P9-CRC-364

ORNITHOLOGY

in Laboratory and Field

Fifth Edition

ORNITHOLOGY
in Laboratory and Field
Fifth Edition

Olin Sewall Pettingill, Jr.
Laboratory of Ornithology
Cornell University

Illustrated by
Walter J. Breckenridge
University of Minnesota
Bell Museum of Natural History

Academic Press, Inc.
(Harcourt Brace Jovanovich, Publishers)
Orlando San Diego San Francisco New York
London Toronto Montreal Sydney Tokyo São Paulo

To the memory of
Alfred O. Gross
(1883–1970)
Professor of Biology at Bowdoin College,
preeminent ornithologist and inspiring teacher.

Academic Press, Inc.
Orlando, Florida 32887

United Kingdom Edition Published by
Academic Press, Inc. (London) Ltd.
24/28 Oval Road, London NW1 7DX

ISBN: 0-12-552455-2
Library of Congress Catalog Card Number: 84-71089
Printed in the United States of America

Contents

Preface

Like its four predecessors, this fifth edition of *Ornithology in Laboratory and Field* is intended for use at the college or university level and assumes that students need have only a background knowledge of general biology. Although the book makes no pretense of covering the realm of ornithology, it treats those major aspects of the science that can be studied during a course in the regular academic year or in the summer.

There are twenty-two chapters, all more or less independent units. Even though they are presented in a fairly logical sequence, they can be taken up in almost any order, and some may be omitted altogether without affecting the value of the others. The instructor may wish to select parts of the book without regard to sequence that will best supplement class work or suit the season of the year during which the course is given.

With few exceptions each chapter concludes with references. While many references are cited in the preceding text of each chapter, many more are included to support or supplement the text, to give views contrary to statements in the text, or to suggest further reading. If the titles are not cited from the text or the titles themselves do not indicate the reasons for their inclusion, they are annotated parenthetically. Students should make a practice of perusing the references after reading the preceding text.

In order to use this book effectively, certain equipment and materials are desirable. The student should have for personal use:

Binocular

Field guide

Key to the bird species in the region where the study is made

Daily field checklists (as many as there will be field trips)

Loose-leaf pocket notebook, preferably with aluminum cover

A set of colored crayons

The institution should make available to students the following:

The American Ornithologists' Union's *Check-list of North American Birds,* Sixth Edition

Annotated checklist of birds in the region of the study

A world atlas

Telescope (preferably the so-called "spotting scope") with either eyepiece or a zoom lens and a tripod

Portable tape recorder

Compound microscope

Meter stick

A Common Pigeon (Rock Dove) and a House Sparrow, properly preserved but not plucked, for external study and dissection

Several nestling House Sparrows preserved in spirit for study of feather tracts

Pigeon skeleton, mounted

Pigeon skeleton, completely disarticulated

Human skeleton, mounted, or a detailed chart of a human skeleton

Human cervical vertebra (fifth)

Hyoid apparatus of a woodpecker

A series of contour feathers illustrating specialized feather types. Some of the more exotic types may be obtained from zoos, which usually save such material for educational institutions.

A semiplume, an adult down feather, a filoplume, a bristle, and a section of the vane of a contour feather mounted on slides for microscopic study.

Parts of feathers with different colors—red, orange, yellow, black, gray, brown, iridescent, blue, green, and white—mounted on slides for microscopic study of color-producing elements

A series of spread wings of a passerine species (e.g., the House Sparrow or European Starling) to illustrate the progress of molt

A collection of bird skins representing all the orders and families of North American birds and all species found in the region of the study. (If possible, the collection should be sufficiently comprehensive to show sex, age, and other constant differences in plumage; abnormal plumage coloration; color phases; eclipse plumage; and plumage changes by wear and fading.) A transparent plastic tube (capped) for each of the smaller skins that will be handled often is recommended in order to prevent damage. (These can be ordered from any biological supply company.)

A record player or tape recorder with reproductions of songs of species occurring regularly in the region of the study

The prefaces to the previous editions of this book tell the story of how it evolved, starting in 1939. In these prefaces I have acknowledged ornithological sources for much of the text. I also acknowledged my many colleagues, former students, numerous friends, and my late wife, Eleanor, who contributed in various ways to the production of all four previous editions. Without the advice or assistance of these generous people I would not have had the framework and much of the authoritative information in this edition.

The cover illustration is by Walter J. Breckenridge and all the illustrations in the text, if not accompanied by credit lines, are his except the following: Figures 1–3, 5–6, and 24 by William Montagna; Figures 4 and 38 by Ray S. Pierce; Figures 8–11 by Robert Gillmor; Figures 12–23, 25–28, 34, and 36 by Robert B. Ewing; Figures 29, 30a, and 31 by Barbara Downs; Figure 33 by Sandra L.L. Gaunt; Figure 41 by Frank A. Pitelka (first published in *The American Midland Naturalist* for 1941); Figures 51–53 by Sidney A. Gauthreaux, Jr.; and Figure 54 by Helen S. Chapman.

For the new chapter "Flight" in this edition, I am grateful to the Laboratory of Ornithology at Cornell University for granting me permission to use the drawings by Robert Gillmor and parts of the accompanying text from the Laboratory's *Home Study Course in Bird Biology,* Seminar IV.

I wish to thank the Bird Banding Laboratory at Patuxent, Maryland, for reviewing in Appendix A the procedures and regulations for obtaining permits to capture, band, and mark wild birds, as well as for providing addresses of state and provincial agencies to which inquiries and applications should be made. I also wish to thank Janet G. Hinshaw, Librarian in the Josselyn Van Tyne Memorial Library at the University of Michigan, for help in checking on, and adding to, the titles of various ornithological journals in Appendix C.

In the preparation of this edition I am deeply indebted to the following authorities in their respective fields of ornithology:

Sidney A. Gauthreaux, Jr., for writing the entire chapter "Migration."

Jack P. Hailman, for writing the entire chapter "Behavior" and for reviewing all the succeeding chapters on the breeding biology of birds.

Peter Stettenheim, for again giving detailed attention to the chapter "Feathers and Feather Tracts," further clarifying or amplifying many aspects of feather growth, structure, coloration, and function.

Andrew J. Berger, for reviewing and updating the "Muscular System," and for taking the time to go over all parts of the book pertaining to internal anatomy and physiology.

Abbot S. Gaunt, for describing the anatomy of the syrinx and its function.

Kenneth C. Parkes, for reviewing and, in some instances, revising or rewriting parts of the chapters "Systematics and Taxonomy" and "Plumages and Plumage Coloration"; also for overseeing the exacting work of Jay Loughlin, Collection Manager at the Carnegie Museum, in updating the synopses of North American orders and families of birds and reconstructing the keys to the same orders and families in accordance with the sixth edition (1983) of *The A.O.U. Check-list of North American Birds*.

Donald E. Kroodsma, for greatly expanding the chapter "Song" and giving the benefit of his expertise in recording bird vocalizations under "Ornithological Field Methods" in Appendix A.

Stephen I. Rothstein, for significantly extending the sections on brood parasitism in the chapters "Eggs, Egg-laying, and Incubation" and "Young and Their Development."

John T. Emlen, for worthy information on bird populations and giving instructions for a new method of measuring populations.

Alan Feduccia, for his critical reading of the ancestry and evolution of birds in the last chapter of this book.

Robert A. McCabe and Ray B. Owen, Jr., for reading and commenting in Appendix A on field techniques applied to capturing, marking, and following wild birds.

Finally, I thank my long-time friend, Edward F. Dana, with his considerable editorial experience, for reading much of the text of this edition when it was in galley proof.

Olin Sewall Pettingill, Jr.

Birds and Ornithology: An Introduction

Birds among all animals offer the most favorable combination of attributes for scientific study. They are numerous, abundantly diversified in form, and easily observed. They are highly organized and responsive with sensory capacities similar to man's and therefore understandable. Pleasing in colors and movements, they are also, with few exceptions, inoffensive in their habits and incapable of physically harming the investigator. Many adapt readily to experimentation. Little wonder that ornithology, the science of birds, boasts so many practitioners, and in turn contributes so significantly to modern concepts of evolution, speciation, behavior, and ecology.

Birds Defined

Birds are unique among all animals in being feathered. Like mammals, they too are warm-blooded, or homeothermous, (capable of regulating their body temperature). And like most of their vertebrate associates, excepting most mammals and a few others, they lay eggs.

Animals move from place to place by running, hopping, walking, crawling, swimming, gliding, and flying. Among birds, flight is the principal means of locomotion, even though some forms—for example, ostriches, kiwis, and penguins—in the course of evolution have lost their ability to fly. Therefore, one recognizes birds as birds because they are formed to fly.

The modern bird, like an airplane, is structurally and functionally efficient. A bird must be able to take flight, to stay aloft, and to reach its destination under the most adverse conditions.

Achievements for Flight

Several achievements have contributed to the bird's mastery of the air.

Lightness Achieved by a covering of feathers—"the strongest materials for their size and weight known"—instead of a thick skin; by the loss of teeth and the heavy jaws to support them; by a reduction of the skeleton and by the hollowing, thinning, and flattening of the remaining bones; by a radical shortening of the intestine and the elimination of the urinary bladder; and by air spaces in the bones, body cavity, and elsewhere.

Streamlining Also achieved by the feathers, overlapping and smoothing the angular, air-resistant surfaces and providing bays, wherein the feet may be withdrawn.

Centralization and Balance Achieved by positioning all locomotor muscles toward the body's center of gravity—leaving the wings, like puppets, controllable by tendinous strings; and by positioning the gizzard, the avian substitute for teeth, and other heavy abdominal organs in the center of the body.

Maximum Power Achieved by the combination of an exceptionally high, steady body temperature for aerial maneuvers in all extremes of climate and weather; by feathers, which aid in conserving the heat; by increased heart rate, more rapid circulation of the blood, and greater oxygen-carrying capacity of the blood stream; by a unique respiratory system, which permits a double tide of fresh air over the lung surfaces, synchronizes breathing movements with flight movements, cools the body internally, and eliminates excess fluids; and by a highly selective diet of energy-producing foods, which contain few indigestible substances to cause excess weight.

Visual Acuity and Rapid Control Achieved by large eyes with a wide visual field and remarkable distance determination, and by a brain whose greatly enlarged visual and locomotor centers are capable of recording and transmitting nerve impulses with the reactions of a seasoned pilot.

Range in Size

Birds range widely in size. The Ostrich *(Struthio camelus)*, standing between 8 and 9 feet tall (2.44 and 2.74 m) and weighing nearly 350 pounds (159 kg), is the largest. But it is, of course, flightless. Among the largest flying birds are the Royal and Wandering Albatrosses *(Diomedea epomophora* and *D. exulans)* and the Andean and California Condors *(Vultur gryphus* and *Gymnogyps californianus)* with wingspans approximating 10 feet (3 m). The Marabou Stork *(Leptoptilos crumeniferus)* may be the largest flying bird, if, as reported, its wingspan measures over 12 feet (3.7 m).

The smallest birds include numerous species of hummingbirds, the extreme being the Cuban Bee Hummingbird *(Mellisuga helenae)* that measures 2.25 inches (5.72 cm) from bill-tip to tail-tip and weighs less than 2 grams. Fourteen Bee Hummingbirds would weigh no more than an ounce (28.35 g).

Within a species there is often sexual difference in size, the males averaging slightly larger. In some species sexual dimorphism is very marked with the male slightly more than twice as large, as in the Wild Turkey *(Meleagris gallopavo)* and in the largest of all grouse, the Capercaillie *(Tetrao urogallus)*, or with the female a third larger, as in the Sharp-shinned and Cooper's Hawks (*Accipiter striatus* and *A. cooperii*).

There are limits to the size that flying birds may attain. They cannot be as small or as large as many other animals. Because they have a high rate of metabolism for supporting a high body temperature and flight movements, birds need sufficient food to maintain this rate and at the same time compensate for heat loss from body surfaces.

Theoretically, the smaller the bird, the greater is its relative body surface in relation to weight and the greater its heat loss. Consequently, the smaller the bird, the more it must eat in proportion to size. Again, theoretically, a bird smaller than kinglets and chickadees would have to eat all the time, night and day. Hummingbirds exist, small as they are, because they lower their body temperature—that is, become torpid—at night or at other times when they cannot eat. Thus they conserve energy.

The larger the bird, the faster it must fly to stay airborne. It needs bigger flight muscles for greater speed. This, in turn, means greater weight because flight muscles are heavy.

The larger birds have attained their size while retaining their ability to fly by developing a dependence on air currents. Albatrosses and condors practically require winds and updrafts in order to fly at all.

Ornithology Defined

Ornithology, simply defined, is the science of birds. For a descriptive definition, there is none more suitable than the one written by Elliott Coues, the perceptive American ornithologist, nearly a century ago:

> Ornithology consists in the rational arrangement and exposition of all that is known of birds, and the logical inference of much that is not known. Ornithology treats of the physical structure, physiological functions, and mental attributes of birds; of their habits and manners; of their geographical distribution and geological succession; of their probable ancestry; of their every relation to one another and to all other animals, including man. [In *Key to North American Birds*, 5e (Boston: Dana Estes and Co., 1903), p. 58.]

One must study ornithology in both laboratory and field because a knowledge of birds "in the hand" is incomplete without a knowledge of birds "in the wild," and vice versa.

Form, Structure, and Physiology

Basic to the study of ornithology is an introduction to the form, structure, and physiology of birds. This the student can best accomplish by making direct observations on the physical make-up of a "generalized" bird, such as the Rock Dove or Common Pigeon *(Columba livia)*, and by learning from a text book the role of each organ system in the bird's way of life. Attention must be centered on those features that will particularly enhance an appreciation of birds as biological entities. Frequently, certain features must be compared to their homologues in man, thereby making them more understandable.

The logical sequence to such an introduction is, first, the identification of the different parts of the bird's topography, followed by a study of the bird's feather covering—how the feathers are structured and variously modified, how they develop, how they are colored, and how they are arranged on the body. A detailed knowledge of these exterior features is indispensable not only in describing birds and their actions but also in accounting for many of their adaptations.

With this knowledge, the student is then prepared to investigate the internal organ systems. Avian anatomy and physiology offer many opportunities for research. Indeed, an increasing number of ornithologists specialize in one or both of these fields, dealing particularly with the adaptive and comparative aspects among different species of birds.

Species and Speciation

Although uniformly specialized for flight, birds have nonetheless changed widely in form and action in order to live in particular environments.

Consider, for example, the adaptations for locomotion and feeding. Some species customarily fly swiftly; others fly slowly. Some hover; others soar. Some swim and dive; others wade. Some walk or hop; others climb. To get food, some species probe in the soil, others dabble in shallow water, scratch the ground, chisel holes in trees, make flying sorties, or hunt for prey in any number of different ways.

These adaptations and others, always in complex combination, account for the different shapes of wings, tails, bills, and feet and differences in body shape, plumages and coloration, breeding habits, seasonal movements, and general behavior. Or, to put it another way, thanks to adaptive radiation operating so vigorously in the descent of birds, there are some 9,000 different species today.

In studying birds one naturally thinks of them in terms of species. Therefore, the logical sequel to a knowledge of their form, structure, and physiology is an acquaintance with the many different species in the student's immediate area. This requires understanding the concept of species and speciation and the methods of classifying, naming, and identifying.

Gaining a thorough acquaintance with the 150 to 300 species regularly occurring in the average study area of temperate North America demands a knowledge of the taxonomic characters and other means of recognizing species in both laboratory and field, together with an understanding of changes in plumage and plumage coloration among different species.

The identification of species is not an end in itself but a stepping stone to investigations of many aspects of bird life, or of biological problems in which birds play a role. Some students find speciation per se a challenging field since there is still much to be learned about the origin, status, and interrelationships of species.

Distribution

Although most modern species of birds can fly and thus can rove the earth, each species is confined to a particular geographical range, which may be from several hundred acres, as on a sea island, to one or more continents in size.

The ranges of species overlap so that in any one area there is an aggregation of species—an avifauna. Because the ranges of species are rarely or never identical, avifaunas vary markedly. Students over the years have given attention to the composition, comparison, and origin of avifaunas, yet there is much about them that remains to be investigated.

Geographical ranges are unstable due partly to the tendency among species to invade new areas. Cyclonic storms may help or hasten resettlement by moving individuals to a different place, where they survive and reproduce if the environment suits them. Man has a part in it, too, when, for example, he transports birds on his ships. House Sparrows *(Passer domesticus)* reached the Falkland Islands in the South Atlantic on ships that first stopped at Montevideo, Uruguay, where the birds, attracted to sheep-pens on deck, came aboard and remained until the ships reached the islands.

Any student, after having observed birds in a given area for a few years, is certain to note shifts in ranges and ponder the reasons. Modern ornithologists pay considerable attention to local distribution as the abundant literature on the subject clearly indicates.

Within its geographical range a species, if normally migratory, is seasonally distributed, appearing in one part of its range in one season, in another part in another season.

Within its geographical range a species is also ecologically distributed. It usually occupies a particular environment or habitat and shares this habitat with other organisms—plant and animal—all of which are adapted to the prevailing conditions of soil, air temperature, moisture, and light. All the organisms in a given habitat collectively comprise a biotic community, since they show relationships to one another.

When any two communities meet, more often than not, there is an area of mixture and overlap, or ecotone, in which the birds and other living forms characteristic of these communities are intermixed and in which are additional forms that, preferring this ecotone, seldom occur elsewhere.

Students soon become aware of the importance of habitat or community in accounting for the presence or absence of species and, before long, learn to associate different species with particular environments—the Red-eyed Vireo *(Vireo olivaceus)* with the deciduous forest, the Horned Lark *(Eremophila alpestris)* with the short-grass prairie, and the Verdin *(Auriparus flaviceps)* with the scrub desert. When students travel northward on the continent of North America or climb a high mountain, they expect a sequence of species as they pass through one environment after another—the Olive-sided Flycatcher *(Contopus borealis)* in the coniferous forest, the White-crowned Sparrow *(Zonotrichia leucophrys)* at the timberline ecotone, and ptarmigan (*Lagopus* spp.) on the tundra.

At the same time students become conscious of several significant aspects of ecological distribution. Rarely do they find one species throughout a community, even though it may be characteristic of that environment. As a rule, it occupies merely a niche and is adjusted to this position in structure, function, and behavior as no other species in the same community. In the forest community, for example, the Red-eyed Vireo occupies a treetop niche and is adjusted to this position in structure, function, and behavior just as the Ovenbird *(Seiurus aurocapillus)* occupies the forest floor. It would be unusual to see the Red-eyed Vireo on the ground or the Ovenbird in the treetops. While a species may appear to share its niche with other species, not one behaves exactly as another does or requires the same food and the same nesting site.

Bird species are of greater variety and density in ecotones than in the pure communities that border them. This phenomenon, called edge effect, is important to anyone wishing to see larger numbers of birds.

Edge effect results in a greater variety of vegetation—grasses, shrubs, and trees—providing a greater variety of food and cover for birds. For example, ecotones where field and forest merge have the plants characteristic of both field and forest and many additional shrubs. Thus they bring together birds of both field and forest and also attract species that require either shrublands or a combination of trees, shrubs, and grasses.

Some bird species are adapted so strongly to a special niche that they cannot live in a different situation. If an element in the niche on which they depend is destroyed or seriously altered, they are more likely to disappear than to make an adjustment. The Snail Kite *(Rostrhamus sociabilis)* probably would disappear in Florida were disaster to befall the big freshwater snail, *Pomacea palludosa,* on which it feeds exclusively. It is likely that the Kirtland's Warbler *(Dendroica kirtlandii)* would disappear in northern Lower Michigan, where it breeds exclusively, if there were no more jack pines 6 to 18 feet high (2 m to 6 m) under which it almost invariably nests.

A good many bird species, on the other hand, are much more adaptable. Sometimes they are so widely tolerant of different situations that their precise niches are unrecognizable. The Blue Jay *(Cyanocitta cristata),* Black-capped Chickadee *(Parus atricapillus),* and Cedar Waxwing *(Bombycilla ced-*

rorum) are so adaptable that one may find them almost everywhere in wooded areas through their ranges.

The species that restrict themselves to narrowly prescribed niches generally have small populations within correspondingly small ranges. The species tolerant of environmental changes and variations are mainly the inhabitants of the ecotones; they have large populations and often range widely.

The underlying factors accounting for the ecological distribution of many species still remain to be determined. Here is a study with a degree of urgency. As man steadily destroys the natural environments, an understanding of a species' ecological requirements is the first step in preventing its decrease. The next step is to see that its requirements are maintained through intensive management and conservation practices.

Behavior

The behavior of birds attracts scores of investigators. Birds are ideal animals for behavioral studies. Each species has an impressive repertoire of innate behaviors and, at the same time, its ability to learn compares favorably with that of most mammals. Thanks to a rich variety of bird species, each with a different mode of life, investigators have available for study a correspondingly rich variety of behaviors.

An understanding of the principles of bird behavior is essential for any beginning student, helping as it does to explain the basis of many avian activities. Even more important, an understanding of bird behavior illuminates many of the basic ethological principles applied to human life. Modern psychologists are now paying attention to such phenomena as individual distance and dominance relationships (first noted in birds!) that are so evident in urban societies. Continued, in-depth studies of avian behavior will, almost certainly, further sharpen man's perception of his own social problems.

The procedure in the study of behavior is to identify, describe, and name the behaviors of a species and then to determine what each behavior accomplishes, its significance to the species' survival, its causes, how it has evolved, and whether it is innate, learned, or both innate and learned. Many mating displays are actually derived from such maintenance activities as preening or scratching; or from displacement activities—for example, when a bird breaks off fighting and pecks at some object; redirected activities—when a bird redirects its attack to an object other than one which elicited the response; and intention movements—when a bird makes a move to fly but fails to do so, thereby performing an incomplete act.

Inherited behavior predetermines the extent to which learned behavior may develop. Learned behavior is actually adaptive behavior resulting from experience. A bird inherits the ability to fly, yet it must learn by experience to take off *into* the wind rather than *with* it and to choose the perch that will best accommodate its feet. This is called learning by trial and error. Other forms of learning are by habituation and by imprinting. A few birds show ability to learn by insight. The different methods of learning among birds demand much more research.

Investigators often give considerable attention to social behavior since most birds are by nature gregarious and have consequently developed many kinds of interactions related to attack, escape, defense, flocking, and reproduction. Although the literature on social behavior in birds is already enormous, the subject is still a fertile field for study.

Migration

No aspect of bird life has so excited man's interest down through the centuries as the withdrawal of birds from an area in the colder seasons and the return to the same area when the seasons become warmer. In spite of a great store of knowledge on the initiation and procedure of migration among modern birds, the question of how and when migration originated still remains speculative—an ever-present challenge to one's thinking.

Experimental studies started over 50 years ago demonstrate that a specific day length in the spring stimulates the activity of a bird's endocrine glands, and this stimulation brings the bird into a migratory state. Some external factor then releases migratory behavior. In the fall, with a regression of endocrine activity, the bird reaches another migratory state ready for triggering by an outside cause.

The present wealth of information on the process of migration—starting and stopping times, rate, duration, distances covered, routes, and relation to weather—is due in large measure to direct observations and record-keeping by hundreds of persons and to returns from many millions of banded birds.

Radar and radiotelemetry are useful tools in fathoming some of the "mysteries" of night migration and determining the speed, direction, and elevation of migratory flights.

Migrating birds have obvious navigational ability, otherwise they could not return as they do to their nesting grounds after the winter spent hundreds, sometimes thousands, of miles away. Just how migrant birds orient themselves has been the object of numerous experiments. By using caged birds that display migratory activity by "fluttering" in the direction of migration in the wild, some investigators have demonstrated that birds migrating on clear days may be guided by the sun and on clear nights by star patterns. These and other experiments, although convincing, do not explain orientation by all birds under all circumstances. Undoubtedly different birds use different cues or different combinations of cues, depending on where and when they migrate and the prevailing weather conditions. The whole subject of orientation, complex and fascinating, beckons for continued research.

The Reproductive Cycle

The main stages of the reproductive cycle of most bird species are the establishment of territory, the coming together of the sexes, nest-building, egg-laying, incubation, hatching of the eggs, and the development and care of the young. Involved in the establishment of territory and the coming together of the sexes are two prominent activities—singing and mating displays.

In the past 75 years many investigators have studied the reproductive cycle of different species, resulting in the accumulation of a vast amount of data. Yet, surprisingly, detailed, comprehensive information is available on relatively few species. For only about 5 percent of North American species is the size of territory known; for about 10 percent, the average length of nestling life; for about 20 percent, the average incubation period; for about 30 percent, the full description of songs and mating displays.

Anyone beginning a study of birds should carefully observe the reproductive cycle of at least one species from territory establishment to fledging and dispersal of the young. The more detailed information one can obtain, so much the better. Ideally, students will contribute to knowledge of the species, but whether they do or not, they are almost certain to profit by gaining an intimacy with the living wild bird, its behavior and problems of survival.

Longevity, Numbers, and Populations

How long do birds live? How many birds are there in given areas? What are the factors controlling the numbers of birds? These are questions that always fascinate anyone studying ornithology, and the answers continue to be unsatisfactory in scope and often controversial.

Students should familiarize themselves with these questions and gain some first-hand experience in estimating numbers of birds.

Direct counting of individuals of most species generally is futile because they are so numerous and widespread. Time is better used in measuring the populations of all species in a given area and understanding how their populations are controlled. This is a complex undertaking. It includes determining their reproductive rates; the ratio of age groups and sexes; the annual fluctuations of their respective populations because of varying physical factors of the environment (air temperature, precipitation, and others) and biological factors (predation, diseases, food supply); and ways in which their populations are controlled over long periods of time. Although populations normally fluctuate in numbers of individuals per year, they are remarkably stable over a period of, say, 50 years if their habitat is unchanged. Annual fluctuations are scarcely more than wrinkles in the long history of a population.

The study of populations has endless opportunities for investigation. It is of vital—"vital" meaning life-or-death—importance at the present time as man hastens his encroachment upon and destroys the natural environment. Determining when certain populations are showing a sharp decline provides the basis for informing conservation agencies and urging remedial action.

Evolution

Where did birds come from? The story goes back to the Triassic Period some 200 million years ago, when birds arose from a somewhat specialized group of reptiles that had long hindlimbs. The avian line from this reptilian specialty may have begun as tree-climbing forms, which first jumped from branch to branch by using membranes stretched between the sections of their shorter and slightly flexed forelimbs.

As they gradually evolved the ability to fly farther, these arboreal forms acquired greater sailing surface through expansion and modification of the scales on the trailing edges of their forelimbs and along the outer edges of their long tails. At this point birds came into being, for of all the physical features of birds, none distinguishes them more sharply from all other creatures than these outgrowths of the skin.

The remarkable fossil *Archaeopteryx lithographica* possessed feathers and is thus recognized as the earliest known bird. This creature of the Jurassic Period, some 140 million years ago, may have been one of several kinds of similarly primitive birds already existing. Nobody knows. But in any case, one such primitive species, probably of either Eurasian or African origin, acquired the power of flight—that is, the ability to sustain itself in the air for indefinite periods by flapping its wings. And from this stock many species began to emerge as they spread out and filled more habitats and niches.

This evolutionary process, commonly called adaptive radiation, was slow at first but steadily quickened during the next 139 million years, through the Cretaceous and Tertiary Periods. Birds in time inhabited all the earth's great land masses and occupied most of the primitive environments.

But as the continents separated, merged, and separated, as mountain ranges rose and were worn away, as the climates shifted, and as plant forms evolved, flourished, and vanished, so did habitats for birds. The species so precisely adapted to one habitat that they could live in no other disappeared when the habitat disappeared. More species were always evolving, however, to fill new niches.

The primitive birds became extinct through the Cretaceous Period. The "new" birds began to look more and more like modern species, and many birds were recognizable by the end of the Tertiary as ostriches, pelicans, cranes, nuthatches, thrushes, and so on.

With the coming of the Pleistocene Epoch, or Ice Age, about a million years ago, the abundance of birds in number of species attained a peak that has never been exceeded. This period of prehistory could have been called the Age of Birds, had mammals not already taken the ascendancy in size and aggressiveness to dominate the earthly scene.

Toward the end of the Pleistocene and the start of the Recent Epoch, about 15 thousand years ago, bird species began disappearing more rapidly than they were evolving. The decrease of birds was under way. Man had not yet become a major destructive force in the avian environment. How, and how fast, that destructive force grows may determine how, and how fast, the presently extant 9,000 species of birds disappear.

The first bird species definitely known to have been eliminated by man was the Dodo (*Raphus cucullatus*), in 1681. Since that date, no fewer than 78 species have become extinct over the world, nearly half of them destroyed by man. At this rate of disappearance, the future for bird life appears alarming.

And it is alarming! Whenever investigating the attributes of birds, students of ornithology must keep this in mind, being constantly alert to discover ways and means that will insure the protection of birds and thus assure their survival for centuries to come.

Topography

The various parts of a bird's exterior are mapped out as the **topography.**

For convenience, the description of the topographical parts are grouped below under seven titles: Head, Neck, Trunk, Bill, Wings, Tail, Legs and Feet. While studying each part, refer to specimens of the Rock Dove *(Columba livia),* called the Common Pigeon in this book, and the House Sparrow *(Passer domesticus)* and to the outline drawings (Figures 1–6). Follow the instructions given in the Activities for labeling the drawings.

Place a pigeon specimen on its back and observe its outline or **contour.** The body shape tapering at both ends is streamlined for cleaving the air in flight.

The Head

The upper, or dorsal, part of the head is somewhat curved and composed of an anterior (foward) and a posterior (rear) part: the anterior part, the **forehead,** extends up and back from the bill to an imaginary line joining the anterior corners of the eyes; the remainder of the upper part of the head, the posterior, is the **crown.** (Some authorities call the sloping posterior portion of the crown the **occiput** or **hindhead.**) Below the lateral boundary of the forehead and crown is the **superciliary line,** distinctively colored in some birds but not in the Common Pigeon or male House Sparrow.

The side of the head is rather flat and divided into the **orbital** and **auricular regions.**

The orbital region includes the **eye, eyelids,** and **eye-ring.** The eye, as revealed through the circular eye-opening, consists only of the dark **pupil** and pigmented **iris.** (The eyeball, actually of great size, can be felt under the skin.) Note that the pigeon's iris is bright orange or yellow. What is the color of the sparrow's iris? The two eyelids are skin-folds, one above the eye and one below. In the pigeon they are unfeathered and red. How do they differ in the sparrow? In all birds, as in mammals and in many reptiles and amphibians, the lids close the eye; only in birds, however, do the lids close the eye at death. Observe that at closure, in the pigeon and sparrow, the lower lid comes up more than the upper lid comes down. This is the rule among most diurnal birds; in most nocturnal species (e.g., owls and goatsuckers) and in a few others the upper lid is the more mobile, as in mammals and alligators. The anterior corner of the eye (toward the nostril), where the eyelids come together, is the **nasal canthus;** the posterior corner (near the temple in man) is the **temporal canthus.** The **nictitating membrane,** sometimes called the "third eyelid," is a translucent, vertical fold under the lids on the side of the eye toward the bill. If the eye of a living bird is touched, the nictitating membrane—just before the lids close—slips obliquely across the exposed surface of the eye. Ordinarily, the membranes of both eyes, and the lids of both, act together (consensually) even when only one eye is touched. Birds, like mammals, blink periodically. The pigeon and a few other species blink with both the nictitating membrane and the lids, but most species, including the House Sparrow, blink with the nictitating membrane alone, the lids—usually the lower—closing only in sleep or when the eye is menaced by foreign objects. In some birds (though not in the pigeon or sparrow) the feathers immediately around the eyelids are distinguished from the surrounding feathers by different color and are called collectively the eye-ring.

The auricular region is around the ear opening, concealed by a patch of feathers, the **auriculars** ("ear coverts"). The temporal region, between the auriculars and the orbital region, is small. Generally, it is considered part of the auricular region and not used in describing birds. The area between the eyelid (or the eye-ring in birds having one) and the base of the upper part of the bill is the **lore.**

The side of the head from the base of the lower part of the bill to the angle of the jaw (found by feeling for a bony prominence behind and below the ear) is the **malar region** (cheek). It is bounded above by the lore, orbital region, and auricular region and below by the edge of the lower jaw.

The under (ventral) part of the head is flat and divisible into an anterior part, the **chin,** a feathered area in the fork of the lower part of the bill; and a posterior part, the **gular region,** a continuation of the chin to an imaginary line drawn between the angles of the jaw.

The Neck

The neck extends from the posterior margin of the crown to the trunk and is divided into four regions: **nape, jugulum,** and **sides.** The upper, or dorsal, part is the nape. The lower, or ventral, part is the jugulum. (The term "throat," frequently used in descriptions of birds, includes the gular region of the head and the jugulum of the neck.) The sides of the neck extend between the nape and jugulum and from the posterior borders of the auricular and malar regions to the trunk.

The Trunk

The trunk is divided into two surfaces: the **upper parts** and the **under parts.** The upper parts include all the trunk above an imaginary line drawn from the shoulder joint to the base of the outermost tail feathers. The under parts include all the trunk below this line. (Sometimes the terms "upper parts" and "under parts" are used to include the dorsal and ventral surfaces of the wing and tail, as well as those of the trunk.) The upper parts of the trunk are made up of the **back** and **rump.** The back is the anterior two-thirds of the area between the base of the neck and the base of the tail; the rump is the posterior one-third. The under parts are divided into **breast, abdomen, sides,** and **flanks.** The rounded portion of the under parts, beginning at the lower border of the jugulum, is the breast; the flatter portion ending in an imaginary line drawn across the vent is the abdomen. The breast and abdomen curve upward, forming the sides of the body. The parts lying between the posterior half of the abdomen and the rump are frequently termed the flanks. Although technically the sides of the body belong to both the upper parts and the lower parts, the imaginary line separating the two surfaces is so high on the trunk that the sides of the body are generally considered under parts only.

Activity

On Figure 1, label the following: eyelids; iris; pupil; nasal canthus; temporal canthus; nictitating membrane (approximate position). Write all labels on this and subsequent figures outside the drawings and parallel to the top of the page. Use dotted (i.e., broken) leader lines.

On Figure 2a, label the following: head; neck; side of body; flanks; breast; abdomen.

On Figure 2b, label the following: forehead; crown; occiput; superciliary line; auriculars; lore; malar region; chin; gular region; nape; jugulum; side of neck; back; rump; side of body; flanks; breast; abdomen.

Figure 1 **Common Pigeon**

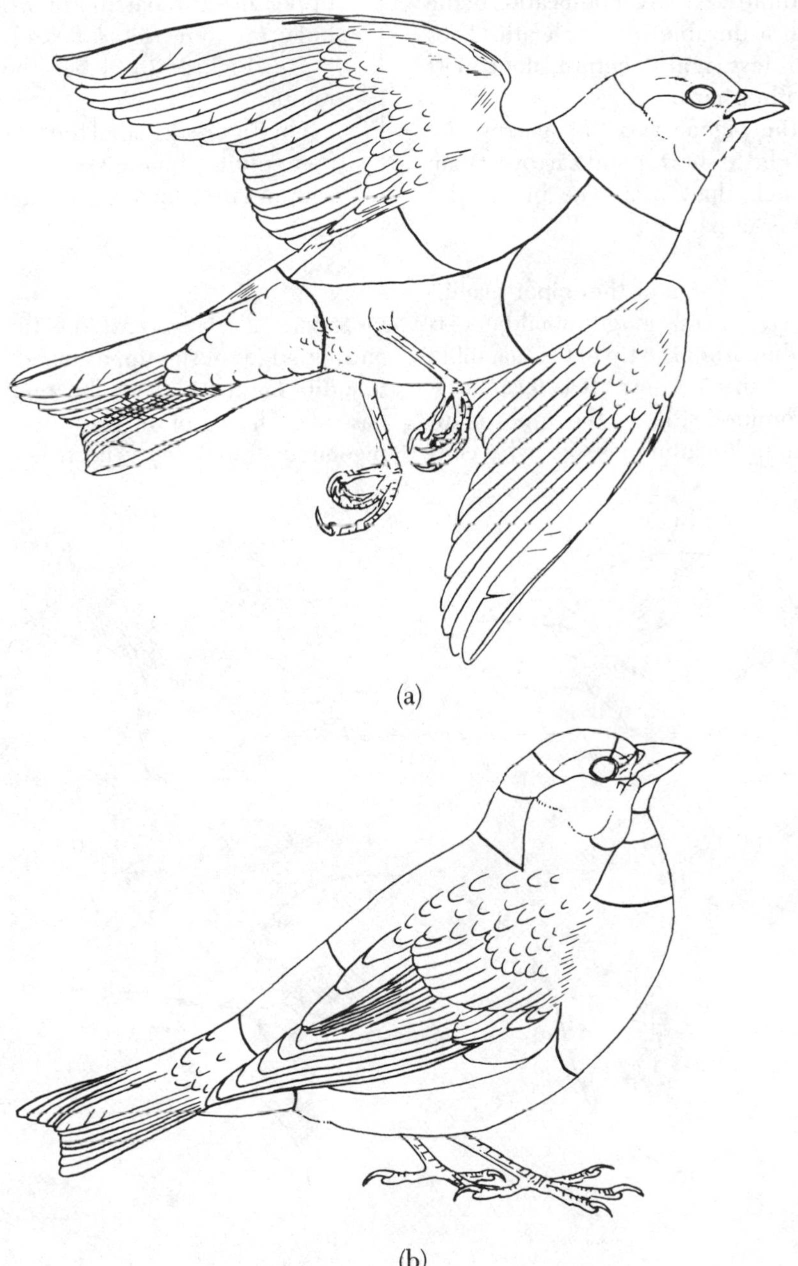

(a)

(b)

Figure 2 **House Sparrow**

The Bill

The bill consists of an **upper** and a **lower mandible,** lying, as their names indicate, above and below the mouth. Each mandible is a bony modification of the skull covered with a durable horny sheath. Thus the bill is a more or less rigid structure; along most of its length it is rather hard.

Notice that in the pigeon and the sparrow the lower mandible is a little shorter and narrower than the upper and much shallower. The bill is also mapped in a number of parts.

Activity

On Figure 3a, label the following: culmen, upper mandibular tomium, lower mandibular tomium, rictus, nasal fossa.

On Figure 3b, label the mandibular ramus.

On Figure 3c, label the following: gonys, operculum, commissural point, nostril, commissure, mandibular ramus.

Upper Mandible The ridge of the upper mandible—the uppermost, central, longitudinal line—is the **culmen,** extending from the tip of the mandible back to the bases of the feathers. It is formed by fusion of the two rounded **sides of the upper mandible.** Seen in profile, the culmen is somewhat convex, particularly toward the tip of the bill. The cutting edges of the upper mandible are the **upper mandibular tomia** (singular, **tomium**). Toward the base of each side of the mandible is a **nostril.** In the pigeon, overarching the nostril posteriorly, is a soft,

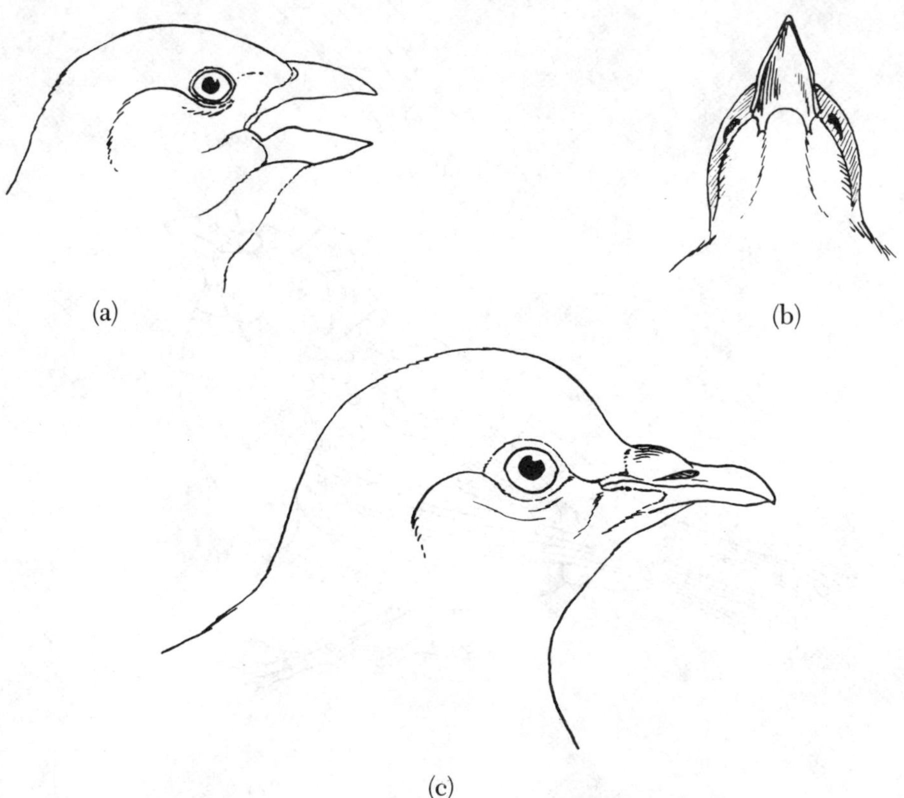

Figure 3 **Common Pigeon and House Sparrow**
(a) Sparrow head, lateral view. (b) Sparrow head, ventral view. (c) Pigeon head, lateral view.

noticeably swollen structure characteristic of pigeons called the **operculum.** The sparrow's nostril opens into a depression, the **nasal fossa,** common in the majority of small birds.

Lower Mandible The cutting edges of the lower mandible are the **lower mandibular tomia.** They are overlapped slightly, when the bill is closed, by the upper mandibular tomia. Viewed from below, the bill has a prong-like projection extending posteriorly on each side of the jaw. This is the **mandibular ramus.** The lowermost ridge of the lower mandible is the **gonys,** formed by an anterior fusion of the rami. Like the culmen, its profile is somewhat convex. The **sides of the lower mandible** include not only the surfaces between the gonys and tomium but also the surfaces of the rami.

Several parts of the bill are evident when the two parts of the mandible are considered in relation to each other. The line along which the mandibles come together is the **commissure** or **gape.** (The term "commissure" is preferred to "gape," which often means the space between the opened mandibles.) The point on each side where the mandibles meet posteriorly is called the **commissural point** (angle of the mouth). The tomium of each mandible has two parts: the **tomium proper**—the hard cutting edge of the mandible; and the **rictus**—the softer, more fleshy, part of the tomium near the commissural point. The rictus is more prominent in the House Sparrow than in the pigeon.

The Wings

The wings are the appendages arising from the shoulder or pectoral girdle. Though homologous to the forelimbs of man and other vertebrates, they are specially adapted to flight, having a peculiar shape and a series of feathers arranged in a definite fashion.

Spread out the wings of the pigeon and note that the feathers belong to two main groups: the flight feathers, or **remiges** (singular **remex**), are the long stiff quills projecting posteriorly; the **coverts** are the smaller feathers overlying the bases of the remiges and covering the rest of the wing. Other groups of feathers are the **alular quills, scapulars, tertiaries,** and **axillars.**

To identify these groups of wing feathers, it is necessary to know from what parts of the wing they arise, and to understand the skeletal framework and external anatomy of the unfeathered wing.

Examine either an articulated human skeleton, or a detailed chart of one, and locate the following bones of the forelimb: humerus (upper arm bone), radius and ulna (forearm bones), 8 carpals (wrist bones), 5 metacarpals (hand bones), and 14 phalanges (digit or finger bones).

The bird's wing has the same skeletal plan and terminology as the forelimb (pectoral appendage) of man and other vertebrates, yet it shows certain striking differences. The skeleton of the bird's wing, like many other parts of its anatomy, is highly specialized for flight. Some of the bones in the human forelimb are lacking in the wing or fused with others; the movements of the various bones upon one another differ markedly. From the wrist outward the skeleton of the bird's wing is especially at variance with the skeleton of the human forelimb.

Examine a prepared articulated skeleton of a pigeon's wing. (See Figure 4a.) Count the number of bones. What is the difference between the number in the wing and the number in the human forelimb? Identify the following bones in the pigeon wing:

Humerus A relatively short, thick bone that articulates, by means of a vertically elongated head, with the shoulder girdle. On the proximal ventral side is the opening of the **humeral pneumatic cavity,** which receives one of the air sacs. The humerus widens out toward its distal end to form two large condyles that articulate with the radius and ulna.

Radius A slender, rather straight bone that articulates with the external condyle of the humerus by a cup-like structure on its proximal end. Its outer posterior margin articulates with the ulna; its distal end fits into one of the carpals.

Ulna A stouter bone than the radius and decidedly more curved. On the outer side is a row of small prominences, the points of attachment of the remiges. Proximally the ulna articulates with the internal condyle of the humerus and ends in an **olecranon process** to form the point of the elbow. Distally the ulna articulates with the two carpal bones.

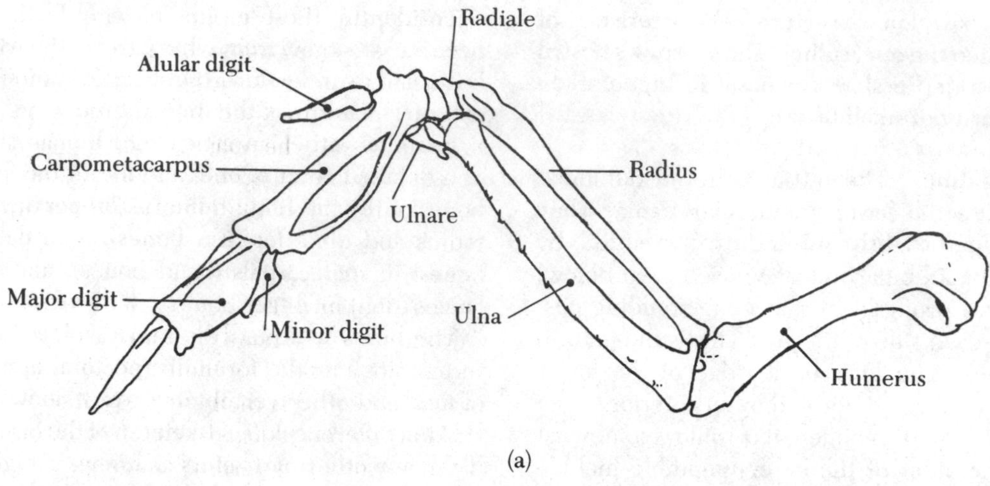

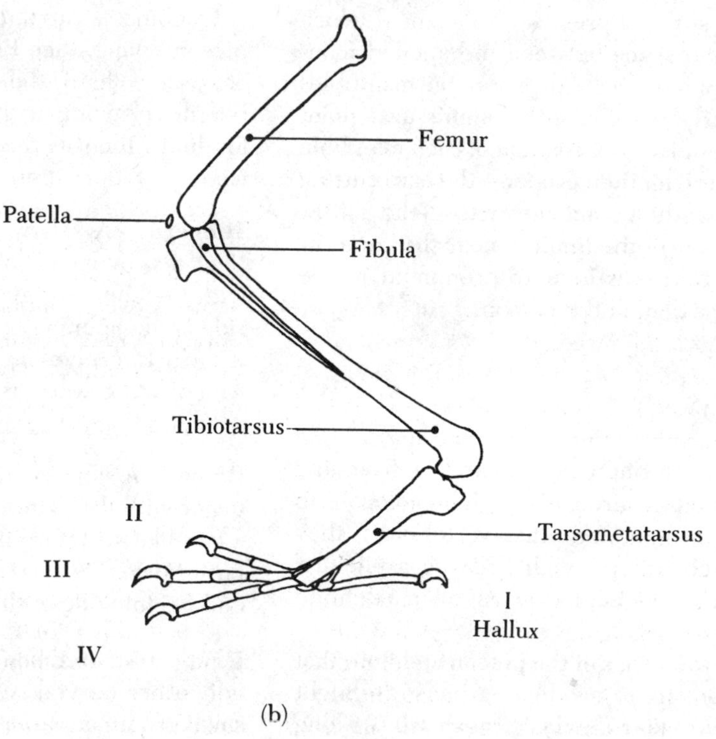

Figure 4 **Common Pigeon**
(a) Skeleton of wing. (b) Skeleton of leg and foot.

Carpals Two somewhat squarish bones. One, the **radiale,** is at the end of the radius; the other, the **ulnare,** is at the end of the ulna. They articulate proximally with the radius and ulna. The additional carpals, found in other vertebrates, are fused with the radiale and ulnare and with the adjoining metacarpals. Although present in the embryo of a bird, they are not distinguishable in the adult.

Metacarpals The first and fifth metacarpals are wanting in the bird, but the second, third, and fourth persist, fusing with vestigial carpals to form the large composite bone called the **carpometacarpus,** which articulates with the radiale and ulnare. The third, or median, metacarpal constitutes a large part of the carpometacarpus. Fused to its proximal end on the radial side is the small remnant of the second metacarpal. On the opposite side is the remnant of the fourth metacarpal, a slender bone nearly as long as the third metacarpal to which its ends are fused; the two bones are sometimes also joined laterally by a thin, bony membrane.

Phalanges The pigeon has only four phalanges. They compose the skeletal structure of the three digits or fingers that persist in the bird. There has been much controversy about the numbering of the bird's fingers—i.e., should they be considered the first, second, and third or the second, third, and fourth. In the absence of proof and in order to attain uniformity, both in osteological and myological studies, the *Nomina Anatomica Avium* (Baumel et al., 1979) proposed that the bird's fingers be called **alular digit** (formerly the pollex), the **major digit** (typically with two phalanges), and the **minor digit** (always with a single phalanx). The proximal phalanx of the major digit is very much flattened and its posterior margin sharply edged. The fourth metacarpal bears the single phalanx of the minor digit and is somewhat triangular in shape. A few birds, belonging mostly to the more primitive orders, have a claw on the alular digit and sometimes on the major digit. Wing claws (Fisher, 1940) are better developed in newly hatched birds than in adults—an indication that these structures are relics of the bird's reptilian ancestry. The young of the Hoatzin *(Opisthocomus hoatzin),* a tree-inhabiting gallinaceous bird of South America, use the claws on their alular and major digits for climbing.

Pluck one wing of the pigeon, leaving only the remiges. Note that the wing has two prominent angles giving it the shape of the letter Z written backwards. The angle nearest the trunk, pointing toward the tail, is the **elbow.** The portion of the wing between the trunk and the elbow is the **brachium.** The angle pointing forward is the **wrist,** or **bend of the wing.** The portion of the wing between the elbow and the bend of the wing is the **forearm,** or **antebrachium.** The entire portion of the wing beyond the bend is the **hand** or **manus.**

Locate on the plucked wing the position of each of the bones just described. Find the rudimentary second finger emerging just beyond the bend of the wing, on the anterolateral surface. The fold of skin extending from the upper arm to the entire antebrachium is the **patagium** (plural, **patagia**). The smaller fold of skin extending from the brachium to the trunk is the **humeral patagium.**

A comparative study of the mechanisms of the bones of the human arm and hand and those of the pigeon's wing reveals many striking differences. Whereas the human shoulder joint is free, permitting the humerus to swing about, the pigeon's is restricted and limits the humerus almost completely to movements up and down and to and from the body. There is no rotary motion (like that in man) of the radius around the ulna and no movements between the bones in the manus. Both antebrachium and manus thus form a firm support for the flight feathers. The wrist joint in the bird does not permit the manus to swing about as in man; the manus can move only to and from the antebrachium and in the same plane. The only joint that allows the same (and no other) motion in man and pigeon is the elbow joint: it permits the antebrachium, or forearm, to move to and from the brachium, or upper arm, in the same plane as the brachium. The most plausible explanation for the sharp differences between the forelimb of man and that of the pigeon (and other birds) is that man's forelimb is "generalized" for a variety of functions, whereas the pigeon's is "specialized" for one function, namely flight.

Having studied the structure of the pigeon's wing, now identify the several sets of feathers, or topographical regions, of the wing. Turn to the unplucked wing of the pigeon and find:

Primaries The remiges attached to the manus. They are counted and numbered from the inside out. How many are there? Which primary is the longest?

Secondaries The remiges on the antebrachium and elbow; all are attached to the ulna. They are counted and numbered from the outside in. How many are there?

Tertiaries Sometimes in descriptive ornithology, the feathers growing upon the adjoining portion of the brachium are called the tertiaries. They are not remiges. In certain species of birds the tertiaries are greatly modified and differ considerably from the secondaries.

Scapulars A group of prominent feathers arising from the shoulder and adjoining portion of the upper surface of the brachium. They slightly overlap the tertiaries.

Alular Quills Three feathers, stiffened like the remiges, springing from the alular digit. Collectively, they are known as the **alula** (pronounced al'-you-la).

Wing Coverts Feathers overlying the remiges on both the upper and under surfaces of the wing. They include all the feathers of the wing except the remiges and the alular quills. The upper wing coverts are as follows:

Greater Primary Coverts The feathers overlying the bases of the primaries. There is one covert for each primary.

Median Primary Coverts The shorter, less exposed feathers overlapping the greater primary coverts. There is one row of them. (Lesser primary coverts are wanting in the pigeon.)

Greater Secondary Coverts The single row of feathers overlying the secondaries.

Median Secondary Coverts A row of shorter, less exposed feathers overlapping the greater secondary coverts.

Lesser Secondary Coverts Even shorter feathers lying in two or three rows directly over the median secondary coverts.

Alular Quill Coverts Three small feathers, each one overlapping an alular quill at its base.

Marginal Coverts The remaining coverts of the upper surface of the wing. They arise immediately anterior to the lesser secondary coverts and are indistinguishable from them. They are densely inserted on the patagium and along its extreme anterior border. They are also inserted along the outer surface of the manus and extend distally to the outermost median primary covert.

The under wing coverts are as follows:

Greater Primary Coverts Overlying the primaries at their bases.

Greater Secondary Coverts Overlying the secondaries at their bases.

Median, Lesser, and Marginal Coverts Generally

Activity

On Figures 5 and 6, identify and label the several sets of wing feathers (except the scapulars and median primary coverts, which are not shown). Number the primaries and secondaries. Color the following regions of the upper wing: alular quills (alula) and alula quill coverts—purple; greater primary coverts—green; greater secondary coverts—yellow; median secondary coverts—red; lesser secondary converts—blue; marginal coverts—brown.

On Figure 1, color as above and label the following: primaries; secondaries; alula; greater primary coverts; greater secondary coverts; median secondary coverts; lesser secondary coverts; marginal coverts.

On Figure 2a, label the following: axillars; lining of the wing.

(a)

(b)

Figure 5 **Common Pigeon Wing**
(a) Upper surface. (b) Under surface.

(a)

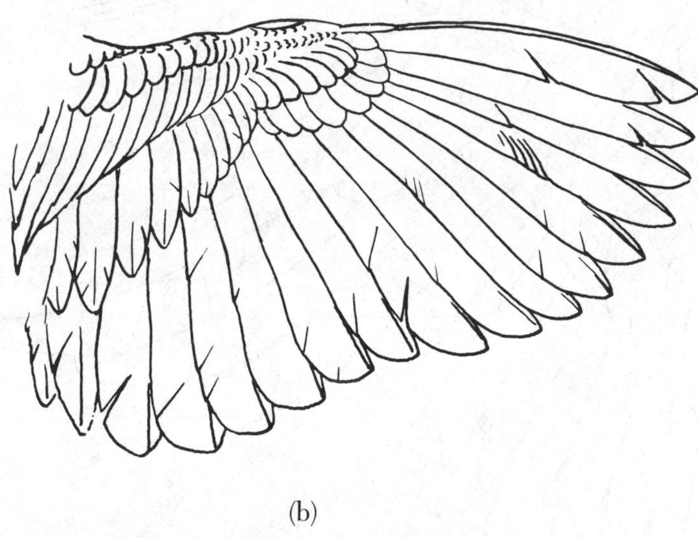

(b)

Figure 6 **House Sparrow Wing**
(a) Upper surface. (b) Under surface.

all of these coverts are referred to as the **lining of the wing.** They overlap each other in much the same fashion as the corresponding coverts on the upper surface of the wing. But the feathers of the three groups are much more alike and less distinctly arranged in rows. Therefore no attempt is made here to distinguish between them.

Axillars. These are under wing feathers lying close to the body in the axilla or "armpit." They are white and both longer and stiffer than the coverts. In certain species of birds the axillars are even more peculiarly modified.

Study the wing of the House Sparrow. Note that the tenth (outermost) primary, rudimentary and quite concealed, is a minute, narrowly pointed feather lying on the outside of the wing next to the outermost greater primary covert. How many secondaries are there?

Certain species of birds lack the tenth primary altogether, hence the outermost primary is the ninth. Certain species also have fewer inner secondaries. This explains why the primaries are counted from the inside out and the secondaries from the outside in.

Consider now the wing as a flying mechanism.

Spread the unplucked pigeon wing to its full extent and notice that the anterior part containing the bones, muscles, and tendons is thicker than the posterior part, which bears the remiges. The basic structure is, therefore, roughly tapered or cambered from the leading edge to the trailing edge. Notice also that the remiges are supported by the wide membrane (mostly skin) from which they emerge and that, except for the tertiaries, they are further supported by being attached to the bones. On the spread wing the remiges point in different directions: the primaries outward, away from the body; the secondaries mostly backward; and the tertiaries toward the body. The scapulars, on the shoulder and brachium, are directed outward from their skin support and somewhat overlap the tertiaries. Each time the wing is spread, the remiges and scapulars, controlled by their supports, automatically take these directions. Thus during flight the wing has a broad, continuous surface from its origin on the body to the tip.

Spread the unplucked wing and examine the upper surface. The marginal coverts arise straight up, then bend posteriorly to overlap one another. The posteriormost marginal coverts bend over the bases of the lesser coverts, and the lesser coverts bend over the median coverts, the median coverts over the greater coverts, and the greater coverts over the remiges in a similar manner. This arrangement, giving an even curve to the tapered structure of the wing, streamlines the wing. Since air has weight and exerts pressure, streamlining is essential for flight. It allows air to flow smoothly over the wing surface and at the same time prevents excessive pressure from building up in front, reduces pressure on the upper and lower surfaces, and lessens vacuum and turbulence behind.

The wing supports the bird in flight and moves it forward in the air. The proximal part of the wing, from body to wrist joint, supplies the principal support by giving **lift**; the distal part provides the forward motion by propulsion, acting in a manner analagous to a propeller.

Note that the proximal half of the extended wing is tilted upward from trailing to leading edge so that more of the under surface will face the direction of flight. During ordinary flapping flight the under surface meets the pressure of the oncoming air stream, deflects it downward, and prevents it from flowing over the upper surface. (See Figure 7a.) This gives lifting force because the pressure on the under surface is greater than the pressure on the upper surface, creating a "suction effect." The wing can increase the lifting force by increasing the tilt, but only up to a certain limit. If the tilt is too great, none of the air stream can slip over the upper surface (see Fig. 7b); thus the pressure on the upper surface is so diminished as to create a partial vacuum into which the surrounding air rushes and swirls, destroying the lift and causing a **stall.** At the trailing edge, eddies of air extend forward along the upper surface; at the wing tip, other eddies (collectively called the **tip vortex**), whirling at still greater speed, move inward along the upper surface. The explanation for this intrusion upon the upper surface is that the air, at higher pressure under the wing, tends to swirl up and over the upper surface where the air pressure is lower.

Turbulence of this sort not only causes **drag** on the wing but also interferes with the smooth passage of air over the wing by extending up and over the upper surface. The effects of turbulence are

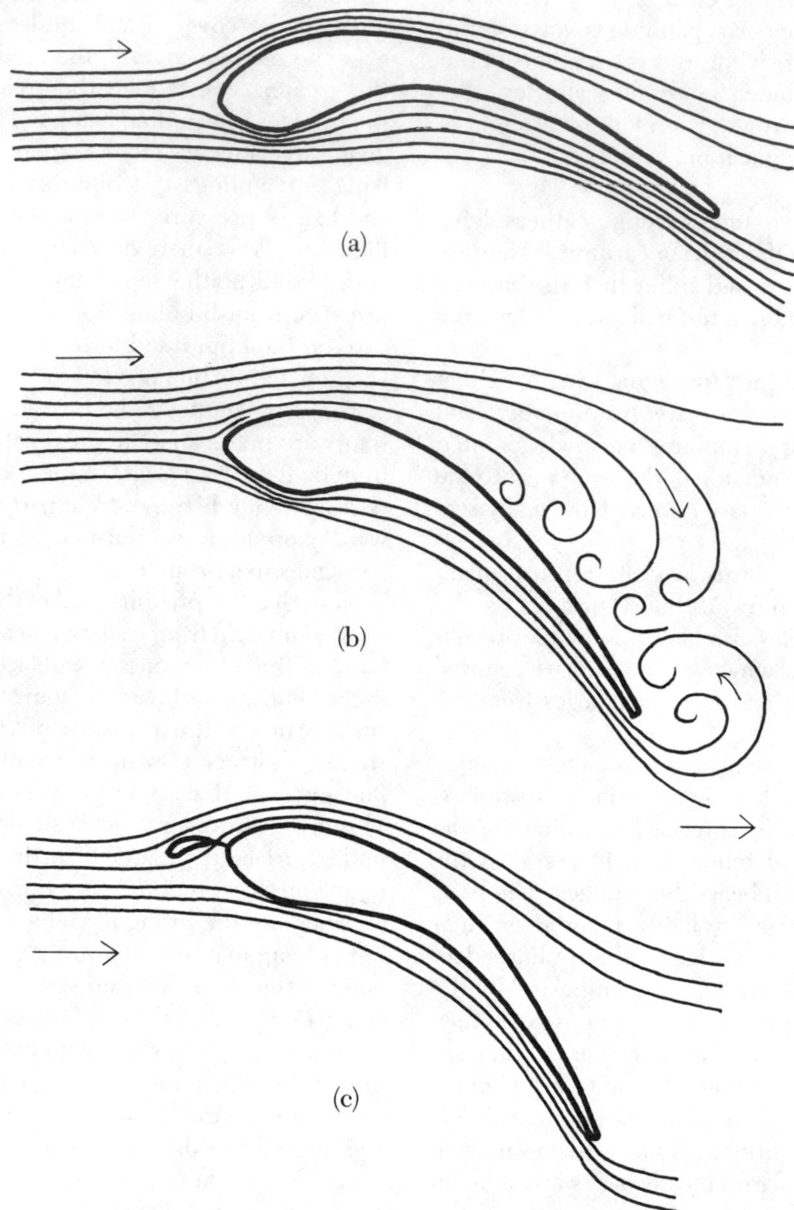

Figure 7 **The Bird's Wing in Flight**
Diagrams showing how a wing meets an oncoming air stream—(a) during ordinary flapping flight; (b) when the wing is greatly tilted; (c) when the wing is greatly tilted and slotted. (Adapted from Jones, *Elements of Practical Aero-dynamics*, John Wiley & Sons, 1942.)

partially offset by the wing's **aspect ratio**—proportion of length to breadth. In other words, a wing is long and broad enough to allow for disturbances and yet has sufficient surface for lifting purposes.

The distal half of the wing, "the propeller," moves in semicircles. In ordinary flapping flight the propeller moves in a half circle, forward and downward, then upward and backward, the tip describing a course like a figure eight. Hold the proximal half of the wing in one hand and move the distal half with the other in the manner described. The forward-downward **thrust** "pulls" the bird along; the upward-backward thrust, which is much quicker, presses against the air and "pushes" the bird along. In small, fast-moving birds, the downward-forward stroke is the main source of forward motion, while the upward-backward stroke is for recovery only and is made by partially folding the wing and separating the remiges so as to reduce pressure on the upper surface. The proximal half of the wing moves only slightly during the propeller's operation, since its primary function is to give lift and act as a shaft for the propeller. It thus has a steadying influence by preventing the bird's body from bounding up and down with the thrusts of the propeller.

Spread the unplucked wing so that the leading edge is on a straight line perpendicular to the body, and force the alula to stand out slightly from the wing. Observe that there are several apertures or "slots" between the tips of the outermost primaries and another between the alula and wing margin. Broad-winged, soaring birds such as eagles and cathartid vultures show slots that are very much larger. Slots prevent stalling and increase lift by making air flow fast and evenly over the upper surface, thus reducing turbulence. (See Figure 7c.)

The wings of both the pigeon and the sparrow, relatively short and broad, consequently have a low aspect ratio. With so small a wing area in relation to body weight, the pigeon must flap its wings rapidly in order to gain sufficient lift. The Herring Gull *(Larus argentatus)* with long, narrow wings (high aspect ratio) need not move its wings as fast. In normal, unhurried flight the pigeon flaps its wings an average of 3.0 times per second, the Herring Gull only 2.3 times per second (Blake, 1947). The sparrow may attain a velocity of 39 miles per hour (63.55 km/h) (Schnell, 1965); the pigeon, 47 mph (75.62 km/h) (Meinertzhagen, 1955); and the Herring Gull up to 49 mph (78.84 km/h) when flying with a strong wind (Schnell, *op. cit.*). For flight-speeds attained in a wide variety of birds, see Terres (1980).

The Tail

The bird's tail is actually a small bony and fleshy structure hidden by the feathers. In descriptive ornithology, however, the term "tail" has come to mean the feathers that arise from this structure.

Spread out the pigeon's tail and observe that it is fan-shaped, with a rounded posterior margin due to the graduated lengths of the feathers. Notice that the tail feathers are of two kinds, the **rectrices** (singular **rectrix**) and **tail coverts:**

Rectrices The strong, conspicuous feathers whose outer ends form the posterior margin of the tail. They are the flight feathers of the tail and correspond to the remiges of the wing. They are paired, the number of rectrices growing on each side of the tail being equal.

How many rectrices are there in the pigeon? Note that the rectrices partially overlap each other. Which one is not overlapped? How many rectrices are there in the sparrow?

Tail Coverts Similar in appearance to the coverts of the wing. They overlie and underlie the flight feathers in much the same way. In descriptive ornithology they are not divided into separate rows or groups. The under tail coverts, sometimes collectively known as the **crissum,** are separated from the feathers of the abdomen by an imaginary line drawn transversely through the vent. The **upper tail coverts** are less clearly marked off from the feathers of the rump. They may be considered to

Activity

On Figure 1 and Figure 2b label the following: upper tail coverts; rectrices. On Figure 2a, label the crissum.

end at an imaginary line drawn transversely through a point on the upper surface directly over the vent.

During flight a bird's tail, especially when spread, serves a variety of purposes. It supplements the lifting surface of the wings and forms a slot in conjunction with the trailing edge of the wings. It can serve as a rudder by steering the bird to left or right; as an elevator, by directing it up or down; or as a brake by retarding its forward speed.

The Legs and Feet

The legs and feet of the bird are less specialized than the wings and show a greater diversity in structure corresponding with the varied habits of different species.

Examine either an articulated human skeleton, or a detailed chart of one, and locate the following bones of the hindlimb (pelvic appendage): femur (thigh bone), patella (knee-cap), tibia and fibula (shin bones), 7 tarsals (ankle bones), 5 metatarsals (foot bones), and 14 phalanges (toe bones).

The skeleton of a bird's pelvic appendage, like the skeleton of a bird's wing, is built upon the same plan as its homologue in man. But since it is more specialized than the human limb, the bones are less easily recognizable.

Examine now a prepared articulated skeleton of a pigeon's pelvic appendage. (See Figure 4b.) Count the separate bones. (The patella may be missing, having been lost when the skeleton was prepared.) What is the difference in the number of bones in the bird and in man? Identify the following bones in the pigeon's pelvic appendage:

Femur A stout, cylindrical bone whose proximal part bends inward and has a prominent head that is received by the pelvic girdle. An irregular projection, the **trochanter,** extends beyond the shaft. The distal extremity of the femur has a pulley-shaped surface, which receives the patella, and two convex condyles, which articulate with the tibiotarsus and fibula.

Patella A small bone found in front of the knee joint.

Tibiotarsus *and* Fibula Two bones running parallel to each other. The tibiotarsus, a composite bone (see following explanation), is expanded at its proximal end, where it articulates with the inner condyle of the femur. It extends to the heel. The fibula articulates with the outer condyle of the femur but is a poorly developed bone extending only two-thirds of the way to the heel. Closely pressed to the outside of the tibiotarsus as a slender spicule, the fibula is partly fused with the shaft of the tibiotarsus.

Tarsals The tarsals, or ankle bones, do not occur in the pigeon as separate elements. Some have fused with the lower end of the tibia and, because of this fusion, the tibiotarsus (tibia + tarsal elements) gets its name. Other tarsals have fused with the next bone of the foot to be considered.

Metatarsals The second, third, and fourth metatarsals fuse to form one bone that is homologous to the human instep. To the proximal end of this composite bone are fused, as previously mentioned, certain tarsals. This bone is, therefore, properly called the **tarsometatarsus,** although it is more commonly referred to as the **metatarsus.** The proximal end of the metatarsus is very irregular and is provided with two concavities that articulate with the tibiotarsus. The three metatarsals making up this composite bone remain distinct at the distal end as three articular projections for the anterior toes. A rudimentary first metatarsal, sometimes known as the accessory metatarsal, is connected by a ligament with the inner and posterior aspect of the distal part of the metatarsus. The fifth metatarsal is absent in adult birds.

Phalanges There are four toes or digits, three anterior and one posterior. Each is made up of a series of phalanges placed end to end; the proximal phalanges are nearest to the metatarsus, and each of the distal phalanges terminates in a strong, curved **claw.** The toe projecting posteriorly is the first toe, or **hallux** (Toe No. 1). It is articulated with the accessory metatarsal and has two phalanges; it is homologous to the "big toe" in man. The three toes extending anteriorly are articulated with the three projections on the distal end of the metatarsus. The innermost of these three toes has three phalanges;

the middle toe, four; and the outermost toe, five. They are, respectively, the second (Toe No. 2), third (Toe No. 3), and fourth (Toe No. 4) toes and are homologous to the second, third, and fourth toes of man. A fifth toe is not found in birds.

Turn now to the pigeon and pluck the feathers from the leg. Find the position of the bones studied; then become familiar with the following parts of the pelvic appendage:

Thigh The proximal segment of the leg containing the femur. It was, prior to plucking, entirely hidden.

Crus The distal segment of the leg containing the tibiotarsus. Sometimes called the **shank** and more popularly known as the "drumstick," it is entirely feathered.

Knee The junction of the thigh and crus. It bends forward as in man.

Foot The remaining portion of the pelvic appendage. It is divisible into two parts:

Tarsus. The third segment of the pelvic appendage, between the crus and the bases of the toes. It is noticeably scaled except at the proximal end, where it is feathered. It contains the metatarsus and accessory metatarsal.

Toes Four in number.

Heel The junction of the crus and foot. It always bends backward. The bird is, therefore, **digitigrade,** walking on the toes with heels in the air. Man is plantigrade, walking on the soles of the feet with heels on the ground.

Examine the unplucked leg of the pigeon. Locate the crus. In descriptive ornithology the crus, together with the feathers, is called the **tibia** and will be thus designated in later sections of this book.

Activity

On Figure 1 and Figure 2a, label the following: hallux; tarsus; tibia. On Figure 1, number the toes of one foot.

Flex the pigeon's pelvic appendage at the knee and heel, drawing the tarsus to the crus and the crus to the thigh; then extend the appendage and repeat the procedure. Note that when the appendage is flexed the toes assume a grasping position and that when the appendage is extended the toes straighten out. This action of the toes is brought about by tensions exerted upon them by muscles in the thigh and crus. When the appendage is flexed, the tension brought to bear on the tendons causes the toes to bend. By means of this arrangement the toes automatically grasp and hold fast to a perch while the bird squats during rest or sleep.

References

Baumel, J.J.; King, A.S.; Lucas, A.M.; Breazile, J.E.; and Evans, H.E., eds.
1979 *Nomina anatomica avium: An annotated anatomical dictionary of birds.* London: Academic Press.

Blake, C.H.
1947 Wing-flapping rates of birds. *Auk* 64:619–620.

Fisher, H.I.
1940 The occurrence of vestigial claws on the wings of birds. *Amer. Midland Nat.* 23:234–243. (A study of 227 genera, the majority North American.)

Meinertzhagen, R.
1955 The speed and altitude of bird flight (with notes on other animals). *Ibis* 97:81–117.

Norris, R.A.
1972 Data on nictitating rates in birds. *Bird-Banding* 43:289–290. (In 257 individual birds, representing 44 species, mostly passerine, the average rate of "blinking" with the nictitating membrane was 17.9 per minute.)

Rand, A.L.
1954 On the spurs on birds' wings. *Wilson Bull.* 66:127–134. (The occurrence of wing spurs is noted for all species of screamers, Family Anhimidae, some plovers, two jacanas, and two ducks. They occur on different parts of the wing and involve the radius, the radiale, or the fused metacarpals, depending on the species. The structures are used in fighting. Wing spurs should not be confused with vestigial claws. See Fisher, 1940.)

Schnell, G.D.
1965 Recording the flight-speed of birds by Doppler radar. *Living Bird* 4:79–87.

Terres, J.K.
1980 *The Audubon Society encyclopedia of North American birds*. New York: Alfred A. Knopf.

Thomson, A.L., ed.
1964 *A new dictionary of birds*. New York: McGraw-Hill. (Various external parts of birds are listed alphabetically and described.)

To see how a knowledge of topography is applied, examine at least one authoritative work in which birds are described by topographical parts. Among the several suitable works, the following series is usually available in most college and university libraries.

Ridgway, R.
1901–1919 The birds of North and Middle America: A descriptive catalogue of the higher groups, genera, species, and subspecies of birds known to occur in North America, from the Arctic lands to the Isthmus of Panama, the West Indies and other islands of the Caribbean Sea, and the Galapagos Archipelago. Parts 1–8. *Bull. U.S. Natl. Mus*. No. 50.

Ridgway, R., and Friedmann, H.
1941–1946 The birds of North and Middle America. Parts 9–10. *Bull. U.S. Natl. Mus*. No. 50.

Friedmann, H.
1950 The birds of North and Middle America. Part 11. *Bull. U.S. Natl. Mus*. No. 50.

Flight

Flight is motion through air. Birds achieve this primarily by using their wings. When in flight birds must contend with two forces: ever-present gravity that pulls them down and air resistance that causes drag. To offset these forces, birds must provide lift and thrust. Read how the bird's wing is structured and functions to meet these two basic needs in "Topography," pages 19–21.

There are three general types of flight: gliding, soaring, and flapping.

Gliding Flight

Gliding flight is moving straight forward in still air with wings spread to create lift equal to the bird's weight. If lift decreases in relation to weight, the bird must either flap its wings a few times, or dive, in order to maintain forward motion and avoid a stall. When gliding downward, the bird partly closes ("sets") its wings to reduce drag, while gravity pulls the bird down against the lift. The steepness of the descent depends on the gliding angle and the **sink rate**—i.e., the rate of downward speed in relation to the surrounding air. The steeper the bird descends, the steeper the bird must adjust its gliding angle to allow for a greater pull of gravity against the drag.

Soaring Flight

Soaring flight is gliding through air without loss of altitude. There are essentially two kinds of soaring: static soaring and dynamic soaring.

Static Soaring This type of soaring entails gliding in currents of air rising at a speed greater than the sink rate. Some vertical currents are warm air, or **thermals,** rising from open land that heats faster than adjacent cooler areas. Thermals begin rising in columns above the cooler air, then expand in huge bubbles free of the ground. Warm air within each bubble circulates upward in the center and downward on the outside, producing a revolving ring of warm air somewhat like a smoke ring. To soar, the bird enters the top of the bubble and circles on the rising air from the center. There is no need for wing-flapping as long as the vertical currents of air are strong enough to equal the bird's sink rate. Other vertical currents of air are **obstruction currents,** currents caused by the upward deflection of horizontal winds striking obstacles or by horizontal winds passing down over obstacles and eddying upward.

Birds that soar successfully on thermals are the broad-winged hawks, eagles, and cathartid vultures, all large birds with light **wing-loading**—the relation of the total area of their wings to their weight—and wings with low aspect ratios (see page 21) that are cambered and slotted and tails that are wide. Birds that soar on obstruction currents include gulls that ride on updrafts along sea cliffs or on updrafts caused by ships. During their migrations, hawks, eagles, and other usually broad-winged birds pass along mountain ranges that run north-south, taking advantage of the easier flight afforded by the updrafts from obstructed winds.

Dynamic Soaring The second type of soaring uses horizontal airstreams or winds of different heights and velocities by alternately gliding up and down between them. Dynamic soaring is a way of life for some pelagic birds that spend much of their time in the air above the sea. Prevailing winds over the sea blow with the ocean currents at a minimum rate near the surface, where they are slowed by friction with the waves, and at gradually greater velocity at higher altitudes, to a maximum velocity at 50 feet (15.24 m) or more. The birds exploit these successive variations in speed by circling swiftly down with the airstream to gain momentum, then circling up against the airstream until the momentum is exhausted, then circling down again to renew the momentum, and so on.

Birds that soar dynamically over the sea are the larger pelagic birds—the albatrosses, for example. They have light wing-loading, but their wings have a contrastingly high aspect ratio and practically no camber and slotting. Lift is as mandatory for them as speed, which is all-important in dynamic soaring.

Flapping Flight

Flapping flight requires lift, as in gliding flight, and thrust from the distal half of each wing, the "propeller," to maintain speed and equal drag. Under "Topography," page 21, first review the function of the bird's wing in ordinary level flapping flight. Then study the progressive action in forward flapping flight, as shown in Figure 8.

A bird taking off from an elevation may drop downward to gain sufficient speed for flapping flight. In taking off from the ground or water, the bird may run or paddle into the wind to gain a sufficient speed, relative to the air, to leave the surface in flapping flight. If, however, there is no wind and the bird cannot run or paddle fast enough for takeoff, it needs extra thrust. It gains this by a more energetic and exaggerated action of the wings, with some additional movement of the proximal half of each wing that usually remains stationary in ordinary flapping flight. (See Figure 9.) A heavier bird may use the upstroke to provide some forward thrust by rotating the whole wing at the shoulder so that, as the wing moves backward and upward, it pushes more air backward and increases the lift.

When landing, a bird must reduce its speed. It accomplishes this by flying into the wind, raising its body vertically, flapping its wings against the direction of flight with vigorous upward and backward thrusts, and extending its legs and feet forward to cushion the impact with the perch or ground. When landing on water, a bird need not slow the forward momentum as much since the water helps to soften the impact. When landing with no wind, the bird exaggerates the forward and backward thrusts of the wings. (See Figure 10.)

Some birds can modify flapping flight for one reason or another. Tree-climbing birds produce an

Figure 8 **Progressive Action of the Wings in Flapping Flight**
(a) The downstroke is just under way, with primaries overlapped and curved upward from pressure against the air. As the wings continue downward, the primaries act as propellers by pulling the wings forward and the whole bird with them. The secondaries meanwhile, provide lift. (b) The downstroke is completed, with the primaries reaching forward and downward to the maximum extent. (c) The upstroke is under way, with the primaries separated and drawn toward the body. During this recovery stroke, the primaries may push backward slightly against the air to propel the bird forward while the secondaries provide lift. (d) The upstroke nears completion as the primaries begin reaching far backward and upward. (e) The upstroke completed, the wings are about to begin another downstroke.

Figure 9 **Taking Off with No Wind**
To gain momentum for lift, the scaup paddles vigorously, picking up speed with its feet while reaching far forward with its wings to obtain the maximum pull and far backward for the maximum push.

(a) Bluebird

(b) Scaup

Figure 10 **Landing with No Wind**
When descending to perch or ground (a), or water (b), the bird starts reducing momentum by braking against the direction of flight—positioning its body upright and spreading its tail to resist the air, and "back-stroking" by stretching its wings far backward to pull strongly against the air. Meanwhile the bird brings its feet forward, ready for impact.

Figure 11 **Hovering**
When hesitating in midair for one reason or another, most birds (for example, the kingfisher at right) simply hover by positioning the body vertically and flapping the wings forward and backward to provide lift while preventing thrust. But a hummingbird (at left), when hovering, with body motionless, to take nectar from a flower, sculls its wings forward and backward. In the forward stroke the leading margin slightly precedes the rest of the wing, which is held upright enough to prevent thrust; in the backward stroke the wing swivels so that the leading margin again precedes the rest of the wing to achieve the same effect.

undulating flight by alternately propelling themselves upward with quick wing-strokes, then setting their wings and dipping downward. This sort of action enables them to swoop upright against a tree trunk. Many birds may **hover** in midair, without moving forward or backward, upward or downward, usually to center their attention on something below. For examples of hovering, see Figure 11.

In flight the bird's tail is so much less important than the wings that most birds can fly when their rectrices are removed. The primary function of the tail is steering, especially at low speed. To change direction, the bird spreads and twists its tail in the chosen direction. (The bird may also steer by tilting and turning the body and lowering one wing in the chosen direction.) When spread and lowered, the tail becomes a brake in landing or a deflector to increase lift in take-off. Flicks of the tail upward and downward help stabilize the bird just as it perches or while it is perched in a strong wind.

For instructive information on various aspects of flight, read G. Rüppell (1977) and "Flight" by J.K. Terres (1980).

References

Rüppell, G.
1977 *Bird flight.* New York: Van Nostrand Reinhold Company. (Chapter 2 includes feather structure in relation to flight.)

Terres, J.K.
1980 Flight. In *The Audubon Society encyclopedia of North American birds.* New York: Alfred A. Knopf.

Feathers and Feather Tracts

Feathers are peculiar to birds and constitute their principal covering. Like the sheath of the bill, the scales on the feet, and the claws on the toes, feathers are horny, keratinized growths of the skin or integument. They develop from tiny pits or follicles in the skin, just as do the hairs of mammals.

The scales on the feet of birds are clearly of the reptilian type and feathers probably evolved from comparable scales, becoming lengthened, elaborated, and diversified. Except in initial development, scales and feathers bear little resemblance to each other.

The two most important functions of the bird's feathers are to provide insulation—thus reducing loss of body heat—and to make flight possible by giving a streamlined contour and increasing the surface of the wings and tail. Through their coloration, feathers also aid certain species in concealment, in sex and species recognition, and in numerous displays. For additional functions of feathers, see Stettenheim (1976).

Structure of a Typical Feather

Remove from the unplucked wing of a Common Pigeon *(Columba livia)* one of the best-developed primaries. Look for the parts described below and locate them on Figures 12 and 13. The terms "dorsal" and "ventral" refer to the upper and under surfaces of the feather itself without regard to the feather's position on the bird.

Shaft

The **shaft** is the axis of the feather and has two parts:

Calamus The proximal (lower) part of the shaft, without vanes. Remaining almost entirely in the skin follicle from which the whole feather developed, the calamus is a tubular and somewhat transparent barrel, circular in cross section, and tapered at the basal end. At this point is an opening, the **inferior umbilicus,** where the nutrient pulp entered during the growth of the feather. The distal end of the calamus is marked by the **superior umbilicus,** a minute opening on the ventral side of the shaft—i.e., the side that faces toward the body of the bird when the wing is closed—between the points where the vanes begin. The superior umbilicus is the remnant of the open, upper end of the tube of epidermis that formed the growing feather.

With a sharp scalpel, slice lengthwise through the covering or **cortex** of the calamus. Note within the hollow interior a series of downward-projecting, cup-like structures seemingly fitted one into the other. Commonly called **internal pulp caps,** they were formed from the cornification at regular points of the layer of epidermis enclosing the pulp, long since resorbed. Each cap consists of a dome and a side-wall that adheres to the calamus next to it.

Rachis The distal part of the shaft supporting the vanes. It is a continuation of the dorsal side of the feather tube above the calamus, from the superior umbilicus to the tip of the feather. Roughly quadrangular in cross section, it has two layers: an outer

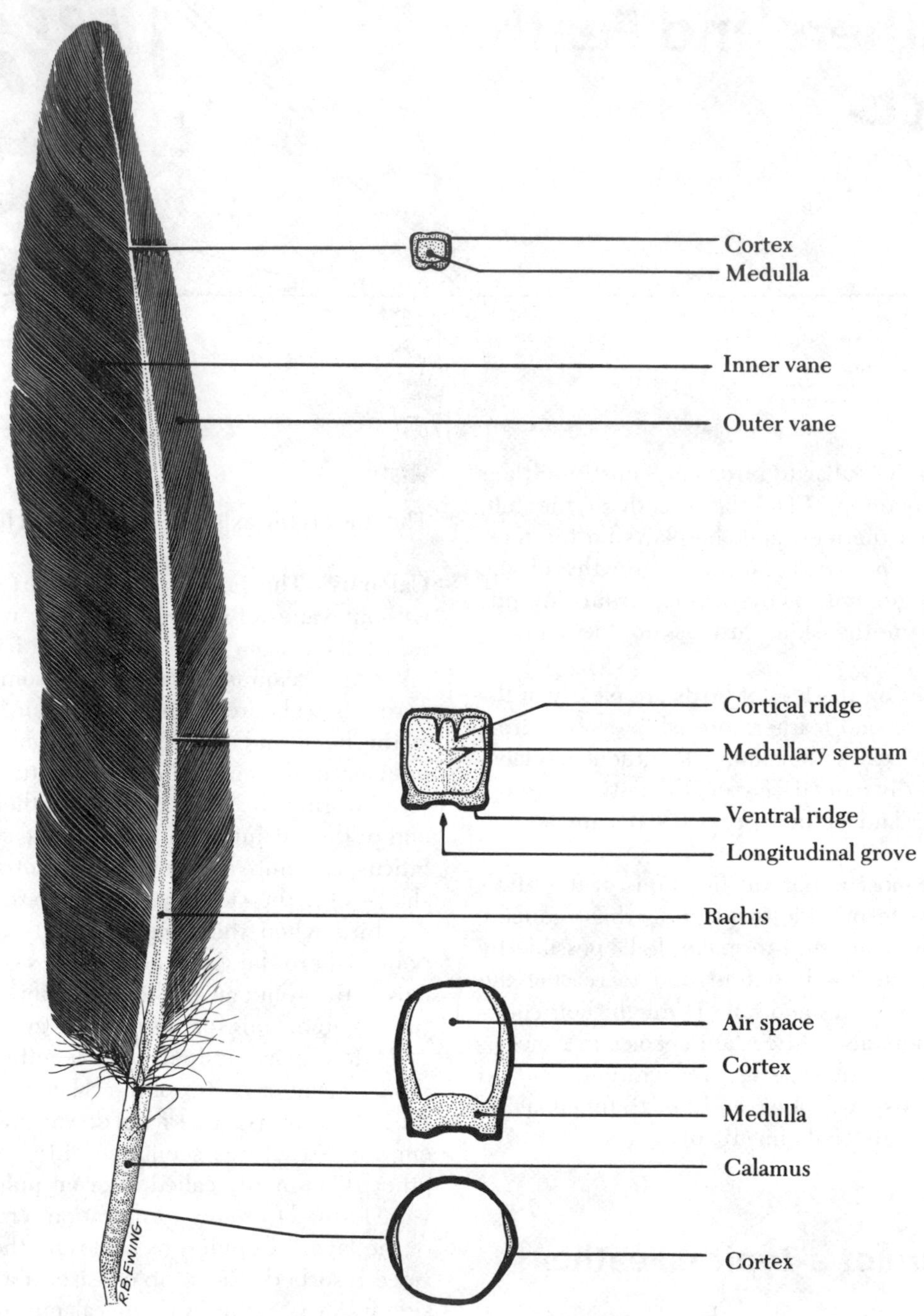

Figure 12 **Primary of Common Pigeon**
Whole feather with cross sections of the shaft, ventral view.

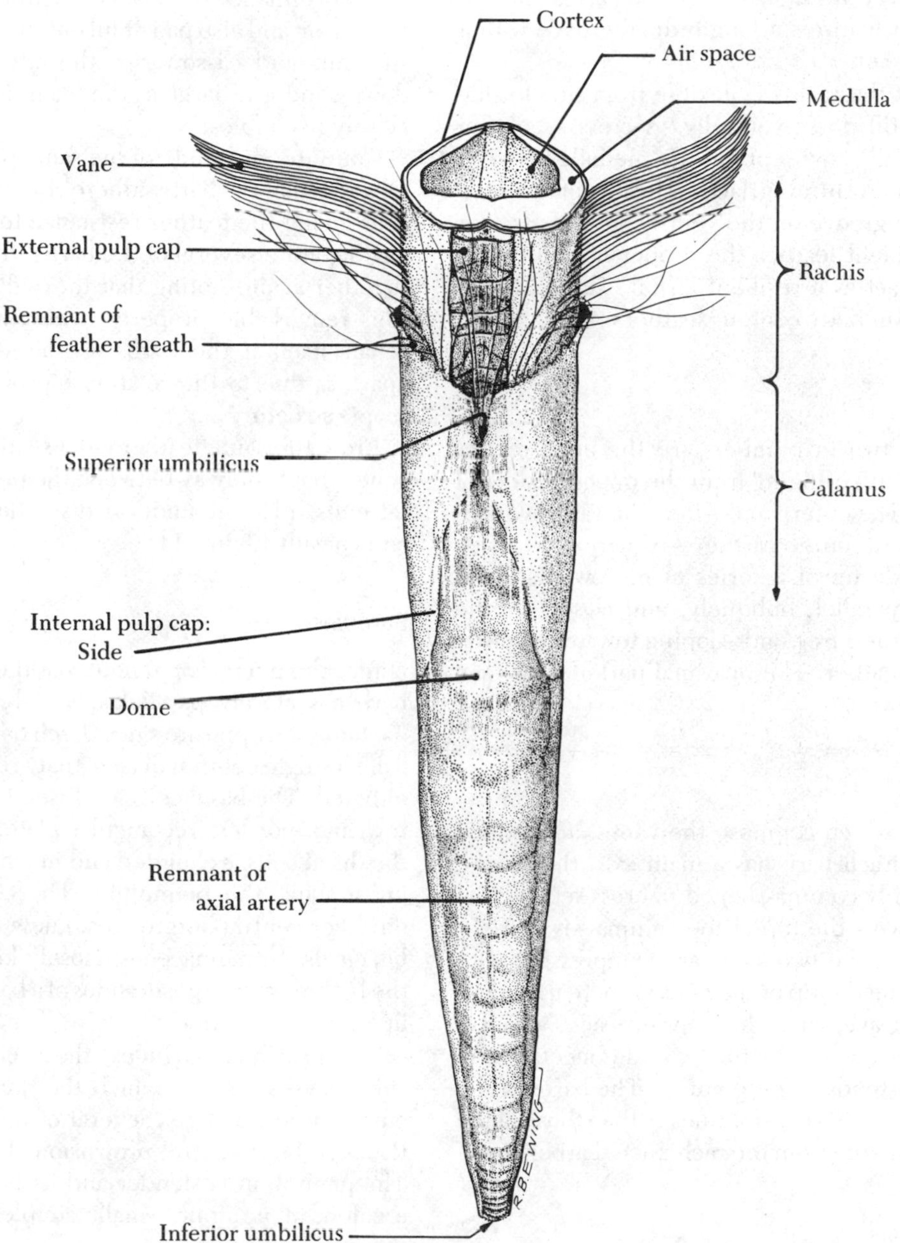

Figure 13 **Primary of Common Pigeon**
Calamus and proximal end of rachis, oblique view.

cortex, thin and transparent, with ridges projecting inward from the dorsal surface; and an inner **medulla** comprised of pithy tissue, firm and opaque. Examine the ventral surface of the rachis and observe that it features a longitudinal groove with a ventral ridge on each side.

Note that the rachis is flexible from side to side but quite stiff dorso-ventrally. A cross section of the rachis will show **septa** in the medulla between the dorsal and ventral surfaces. These septa plus the longitudinal groove on the ventral surface assist in giving the flight feather the proper flexibility and stiffness to act as a resilient airfoil. The septa are not present in body contour feathers.

Vanes

The **vanes,** two in number, are the more or less flexible sheets springing from the opposite sides of the rachis. The outer vane—i.e., the vane overlapping the next outer feather—is narrower. Each vane is made up of a series of narrow, flattened plates, set parallel, obliquely, and closely on the rachis with their free ends sloping toward the distal end of the feather. The proximal part of each vane is rather downy.

Barbs

The plates, which compose the vanes, are termed the **barbs.** Each barb has a main axis, the **ramus,** that is roughly comma-shaped in cross section: the dorsal surface—the top of the comma—is more or less rounded; the two sides are compressed, with the side facing the tip of the feather more flattened, or even concave, than the opposite side, which is always convex; ventrally the two sides meet to form a ridge that tends to be pointed. The barbs of the outer vane are shorter and thicker than those of the inner and emerge from the rachis at a sharper angle.

Afterfeather

The **afterfeather,** virtually absent in the pigeon, is any structure on the rim of the superior umbilicus. It may consist of a row or tuft of barbs, or a shaft, the **aftershaft,** that bears barbs. Examine the large afterfeather on the abdominal feather of a grouse or pheasant and note the aftershaft from which barbs emerge just as they do from the shaft of the main feather. The afterfeather is generally downy in texture and shorter than the main feather. Afterfeathers enhance the insulative property of a bird's feathering and also pad or fill out the body contours. In emus and cassowaries the afterfeather is not downy and is as large as the main feather, which it closely resembles.

Continue the study of the same pigeon primary. Observe that the barbs adhere closely to each other, thus giving the feather resistance to the passage of air. Separate several of the barbs, then press them together again, noting that they adhere as before. This remarkable property, wherein the vane can repair itself if the barbs are accidentally pulled apart, is due to the relationship of certain microscopic structures.

From the same feather cut a section of the inner vane, about midway between the proximal and distal ends, place it under a dissection microscope, and consult Figure 14.

Barbules

Under the microscope, note that the ramus of each barb has closely parallel sets of branches, called **barbules,** on opposite sides. Each barbule is formed from a single column of cells that are serially differentiated. The basal cells are fused into a long, narrow, more or less rectangular plate, the **base,** and the distal cells are jointed and much smaller, forming a stalk, the **pennulum.** The cells often bear dorsal or ventral outgrowths, known collectively as **barbicels.** In pennaceous (closely knit, flat) vanes, the barbules on opposite sides of a barb are different in form and function.

The **proximal barbules,** those on the proximal side, have a base in which the dorsal edge is recurved into a **flange.** Several of the outer cells of the base bear ventral protrusions known as **teeth.** The pennulum is slender and its barbicels, if any, are few in number, small, simple, pointed processes, known as **cilia.**

The **distal barbules,** those on the distal side of a ramus, have a base in which the dorsal edge is thickened but not recurved. Their teeth are more elaborate and variable than those of proximal barbules. Their most distinctive parts are long barbicels with hooked tips, the **hooklets,** situated on the

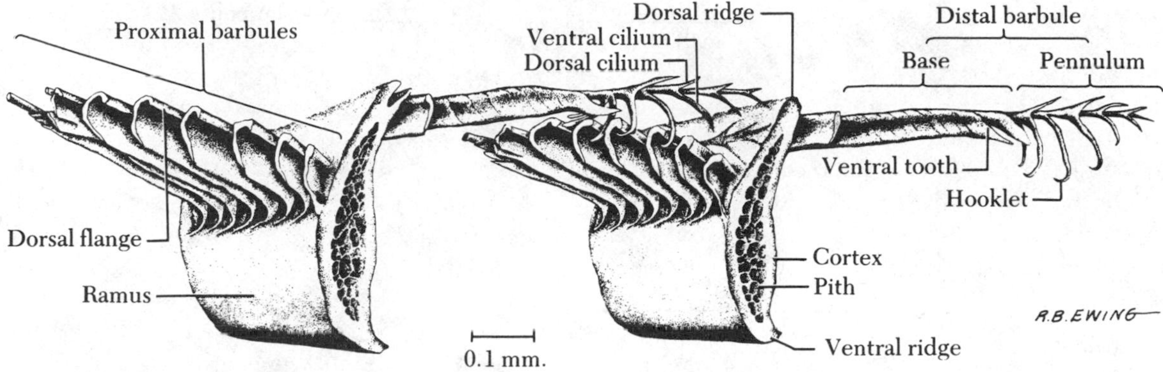

Figure 14 A pennaceous vane showing two adjoining barbs with proximal and distal barbules interlocking. (From Lucas and Stettenheim, 1972.)

ventral side of the inner cells of the pennulum. Beyond the hooklets are dorsal and ventral **cilia,** which tend to be larger and more numerous than those on the pennula of proximal barbules.

Functionally, the distal barbules of one barb cross over one to four proximal barbules of the next higher barb, the hooklets of one grasping the dorsal flanges of the other. The other barbicels maintain proper spacing between the barbules and restrain the hooklets from sliding off the flanges. This arrangement creates a flexible yet strong and self-adjusting mechanism that holds the barbs together.

Pluck a typical feather from the body of the pigeon and study one of the downy barbs from its basal part. Notice that the ramus is more slender and flexible than that of the barb of the pigeon primary. The barbules are of the plumulaceous type, each with a relatively short, strap-like base and a long slender pennulum. Plumulaceous barbules may become loosely entangled, but they do not interlock. They create downy (fluffy) texture in a feather, which varies according to the length, orientation, and branching of the barbs and the size and orientation of the barbules. The pennulum has swellings along its course, called **nodes,** and thus resembles a stalk of bamboo. The nodes are shaped in various ways that are characteristic of different orders of birds. (See Figure 15, showing the downy barbule of the Common Pigeon.) Partly for this reason, downy barbules have been used more successfully than pennaceous barbules in identifying feather remains as in archaeological finds and in the stomach contents of mammalian predators (see Day, 1966).

Feathers of Adult Birds

All feathers on adult birds are called **teleoptiles.** There are five kinds on the pigeon; a sixth kind occurs on many other groups of birds.

1. Contour Feathers

Contour feathers, which shape the body, have vanes that are at least partly pennaceous (closely knit, flat). In flight feathers (remiges and rectrices) the vanes are almost entirely pennaceous whereas in body feathers only the distal half or less has such texture. Also, the afterfeather, if any, is better developed on body feathers than on flight feathers.

Examine the contour feathers on different parts of the pigeon, observing especially the variation in size, shape, and texture. Compare them with the drawing of a typical contour feather of the body (Fig. 16). In all flight feathers the vanes are of unequal width, the outer one being the narrower; from the innermost remiges and rectrices to the outermost, the outer vane becomes gradually narrower. The contour feathers making up the body plumage generally have vanes of equal width. Noteworthy is that the coverts of the flight feathers show a complete transition in structure from a flight feather to a body feather. Thus a greater covert is more like a flight feather than a body feather, while

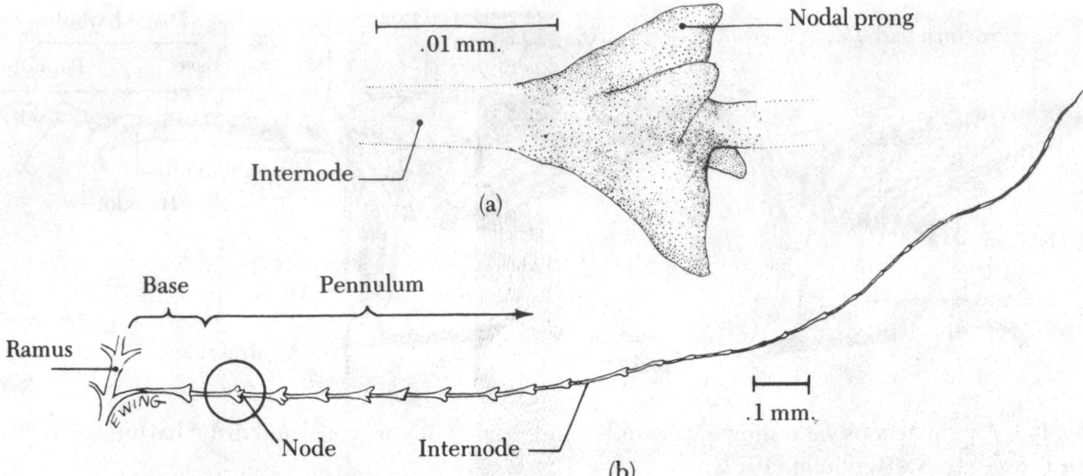

Figure 15 **Downy Barbule of Common Pigeon**
(a) Magnified view of node encircled in (b). (b) Whole barbule.

a lesser covert is more like a body feather than a flight feather.

Among other species of birds are many peculiarly modified contour feathers:

Modified Remiges The primaries and secondaries may be modified in overall shape and structure, in their barbs, and in microscopic structure of their barbules. Modifications are most commonly for flight—e.g., emarginate vanes; stiffening of outer vanes; and the glistening flap, the **tegmen,** on the ramus. In some birds the remiges may be modified for sound production—e.g., curved, stiffened, or narrowed, as the case may be, in the American Woodcock *(Scolopax minor),* male Cock-of-the-Rock *(Rupicola rupicola),* Lesser Whistling-Duck *(Dendrocygna javanica),* most guans, and in the highly modified inner secondaries of certain male manakins. Remiges are also modified for quieting flight—the comb-like leading edge of the outermost primaries of most owls; for physical protection—the bare quills of the cassowary wing (that cannot be identified as either primaries or secondaries); and for visual signals—the innermost secondaries of the Mandarin Duck *(Aix galericulata),* spangled tips of waxwing secondaries, primaries of the Hawfinch *(Coccothraustes coccothraustes),* and the elongate remiges or their coverts in three species of male nightjars.

Modified Rectrices In woodpeckers, woodhewers, and creepers, which use their stiffened tails as supports while climbing, the barbs continue to the tip of the rachis. The terminal barbs have thick, laterally compressed rami without barbules. More important than the lack of interlocking is the arrangement whereby the barbs lie nearly parallel to the tip of the rachis and appear to reinforce it. Some swifts—e.g., the Chimney Swift *(Chaetura pelagica)*—whose tails function as supports in vertical roosting, have rectrices with shafts free of distal barbs and consequently projecting beyond the vanes as stiff spines. Some rectrices are racket-shaped. Those of certain kingfishers (*Tanysiptera* spp.) and drongos *(Dicrurus paradiseus)* grow in their special shape, whereas those of motmots lose the subterminal barbs after the tail is formed. Rectrices may be modified to create sounds by vibration in courtship flights—e.g., those of the Black Turnstone *(Arenaria melanocephala)* and Common Snipe *(Gallinago gallinago).* Probably no birds show more remarkable modifications in all their rectrices than the male Superb Lyrebird *(Menura novaehollandiae):* his outermost pair of rectrices, broad and S-shaped, produce the lyre-like form; the innermost pair, extremely narrow, cross each other soon after emerging from the skin and then curl forward near their tips; the remaining 14 pairs, delicate throughout, with hair-like barbs wide apart

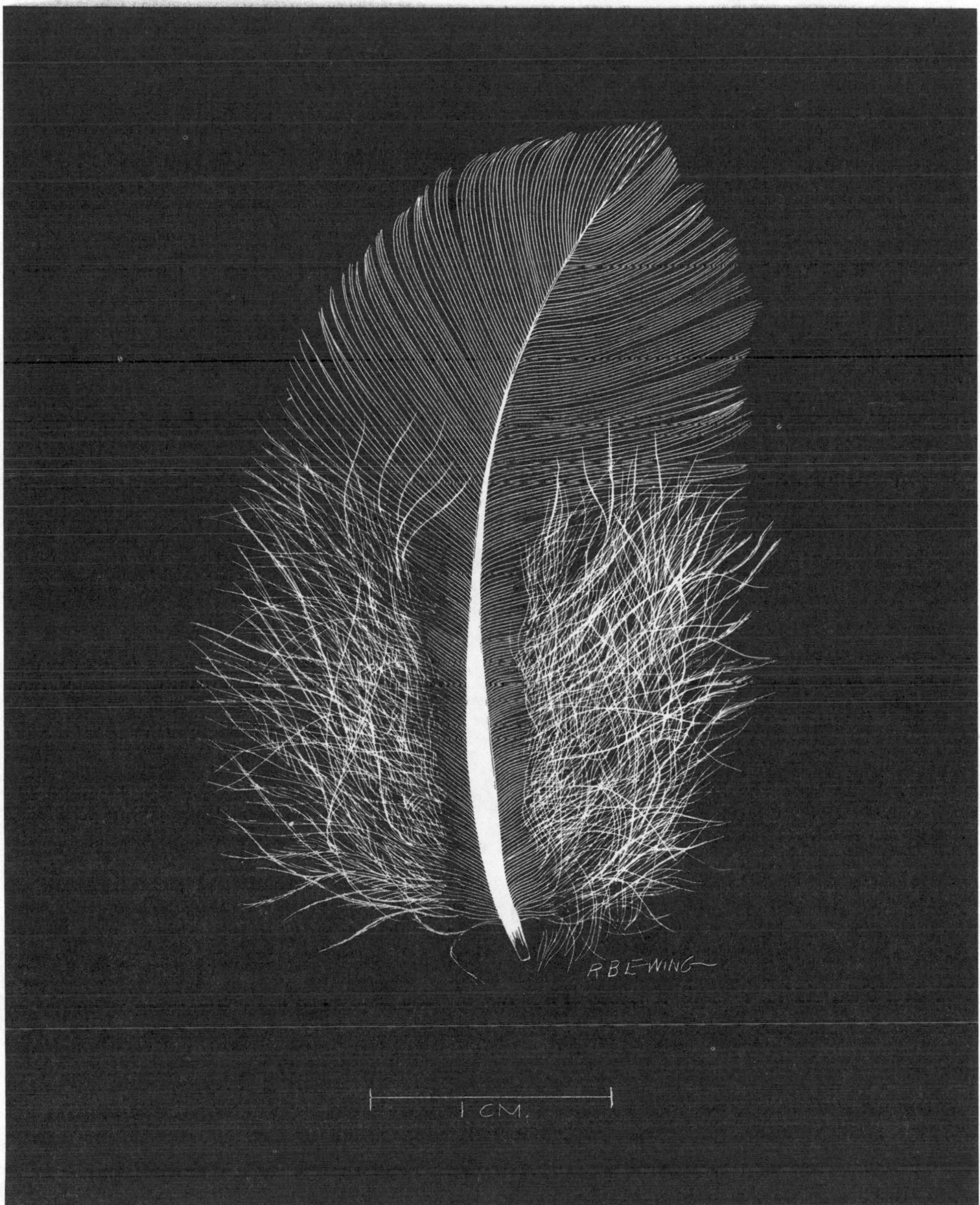

Figure 16 **Typical Body Contour Feather of Common Pigeon**

and without barbules, give a lacy effect to the tail when it is fully displayed. The male lyrebird must pass through five to six years of molts and plumages before his tail attains the adult form.

Other Modified Contour Feathers The auriculars (ear coverts) in most birds are longer around the anterior margin of the ear opening than around the posterior margin. They extend posteriorly over the ear opening and, with their widely separated barbs and short barbules, these feathers form a screen that prevents intrusion of foreign particles. (Feathers of a similar structure grow densely over the nostrils of such birds as crows, forming nasal tufts for the same function.) Auriculars also improve the hearing ability in nocturnal owls, certain parrots and hawks, and owlet-frogmouths. Like the external ear of mammals, they collect sound energy and funnel it into the ear. In owls, the auriculars are mounted on large movable flaps of skin.

Contour feathers with a short, widened bare tip to the rachis arise on the front of the head in certain grebes and rails. The bare tips appear to prevent abrasion of feathers while the birds push through aquatic vegetation. The scapulars and rectrices of anhingas are corrugated, owing to modifications of the barbs. The resulting ripples are believed to have solely a visual effect and do not play any role in swimming.

The contour feathers of many birds are variously developed as plumes for display, and they may arise from almost any part of the bird except the wings. In egrets the plumes are principally the back feathers that extend over the tail; in some species of birds-of-paradise they are long, flowing feathers that project from the flanks. In both egrets and birds-of-paradise the plumes are of soft texture with long, widely spaced barbs and reduced barbules. Sometimes birds have the coverts of their flight feathers enormously developed. Thus in the male Quetzal *(Pharomachrus mocino)*, a trogon, the median coverts of the remiges are drooping plumes, while the upper tail coverts are long streamers, the central pair being four times the length of the tail; in the male peafowl (*Pavo* spp.) the lower rump feathers, as well as the upper tail coverts, are of striking form, owing to the uneven arrangement of the barbs and barbules, and are extended back over the rectrices as a long "train." Numerous species have plumes arising from the head as crests with varying conditions of shaft, barb, and barbule development. Both sexes of the peafowl, for instance, have crests of delicate feathers, each scantily barbed except at the tip, where there are wide vanes of interlocked barbs.

Contour feathers on the belly of several species of sandgrouse *(Pterocles)* have peculiarly coiled barbules that serve to retain water by interfacial tension. Male sandgrouse wade into water holes, soak their belly plumage, and fly back to the nest, sometimes far away in the desert. When they arrive, the chicks run out, cluster under the male and "strip" the water from his feathers.

Activity

Inspect a series of modified contour feathers of various birds and note the structural differences between these feathers and the primary of a pigeon.

2. Semiplumes

These are small, white feathers hidden beneath the contour feathers. On the pigeon (Fig. 17) they are on the abdomen, undersides of the wing and tail, and the sides of the body. On other birds they are often along the margins of the feather tracts. Search for them on the pigeon and pluck several for inspection. The semiplume has a downy texture—no interlocking barbs—and its rachis is longer than the longest barb. Structurally, the semiplume is intermediate between the contour feather and the next kind of feather to be considered; functionally, the semiplume assists in entrapping air for thermal insulation.

3. Adult Down Feathers

These feathers—to be distinguished from nestling down feathers described later—are uncommon on the adult pigeon, but they may be found by careful searching under the contour feathers on the sides of the body as far back as the tail. The down feather

Figure 17 **Semiplume of Common Pigeon**

appears as a fluffy tuft (Fig. 18). Pluck one, mount it on a slide, and examine it under the microscope. Soft throughout, the down feather differs from the semiplume in having a rachis that is either absent or relatively short. Down feathers may grade into semiplumes, depending on the relative length of the rachis and the longest barbs. Like the semiplumes, down feathers assist in thermal regulation. They are especially abundant on waterfowl and other aquatic birds, and they possess afterfeathers in birds that have afterfeathers on their contour feathers.

4. Filoplumes

These are very slender, hair-like feathers (Fig. 19a) found all over the body of the pigeon, always accompanying other feathers, including semiplumes

Figure 18 **Adult Down Feather of Common Pigeon**

and down feathers. Typically, there are one or two filoplumes with nearly every body feather, situated laterally and medially rather than dorsally at its base. There may also be as many as eight filoplumes around the rim of the follicle for each remex and rectrix. Pluck a filoplume, being careful that it does not break off at the superior umbilicus, mount it on a slide, and examine it under the microscope. Note that it has a distinct calamus and rachis, with a few barbs growing from near the tip of the rachis. A filoplume, broken at the superior umbilicus, might look as though it consists only of a rachis.

Filoplumes are somewhat longer, less slender, and consequently easier to find on bigger birds, particularly at the bases of the remiges. In a few passerine birds, filoplumes extend beyond the tips of the contour feathers on the back of the neck. Exposed filoplumes are unusually conspicuous in cormorants; white in males and brownish in females, they appear on the back, rump, thighs, and less frequently on the under parts of the body.

Filoplumes may transmit vibrations or pressure changes to sensory nerve corpuscles close to the feather follicle in the skin. Reflex stimuli from the brain then actuate the muscles to the feathers.

5. Bristles

Bristles are characterized by a stiff, tapered rachis with barbs along the proximal portion, never at the tip. Such feathers are absent in the pigeon, but they are present in many birds, almost exclusively on the head and neck—about the base of the bill, on the lores, on the eyelids, around the eyes, and on the malar and gular regions. Where bristles are situated around the rictus at the base of the bill, they are designated rictal bristles simply because of their location (Fig. 19b). Bristles appear to perform various functions according to their structure and location. Those screening the nostrils, eyes, or ear openings keep out insects, dirt, and other foreign matter. Large bristles on the lores and rictus of kiwis, owls, Whip-poor-wills *(Caprimulgus vociferus)* and other goatsuckers are presumed tactile receptors—analogous to mammalian whiskers—conveying vibrations or pressure to the special sen-

Figure 19 (a) Filoplume of Common Pigeon (14 mm in actual length). (b) Rictal bristle of American Robin (*Turdus migratorius*) (7 mm in actual length).

sory corpuscles surrounding the feather follicle. Short bristles in place of contour feathers on the head and neck of hawks and vultures may keep the head from being smeared and matted with blood from the gory food. In ostriches, rheas, and cassowaries, bristles may expose the skin for heat radiation. For more information about bristles, see Stettenheim (1973).

6. Powder-down Feathers

In many birds the plumage is dusted by a talc-like powder that comes from the downy elements of contour feathers, semiplumes, and ordinary downy feathers. But the powder comes chiefly from specially modified down feathers, considered by some authorities as a separate kind, the powder-down feathers. They are notably well developed on such birds as herons and bitterns, being clustered in areas known as powder-down patches. Examine the breast of an American Bittern *(Botaurus lentiginosus)*, pushing aside the contour feathers. Here are two large, thick patches of yellowish feathers, the powder-downs. Their growth and the accompanying production of powder are said to be continuous.

Powder-down feathers are present on the pigeon, the Northern Harrier *(Circus cyaneus)*, some parrots, and many other birds. On the pigeon they have the structure of contour feathers, semiplumes, or downs and grade into ordinary feathers that shed a small amount of powder. They are distributed on the sides of the trunk from the axilla to the base of the tail, being most abundant and most highly modified anterior to the thigh and antero-lateral to the tail. The powder is formed by the proliferation and keratinization of cells that surround the barbs in the feather germ and is released while and after the feather emerges from its sheath. Because it has a waterproof quality, some authorities regard the powder as supplementing the function of the oil gland in providing a dressing for the feathers.

Feathers of Newly Hatched Birds

All kinds of birds have natal down feathers, called **neossoptiles,** present either at hatching or within a few days. They may be thickly or sparsely distributed. Not long after hatching they are pushed out from their follicles by the ensheathed tips ("pinfeathers") of the next generation of feathers. The neossoptiles may remain attached to the tips of the pinfeathers for a short period only; by the time the pinfeathers have completely unfolded from their sheaths, the neossoptiles will have been dislodged.

Neossoptiles differ mainly from adult down feathers in two respects: their barbules are shorter and less distinctively shaped; and the tips of their central barbs are without barbules. Neossoptiles, nevertheless, vary widely in such details as the size or absence of the rachis, the condition of the calamus, and the placement of the barbs—all depending on the particular kinds of birds and the contour feathers, semiplumes, and adult down feathers that the neossoptiles precede.

Feather Development

The skin or integument of birds consists of two major layers: the dermis and the epidermis. The **dermis,** the inner part, is the nutrient tissue, carrying blood vessels. The **epidermis,** the outer part, consists of numerous cell layers: the innermost layers comprise the stratum germinativum, divisible (from the outside in) into the stratum transitivum, stratum intermedium, and stratum basale. These layers continually proliferate new cells outward. The outermost layers constitute the stratum corneum—horny tissue consisting of cells that have become flattened and forced outward by the new cells from the stratum germinativum. In the formation of a feather the dermis gives rise to the pulp whose blood vessels supply the nutrition for growth. The stratum corneum and stratum transitivum form the sheath of the feather and the outer layers of the calamus. The stratum intermedium is the source of nearly all parts of the feather, including the inner layers of the calamus. The stratum basale forms the germinating ring at the base of the feather, the walls of the pulp caps, and a small amount of material that is discarded when the feather unfurls from its sheath.

The growth of a feather or scale starts with a papilla that pushes up the overlying epidermis. (See Figure 20a.) Thereafter the gross similarity between a feather and a scale ceases altogether. If destined to form a feather, the growth becomes elongated and tubular; if a scale is to result, it soon becomes flattened and plate-like.

Development of a Neossoptile

The feather germ, which constitutes the dermal papilla or pulp and its epidermal covering or cap (like a finger bearing a thimble), continues its outward growth and at the same time sinks into a pit, the future feather follicle. (See Figure 20b.) Meanwhile the stratum germinativum of the overlying epidermis, through the outward proliferation of cells by the germinating ring around the base of the feather germ, produces (1) a series of columns of barb ridges (the future barbs; also the rachis, if there is to be one, as in ducks) that run parallel from the ring to the tip of the germ and are closely applied to the inner pulp, and (2) an outer, cone-shaped sheath. As a result of rapid growth, barb ridges each differentiate within the sheath into ramus and barbules, and the distal end of the feather soon projects from the follicle above the surface of the skin; the proximal end stays in the follicle as the calamus. (See Figure 21a.) The production of keratin, a specific protein within the cells, gradually hardens the feather parts. Eventually the sheath of the feather splits, beginning at the tip; the barbs separate and extend free, and the pulp disappears

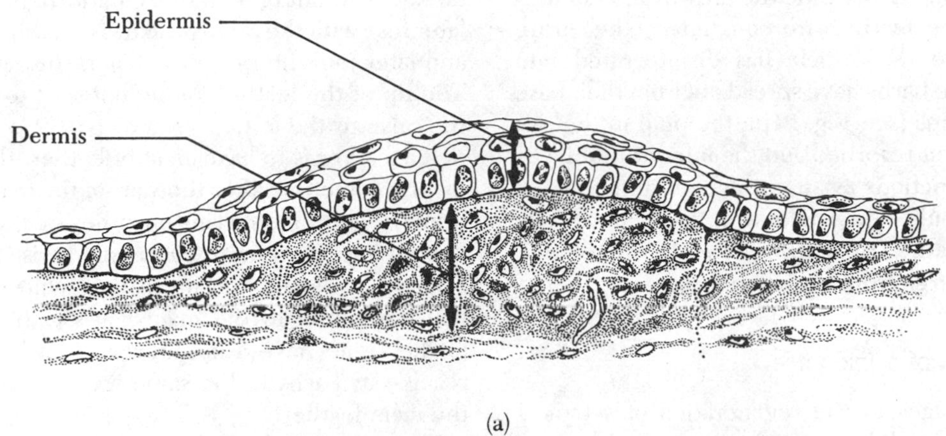

(a)

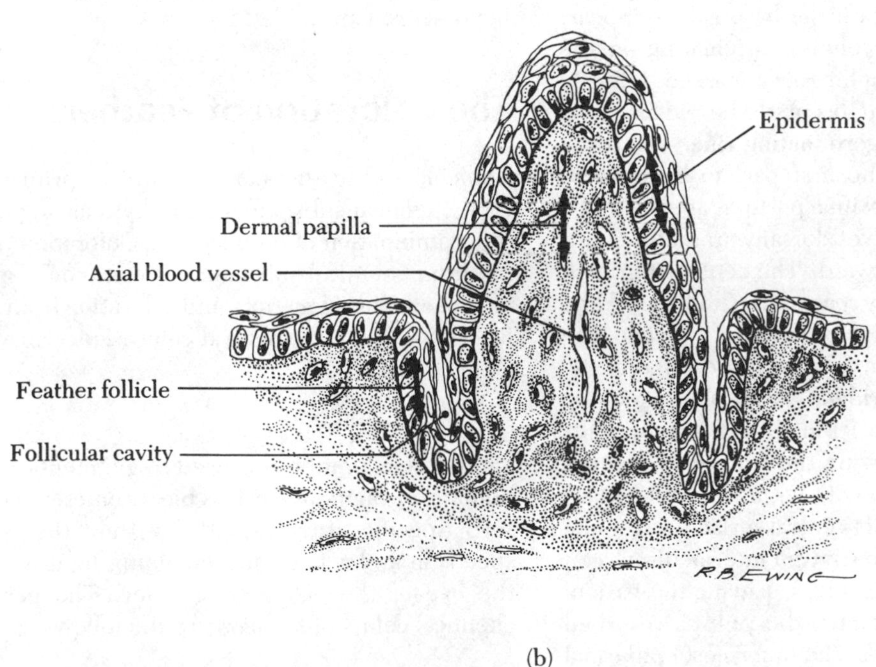

(b)

Figure 20 **Formation of a Feather**

(a) Initial stage in embryonic formation of a feather germ. (b) Later stages in the formation of a feather germ. The dermal papilla consists of pulp and blood vessels, all temporary. The epidermis forms the feather.

by resorption. By the time the growth and keratinization of the feather are completed, the entire sheath above the follicle has disintegrated into flakes and the barbs have spread out from their base on the calamus (see Fig. 21b); the pulp in the calamus has been resorbed and the inferior umbilicus no longer functions as an entrance for pulp and is closed by a pulp cap. The base of the calamus rests on the ensheathed tip of the next feather generation that will emerge from the same follicle.

Development of a Teleoptile

The early stages in the regeneration of a typical contour feather differ from those in initial embryonic stages in that both the follicle and the feather germ are already formed. First, a new epidermal cap must grow over the dermal papilla—in case the previous cap was torn or pulled away by plucking. Then, the feather germ grows outward and regenerates. Within the sheath the barb ridges appear, as before, in parallel columns originating on the germinating ring; but after this a marked differentiation in development occurs. The ridges grow tangentially from the germinating ring. The distal end of each barb is the first part to be formed. Progressively more proximal portions are laid down in a half-spiral that curves dorsally around the pinfeather as it grows outward. The center of growth for each barb eventually reaches the mid-dorsal line where it meets the ridge of the future rachis. The bases of the barbs then fuse with the ridge of the rachis, "the rachidial ridge," at an oblique angle. (See Figure 22a and b.) Finally the sheath splits and disintegrates, allowing the feather (which has been rolled up in the sheath with its dorsal surface against the sheath and its ventral surface next to the pulp) to flatten out. The stratum intermedium soon stops proliferating barb ridges, leaving the base of the feather as the calamus; the pulp is resorbed and caps are formed from the innermost epidermal layers surrounding the pulp.

The parts of a barb—barbules, flanges, hooklets, etc.—develop by differentiation of the barbule cells and ramus cells while a feather is growing. The feather does not grow like a tree by sending out branches and twigs. Instead, it grows by cell division mostly at the base of the feather germ, and the parts differentiate as they move upward. Barbules do not grow out of the ramus but form in place and later fuse with the ramus. Likewise, barbs form first and later join the rachis or hyporachis (aftershaft). The tip of the feather is the first part to form and the calamus the last.

If a feather is to have an afterfeather, this begins to form sometime after the start of the main feather from the germinating ring on either side of the mid-ventral line. New barbs continue to arise as before; some will make up the afterfeather while the others will contribute to the main feather. If the afterfeather is to have a hyporachis, it will arise and receive its barbs in the same way as the rachis of the main feather.

Examine a pinfeather 10 to 30 mm long from a molting bird. Carefully dissect it by cutting it free at the base of the follicle. Make an incision on one side along its full length. Unroll the feather tube, remove the core of pulp, and observe the barb ridges, the developing rachis, and the developing hyporachis if any.

The Coloration of Feathers

The colors of feathers are due to two primary factors: chemical substances and physical properties, or a combination of both factors. Coloration resulting from chemical substances is commonly spoken of as **chemical coloration,** and coloration from physical properties as **structural coloration.**

Chemical Coloration

Chemical coloration is caused by pigmentary compounds, or pigments, called **biochromes,** which absorb specific wave lengths within the visible spectrum and reflect the remaining light waves to the eye of the observer as color. The principal chemical colors of feathers are the following:

Red, Orange, and Yellow Produced in most instances by carotenoids (formerly called lipochromes), which are fat-soluble pigments appearing in a diffused state rather than in discrete granules. Carotenoids occur in fat deposits, egg yolk, secretion of the oil gland, and bare skin, as well as in feathers. They are primarily synthesized by plants and appear in birds only after being modified from

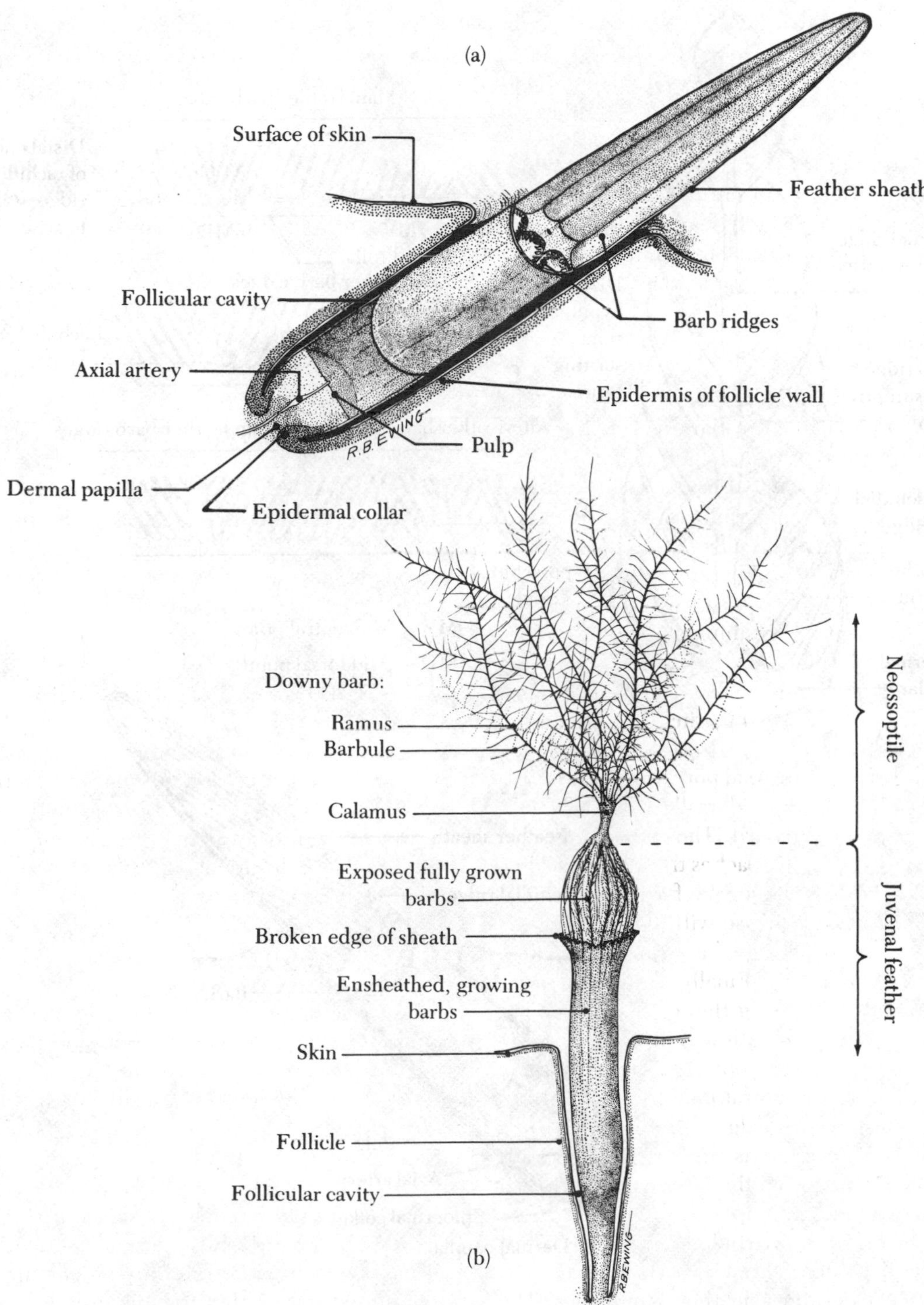

Figure 21 **Generalized Structure of Growing and Mature Natal Down Feathers (Neossoptiles)**
(a) Cutaway diagram of a growing natal down feather. (b) Mature neossoptile, with juvenal feather (teleoptile) forming below it.

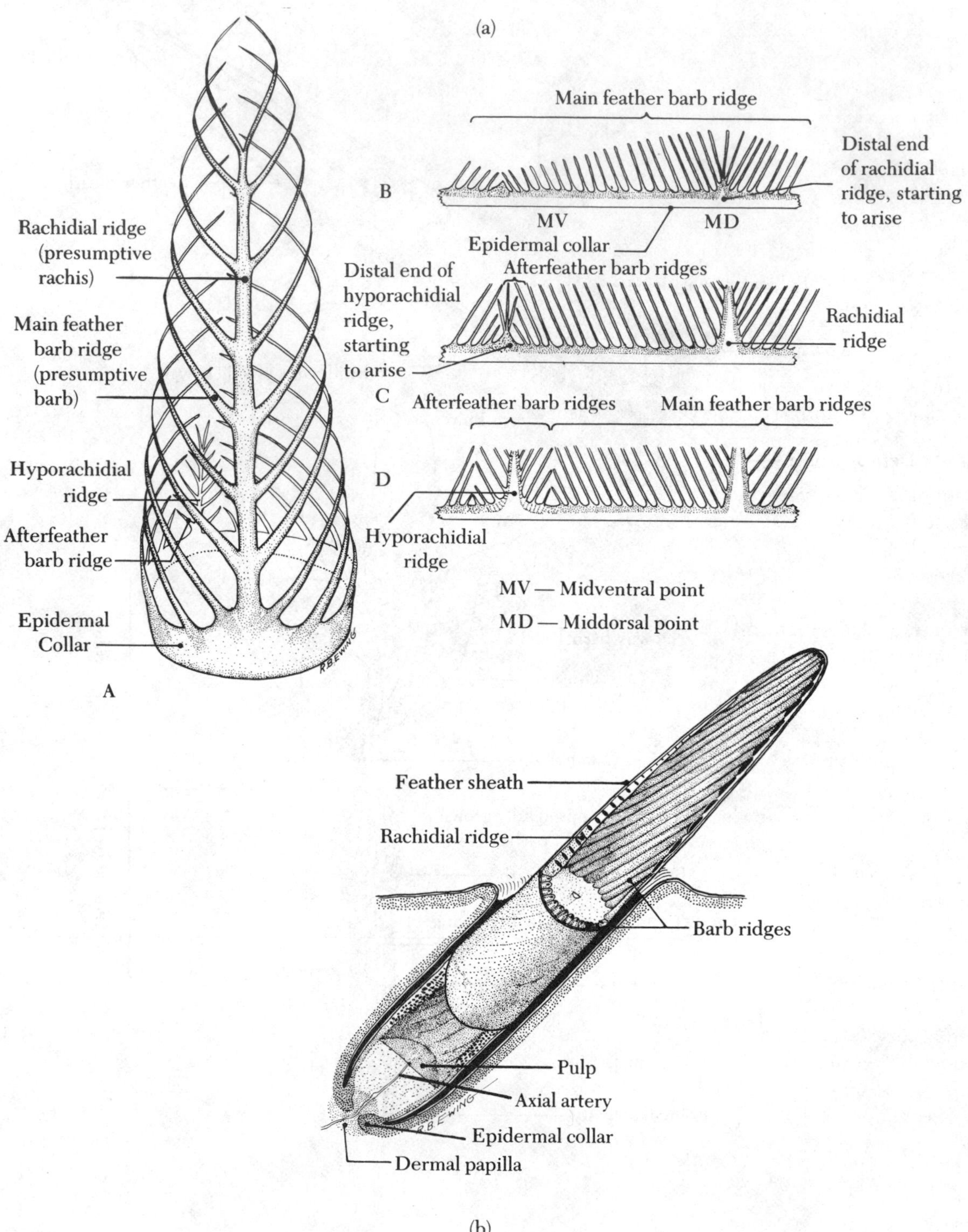

Figure 22 **Generalized Structure of Growing Contour Feathers (Teleoptiles)**
(a) Diagrams of growth of barb ridges. A, dorsal view of feather without sheath or pulp. B, C, and D, series of split feather preparations. (b) Cutaway diagram of a growing feather in its follicle.

ingested food. Carotenoids within a feather occur primarily near the tip of the rachis and the rami of pennaceous barbs, and they may be removed by various organic solvents such as alcohol or ether.

Black, Gray, and Brown Produced by melanins, which are relatively insoluble pigments appearing as granules. Certain melanins can also produce brownish-yellow, reddish brown, and chestnut red. Unlike carotenoids, melanins are synthesized in special pigment cells (melanocytes) from amino acids. (See Figure 23.) In feathers melanins occur in the rachis and all parts of the barbs except the barbicels; elsewhere in the body they occur in the skin, the horny covering of the bill, scales of the feet, and certain internal organs. The differences in the melanistic colors depend in part on the amount of pigment deposition and the size of the pigment granules. For example, black is due to a large amount of pigment, which absorbs all light waves; gray to a smaller amount, which absorbs fewer light waves, etc. Melanins may mask carotenoids, making them invisible, or combine with them visually (not chemically) to create certain colors. Quite apart from its color-producing function, melanin serves to increase a feather's tensile strength and resistance to wear. Thus a black or brown feather is less subject to abrasion than a white or brightly colored one; the black or brown part of a feather may remain intact after the white or brightly colored tip has worn away.

Green Produced in most cases by structural conditions and combinations of structural colors and yellow biochromes. The best known green pigment is turacoverdin, a porphyrin derivative, which gives green in the feathers of touracos (Musophagidae). The other green pigments are thought to be carotenoids. Typical porphyrins create red, brown, or buff colors in feathers of birds in 13 orders; they are most common in owls and bustards.

Structural Coloration

Structural coloration is caused by the presence of structural elements within the cells that are capable of modifying or separating white light by interference. There are two kinds of structural coloration: iridescent coloration, which changes according to the angle at which feathers are viewed, and non-iridescent coloration, which does not change. Biochromes, especially melanins, may combine with structural elements to modify or intensify structural coloration. Some examples of structural coloration are the following:

Iridescent Coloration Produced by barbules that are modified in various ways among birds by the broadening and flattening of the base and/or the pennulum wholly or in part. This enlarges the visible surface of the barbs. Iridescence is created by interference of light waves within each barbule. The cortex on the enlarged, exposed portion of the barbule contains a layered arrangement of melanin particles in a keratin matrix. This structure differs widely among iridescent birds. For example, in the peafowl the melanin particles are long, slender, solid rods, separated by air vacuoles, the whole constituting a three-dimensional space lattice. Trogons and hummingbirds have a variety of conditions involving layers of melanin in the form of air-filled, oval platelets ("foamy pancakes") or tubes. In all cases the visual effect arises from the interference of light waves reflected from surfaces within the lattice. Colors and patterns result from small variations in the reflective indices of keratin and melanin, the shape and dimensions of the melanin bodies and air vacuoles, and the spacing between the melanin bodies. All these factors are genetically controlled with a very high degree of precision.

Blue Coloration Produced by the rami of the barbs. Each ramus, like the rachis, is composed of a thin, dense, transparent outer layer, the cortex, surrounding a thick, porous, opaque core, the medulla. In the medulla are cells completely filled with a "spongy structure," except for some large nuclear vacuoles and small cavities. Blue results from interference of light waves reflected from the surfaces of the spongy structure and the absorption of other light by the melanin pigmentation in the core of the medulla.

Green Coloration Feathers may appear green instead of blue when the cortex above the spongy medullary cells is thicker and has yellow carotenoids. Green may also be due to differences in the cross-sectional size and shape of the ramus, the spongy structure, and the size and shape of the barbules. If the vane of a structurally green feather

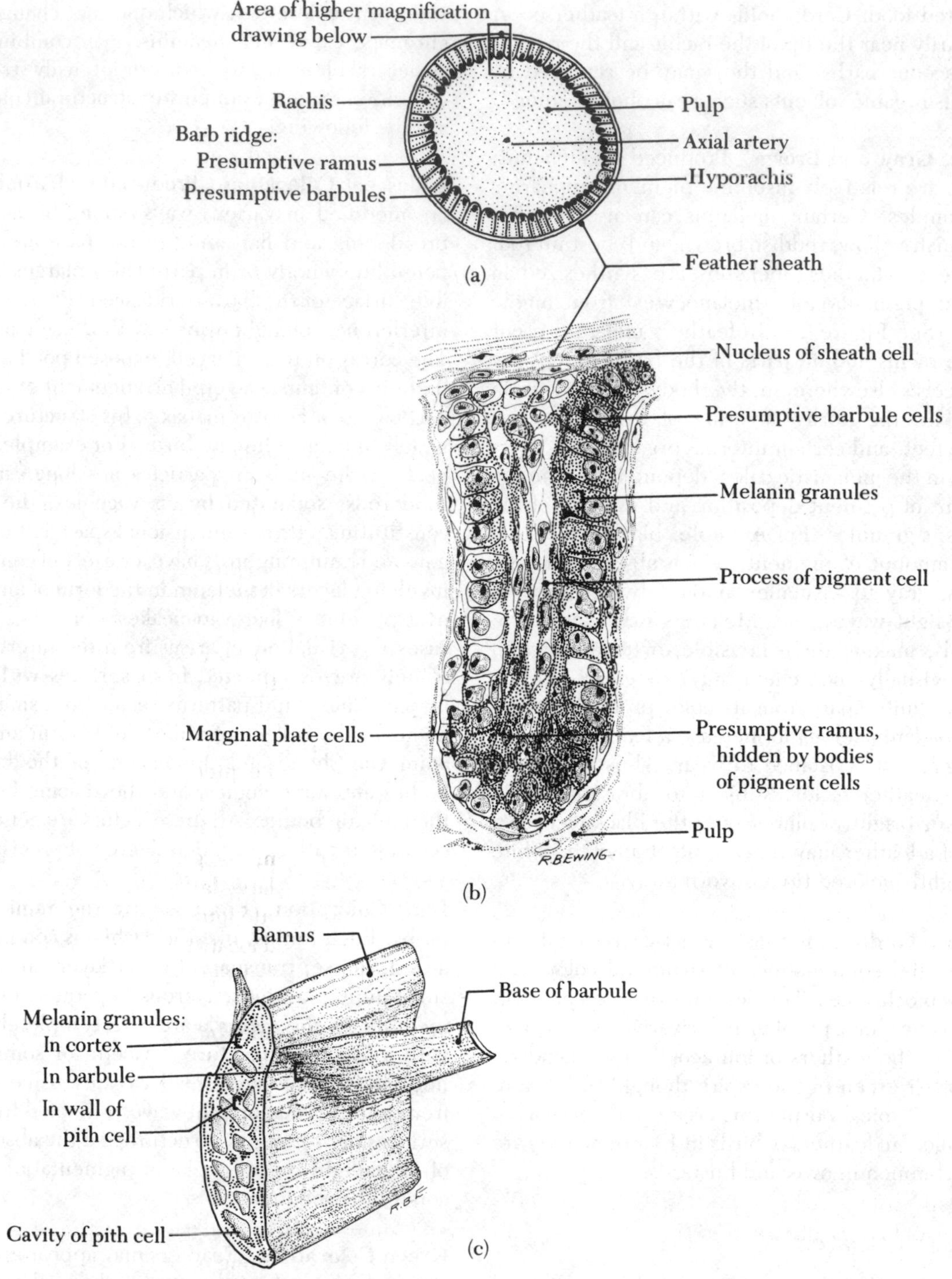

Figure 23 **Melanin in a Contour Feather**

(a) Cross section of a growing feather showing the arrangement of melanin. (b) Higher magnification of a part of the same feather showing the pigment cells in a barb ridge. (c) Segment of a pennaceous barb showing the distribution of melanin granules.

Activity

Place under the microscope small feathers, or portions of feathers, mounted on slides, which show the colors discussed above. Include several feathers with iridescent colors. Study the colors, using first reflected light (i.e., light falling on the feathers from the side and reflected from them to the eye) and then transmitted light (i.e., light passing from the mirror and through the feathers to reach the eye).

is scraped with a sharp scalpel, thereby removing the yellow cortex from the barbs, the vane will appear blue.

White Not a color. Feathers white or partly white owe their appearance to the scattering of light in two ways: by reflection from the vanes and rachis, chiefly from the surfaces of unpigmented barbules, or from light passing through the colorless cortex of the rachis and rami and reflecting off of the internal medullary cells or pith; or by translucency of the vanes and calamus, resulting from the scattering of light by the keratinized cells of the cortex. If a white feather is immersed in a liquid with the same refractive index as keratin (approximately 1.54), such as Canada balsam, microscope immersion oil, or xylene, reflecting surfaces will disappear and the vanes will become transparent while the whiteness of the calamus will not change.

Feather Tracts

Penguins, ostriches, rheas, cassowaries, emus, and a few other flightless birds have feathers rather uniformly distributed over the body. The same condition would superficially appear to be true of all flying birds, such as the Common Pigeon and House Sparrow, but close inspection reveals that their contour feathers arise from tracts or **pterylae** (singular **pteryla**) on the skin and that there are certain areas or **apteria** (singular **apterium**) where contour feathers do not occur. This does not mean that apteria are necessarily bare, as they may have down feathers and semiplumes. The apteria are normally concealed because they are overlapped by the adjoining pterylae. The arrangement and distribution of feathers in tracts is technically spoken of as the **pterylosis.**

Begin a study of the pterylosis of the House Sparrow. In this bird the apteria are proportionately larger than in the pigeon; consequently, the pterylae are somewhat narrower and more distinct. If possible, use a nestling sparrow with feathers only partially developed; the tracts show clearly without need of special preparation. If a nestling sparrow is not available, use an adult bird. With scissors, closely clip the entire plumage with the exception of one wing. Note that the contour feathers do not simply "fill" the different tracts; they are evenly spaced in regular rows. Determine the following pterylae and their associated regions and apteria:

Capital Tract The pteryla extending over the top of the head from the base of the upper mandible posteriorly to the point where the head and neck join. It is bounded on each side by an imaginary line passing from the mandibular ramus to the angle of the jaw.

Dorsal Tract The pteryla extending posteriorly from the capital tract to the upper tail coverts. Along the neck it is bordered on each side by a **cervical apterium;** along the trunk it is bordered on each side by a large **lateral apterium.** The dorsal tract is divisible into four regions identified mainly by their shape and locations. Thus the narrow **cervical region** extends from the head to the trunk; the narrow **interscapular region** extends posteriorly between the shoulder blades (scapulae); the saddle-shaped **dorsal region** extends from the shoulder blades to a point approximately halfway to the tail; the broad **pelvic region,** lying between the hips and extending from the dorsal region to the tail coverts, completes the dorsal tract.

Scapulohumeral Tract A narrow pteryla found on each shoulder running obliquely backward on the brachium from the anterior part of the shoulder, where it barely merges with a feather tract below. The feathers arising from this pteryla are the scapulars.

Femoral Tract A narrow pteryla that extends

Activity

On Figure 24 (the two drawings on the plucked sparrow) indicate the feather tracts by stippling. Use approximately as many dots as there are feathers. Take care, particularly in the case of the ventral tract, to show how the feathers are arranged in rows. Then label all pterylae, regions of the dorsal pteryla, and apteria.

along the outer surface of each thigh from a point near the knee to the vent. When the leg is drawn up, this tract is almost parallel to the dorsal tract.

Crural Tract The remaining feathers of the leg. They are separated from the femoral tract by a narrow apterium.

Ventral Tract The pteryla beginning at the junction of the mandibular rami and extending posteriorly to the circlet of feathers around the vent. It encloses a longitudinal **mid-ventral apterium.** It is bounded by the capital tract and the cervical and lateral apteria. In some birds the ventral tract is not continuous as in the sparrow; instead it may be more or less divisible into such tracts as the cervical, pectoral, sternal, and abdominal.

Caudal Tract Includes the rectrices, the upper and under tail coverts (see "Topography," p. 21) and also the circlet of feathers around the vent. The oil gland, which is located dorsally at the base of the tail, is not feathered in the sparrow. When feathers do occur on it, as in certain kinds of birds, the feathers are considered part of the caudal tract.

Alar Tract Includes the remiges, all their coverts, and the other feathers arising on the wing except the feathers of the scapulohumeral tract, which is separated by a narrow, unnamed apterium. See "Topography," page 15, for details on the arrangement of the feathers on the wing.

The pterylosis of the House Sparrow must not be considered the standard or typical pattern for all birds as there is no typical pattern. If possible, study

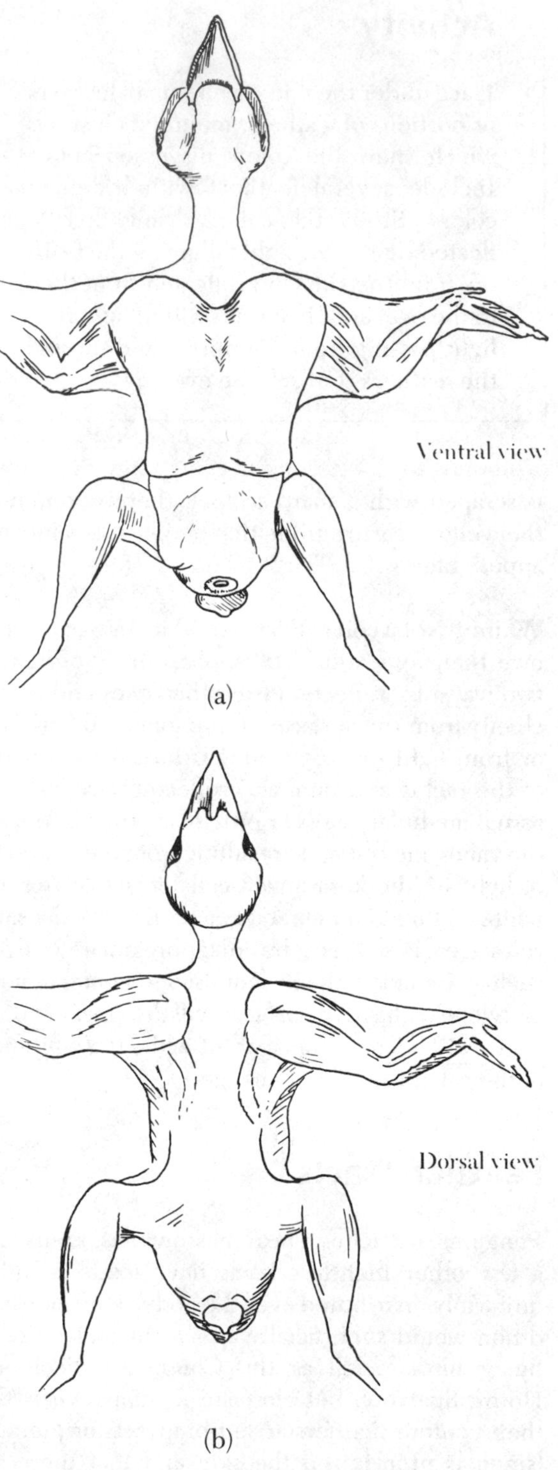

Figure 24 **Plucked House Sparrow**

and compare the pterylosis of several widely different species of birds, adult as well as natal forms. Some suggested references: Clench (1970); Humphrey and Clark (1961); Lucas and Stettenheim (1972, chapters 2 and 3); Morlion and Vanparijs (1979); and Wetherbee (1957).

Among different species, and between groups of species, a considerable variation exists with respect to the size and shape of pterylae, presence of certain apteria, the density of feathers on certain pterylae, and the number of remiges and rectrices and their coverts. Within a species, however, or a group of species, the pterylae and apteria show generally constant characteristics.

The study of pterylae, **pterylography**, is useful in taxonomic investigations. By comparing the pterylosis of apparently similar species, it is often possible to determine whether or not a close relationship actually exists. The works by Ames, Heimerdinger, and Warter (1968), Berger (1960), Compton (1938), and Heyman and Morlion (1980) illustrate how pterylography can be applied to taxonomy. Through the study of pterylosis, researchers can identify particular feathers, individually and in groups. Then they can follow a series of plumages of birds of known ages which, in turn, enables researchers to develop keys for aging wild-caught specimens by plumage characters. Ultimately, the procedure enables researchers to ascertain the age composition of a population at certain times of the year. See, for examples, the papers by Aldrich (1956), Foster (1967), Holmes (1966), Mewaldt (1958), and Ohmart (1967).

The Numbers of Contour Feathers on a Bird

Birds of the same species in the same area and season presumably show only slight individual variation in number of contour feathers. The main exceptions are species in which the sexes differ greatly in size and those in which the males have a more elaborate feather arrangement; in these species a correlation between the number of feathers and sex may be expected. Sharp seasonal differences in the number of feathers on birds of a species occur when the individuals inhabit a region with marked seasonal changes. As a rule, the number of feathers for insulation is much greater in winter than in summer. For example, the contour feathers of three House Sparrows collected in Michigan during January and February numbered 3,546, 3,615, and 3,557, respectively; on two taken in July in the same area the totals were 3,138 and 3,197. The authority for these figures (Staebler, 1941) calculated that there was a loss from winter to summer of 11.5 percent of the feathers. By contrast, in Pretoria, South Africa, where the year-round temperature is milder than in Michigan, the counts of feathers on 11 specimens of the Laughing Dove *(Streptopelia senegalensis)* taken in different seasons showed no apparent variation (Markus, 1965).

Species of birds differ widely in the number of contour feathers among species, a fact borne out in counts on birds from Michigan by Ammann (1937), from Florida by Brodkorb (1949), and from the vicinity of Washington, D.C., by Wetmore (1936). The lowest number found was 940, on a Ruby-throated Hummingbird *(Archilochus colubris)* collected in June (Wetmore), while the highest was 25,216 on a Tundra Swan *(Cygnus columbianus)* taken in November (Ammann). On 74 species of passerine birds (flycatchers, jays, chickadees, vireos, blackbirds, etc.) the numbers ranged from 1,119 (a Ruby-crowned Kinglet, *Regulus calendula*, taken in October—Wetmore) to 4,607 (an Eastern Meadowlark, *Sturnella magna*, taken in February—Brodkorb). Some of the counts on nonpasserine birds were as follows:

15,016	Pied-billed Grebe *(Podilymbus podiceps)*	December	Brodkorb
3,867	Least Bittern *(Ixobrychus exilis)*	May	Brodkorb
14,914	Northern Pintail *(Anas acuta)*	January	Brodkorb
7,224	Clapper Rail *(Rallus longirostris)*	April	Brodkorb
13,913	American Coot *(Fulica americana)*	November	Brodkorb
4,480	Least Sandpiper *(Calidris minutilla)*	April	Brodkorb
2,635	Mourning Dove *(Zenaida macroura)*	June	Wetmore
9,206	Barred Owl *(Strix varia)*	June	Brodkorb
3,332	Common Nighthawk *(Chordeiles minor)*	April	Brodkorb
3,665	Red-bellied Woodpecker *(Melanerpes carolinus)*	April	Brodkorb

It is not surprising that the diminutive hummingbird has relatively few feathers and that a big bird, such as a swan, has so many, or that small birds generally have fewer feathers than large birds. But body size is not the sole criterion in the number of feathers. Ammann found that nearly 80 percent of the feathers on the Tundra Swan were on the head and neck, indicating that a species' physical peculiarity—e.g., an exceptionally long neck—may account for a large number of feathers. The number of feathers may be attributed also to the structure and size of feathers and the uses that feathers serve. As an illustration, the Northern Pintail has short feathers with tightly interlocking barbs to resist water, and the Barred Owl has long, loose-textured feathers to enable silent flight. Though the two birds have nearly the same body size, the duck has more than double the number of feathers.

Two investigators, Hutt and Ball (1938), demonstrated that a small bird has more feathers per unit of body surface than a large bird and that among land species the number of feathers per unit of body surface actually increases with decreasing body weight. The basis for this phenomenon is that the amount of heat lost by a bird, or any warm-blooded animal, is directly proportional to the surface area of its body. Because the surface area per unit of weight is much greater, a small bird, as a consequence, has greater difficulty in maintaining a body temperature above the environment than a large one. A small bird must, therefore, have relatively more feathers for insulation. The weight of plumage in relation to body weight shows a similar trend. Turček (1966) found that the relative weight of the plumage is lighter in heavier birds because they need more feathers.

Anything approaching a thorough investigation of the number of feathers among all species and on individuals within a species has not yet been undertaken. Do the Hutt-Ball findings apply to aquatic birds? Does the presence of afterfeathers in any way affect the number of feathers per unit of surface area or weight? Does the number of down feathers (teleoptiles) increase or decrease in proportion to the number of contour feathers? Does the number of filoplumes increase or decrease in the same way? These are only a few of the questions to be answered by additional feather counts.

Feather Parasites

All birds are hosts to parasites on their feathers. While this is a sweeping statement, it is undoubtedly true, as investigations have borne out.

There are two principal groups of feather parasites: the feather lice, order Mallophaga, of the arthropod class Insecta; and the feather mites of the arthropod class Arachnida. The feather lice are the more prominent in numbers of species. Many are parasitic on particular species, or closely allied species, of birds. Considering the extent of their distribution and their host specificity, feather lice are believed to have become parasitic on birds at a very early stage in their evolution and therefore are useful to ornithologists in determining the taxonomic relationships of birds. Feather mites are much less well known since their study is in its infancy. For further information on feather parasites, refer to Beer (1970) and Rothschild and Clay (1952). For the effects of fungi on feathers, see Pugh (1972).

References

Aldrich, E.C.
1956 Pterylography and molt of the Allen Hummingbird. *Condor* 58:121–133.

Ames, P.L.; Heimerdinger, M.A.; and Warter, S.L.
1968 The anatomy and systematic position of the Antpipits *Conopophaga* and *Corythopis*. *Peabody Mus. Postilla* No. 114. (Includes the use of pterylosis.)

Ammann, G.A.
1937 Number of contour feathers of *Cygnus* and *Xanthocephalus*. *Auk* 54:201–202.

Baumel, J.J.; King, A.S.; Lucas, A.M.; Breazile, J.E.; and Evans, H.E., eds.
1979 *Nomina anatomica avium: An annotated anatomical dictionary of birds*. London: Academic Press.

Beer, R.E.
1970 Ectoparasites of birds: A brief review. In *Ornithology in laboratory and field*, Appendix I, by O.S. Pettingill, Jr. Minneapolis: Burgess.

Berger, A.J.
1960 Some anatomical characters of the Cuculidae and the Musophagidae. *Wilson Bull*. 72:60–104. (Includes pterylosis.)

Brodkorb, P.
1949 The number of feathers in some birds. *Quart*.

Jour. Florida Acad. Sci. 12:1–5.
1955 Number of feathers and weights of various systems in a Bald Eagle. *Wilson Bull.* 67:142. (The total count of contour feathers was 7,182).

Brush, A.H.
1978a Feather keratins. In *Chemical zoology*, Vol. 10. A.H. Brush, ed. New York: Academic Press.
1978b Avian pigmentation. In *Chemical zoology*, Vol. 10. A.H. Brush, ed. New York: Academic Press.
1978c Structural aspects of the speculum of Mallard *Anas platyrhynchos*. *Ibis* 120:523–526.

Chandler, A.C.
1916 A study of the structure of feathers, with reference to their taxonomic significance. *Univ. Calif. Publ. in Zool.* 13:243–446. (One of the first important studies on the subject.)

Clench, M.H.
1970 Variability in body pterylosis, with special reference to the genus *Passer*. *Auk* 87:650–691.

Compton, L.V.
1938 The pterylosis of the Falconiformes with special attention to the taxonomic position of the Osprey. *Univ. Calif. Publ. in Zool.* 42:173–212. (Another good example of a comparative study of feather arrangement.)

Conover, M.R., and Miller, D.E.
1980 Rictal bristle function in Willow Flycatcher. *Condor* 82:469–471.

Day, M.G.
1966 Identification of hair and feather remains in the gut and faeces of stoats and weasels. *Jour. Zool.* 148:201–217.

Dyck, J.
1971a Structure and spectral reflectance of green and blue feathers of the Rose-faced Lovebird *(Agapornis roseicollis)*. *K. Danske Vidensk. Selsk. Biol. Skr.* 18(2):1–67.
1971b Structure and colour-production of the blue barbs of *Agapornis roseicollis* and *Cotinga maynana*. *Ztschr. f. Zellforsch.* 115:17–29.

Foster, M.S.
1967 Pterylography and age determination in the Orange-crowned Warbler. *Condor* 69:1–12.

Fox, D.L., ed.
1976 *Animal biochromes and structural colours: Physical, chemical, distributional & physiological features of coloured bodies in the animal world.* 2nd ed. Berkeley: Univ. of Calif. Press.

Greenewalt, C.H.; Brandt, W.; and Friel, D.D.
1960 Iridescent colors of hummingbird feathers. *Jour. Optical Soc. Amer.* 50(10):1005–1016.

Heyman, R., and Morlion, M.L.
1980 The pterylosis in the genera *Pycnonotus* and *Andropadus*. Gerfaut 70:225–244.

Höhn, E.O.
1977 The "snowshoe effect" of the feathers on ptarmigan feet. *Condor* 79:380–382.

Holmes, R.T.
1966 Molt cycle of the Red-backed Sandpiper *(Calidris alpina)* in western North America. *Auk* 83:517–533.

Humphrey, P.S., and Clark, G.A., Jr.
1961 Pterylosis of the Mallard Duck. *Condor* 63:365–385.

Hutt, F.B., and Ball, L.
1938 Number of feathers and body size in passerine birds. *Auk* 55:651–657.

Joubert, C.S.W., and MacLean, G.L.
1973 The structure of the water-holding feathers of the Namaqua Sandgrouse. *Zoologica Africana* 8:141–152.

LaBastille, A.; Allen, D.G.; and Durrell, L.W.
1972 Behavior and feather structure of the Quetzal. *Auk* 89:339–348.

Lucas, A.M., and Stettenheim, P.R.
1972 Avian anatomy/integument. 2 vols. *Agriculture Handbook* 362. Washington, D.C.: U.S. Department of Agriculture, (An indispensable work on the structure of the skin and all its derivatives, with a comprehensive list of references.)

Maderson, P.F.A.
1972 On how an archosaurian scale might have given rise to an avian feather. *Amer. Nat.* 106:424–428.

Markus, M.B.
1965 The number of feathers on birds. *Ibis* 107:394.

Mayaud, N.
1950 Teguments et phanères. In *Traité de zoologie*, ed. P.-P. Grassé. Vol. 15. Paris: Masson et Cie. (Highly useful information on the integument, structure, development, varieties, and coloration of feathers, and on pterylosis; with excellent illustrations.)

Mewaldt, L.R.
1958 Pterylography and natural and experimentally induced molt in Clark's Nutcracker. *Condor* 60:165–187.

Miller, W.DeW.
1924 Variations in the structure of the aftershaft and their taxonomic value. *Amer. Mus. Novitates* No. 140:1–7.

Morlion, M.
1964 Pterylography of the wing of the Ploceidae.

Gerfaut 54:111–158. (A helpful paper for studying the pterylosis of the House Sparrow.)

Morlion, M.L., and Vanparijs, P.
1979 The pterylosis of five European corvids. *Gerfaut* 69:357–378.

Nitzsch, C.L., and Burmeister, C.C.H.
1840 *System der pterylographie.* English translation by W.S. Dallas; edited by P.L. Sclater and published in 1867 by the Ray Society, London. (The classic work on the subject of feather arrangement.)

Ohmart, R.D.
1967 Comparative molt and pterylography in the quail genera *Callipepla and Lophortyx. Condor* 69:535–548.

Pitelka, F.A.
1945 Pterylography, molt, and age determination of American jays of the genus *Aphelocoma. Condor* 47: 229–260.

Pugh, G.J.F.
1972 The contamination of birds' feathers by fungi. *Ibis* 114:172–177.

Richardson, F.
1942 Adaptive modifications for treetrunk foraging in birds. *Univ. Calif. Publ. in Zool.* 46:317–368. (Contains a discussion of the stiffening of tail feathers for climbing.)

Rijke, A.M.
1970 Wettability and phylogenetic development of feather structure in water birds. *Jour. Exper. Biol.* 52:469–479.

Rothschild, M., and Clay, T.
1952 *Fleas, flukes & cuckoos: A study of bird parasites.* London: Collins.

Rüppell, G.
1977 *Bird flight.* New York: Van Nostrand Reinhold. (Chapter 2 includes feather structure in relation to flight.)

Schmidt, W.J., and Ruska, H.
1962 Über das schillernde federmelanin bei *Heliangelus* und *Lophophorus. Ztschr. f. Zellforsch.* 57:1–36. (Modifications for iridescent coloration.)

Schüz, E.
1927 Beitrag zur kenntnis der puderbildung bei den vögeln. *Jour. f. Ornith.* 75:86–224. (A comprehensive treatise on powder-down feathers.)

Sengel, P.
1971 The organogenesis and arrangement of cutaneous appendages in birds. *Advances in Morphogenesis* 9:181–230.

Simon, H.
1971 *The splendor of iridescence: Structural colors in the animal world.* New York: Dodd, Mead.

Staebler, A.E.
1941 Number of contour feathers in the English Sparrow. *Wilson Bull.* 53:126–127.

Stettenheim, P.R.
1973 The bristles of birds. *Living Bird* 12:201–234.
1976 Structural adaptations in feathers. *Proc. 16th Internatl. Ornith. Congr.* pp. 385–401.

Turček, F.J.
1966 On plumage quantity in birds. *Ekolog. Polska (Ser. A)* 14:617–633.

Watson, G.E.
1963 The mechanism of feather replacement during natural molt. *Auk* 80:486–495.

Wetherbee, D.K.
1957 Natal plumages and downy pteryloses of passerine birds of North America. *Bull. Amer. Mus. Nat. Hist.* 113:341–436.

Wetmore, A.
1936 The number of contour feathers in passeriform and related birds. *Auk* 53:159–169. (The results of painstaking work in counting feathers in many species.)

Anatomy and Physiology

A study of the organ systems of the bird, except the integumentary system already treated at length, is undertaken in the ensuing pages. The first half of this chapter deals with the anatomy of the Common Pigeon *(Columba livia)* as revealed by observation and dissection. In the directions for study are a few comments relating to the function of certain organs or their parts. Some of the more important aspects of avian anatomy and physiology are discussed later in the chapter, with a view to stimulating an interest in further reading and investigation.

The Anatomy of the Pigeon: Skeletal System

The skeletal system of a bird has two notable characteristics: a strong tendency toward fusion of adjacent bones; and a lightness resulting from the pneumaticity of many of the bones. As in other vertebrates, the skeleton of a bird is divisible into the axial skeleton (bones lying along the central axis of the body—skull, vertebral column, ribs, and sternum) and the appendicular skeleton (bones of the pectoral and pelvic girdles and their limbs). For convenience, the skeleton will be considered here as made up of three parts: bones of the limbs; bones of the trunk; and bones of the head.

The Bones of the Limbs

Since the bones of the limbs were studied in connection with the external anatomy of the wings, legs, and feet, they will not be considered again.

The Bones of the Trunk

The trunk comprises the vertebral column, ribs, sternum, and the pectoral and pelvic girdles. Use two prepared skeletons of the pigeon, one in which the bones are completely articulated and one in which the bones of the trunk and limbs are disarticulated and may be handled at will. Use a human skeleton, or a detailed chart of one, for comparative purposes.

Vertebral Column

The vertebral column provides a base for the bones of the trunk and limbs and is the main support of the head. It is made up of a chain of bony elements called **vertebrae.**

Vertebrae conform to one general plan in all animals bearing them. Study a human cervical vertebra, preferably the fifth. (See Figure 25a.) Viewed from the front, there is a body or **centrum** (plural, **centra**) surmounted on the dorsal side by a **neural arch,** which surrounds a **neural canal** for the passage and protection of the spinal cord. The neural arch is in reality a composite structure made up of two plate-like masses meeting at a median line above the canal to form a **neural spine.** Vertebrae commonly have seven processes. The neural spine, already mentioned, is one. The **transverse processes** are two. They are located on either side of the neural arch and project laterally; at the base of each transverse process is the **vertebrarterial canal.** The remaining four processes bear sufaces for articulation with adjoining vertebrae: two, the

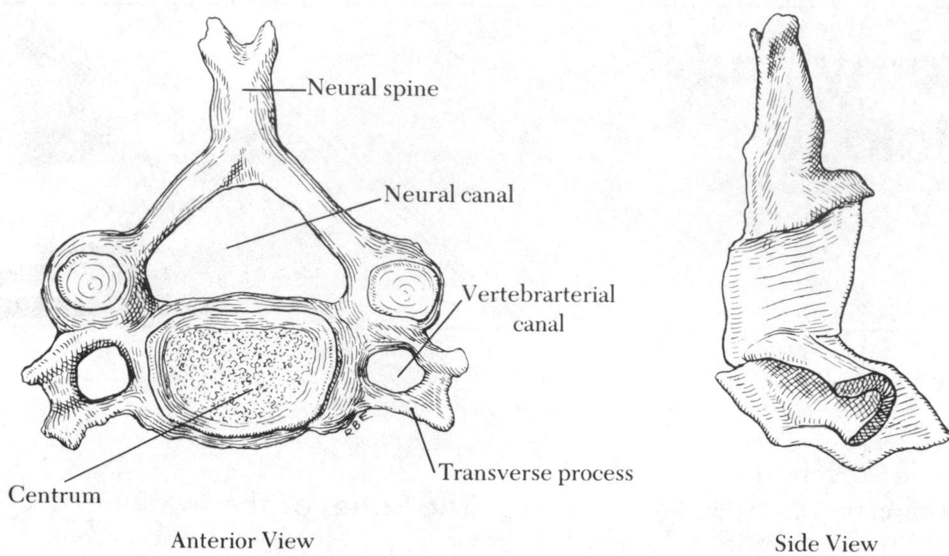

(a)

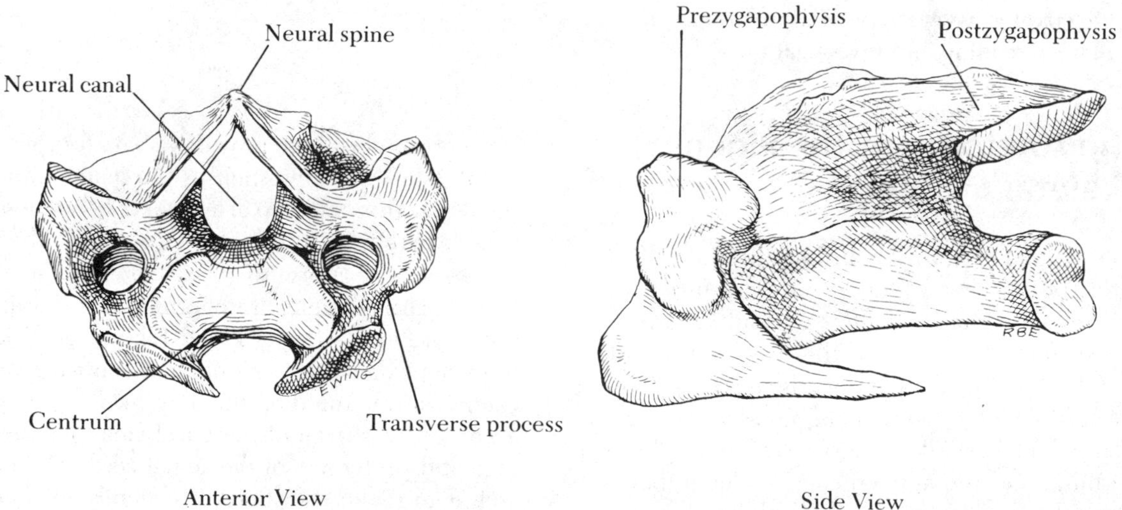

(b)

***Figure* 25 Vertebrae**
(a) Cervical vertebra of man. (b) Cervical vertebra of Common Pigeon.

prezygapophyses, extend anteriorly from either side of the arch and face upward; two, the **postzygapophyses,** extend posteriorly from either side of the arch and face downward. The prezygapophyses of any given vertebra rest on the corresponding postzygapophyses of the vertebra next in front, thus allowing a certain amount of movement between them.

The vertebrae have different characteristics in the different regions of the body in which they occur. They are, therefore, classified into: **cervical** (neck), **thoracic** (chest), **lumbar** (loin), **sacral** (pelvis), and **caudal** (tail) **vertebrae.** In man and other mammals the vertebrae in these groups are relatively distinct, but in birds they have undergone considerable modification, certain groups having been reduced, crowded, and even fused together. All the cervical vertebrae remain completely distinct and freely movable. They warrant special study here, for they show many significant specializations.

Cervical Vertebrae

The cervical vertebrae lie between the skull and the first vertebra that is connected with the sternum by a pair of complete ribs. In the pigeon there are fourteen such vertebrae, but this number is not constant among birds; it ranges from thirteen to twenty-five. Man and most mammals have but seven cervical vertebrae.

Study and compare the sixth cervical vertebra of the pigeon (see Figure 25b) with the fifth cervical vertebra of man. Observe particularly the following parts:

Centrum In the bird the anterior end of the centrum is saddle-shaped; that is, it is convex dorsoventrally and concave from side to side. The posterior end shows the reverse condition. A centrum of this sort is said to be **heterocoelous.** The human centrum is acoelous, for its anterior and posterior ends are flattish. From each side of the pigeon centrum, near the base of the transverse process, there projects posteriorly, and somewhat medially, a process which, with its fellow of the other side, forms on the forward end of the centrum a groove for the carotid artery. Does it occur in man?

Transverse Process The transverse process in both bird and man is pierced by an opening, called the **vertebrarterial canal,** through which the vertebral artery and vein pass.

Neural Spine In man the neural spine is bifurcated. Does it differ in the bird?

The other cervical vertebrae have many features in common with the ones just studied. The first and second vertebrae in both bird and man are more specialized than the others and are called respectively the **atlas** and **axis**. The atlas is a ring-like bone without a centrum. In the bird the atlas is in contact with the skull by one ball and socket joint formed by the occipital condyle of the skull (described later) and the cupped articular surface of the atlas. How does the contact between skull and atlas differ in man? The axis possesses the **odontoid process,** upon which rotates the atlas bearing the skull. Some of the cervical vertebrae of the bird have small median spines, **hypapophyses,** extending ventrally from their centra. Others have small ribs which do not reach the sternum. Which vertebrae have hypapophyses? Which have incomplete ribs?

Many of the features of the avian cervical vertebrae that have just been studied—e.g., the large number of vertebrae, the heterocoelous centrum, and broad, overlapping zygapophyses—are specializations for greater flexibility. The neck in the pigeon and most birds has three functionally distinct segments: anterior, for bending below but not above a straight line; middle, for bending above and below; and posterior, for bending slightly below but mostly above. These properties are determined mainly by the angles of the pre- and postzygapophyses.

Thoracic, Lumbar, Sacral, and Caudal Vertebrae

In the pigeon there are five thoracic vertebrae. They succeed the cervical vertebrae posteriorly and are fused together by centra and zygapophyses, and by two processes: the neural spines that form a dorsal ridge and the transverse processes that form thin lateral plates. The thoracic vertebrae each possess a pair of complete ribs passing to the sternum. The side of each vertebra thus has two articular sufaces for a rib, one on the transverse process and one on the centrum. A variable number of thoracic, lumbar, and sacral vertebrae fuse to form the **synsacrum.** No attempt will be made either to distinguish between them or to note their individual modifications. It will suffice to point out the long dorsal ridge, formed by the fusion of the neural spines, and the crossbars seen from below, which represent transverse processes emerging from their closely fused vertebrae and joining the inner walls of the pelvic girdle. The six caudal vertebrae are freely movable, with unfused neural spines and transverse processes. The terminal vertebra represents the fusion of several caudal vertebrae. In the shape of a ploughshare, it is appropriately called the **pygostyle.** It serves as a base for the rectrices.

How many thoracic vertebrae does man have? Lumbar? Sacral? Caudal (coccygeal)?

Ribs

A typical rib, such as that found in man, is a flattened arch of bone attached to the transverse process and centrum of a vertebra by two heads and to the sternum by one cartilaginous extremity. In man how many ribs are there? How many are unattached to the sternum?

In the pigeon there are seven ribs on each side. (See Figure 26.) The first two are articulated with the cervical vertebrae and do not reach the sternum. The next four are articulated with the thoracic vertebrae and sternum. The last is articulated with a thoracic vertebra but has its ventral extremity attached to the ventral end of the rib in front, instead of to the sternum. All but the first two ribs are jointed and are made up of two pieces: the dorsal or **vertebral rib** and the ventral or **sternal rib.** The sternal rib corresponds to the cartilaginous portion of the human rib. From the posterior margins of all but the first and the last vertebral ribs project **uncinate processes**—crossbones that weld the thorax into a firm unit and provide surface for the attachment of costal muscles which, in the bird, play a prominent role in respiratory movements.

In the articulated skeleton of the pigeon observe the relative lengths and positions of the ribs. The third, fourth, fifth, and sixth ribs tend to increase in length successively, and their vertebral and sternal parts tend, at the same time, to meet at decreasingly acute angles. This arrangement permits the sternum to move downward and forward, then upward and backward, in breathing.

Sternum

The sternum or breastbone is one of the most highly specialized parts of the avian skeleton. Two kinds of sterna exist in birds: **ratite** and **carinate.** The ratite sternum has the ventral surface flattened, like the bottom of a raft, and occurs in flightless birds such as the ostrich. The carinate sternum has the ventral surface keeled, like the bottom of a sail boat. Flying birds have this kind; it permits more surface for the origin of the all-important muscles that operate the wings. The pigeon has the carinate sternum. This will vary in size and shape among different birds in accordance with the type of flight and the arrangement of muscles needed for it.

Viewed from above, the sternum of the pigeon is somewhat oblong, with the longer lateral borders more or less parallel but converging posteriorly. Several noteworthy structures are evident. On the anterior border are two deep, smooth-faced grooves that receive the expanded portions of the coracoid bones of the shoulder girdle and thus provide a base for the bones of this region. These articular surfaces are properly called the **coracoidal facets.** Between them a small forked process, the **rostrum,** projects anteriorly. On the anterior lateral borders are small pits, each one of which accommodates the sternal end of a true rib. They are the **costal facets.** Just behind the costal facets appears a conspicuous backward-pointing process, the **posterolateral process,** followed by a deep notch. The posterior border of the sternum is rounded and is formed by the fusion of the posterior ends of two more back-pointing processes, namely, the lateral metasternal process and the median **metasternum.** This fusion forms a bony bar, leaving a window or **fenestra** just anterior to it.

The lateral and posterior borders of the sternum vary considerably in different birds. In some cases all three processes are unfused at their ends, leaving two notches between them. The sternum is then **double-notched.** In other cases all three processes may be fused at their tips, making the sternum **bifenestrate;** or completely fused along their entire lengths, making the sternum **entire.** The pigeon, of course, has a **single-notched uni-fenestrate** sternum.

Viewed from the side, the sternum shows an enormous keel, the **carina,** extending ventrally along the median line. The carina drops down

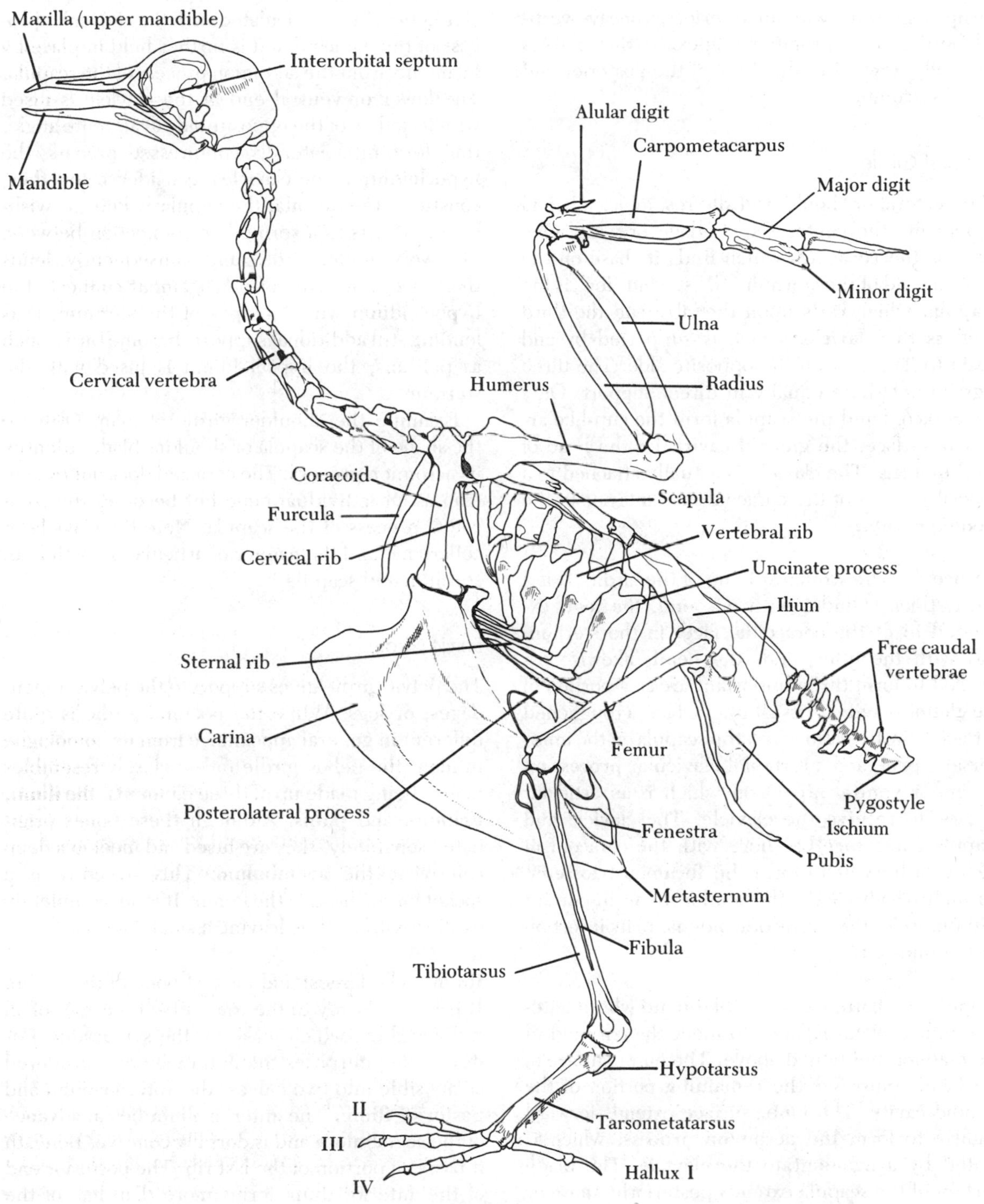

Figure 26 **Skeleton of Common Pigeon**

abruptly in front, with an anteriorly concave vertical border, to a prominent **apex.** It then curves gradually upward to the level of the posterior end of the sternum.

Pectoral Girdle

The pectoral or shoulder girdle resembles a tripod supporting the pectoral appendage or wing. One "leg" is the **coracoid,** which finds its base on the anterior end of the sternum; the second "leg" is the **scapula,** which rests upon the ribs; and the third "leg" is the **clavicle,** which is supported by and fused to its fellow on the opposite side. The three legs do not share equally in direct support. Only the coracoid and the scapula form the cup-like articular surface, the **glenoid cavity,** for the head of the humerus. The clavicle is actually attached to a special process of the coracoid just anterior to the glenoid cavity.

Coracoid The strongest bone of the girdle. It is a short, thick cylinder with one end, the **foot,** expanded to fit the coracoidal facet in the sternum and with the other end, the **head,** likewise expanded to form three important areas: a portion of the glenoid cavity on its lateral surface; a roughened surface for articulation with the scapula on the inner dorsal aspect; and a terminal **clavicular process** on the inner ventral aspect on which is an articular surface to receive the clavicle. The clavicle and scapula come together here with the coracoid in such a fashion as to form the **foramen triosseum** through which passes the tendon of an important flight muscle, the supracoracoideus, to its insertion on the humerus.

Scapula A flattened sabre-like bone whose anterior end or head expands to meet the coracoid in the manner mentioned above. The outer surface of the head comprises the remaining portion of the glenoid cavity. The inner surface extends forward slightly to form the **acromion process,** which is united by a ligament to the clavicle. The blade portion of the scapula extends posteriorly, more or less parallel to the vertebral column, to rest upon the ribs.

Clavicle A thin, rod-like bone whose upper or dorsal end expands considerably to form the **epicleidium.** This is articulated with the clavicular process of the coracoid and is further held in place by ligaments from the acromion process of the scapula. The lower or ventral end of the clavicle is fused with its fellow of the opposite side at an acute angle, thus forming a laterally compressed process, the **hypocleidium.** The clavicles, considered together, constitute the **furcula,** the popularly known "wishbone." It acts as a spring-like connection between the two shoulder girdles and, consequently, lends them necessary support. A ligament connects the hypocleidium with the apex of the sternum, thus lending still additional support. In some birds, such as pelicans, the hypocleidium is fused with the sternum.

Examine the shoulder girdle of man. Observe the shape of the scapula or shoulder blade. Identify its acromion process. The coracoid does not exist in man as an individual bone but becomes the coracoidal process of the scapula. Note the clavicle or collarbone and its manner of articulation with both sternum and scapula.

Pelvic Girdle

The pelvic girdle gives support to the pelvic appendages, or legs. While the pectoral girdle is quite different in general appearance from its homologue in man, the pelvic girdle rather closely resembles man's, being made up of three elements, the **ilium, ischium,** and **pubis.** Although these bones originated separately, they are fused and meet in a deep concavity, the **acetabulum.** This structure is a socket for the head of the femur. It is not completely ossified within, thus leaving a small foramen.

Ilium The largest and longest bone of the girdle. It is joined firmly to the transverse processes of all the vertebrae which make up the synsacrum. For descriptive purposes the ilium may be considered as divisible into two halves, the **anterior ilium** and **posterior ilium.** The anterior ilium lies in advance of the acetabulum and is dorsally concave. Beneath it passes a portion of the last rib. The posterior end of the anterior ilium forms more than half of the wall of the acetabulum. The posterior ilium is dorsally convex.

Ischium An exceptionally thin, plate-like bone that is continuous dorsally with the posterior ilium and completes the side wall of the girdle. The **ischiadic foramen** is an oval opening formed along the line of fusion of the ilium and ischium. It is homologous to the sacrosciatic notch in man. Anteriorly the ischium forms the upper fourth of the acetabulum.

Pubis A slender, needle-like bone whose anterior end completes the lower fourth of the acetabulum. From the acetabulum the pubis passes posteriorly along the ventral border of the ischium and comes to an end quite far behind the side wall of the girdle. Close to the acetabulum a small round opening, the **obturator foramen,** separates the pubis and ischium.

The pelvic girdle, unlike the pectoral girdle, is closely attached and fused to the vertebral column, forming a composite structure.

The Bones of the Head

The bones of the head include all the bones anterior to the atlas of the vertebral column. In the adult bird, they show a remarkably close fusion, with the result that the original lines of demarcation are almost wholly obscured. Examine, if available, the cleaned skeleton of a young bird—or better still, a stained and cleared embryo—in which the different bones still show distinctly.

Only a comparatively few bones of the head are movable. They are the compound lower jaw bones, the quadrates on which they work, the pterygoids of the roof of the mouth, and the bones of the tongue. The upper jaw is also movable or **kinetic** in relation to the bones of the head. Kinesis varies among different birds, depending on the adaptations for feeding. It is moderate in the pigeon, very pronounced in fruit-eating birds such as parrots, which must manipulate their food in the mouth, and markedly reduced in grazing birds—e.g., ostriches and rheas—which must bite off food with their jaws.

The bones of the head that enclose the brain and form the skeletal structure of the face are collectively known as the **skull.** The bird's skull, compared with that of man and other higher vertebrates, shows several remarkable features: (1) The orbits are relatively large and spacious, with only a thin plate of bone, the **interorbital septum,** along the midline of the head separating them. In fact, they are so huge that the bones enclosing the brain must occupy a position posterior to them. (2) Instead of teeth, certain bones are greatly elongated and covered with a horny sheath that partially performs their function. (3) The lower mandible is made up of two compound bones, one on each side of the head.

In an ornithology course a detailed study of the bones of the head is not necessary. Many of the bones are quite small and inconspicuous and have no place in descriptive works. Be able to recognize, however, the bones that are prominent and of paramount importance in the classification of birds.

For convenience, consider the bones of the head in three groups: bones of the cranium; bones of the face; bones of the tongue. Refer to Figures 27 and 28.

Bones of the Cranium

The bones of the cranium enclose the brain and form the so-called brain-case. None is freely movable. Compare the shape of cranium in the pigeon with that in man.

Occipital The bone forming the base of the cranium. A large opening, the **foramen magnum,** permits the passage of the spinal cord to the brain. The occipital is composed of four bones with indistinct boundaries. They are the **supraoccipital,** which forms the upper boundary of the foramen; the **exoccipitals,** which form the lateral boundaries of the foramen and extend forward on each side to the ear opening; and the **basioccipital,** which forms the lower boundary of the foramen and bears the ball-like **occipital condyle** for articulation with the atlas.

Parietals A pair of broad, squarish bones, fused along their medial borders. They continue up over the back of the skull from the occipital. They are bounded anteriorly by the frontals and laterally by the squamosals. The parietals, together with the occipital, roof over the posterior portion of the brain.

Frontals A pair of bones indistinctly fused along their medial borders. They continue forward from the parietals, forming the roof of the anterior portion of the brain and the roof and superior margins of the orbits.

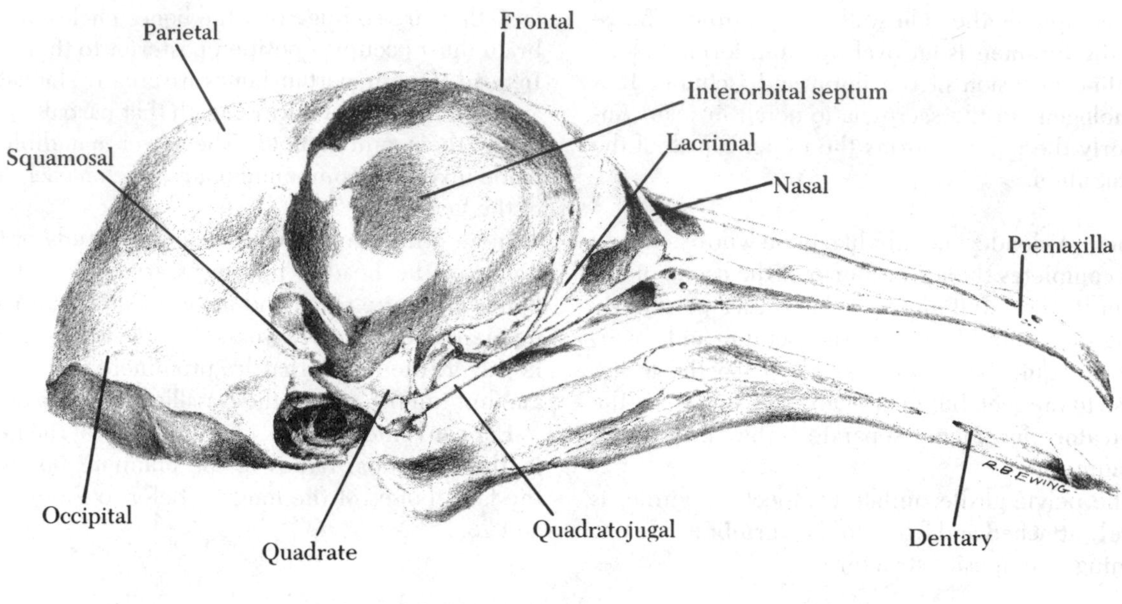

(a)

Parietal
Quadratojugal
Frontal
Culmen
R.B.EWING
Premaxilla
Nasal
Occipital

(b)

***Figure* 27 Skull of Common Pigeon**
(a) Viewed from side. (b) Viewed from above.

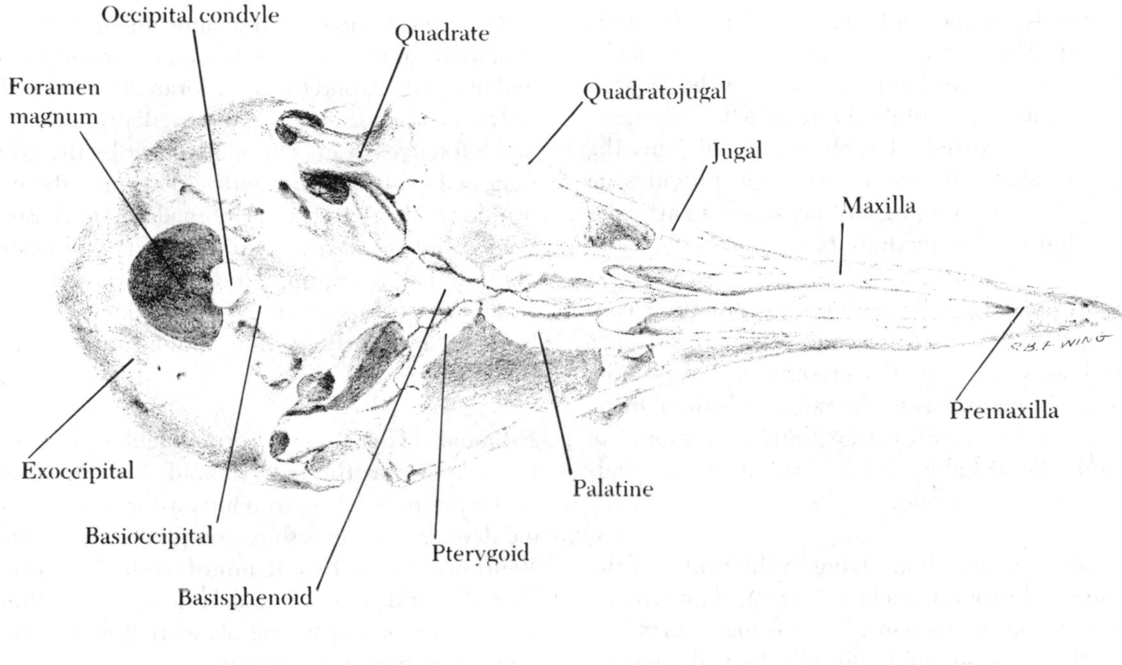

(a)

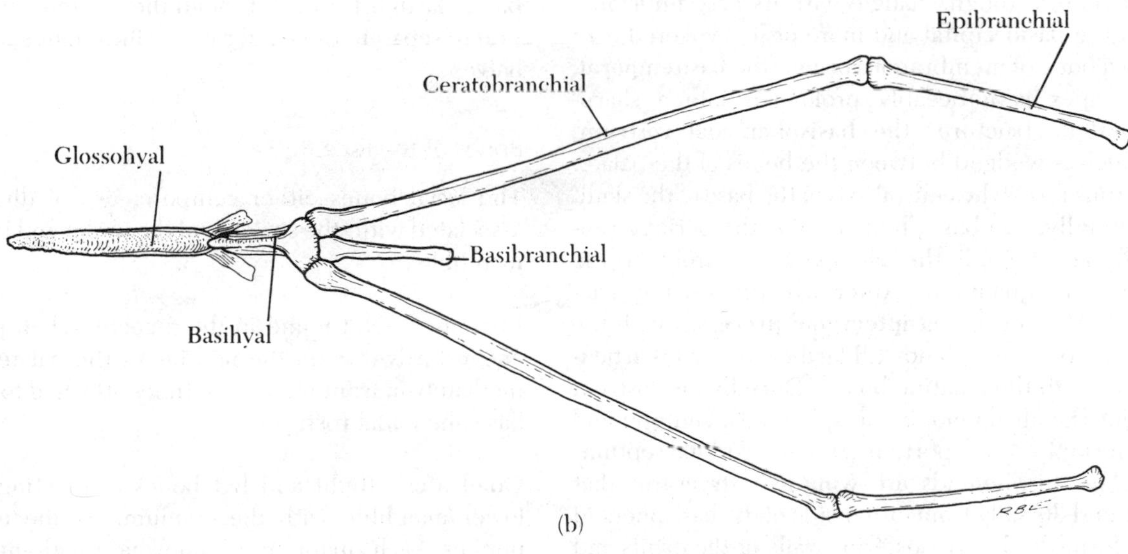

(b)

Figure 28 **Skull and Hyoid Apparatus of Common Pigeon**
(a) Ventral view of skull with lower jaw removed. (b) Hyoid apparatus.

Squamosals A pair of bones, one on each side of the head. They form the posterior margins of the orbits and join the lateral borders of the frontals and parietals to complete the roof of the cranium.

The lower border of each squamosal joins the occipital, while its lowermost angle provides an eave for the ear opening and possesses an articular depression for the quadrate bone.

Periotic Capsules Paired skeletal structures containing the organs of hearing. Each capsule has three bones: the **prootic, epiotic,** and **opisthotic.** They are homologous to the petrosal bone of man.

The periotic capsules are within the cranium and are indistinguishably fused with the exoccipitals and basisphenoid bone.

Sphenoid A large bone lying in the center of the cranium and providing a base for many bones of the skull. It is commonly considered as made up of four parts: the **basisphenoid,** the **alisphenoids (laterosphenoids),** the **presphenoids,** and the **orbitosphenoids.**

The basisphenoid forms the greater part of the cranial floor. Viewed from below, it assumes the shape of a rough triangle with its base implanted on the basioccipital and more or less covered over by a bone of membranous origin, the **basitemporal.** Its apex is noticeably prolonged into a sharp-pointed structure, the **basisphenoidal rostrum,** which is wedged between the bones of the palate. It thus forms the central axis of the base of the skull. Laterally, the basisphenoid joins the periotic capsule and bounds the ear opening in front; it also joins the squamosals. Anteriorly, the rostrum articulates by special **basipterygoid processes** with the pterygoid bones, and still farther forward it articulates with the palatine bones. Dorsally the rostrum joins the alisphenoids, presphenoids, and ethmoid to form the lower portion of the interorbital septum.

The alisphenoids are wing-like structures that extend dorsally from either side of the basisphenoid to form the lower posterior walls of the orbits and to enclose that part of the brain that is not already encased by the bones mentioned above. The alisphenoids pass upward to meet the frontals and squamosals, and far enough forward to form the inner walls of the orbits up to and including the hind margin of the **optic foramen**—the large oval opening for the passage of the optic nerve.

The presphenoids and orbitosphenoids are continuations of the alisphenoids and rostrum forward and upward beyond the optic foramen. Save at their posterior ends they are fused medially, their right and left halves being closed together by the greatly enlarged eyeballs. The central portion of the interorbital septum is a thin plate made up of these two bones. In two places (posterodorsally and posteroventrally) it is incomplete, leaving openings between the orbits. It is bounded dorsally by the frontals, anteriorly by the ethmoid, and ventrally by the rostrum.

Ethmoid This bone, a perpendicular plate sometimes known as the **mesethmoid,** is a continuation of the interorbital septum forward from the sphenoidal elements and thus completes the septum. Ventrally, the ethmoid unites with the rostrum. Dorsally and somewhat anteriorly, the ethmoid sends out a pair of lateral plates that form an **orbitonasal septum** separating the orbits from the nasal cavities. The ethmoid consequently forms a portion of the anterior orbital wall and is bounded anterodorsally by the nasal and frontal bones and laterally by the lacrimal bones. A part of the mesethmoid passes farther forward through the orbitonasal septum to separate the nasal cavities into right and left halves.

Bones of the Face

The facial bones either compose, or are directly associated with, the skeleton of the upper and lower mandible.

Prevomer Not found in the pigeon. When present in birds it is on the midline of the palate immediately in front of, or sometimes attached to, the basisphenoidal rostrum.

Quadrates Right and left bones connecting the lower mandible with the cranium. As the name implies, each quadrate is somewhat quadrangular but with a constriction in the middle making two expanded ends. The end toward the cranium freely articulates with a depression in the lowermost angle of the squamosal, and the basisphenoid just anterior and slightly dorsal to the ear opening. The end of the quadrate pointing in the opposite direction freely articulates with the adjoining side of the lower mandible, the zygomatic bar, and the pterygoid. A process extending from the quadrate toward the orbit is for the attachment of muscles.

Quadratojugals and Jugals The major part of the slender **zygomatic bar** on either side of the head, below the orbit, is composed of these two bones. The posterior portion is the quadratojugal, which articulates with the quadrate. It joins the jugal in front by an oblique suture. The jugal is a very small scale-like bone, which is in turn obliquely sutured in front to the posterior process of the maxillary bone. Frequently a portion of the quadratojugal bone extends far enough forward to be sutured also to the maxillary process.

Maxillae Right and left bones, each one consisting of three processes: a **posterior process,** which unites obliquely with the jugal and quadratojugal and thus completes the zygomatic bar; an **anterior** or **dentary process,** which is sutured laterally to the premaxillary and nasal bones; a broad **ventral process,** sometimes called the **maxillopalatine process,** which descends downward and inward. The ventral process ends blindly and does not meet its fellow of the opposite side. A cleft is therefore formed in this region of the palate.

In man, the maxilla joins with its fellow to form a prominent bone for holding the upper teeth. No cleft normally exists between them.

Pterygoids A pair of short, rather thick, rod-like bones articulating with the quadrates and the palatines. Their long axes are obliquely disposed between these bones. The middle posterior surfaces of the pterygoids articulate with the basipterygoid processes of the basisphenoidal rostrum.

Palatines A pair of bones forming the greater portion of the palate of birds. In the pigeon they are long and slender bones and lie along the midline of the mouth parallel to each other. Anteriorly they unite with the palatal processes of the premaxillary bones; posteriorly they become considerably flattened horizontally and rest on the basisphenoidal rostrum, articulating at their extreme ends with the pterygoids.

Premaxillae These two bones together form the tip of the upper mandible. Each premaxilla composes one-half of it and has three backward-projecting processes that form the main bulk of the mandible. One, the **frontal** or **nasal process,** passes to the frontal bone and fuses with its fellow to form the culmen of the beak. The second, the **dentary process,** extends horizontally to join the anterior process of the maxilla and forms half the tomium of the mandible. The third, the **palatal process,** contributes to the formation of the palate by extending along the roof of the mouth to join with the anterior end of the palatine bone of the same side.

Nasals Paired bones extending forward from the frontal bones and separated by the frontal process of the premaxillary bones. They rest on the ethmoid bone below. Each nasal bone forms the posterior boundary of a nostril by dividing into two processes: one, the **superior process,** passes along the medial side of the nostril to meet medially the frontal process of the premaxilla; the other, the **inferior process,** descends to fuse with the anterior process of the maxilla.

Lacrimals Paired bones (sometimes called the **prefrontals**) situated in the anterior portions of the orbits. Each is attached above to the frontal and nasal bones and descends, somewhat flattened, toward the dorsal surface of the zygomatic bar, bounding the orbit anteriorly.

Lower Jaw Bones The lower mandible of the bird is made up of two jaw bones that fuse anteriorly to form a V-shaped structure. Each lower jaw bone is actually a composite affair, being made up of five bones immovably fused together—the **dentary, splenial, angular, surangular,** and **articular.** No attempt will be made here to distinguish between them. The posteriormost of the five bones, the articular, has a double-cupped superior surface for articulation with the quadrate.

Bones of the Tongue

The bones of the tongue, sometimes called the **hyoid apparatus,** are divisible into two groups: median and paired. Of the median group there are three bones. The anteriormost, the **glossohyal,** serves the skeleton for the main bulk of the tongue. A small piece of cartilage projects from its forward end. Loosely articulated to the posterior end of the glossohyal is the **basihyal.** The third median bone, the **basibranchial,** is closely fused to the posterior end of the basihyal; it is usually recognized as the tapering portion of the two united bones. Coming off near the posterolateral portions of the basihyal are the "horns" of the hyoid apparatus. The first bones are the slender, paired **ceratobranchials.** To their posterior ends are articulated the equally slender **epibranchials.** (One pair of bones not clearly seen are the **ceratohyals,** which emerge from the anterior end of the basihyal. For present purposes they are not important.)

Examine the hyoid apparatus of a woodpecker, which shows extraordinary development of the branchials. These greatly elongated bones and the muscles attached to them are an important part of the mechanism that permits the woodpecker to extend and retract the tongue when getting insects from holes in trees.

Muscular System

by Andrew J. Berger

The muscular system has 175 different muscles, most of which are paired—i.e., each muscle is represented on both the right and left sides of the body. A thorough study of these muscles would require far more time than the average course in ornithology will permit. The aim in the following text is to introduce some of the more important muscles used in systematics and to point out special features of avian musculature. For more descriptive information on all the muscles of the bird, refer to *Avian Myology* by J.C. George and A.J. Berger (New York: Academic Press, 1966). More recently the *Nomina Anatomica Avium,* edited by J.J. Baumel, et al. (London: Academic Press, 1979), was prepared by the International Committee on Avian Anatomical Nomenclature to promote international communication "by establishing an agreed list of terms in a universally accepted language" (Latin). Some of the proposed name changes are listed in Table 1. The names used in the following text follow George and Berger.

Obtain a specimen of a pigeon and prepare to dissect the muscles.

The first step is to remove the skin. As in most small birds, it is relatively very thin. Therefore, be careful when taking off the skin not to cut the underlying muscles.

Place the specimen on its ventral side and make

Table 1 **Proposed Changes in Muscle Names**

Nomina Anatomica Avium, 1979	George and Berger, 1966
M. coracobrachialis caudalis	M. coracobrachialis posterior
M. biceps brachii pars propatagialis	M. biceps slip
Ligamentum humerocarpale	Humerocarpal band
M. latissimus dorsi	M. latissimus dorsi
pars cranialis	pars anterior
pars caudalis	pars posterior
M. tensor propatagialis	M. tensor patagii longus et brevis
pars longa	propatagialis longus
pars brevis	propatagialis brevis
M. scapulohumeralis caudalis	M. dorsalis scapulae
M. ectepicondylo-ulnaris	M. anconeus
M. iliotibialis cranialis	M. sartorius
M. caud-ilio-femoralis	M. piriformis
M. flexor cruris medialis	M. semimembranosus
M. flexor cruris lateralis	
pars pelvica	M. semitendinosus
pars accessoria	M. accessory semitendinosus
M. iliofibularis	M. biceps femoris
M. iliotrochantericus caudalis	M. iliotrochantericus posterior
M. iliofemoralis internus	M. iliacus

a two-inch incision in the dorsal midline of the neck. Then separate the skin from the underlying muscles by inserting the blunt handle of a scalpel (not the blade!) between the skin and the superficial layer of muscles. Lift the skin upward—away from the muscles—and continue the incision posteriorly to the base of the tail. Stay in the middorsal line. This can be readily determined by feeling the neural spines of the vertebrae with one's fingers. Proceed slowly and carefully.

Now, "work" the skin on the right side of the body outward by separating the skin from the underlying muscles with the handle of the scalpel. In most regions the skin is fastened to the **connective tissue** or **fascia** (plural, **fasciae**) covering the muscles by loose fibroelastic connective tissue, the **subcutaneous connective tissue** or **superficial fascia.** A blunt instrument will break these fibers but will not cut into the muscles. While removing the skin, be alert for **dermal muscles,** which insert into the skin. As these muscles will not be studied, cut them from their attachment to the skin by using the blade of the scalpel.

After freeing the skin to about the midlateral line of the body, turn over the specimen. Make an incision in the skin down the ventral midline, beginning at the anterior end of the carina and extending to the vent. The carina can be felt and often seen through the skin. Be careful, posterior to the sternum, not to cut through the abdominal wall because both the skin and the abdominal wall itself are very thin in this region. Separate the skin from the underlying muscle and, posteriorly, the abdominal wall, but on the left side only. When the skin has been freed completely around the side of the body, cut the skin and remove it. Take special care when removing the skin posterior to the humerus, from the elbow to the shoulder. Look for the tiny, shiny tendon of the **expansor secundariorum** (a muscle to be dissected later) between the two layers of skin. Try not to cut it when removing the skin. Examine the feather follicles on the deep surface of the skin. Complete the skinning of the wing and leave the stubs of the primaries and secondaries attached to the bones.

Note that the muscles are surrounded by deep fasciae. These connective tissues invest the individual muscles and some of them serve either to connect or to separate muscles or other organs. Basically, the dissection of musculature is a matter of following fascial planes and of removing fascia—and sometimes fat—in order to expose and define clearly the individual muscles or other structures. As a rule, dissection is done better with the blunt handle of a scalpel rather than the sharp blade, which is likely to cut structures that need to be preserved.

The Muscles of the Wing

For convenience in study, the muscles of the wing are grouped into ventral muscles and dorsal muscles and are considered in that order.

Ventral Muscles

Most of ventral muscles are fundamentally flexor muscles, serving to elevate and depress the wing or to flex the forearm. See Figure 29a.

Pectoralis The largest and most powerful muscle in the bird. Note its general relationship to the body as a whole. Pass a probe from lateral to medial deep to the pectoralis about one inch from its insertion on the humerus. Work the probe posteriorly, cutting the fibers of origin of the pectoralis from the carina and the lateral border of the sternum. Scrape the muscle attachments from the clavicle and from the membrane running from the clavicle to the coracoid. After all the attachments have been cut, reflect the pectoralis outward toward its insertion but do not cut the fibers of the insertion. Lying deep under the pectoralis are two muscles, the supracoracoideus and the coracobrachialis posterior. Try not to cut them as they will be studied next.

Supracoracoideus Note the midline raphe—a whitish, seam-like line. Cut the muscle's fibers of origin from the carina, the body of the sternum, and the coracoclavicular membrane. Trace the belly of the muscle and tendon until they disappear dorsally. The actual site of the insertion will be seen later. The supracoracoideus together with the pectoralis play the principal role in elevating and depressing the wing. Both muscles constitute a considerable portion of total body weight—as much as 34 to 36 percent in some species (see F.A. Hartman, *Smithsonian Misc. Collections,* 143:1–91, 1961).

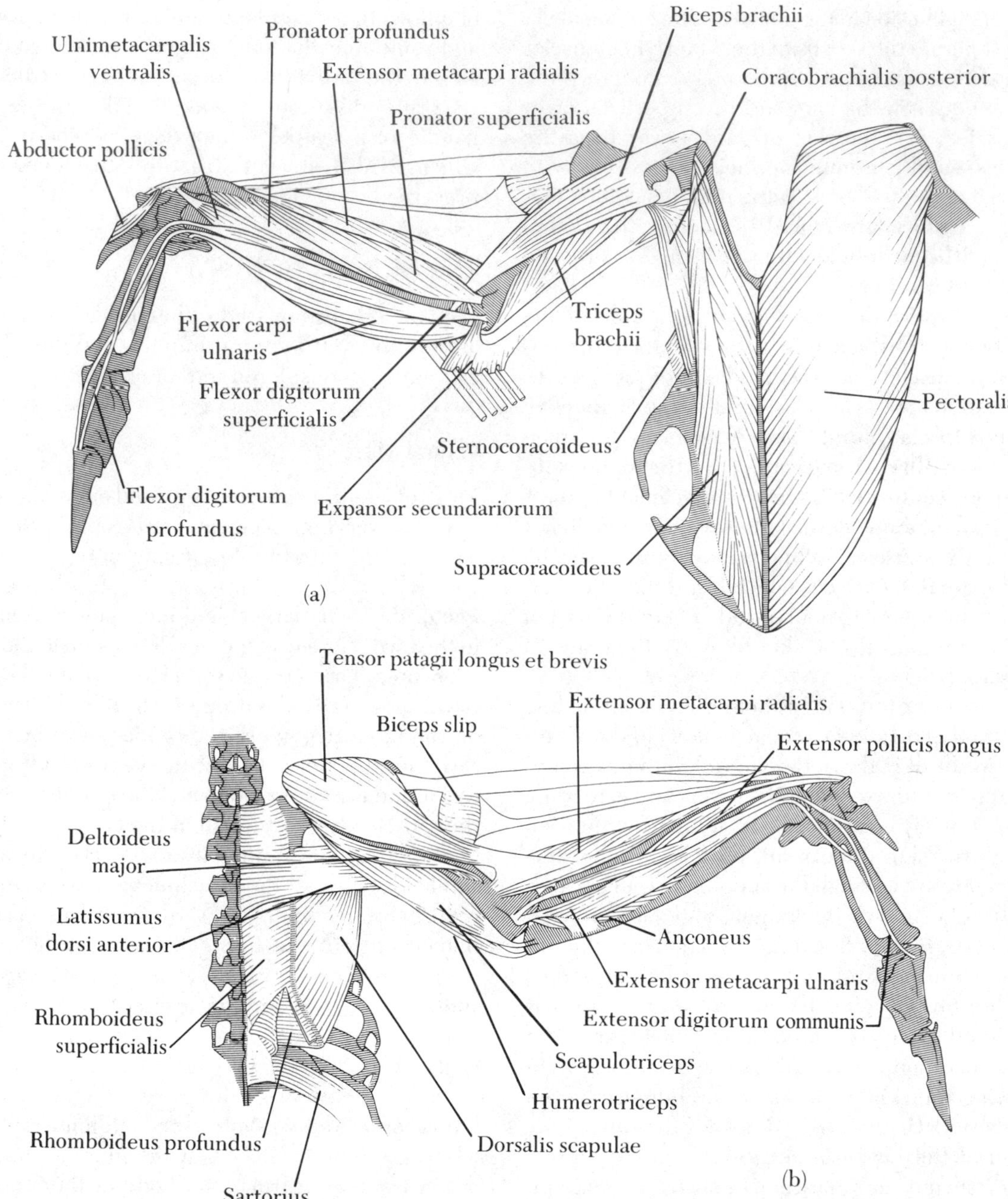

***Figure* 29 Muscles of the Wing, Common Pigeon**

(a) A ventral view of the sternum and right wing. The bird's right pectoralis muscle has been removed to reveal the underlying supracoracoideus and coracobrachialis posterior muscles. The humerocarpal band has been removed to show the relationships of the flexor carpi ulnaris and flexor digitorum superficialis muscles. (b) A dorsal view of the muscles of the right shoulder and wing as seen after the skin and fascia have been removed. A flap has been reflected in the posterior part of the rhomboideus superficialis muscle to show the deeper rhomboideus profundus muscle. The posterior aponeurotic extension of the insertion of the tensor patagii brevis muscle on the belly of the extensor metacarpi radialis muscle has been removed to show better the relationships of the proximal forearm muscles.

Coracobrachialis Posterior Note the relationships of this muscle lateral to the supracoracoideus but do not dissect it. This muscle assists in drawing the humerus posteriorly.

Biceps Brachii Clean the tendon of origin so that it can be seen clearly. It has one attachment to the anterior surface of the head of the coracoid and another to the head of the humerus. Follow the belly downward to the formation of the tendon of insertion on the radius and ulna. The actual insertion is concealed by the bellies of several forearm muscles. The biceps brachii flexes the forearm or antebrachium.

The Biceps Slip Note the relationship of the belly of this fleshy muscle to the biceps brachii, from which it arises, and to the tendon of insertion of a muscle, the **tensor patagii longus.** The presence or absence of the biceps slip in different groups of birds is commonly used as a diagnostic character.

Expansor Secundariorum Look for the belly of this muscle posterior to the elbow. It is a peculiar muscle composed entirely of smooth rather than striated muscle fibers. The insertion is on the calami of several of the secondary feathers. From the proximal end of the belly, trace the fine scapular tendon upward to the armpit or axilla. If necessary, cut the tendon of the biceps and reflect the belly outward. The scapular tendon is attached to the dorsomedial edge of the scapula; it is reinforced by a second tendon which arises from the fascial envelope surrounding the distal portion of the belly of the dorsalis scapulae, to be studied later. The expansor secundariorum also has a short tendinous origin from the distal end of the humerus. The expansor secundariorum serves to draw the proximal secondaries medially and downward.

Triceps Brachii Note the general relationships of this large extensor muscle along the posterior surface of the humerus. The muscle consists of two distinct parts: the **scapulotriceps** and the **humerotriceps.** The scapulotriceps arises from the lateral surface of the scapula just posterior to the glenoid cavity; it inserts on the dorsal surface of the base of the ulna (dissected later). The humerotriceps arises from most of the posterior surface of the humerus. The only indication of two heads for this muscle is found in the pneumatic fossa of the humerus. The humerotriceps muscle inserts by tendinous fibers on the proximal end of the ulna. The triceps brachii extends the forearm.

Other Ventral Muscles Observe the following five muscles on the ventral surface of the antebrachium but do not dissect them except to trace the tendon of the biceps brachii to its insertion: the **extensor metacarpi radialis,** the **pronator superficialis,** the **pronator profundus,** the **flexor digitorum superficialis,** and the **flexor carpi ulnaris.** What do the names tell you about these muscles and their actions? Be alert for the **humerocarpal band** covering (and concealing) parts of the bellies of the flexor digitorum superficialis and the flexor carpi ulnaris.

Dorsal Muscles

Most of the dorsal muscles are extensor muscles serving to extend and elevate the wing. See Figure 29b.

Latissiums Dorsi The most superficial muscle in the back. There is a single belly, the **pars anterior,** in the pigeon. The latissimus arises from the neural spines of the last cervical and the first two thoracic vertebrae. There are four parts in some birds: besides the pars anterior, the **pars posterior, pars metapatagialis,** and **pars dorsocutaneous.** At this time, trace the belly laterally only until it disappears deep to the muscles of the arm. The latissimus serves to draw the humerus posteriorly and medially.

Tensor Patagii Longus et Brevis This complex in the pigeon consists of a single hypertrophied belly; it arises from the apex or epicleidium of the clavicle and from the acromion process of the scapula. (There are two separate bellies in most birds.) The anterior, thinner part of the belly is the tensor patagii longus; the posterior, thicker part of the belly represents the tensor patagii brevis. Clean and study the entire muscle, including the insertion of the brevis tendon on the belly of the extensor metacarpi radialis (see previous explanation). The brevis part of the tensor patagii assists in flexing and elevating the wing. Note the relationship of the longus tendon to the biceps slip and trace the tendon to its insertion at the wrist. The longus part of the tensor patagii together with the biceps slip assist in tensing the patagium (see page 15). Cut the origin of the belly of the tensor patagii from the epicleidium of the clavicle and the acromion process of the scapula; then turn the belly outward. Do not cut the tendons of insertion.

Deltoideus Major This has a small **anterior head** and a large **posterior head.** Note the extent of each. Cut each head at its origin from the anterior end of the scapula and remove the anterior head. Now identify the tendon of insertion of the supracoracoideus. The deltoideus major elevates the brachium and draws it posteriorly.

Scapulotriceps This is a large head of the triceps brachii. Note its extent and relationships, and then cut the muscle at its origin from the inferolateral surface of the scapula and from the inferior margin of the posterior lip of the glenoid cavity. Turn the belly down toward the elbow. Now trace the latissimus dorsi pars anterior to its insertion on the humerus.

Rhomboideus Superficialis and Rhomboideus Profundus Identify these two muscles running between the neural spines and the medial border of the scapula. These muscles serve to stabilize the scapula or to draw it toward the vertebral column. Do not remove them.

Dorsalis Scapulae Note this large muscle, arising from the lateral surface of the scapula. The belly passes forward, ends on a tendon, and inserts on the proximal end of the humerus. This muscle serves to draw the brachium medially and rotates it so that the leading edge of the wing is turned downward. Do not dissect it.

Other Dorsal Muscles Examine, but do not dissect, the following extensor muscles on the dorsal surface of the forearm: **extensor metacarpi radialis, extensor digitorum communis, extensor metacarpi ulnaris,** and **anconeus.**

The Muscles of the Pelvic Appendage

Beginning in the dorsal midline of the synsascrum, remove the skin from the lateral surface of the left thigh and leg; then remove the skin from the medial surface. Remember that the skin is very thin in these areas. For convenience, the muscles are considered in two groups, the superficial muscles and the formula muscles. See Figures 30 and 31.

Superficial Muscles

The superficial muscles must be removed in order to expose the formula muscles.

Sartorius The most anterior muscle on the anterolateral surface of the thigh. Trace the belly from its origin on the anterior end of the ilium to its insertion on the patellar ligament at the proximal end of the tibiotarsus. The sartorius serves to extend both the thigh and the leg. Leave it in place but separate it from the muscle lying immediately posterior to it, the iliotibialis.

Iliotibialis This, the most superficial muscle on the lateral surface of the thigh, is a very thin layer of muscle and aponeurosis (dense connective tissue). Considerable care is required in removing it without damaging the underlying muscles. Cut the aponeurosis of origin from the anterior and posterior iliac crests of the ilium and from the intervening median dorsal ridge of the synsacrum. Pull the aponeurosis and belly outward and downward toward the knee. Observe the three parts of the iliotibialis complex: anterior and posterior fleshy bellies and an aponeurotic central sheet in approximately the distal three-fourths of the muscle. The aponeurotic portion probably will be fused with the underlying muscles and will have to be shaved from them. The muscle ends in a tough aponeurosis, which forms the anterior layer of the **patellar tendon.** It encloses the patella. The iliotibialis serves to abduct or draw the thigh away from the medial axis of the body and to extend the leg.

Iliotrochantericus Posterior This large muscle lies concealed by the aponeurosis of origin of the iliotibialis. Cut the fleshy attachments of the iliotrochantericus posterior from the anterior ilium and reflect the muscle outward to its insertion on the femur just distal to the trochanter. Do not cut the femoral attachment. The iliotrochantericus posterior rotates the lateral surface of the femur forward and inward.

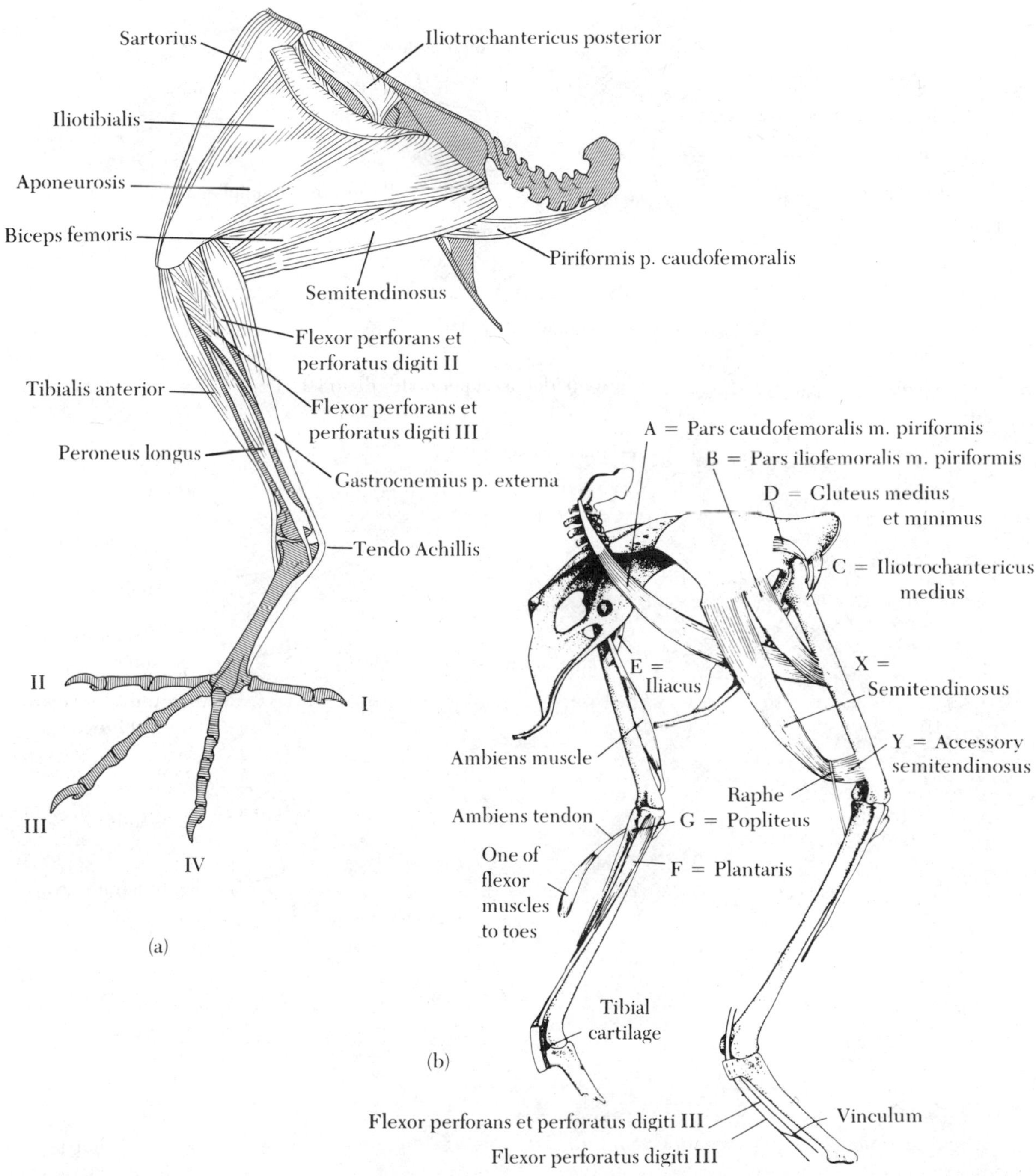

Figure 30 **Muscles of the Leg**

(a) Lateral view of muscles of the left leg of the Common Pigeon as seen after the skin and superficial fascia have been removed. The aponeurosis of origin of the iliotibialis muscle from the anterior iliac crest of the ilium has been cut and reflected downward to show the iliotrochantericus posterior muscle. (b) Drawing of a generalized bird showing all the formula muscles. (From *Avian Myology,* J.C. George and A.J. Berger, New York: Academic Press, 1966.)

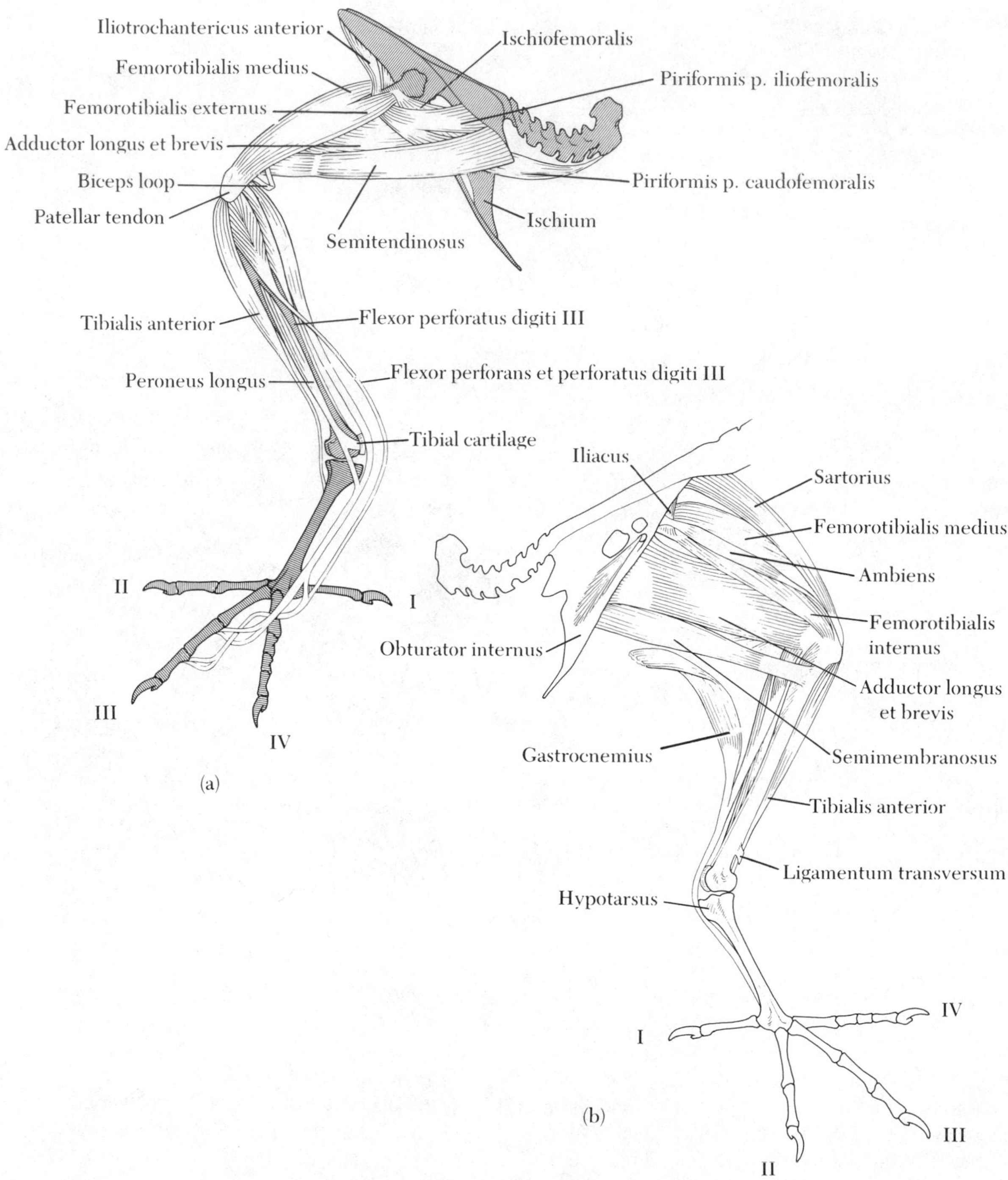

Figure 31 **Muscles of the Leg and Thigh, Common Pigeon**
(a) A deep layer of muscles of the left leg. The following muscles have been removed: sartorius, iliotibialis, iliotrochantericus posterior, biceps femoris; and gastrocnemius. (b) A view of the muscles on the medial surface of the left thigh and leg after the skin and superficial fascia have been removed. The internal head of the gastrocnemius muscle has been reflected backward to reveal the plantaris muscle. The popliteus is not shown.

Gastrocnemius The most superficial muscle mass on the posterior, lateral, and medial surfaces of the crus. Its three separate heads—lateral, medial, and internal—contribute to the formation of a common **Tendo Achillis** a short distance from the intertarsal joint. Identify this tendon and pass a probe deep to it, making sure that the probe lies superficial to all of the other tendons running toward the tibial cartilage. The gastrocnemius serves to extend the tarsometatarsus. Cut the Tendo Archillis and carefully reflect the gastrocnemius upward to the knee joint.

Formula Muscles

These are the eleven muscles used in systems of classification, and they are designated by symbols A, B, C, D, E, F, G, X, Y, Am, and V. (See Figure 30b for a drawing of the formula muscles.) The leg-muscle formula for the Common Pigeon is ABCEFGXYAmV since in this species all but one muscle—D, the gluteus medius et minimus—is represented.

The formula for the Osprey *(Pandion haliaetus)* is ADEBAm; for the Eastern Screech-Owl *(Otus asio)*, ADEG; for the Chimney Swift *(Chaetura pelagica)*, AE. Refer to *Avian Myology* (New York: Academic Press, 1966) by J.C. George and A.J. Berger, pages 233–236, for further explanation on the use of symbols and a list of formulas of selected species.

Iliotrochantericus Medius "C" in leg-muscle formulas. Identify this small muscle running from its origin on the ventral edge of the ilium, just anterior to the acetabulum, to its insertion on the femur dorsal to the iliotrochantericus anterior. Its action is similar to that of the iliotrochantericus posterior.

Iliacus "E" in leg-muscle formulas. This is a tiny, flat, bandlike muscle arising from the ventral edge of the ilium just medial to the origin of the iliotrochantericus medius. The flat belly passes outward to insert on the medial surface of the femur, only 5 mm from the proximal end of the bone. A hand lens or dissecting microscope may be required to find this muscle. It acts to rotate and to adduct or draw the thigh toward the medial axis of the body.

Semitendinosus and Accessorius Semitendinosi This complex lies posterior to the femur. Trace the semitendinosus ("X" in leg-muscle formulas) from its origin on the posterior third of the posterior iliac crest downward to the ligamentous raphe that separates the muscle from the accessorius semitendinosi ("Y" in formulas). Follow the latter muscle to its insertion on the posterolateral surface of the distal end of the femur. The semitendinosus together with the accessory semitendinosus serve primarily to flex the femur (draw it posteriorly). The accessory semitendinosus is absent in some species; both muscles are absent in others.

Cut the fibers of origin of the semitendinosus muscle and reflect the belly downward. This will expose the following muscle.

Piriformis This muscle in the pigeon has two separate bellies: the **pars iliofemoralis** and **pars caudofemoralis** ("B" and "A" respectively in leg-muscle formulas).

The pars iliofemoralis arises from the ventral surface of the posterior iliac crest. The belly passes downward and forward, superficial to the pars caudofemoralis, and inserts on the posterolateral surface of the femur. Cut the muscle at its origin and reflect the belly forward, thus exposing all of the next part.

The pars caudofemoralis arises from the ventral surface of the pygostyle and from a dense aponeurosis associated with the under tail coverts. The strap-like belly passes anteriorly between the semitendinosus (already reflected) and the **semimembranosus** (a muscle consisting of a flattened band of fleshy fibers running from its origin on the postero-inferior surface of the ischium to its insertion on the tibiotarsus; do not dissect it). The fleshy belly of the pars caudofemoralis ends on a small tendon that fuses with the tendon of insertion of the pars iliofemoralis.

The piriformis assists in flexing the thigh and moving the tail laterally and downward.

Ambiens "Am" in leg-muscle formulas. This is the most medial muscle of the thigh. Follow the muscle from its aponeurotic origin on the pubis to the knee. Trace the tendon through a compartment in the patellar ligament to the lateral surface of the knee. The tendon of the ambiens muscle then descends deep to the tendon of insertion of the **biceps femoris** and serves as part of the origin of three of the flexor muscles to Toes No. 2, 3, and 4, thus reinforcing their actions. The pattern varies in other bird species. This muscle is found in some reptiles but not in mammals.

Plantaris "F" in leg-muscle formulas. It may be necessary to cut the internal head of the gastrocnemius in order to study the plantaris muscle. The plantaris arises from the posteromedial surface of the tibiotarsus, beginning just below the proximal articular surface. The belly extends about halfway down the crus, tapers to a flattened tendon, and inserts on the proximal end of the tibial cartilage. The plantaris serves to draw the tibial cartilage proximally and to aid in extension of the tarsometatarsus.

Popliteus "G" in leg-muscle formulas. This is the deepest muscle on the posterior surface, proximal end, of the crus. Carefully cut the origins of the flexor muscles posterior to the head of the tibiotarsus and reflect the bellies distad. Identify the popliteus as a small, rectangular-shaped muscle extending between the head of the fibula and the tibiotarsus. Its weak action is to draw the head of the fibula toward the tibiotarsus.

The Vinculum "V" in leg-muscle formulas. As they pass down the posterior side of the tibiotarsus, the two tendons of the flexor muscles for Toe No. 3 are connected by a tendinous band or vinculum. In order to expose the vinculum and the two tendons, it is necessary to remove the fibers of the gastrocnemius tendon that insert on the posterior surface of the tibiotarsus. Now grasp the Tendo Achillis (already cut) a short distance above the tibial cartilage and pull it downward sharply in order to break the attachments to the bone. Then carefully separate the tendons on the posterior surface of the tibiotarsus and locate the two tendons to Toe No. 3 and the interconnecting vinculum.

Now trace the tendons of the **flexor perforatus digiti III** and **flexor perforans et perforatus digiti III** to their insertion on the phalanges of Toe No. 3 and study the pattern of insertion. A similar general pattern of relationships to both fingers and toes is found in most mammals, including man. The multiple insertions on the several phalanges of the fingers in man, plus his opposable thumb, account for man's great manual dexterity.

Respiratory and Digestive Systems

These two organ systems may be conveniently considered together, since they are closely associated with each other. Both systems in the bird show marked modifications.

Obtain a specimen of a pigeon and prepare to dissect it internally.

Lay the left side of the mouth open by cutting the angle of the jaw and continuing the incision down the side of the neck. Study the following regions.

Mouth The mouth is a cavity, sometimes called the **buccal cavity,** between the upper and lower mandibles. Its roof or **palate** is hard and horny anteriorly and somewhat softer posteriorly, resembling the soft palate of many mammals. A median slit, the **choana,** separates the palate into two longitudinal **palatal folds** with small backward-projecting horny papillae. A cleft palate of this sort is characteristic of birds. Separate the palatal folds and observe the **nasal cavity.** Anterodorsally it is divided by the **nasal septum** into right and left nasal cavities, which communicate directly with the corresponding right and left **anterior nares,** or **nostrils,** to be seen externally at the base of the upper mandible.

The floor of the mouth is occupied by the **tongue,** whose shape conforms generally to that of the lower mandible. It is attached by only a small part of its under surface. Its covering is thick and horny. Anteriorly it is sharply pointed but posteriorly it is forked and bears, like the roof of the mouth, backward-projecting horny papillae. **Taste buds,** said to number between 25 and 60, occur at the base of the tongue, and a few more lie in the softer, posterior part of the palate. Small **salivary glands,** four pairs, empty into the floor of the mouth.

Pharynx The pharynx is a continuation of the mouth posteriorly, beginning at the posterior end of the palatal slit. In contrast to the mouth, its walls are more or less muscular. In the middle of the dorsal wall, and almost continuous with the palatal slit, is a smaller slit-like opening common to the paired Eustachian tubes, which connect the pharynx with the middle ears. In mammals these tubes enter the pharynx separately, but in birds they enter together, In the middle of the floor of the pharynx are two **laryngeal folds,** which bound a relatively narrow, slit-like opening, the **glottis.** Its margins bear horny papillae, but there is nothing to represent the "trapdoor," or epiglottis, which in mammals protects the opening. The posterior dorsal and ventral walls of the pharynx have pairs of membranous folds with horny papillae, the **dorsal** and **ventral pharyngeal folds.**

Hyoid Apparatus The hyoid aparatus has already been studied in detail. However, determine its location and function by carefully dissecting away (1) the covering of the tongue, to find the median bones, and (2) the skin of the malar and auricular regions, to find the paired bones which pass toward the ears. Pull the horns of the hyoid apparatus to see how they work.

Continue the dissection down the side of the neck by pushing aside the skin along the incision. Identify the soft, thin-walled food tube, or **esophagus,** which connects the pharynx to the stomach, and the windpipe, or **trachea,** with the stiffened rings in its wall, which connects the pharynx with the bronchi of the lungs. Study these structures further.

Trachea The rings in the walls of the trachea are bony on their ventral sides but cartilaginous on their dorsal sides. This is in marked contrast to the rings of the human trachea, which are cartilaginous ventrally and membranous dorsally.

The anterior end of the trachea is expanded to form the **larynx.** It is not a sound-producing organ as in mammals. The larynx supports the laryngeal folds in the floor of the pharynx and the glottis opens into it. Dissect out the larynx by cutting around it and freeing it from the pharynx. Then cut it open on one side and identify the cartilages within. Conspicuous ventrally is the large triangular cartilage, the **cricoid,** with lateral processes which bend around dorsally, coming to narrowed ends in back of the larynx. Between these two dorsal ends of the the cricoid is a median piece of cartilage, the **procricoid.** At the base of the procricoid are attached a pair of slender, curved, somewhat bony cartilages, the **arytenoids,** which extend anteriorly along the upper parts of the larynx and form the skeletal structure of the margins of the glottis. Anteriorly, the arytenoid cartilages are attached to the upper, inner surface of the cricoid cartilage. There are no true vocal cords in birds.

Esophagus This is a distensible tube (see Fig. 34, p. 78) lying dorsal to the trachea and following a relatively straight course to the stomach. Trace it posteriorly. Just before it enters the thoracic cavity it becomes dilated into a bi-lobed sac, the **crop.** Here food is detained before it is passed to the stomach. All pigeons and gallinaceous birds possess a crop.

Air Sacs and Body Cavity The lungs of birds feature outpocketings filled with air that extend between various organs and penetrate certain bones; they have few blood vessels and no respiratory surfaces. The sacs are noticeably thin-walled and resemble soap bubbles. In the pigeon one single and four paired air sacs are recognized. To be studied successfully the respiratory system should be artifically inflated by cutting the trachea, inserting a tight-fitting glass tube, blowing through it, and then immediately tying off the trachea to prevent the air from escaping. In this way all the parts of the respiratory system, including the air sacs, will be distended and made more prominent. Another way to study the air sacs is to inject them with Woods Alloy and obtain casts (for the method, see paper by P.W. Gilbert, *Auk*, 56:57–63, 1939). Latex is also a good injection medium (see D.H. Tompsett, *Ibis*, 99:614–620, 1957).

Inflate the respiratory system as directed and identify the **interclavicular** and **cervical air sacs** (see Fig. 32) in the vicinity of the crop. The interclavicular sac is directly dorsal to the furcula, touches the dorsal side of the crop, and surrounds the posterior end of the trachea and bronchi. It is the only single sac of the respiratory system and has on each side a diverticulum that sends branches to the shoulder region and into the sternum, clavicle, coracoid, and humerus. The cervical sacs are dorsal and paired; they are anterodorsal prolongations of the interclavicular sac supplying the cervical and thoracic vertebrae.

With the scissors, extend the incision in the side of the neck ventrally to the vent, keeping it just to the left of the median line. In the breast region cut through the large flight muscles and sternum, keeping close to the keel. In the abdominal region cut

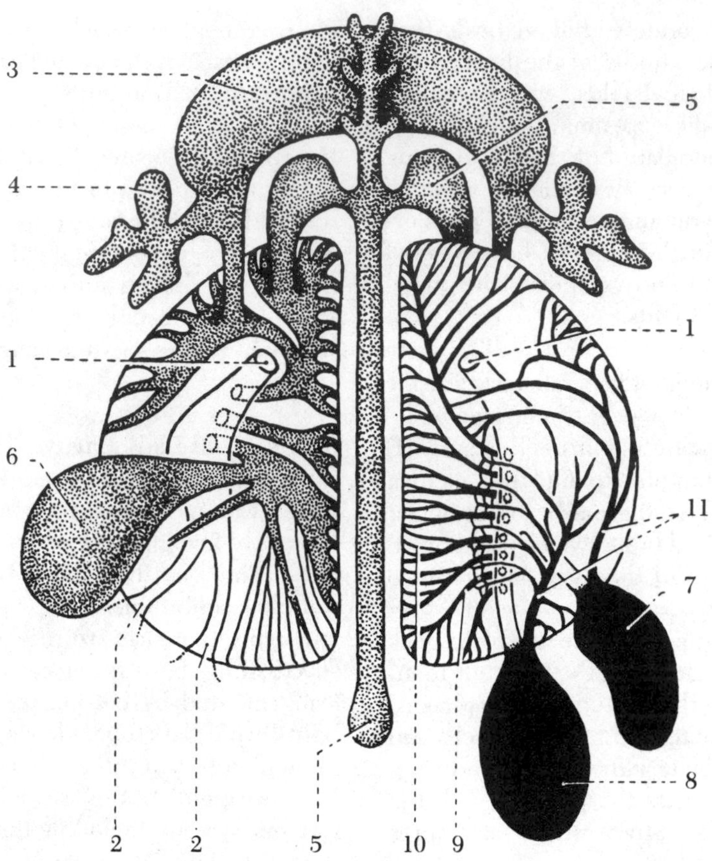

Figure 32 **Diagram of the Lungs and Air Sacs of the Bird** The ventral view is at the left, the dorsal view at the right. The parts of the system concerned with inspiration are in black, the parts with expiration are stippled. 1. Mesobronchi. 2. Opening of the mesobronchi into the air sacs. 3. Interclavicular air sac. 4. Diverticulum of the interclavicular sac to the sternum, coracoid, clavicle, and humerus. 5. Cervical sac. 6. Anterior thoracic air sac. 7. Posterior thoracic air sac. 8. Abdominal air sac. 9. Dorsobronchi. 10. Parabronchi. 11. Recurrent bronchi. (From Portmann, 1950, in *Traité de Zoologie,* edited by Grassé, Vol. 15, Figure 200, Masson et Cie; after Brandes and Hirsch.)

through the thinner muscular layers. Be careful not to cut too deeply, thus injuring the organs below. Spread apart the edges of the incision and examine the body cavity.

As in the higher vertebrates the body cavity is readily divisible into the **thoracic** and **abdominal cavities.** The thoracic cavity is located dorsal to the sternum and contains three smaller divisions. Medially there is the **pericardial cavity**—a space between the prominent **heart** and the thin sac surrounding the heart. This pericardial sac or **pericardium** is in contact ventrally with the inner surface of the sternum and dorsally and laterally with the inner surfaces of the body cavity. Only the posterior part of the pericardium is free. Laterally and somewhat anteriorly are two **pleural cavities** containing the lungs. These fill in the remainder of the thoracic cavity. The abdominal cavity is posterior to the sternum. In it is the large chocolate-colored **liver,** the tightly coiled **intestine,** and, to the left, the enormous **stomach.** Observe that there is no diaphragm separating the thoracic from the abdominal cavity. Instead there is a membranous, double-walled partition extending obliquely backward between the two from the points where the pericardium meets the body walls laterally. This is called the **oblique septum.** The part of the bird corresponding to the diaphragm of the mammal is a thin sheet of muscle arising from the inner surfaces of the ribs and bodies of the vertebrae and closely attached to the ventral surfaces of the pleural cavities.

Lying on each side of the pericardial cavity is a small **anterior thoracic air sac** (see Fig. 32), while between each lung and the liver inside the double-walled oblique septum is a **posterior thoracic air sac.** On each side of the abdominal cavity is a large **abdominal air sac,** which passes between the various organs of the cavity.

Prepare to study in more detail the organs of the thoracic and abdominal cavities. The air sacs may now be punctured.

Return to the trachea and follow it backward to the point where it bifurcates to form the **right** and **left bronchus.** Each bronchus passes directly to the lung on the corresponding side. The **syrinx,** or voice organ (Fig. 33), is at the point where the trachea divides and lies within the interclavicular air sac.

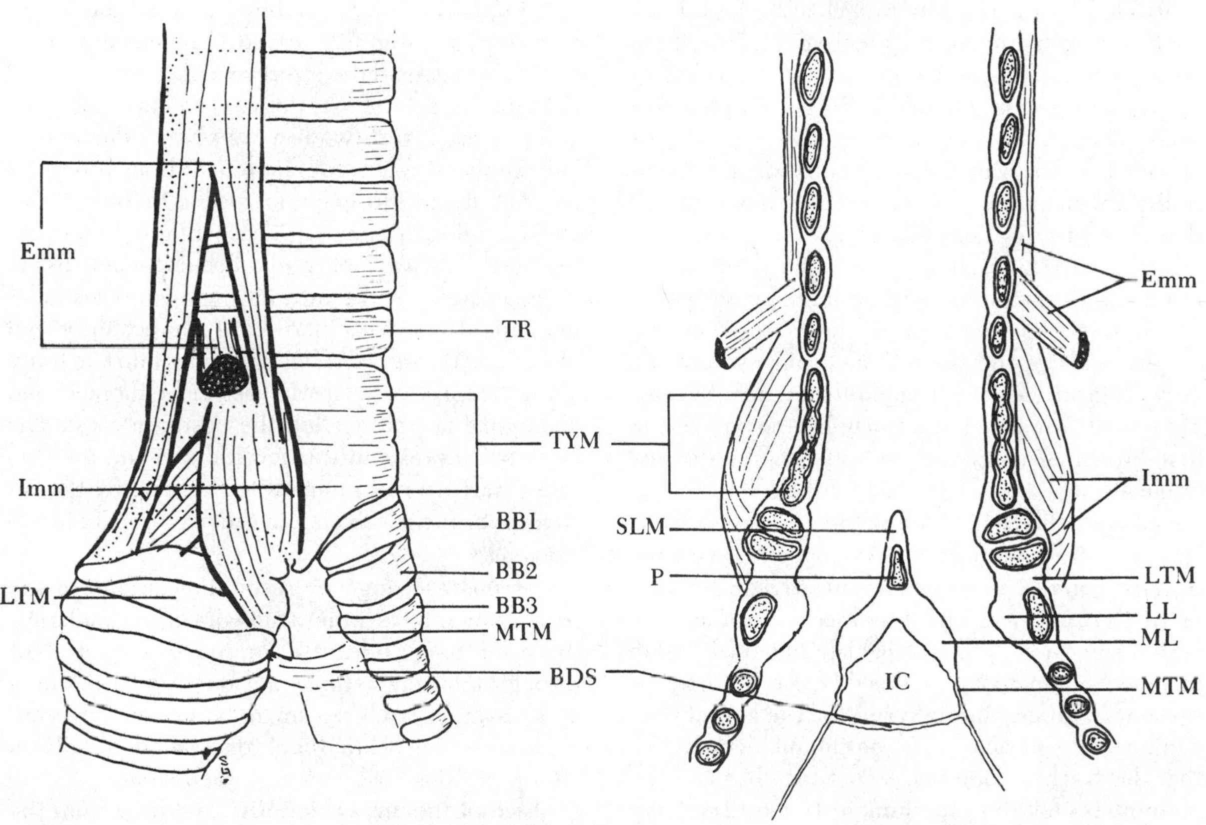

***Figure* 33 Idealized Syrinx of a Songbird**
At left, an external, ventral view with muscles removed from the left side; at right, frontal section through the syrinx. Abbreviations: BB, bronchial bar; BDS, bronchiodesmus; Emm, extrinsic muscles; IC, interclavicular air sac; Imm, intrinsic muscles; LL, lateral labium; LTM, lateral tympanic membrane; ML, medial labium; MTM, medial tympanic membrane; P, pessulus; SLM, semilunar membrane; TR tracheal ring, TYM, tympanum.

Syrinx A pigeon's syrinx is much simpler than that of the songbird (Fig. 33). Indeed, it is simpler than the syringes of most nonpasserines. However, it does show several features of a kind of syrinx that is widely distributed in "lower" birds. Its shape is controlled by two pairs of **extrinsic muscles** only. In pigeons the anterior pair originates on the sternum and passes anteriorly to insert together asymmetrically to the right of the ventral midline of the trachea. The posterior pair arises on either side of the trachea at about the level of the insertion of the anterior pair and passes posteriorly to insert on the **lateral tympaniform membranes.** The syrinx of songbirds contains several pairs of **intrinsic muscles.** These arise on the posterior end of the trachea and insert onto the anterior most **bronchial bars.** The shape of the complex syrinx can be very finely controlled by rotating the bronchial bars, and this control presumably permits increased plasticity in vocal behavior. The lateral tympaniform membranes are stretched between the last two tracheal rings. These are joined in the midline by a cartilagenous bar that prevents their approximation ventrally. In many species the last few tracheal bars that form the anterior end of the syrinx are closely joined, or even fused, and are collectively called the drum or **tympanum.** Large lateral membranes are found in many simple syringes. The anteromedial walls of the bronchi are composed of the very delicate **medial tympaniform membranes.** These are supported anteriorly by the tips of the first bronchial rings and an extension of the last tracheal ring.

Cut open the ventral side of the syrinx in a pigeon and compare it with Figure 33. Pigeons have a bar of dense connective tissue instead of a rigid pessulus of cartilage or bone. The semilunar membrane, the lateral labium, and the medial labium are absent in pigeons. Some or all of these structures may be present in other, simple syringes. The lateral tympaniform membranes are continuous dorsally, so that the tracheal rings can come together and the membranes fold into the lumen. In most birds the posteromedial walls of the bronchi are connected by a broad, sheetlike **bronchidesmus,** but that structure also is absent in pigeons.

Bronchi The bronchi have their outer walls strengthened by half-rings of cartilage; their inner walls are membranous only. Each bronchus enters the ventral surface of a lung and passes through it as the **mesobronchus.** As the mesobronchus proceeds posteriorly the half-rings of cartilage gradually disappear.

Lungs The lungs are covered ventrally by the linings of the pleural cavitites, or **pleura,** and the rudimentary diaphragm. Note that each lung is bright red, owing to its containing a large amount of blood, and that it is somewhat flattened against the dorsal wall of the cavity where it is not invested by pleura. Closer examination will show that the lung fits into the spaces between the ribs and vertebrae so that the impressions of these bones are visible on its surface. How far back does the lung extend?

Within each lung, leading off from the mesobronchus (see Fig. 32), are several **ventrobronchi,** from which extend the interclavicular, cervical, and anterior thoracic air sacs (for convenience called the anterior sacs), and two rows of several **dorsobronchi.** The ventro- and dorsobronchi branch into innumerable **parabronchi** (not visible to the unaided eye) of uniform diameter. These minute tubes connect with one another freely, forming a network of air capillaries, in the meshes of which is a similar network of blood capillaries. At the posterior end of the lung the mesobronchus divides into two tubes going, respectively, to the posterior thoracic and abdominal air sacs (called the posterior sacs). The anterior sacs (except the cervical sac) and the posterior sacs are reconnected to the lung by the **recurrent bronchi,** which join the parabronchi inside the lung.

Gaseous exchange or respiration occurs in the parabronchi. By forming a network of air capillaries, the parabronchi permit a continuous circuit of air through the lung. This is not the case in the lung of a mammal, where small branches, or bronchioles, arising from the bronchi end blindly in alveoli, making such a circuit impossible.

Much of the air, on entering the lung from the bronchus, passes through the mesobronchus (a) to the posterior sacs, (b) back through recurrent bronchi to the parabronchi, (c) then to the anterior sacs by way of the ventrobronchi, and (d) finally out through recurrent bronchi and parabronchi to the bronchus. Some of the air, however, on entering the lung from the bronchus, passes into the mesobronchus and then out through the ventrobronchi to the anterior sacs without making the circuit through the posterior sacs. The direction of the air movement through the lung and air sacs is controlled to a large extent by a complicated valvular system.

The relation of air movement to inspiration and expiration is briefly as follows: when the body cavity is enlarged at inspiration, the posterior sacs receive pure air from the bronchus and mesobronchus and the anterior sacs receive partly vitiated air from the parabronchi and partly pure air from the bronchus, mesobronchus, and ventrobronchi. As the body cavity decreases in volume at expiration, all the air sacs expel air through the recurrent bronchi and parabronchi to the bronchus. During both inspiration and expiration a gaseous exchange takes place in the parabronchi, but the exchange is much less during inspiration because much less air passes through the parabronchi at this time.

Pay particular attention to the following organs in the abdominal cavity. (Refer to Figure 34.) Note that the abdominal cavity is lined, as in other vertebrates, by a thin membrane, the **peritoneum,** and that this membrane is deflected at certain points to cover the organs and to form **mesenteries,** which hold the organs in place.

Liver This organ monopolizes the anterior end of the abdominal cavity. Two lobes are present: the right one extends far back into the abdominal cavity. The forward end of the liver is somewhat ventral to the heart and partly conceals it.

Stomach The stomach is obvious on the left side of the abdominal cavity posterior to and partially covered over by the left lobe of the liver. Two portions of the stomach are recognized: the **proventriculus,** which is the soft, glandular anterior portion continuous with the esophagus, and the **gizzard,** which is the hard, muscular portion. The gizzard is the conspicuous portion of the stomach; the proventriculus appears to be little more than an enlargement of the esophagus before entering the gizzard. A small constriction marks the union of proventriculus and gizzard.

Small Intestine The small intestine emerges from the gizzard near the inner side where the proventriculus enters. The first part of the intestine, the **duodenum,** makes a long, U-shaped loop posteriorly and is easily distinguished for this reason. A thin, lobulated **pancreas** occupies the main area within the loop. Three **pancreatic ducts** pass from the right side of the pancreas into the right side of the duodenal loop. Into the duodenal loop pass also two **bile ducts** from the deep depressions in the dorsal surface of the right lobe of the liver. One enters the left side of the loop just beyond the gizzard; the other enters the opposite side. No gall bladder is present in the pigeon. The remaining parts of the small intestine, the **jejunum** and **ileum,** cannot be distinguished. They are greatly coiled and are suspended from the dorsal wall of the cavity by a mesentery.

Large Intestine The large intestine is relatively reduced in the bird and does not differ markedly from the small intestine. It is merely a continuation of the small intestine, without enlargement, from the middle of the abdominal cavity straight back to a point just ventral to the vertebral column. No attempt is made to distinguish between the colon and rectum.

Ceca Where the small and large intestines merge, appear two lateral pouches, or diverticula, called **ceca** (singular **cecum**). These structures show great variation in birds, ranging from bud-like objects to ones of great length.

Cloaca This is a tubular cavity common to the digestive and urogenital systems and opening exteriorly through the **vent.** It receives the large intestine on its median ventral surface. (See urogenital system later.)

Separate the stomach from the adjacent mesenteries and lift it forward. Running along its dorsal surface to the intestine is the **spleen,** a rather round, reddish organ. Remove the stomach from the body cavity by severing the esophagus and small intestine. Note that the gizzard is flattened and rounded like a bi-convex lens but with one curved surface greater than the others. The glistening effect of the gizzard is due to the many tendons of the outer muscular layer. Cut the stomach in two, anteroposteriorly. The proventriculus, being highly

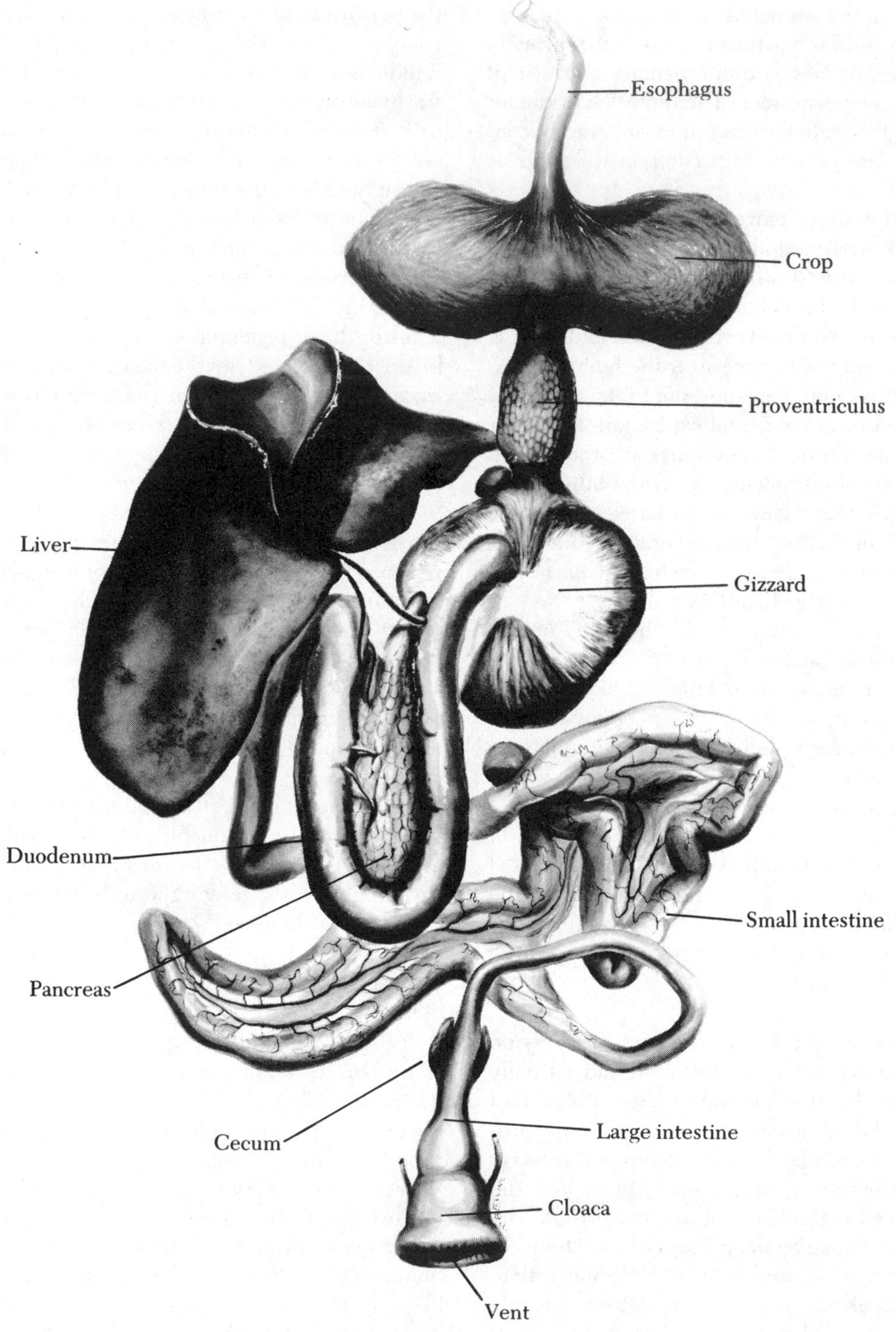

Figure 34 **Digestive System of Common Pigeon**

glandular, presents a spongy appearance within. The gizzard, on the other hand, shows a thick horny lining, raised in hard ridges. It contains many small pebbles that the bird has swallowed. Notice that the walls of the gizzard are not uniformly muscular. Thus the center of each right and left half contains no muscle, while the anterior and posterior ends of each half contain powerful masses of muscle. These are the **lateral muscles.** Observe that their inner horny walls almost meet each other, leaving little space in the gizzard. When food is taken into the gizzard, it is immediately pressed between the walls. By alternate movements of the lateral muscles the food is rubbed against the hard walls and pebbles and thus ground into fine particles.

Remove the intestine. How many times longer than the body of the bird is it? In relation to the body length the intestine is proportionately shorter in the bird than in man.

Circulatory System

The major features of the circulatory system should be observed. Use Figure 35 as a guide.

The heart of the bird is proportionately large and conical, with its apex, in the pigeon and most other species, pointing posteriorly. The chambers of the heart, like those of the mammal, are four in number and entirely separate from one another. Two of the chambers, the **atria** (singular **atrium**), occupy the anterior end or base of the heart; both are thin-walled. The other two chambers, the **ventricles,** comprise the remaining part of the heart and have thick, muscular walls. The greater bulk of the heart is, therefore, ventricular.

The vessels leaving and entering the heart are essentially the same in their form, distribution, and function as the mammalian vessels.

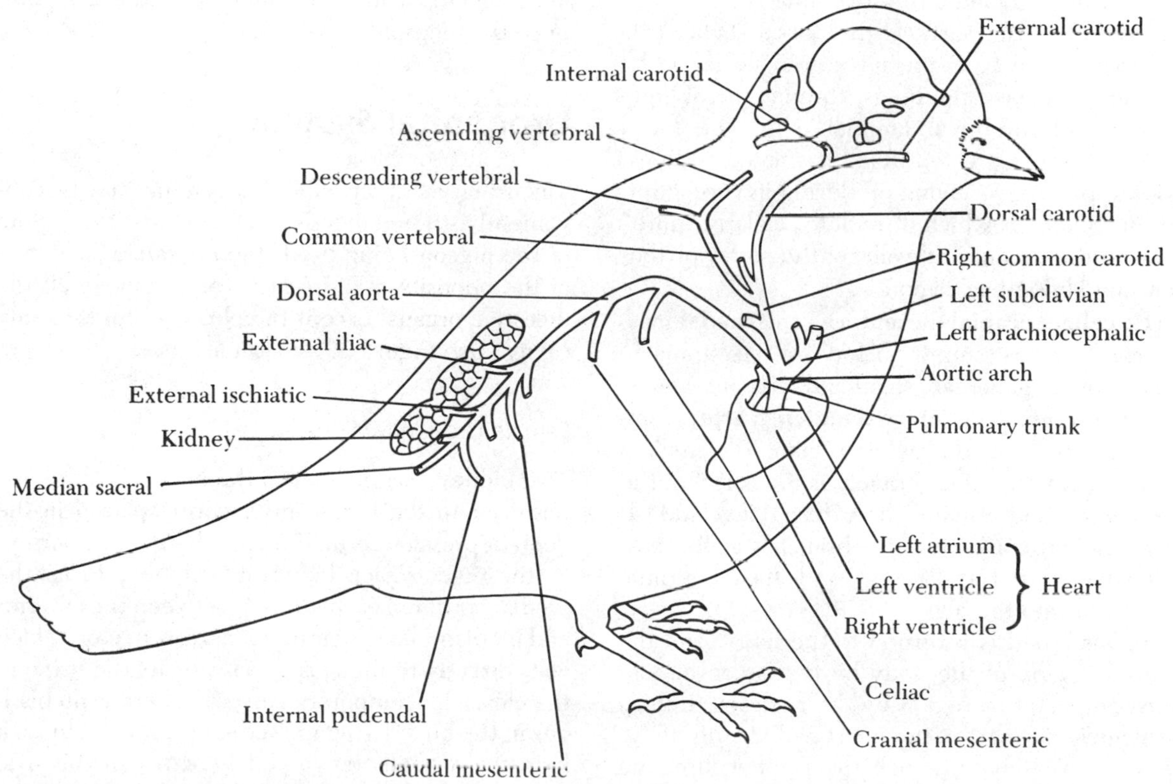

Figure 35 **Schematic Drawing of the Heart and Main Branches of the Aorta**
(Adapted from Van Tyne and Berger, *Fundamentals of Ornithology,* John Wiley & Sons, 1976.)

The **aorta,** on emerging from the left ventricle, turns to the right, instead of to the left as in mammals. The aorta, which is very short, continues as the **aortic arch,** giving off right and left **brachiocephalic arteries.** These large vessels carry oxygenated ("pure") blood to the wings and to the anterior thoracic, neck, and head regions. Follow either one laterally and a little anteriorly to the point where it divides into a **common carotid artery** to the head and a **subclavian artery** to the adjacent wing. In the angle formed by this division and lying against the common carotid artery is the **thyroid gland,** an oval body with a reddish color. Just lateral to the thyroid gland and separate from it are two very small **parathyroid glands.** The thyroid and two parathyroid glands are paired endocrine structures in the same position in the pigeon's right and left sides.

The **pulmonary trunk** emerges from the right ventricle, proceeds to the left, dorsal to the aorta, and then divides into two branches that carry venous ("impure") blood to the two lungs.

Now follow the aorta as the **dorsal aorta.** This leads posteriorly from the aortic arch, dorsal to the heart and between the lungs, through the oblique septum, and into the abdominal cavity. The dorsal aorta carries oxygenated blood to the tail and posterior appendages, giving off during its course numerous branches which distribute similarly "pure" blood to the organs and walls of the posterior thoracic and abdominal regions.

The **celiac, cranial mesenteric,** and **caudal mesenteric arteries** supply blood to the stomach, spleen, liver, pancreas, small and large intestines, and the rectum. The **internal pudental artery** supplies structures in the pelvic region. The **median sacral artery,** the final branch of the dorsal aorta, supplies the tail region. The **external iliac** and **external ischiatic arteries** give branches to the posterior lobes of the kidneys and then continue downward into the legs.

Venous blood is returned to the heart from the anterior region of the body by two **precavae** and from the posterior region by the **postcava.** Lift up the ventricular part of the heart and identify these big veins as they approach the right atrium: the precavae from a right and left direction, respectively, and the postcava from the liver through which it passes. The postcava enters the right atrium between the two precavae.

The pulmonary veins bear oxygenated blood from the lungs to the left atrium. They are small and difficult to find.

Remove the heart from the body cavity by severing the connecting vessels. Make a cut across the ventricles, midway between the atria and apex. Note the crescent-shape of the right ventricle as it tends to overlap the larger, more rounded left ventricle; and note also that only the left ventricle extends to and includes the apex of the heart. The wall of the right ventricle is thinner than that of the left. Lay open the right ventricle further and observe the crescent-shape of the opening into the right atrium, also the muscular fold or flap that acts as a valve. In the mammalian heart the opening into the right atrium is controlled by the more elaborate tricuspid valve. Open the other ventricle and observe two valves, both thin and membranous, which may close the more circular opening into the corresponding atrium. These two valves, present in the mammalian heart as well as the avian, comprise the **bicuspid valve.**

Urogenital System

The urinary and reproductive systems may be conveniently studied together. Dissect first the system in the pigeon being used, then obtain a specimen of the opposite sex. In each case, remove all the digestive organs, except the cloaca, from the body cavity. Use Figure 36 as a guide.

Female Urogenital System

Two **kidneys** are present in the bird. They are just posterior to the lungs and securely placed in the deep depression formed mainly by the synsacrum. In the pigeon each kidney is trilobed. From the medial border of each kidney, between the anterior and median lobes, emerges a narrow **ureter,** which goes directly to the cloaca. This bears the urine to the cloaca for temporary storage. There is no bladder in the bird. Lying immediately anterior to each kidney is a small orange-yellow body, roughly oval in shape. This is the **adrenal gland,** a part of the endocrine system.

Ventral to the left kidney at its anteromedial end, and very near the adrenal gland is a single **ovary,**

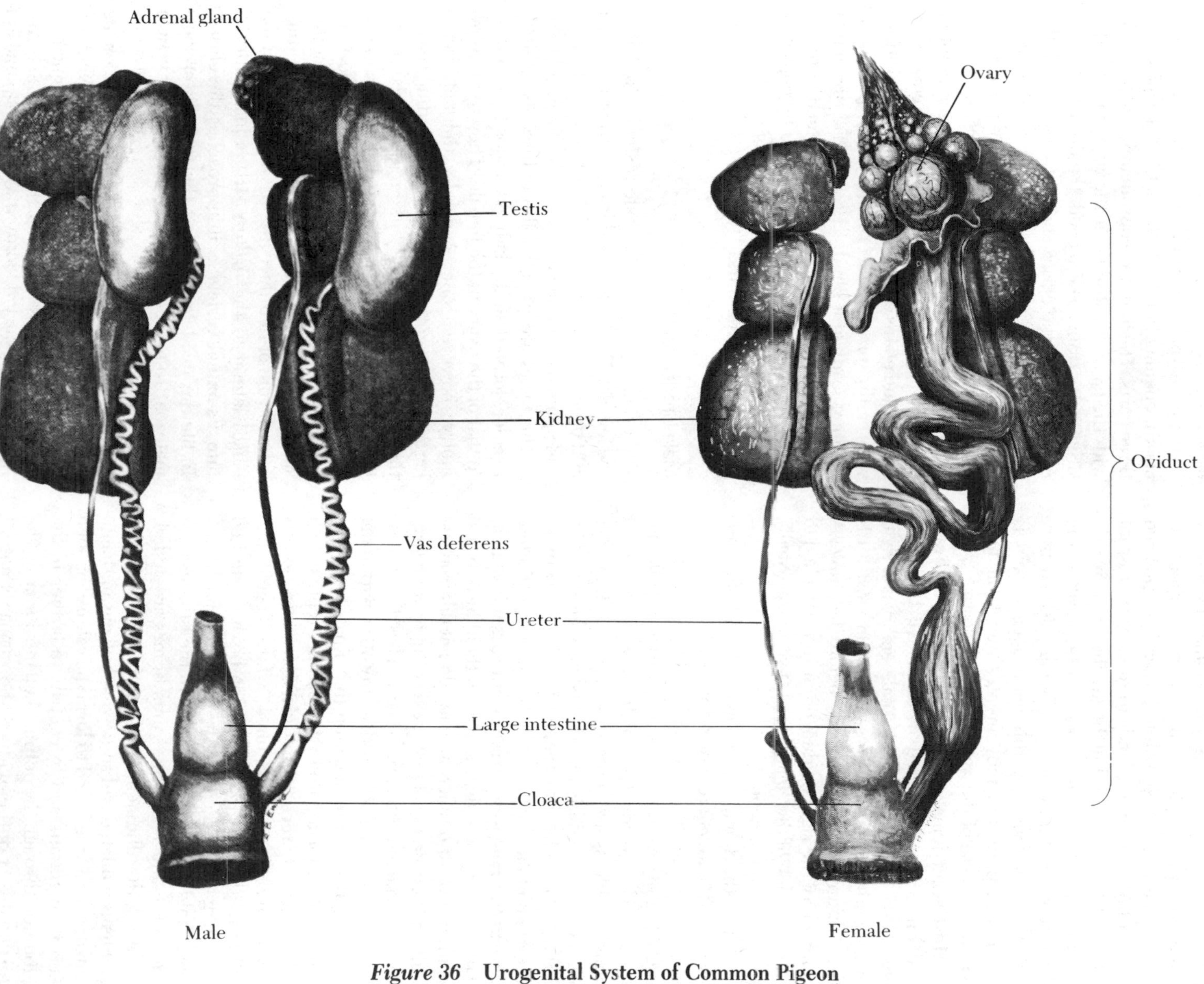

Figure 36 **Urogenital System of Common Pigeon**

whitish in color. It shows many rounded follicles of different sizes containing ova. A small mesentery holds the ovary in place. Slightly posterior and lateral to the ovary a convoluted **oviduct,** supported by another mesentery, leads to the cloaca; it is not attached to the ovary but begins beside it with a funnel-shaped dilation or **ostium** (also called the **infundibulum**). An ovum, on being released by a follicle, is engulfed by this wide opening and is pushed by peristaltic movements through the long winding course of the oviduct to the cloaca.

Most adult female birds possess but one ovary and one fully developed oviduct, and these are always on the left side. In female embryos a right ovary and oviduct are present, but in most birds they disappear or nearly so, before hatching. A remnant of the right oviduct often stays attached to the right side of the cloaca. Such a vestige is usually present in the pigeon.

When the female is in laying condition, both ovary and oviduct are more obvious. Many follicles in the ovary are very big and contain correspondingly big ova holding great quantities of yolk material. The oviduct, which now occupies a considerable space ventral to the left kidney, has its walls considerably lengthened and thickened by enlarged muscles and glands. Besides the ostium, the oviduct has four parts; they are (from ostium to cloaca) the **magnum, isthmus, uterus,** and **vagina.** Though they are distinctive in their histological composition and function, they do not differ markedly in their gross anatomy, except in relative size. As the name indicates, the magnum has the greatest size–always much longer than any other part of the oviduct. It is in the magnum that a passing ovum, having been picked up by the ostium, replete with yolk material, receives by glandular secretion its principal coatings of albumen ("white of egg"). The isthmus, the next part, can often be distinguished by being slenderer than the magnum and having fewer folds in its internal wall. In the isthmus both the outer and inner shell membranes are added, covering the albumen. In the uterus, which is conspicuously dilated and somewhat bulbous in shape, substances are secreted to form the calcareous shell and its pigments and to contribute further to the albumen through the porous shell membranes. The terminal part of the oviduct, the vagina, is noticeably constricted and has no known function other than to direct the completely formed egg to the cloaca.

Make an incision along the right side of the cloaca, opening it fully. Note that it is divisible into three parts. The first, the **coprodeum,** is continuous with the large intestine. It is the largest part and is situated ventrally. The second, the **urodeum,** is the middle part of the cloaca and is situated above the coprodeum, being separated from it by a membranous fold. Through its lateral walls the right and left ureters enter, while above the left ureter the oviduct enters. The third, the **proctodeum,** is more posterior and somewhat dorsal to the second; it is smaller and opens directly to the vent. (In the dorsal wall of the proctodeum of the young pigeon an opening leads to a blind, unpaired sac, the **bursa of Fabricius.** Though it seems to be active in the early life of the bird, its function is not completely known. (See Appendix A, page 377, for further information on the bursa.)

Male Urogenital System

The urinary system of the male is similar to that of the female. The preceding description of the system as it occurs in the female should, therefore, be read and applied to the male.

The **testes** are paired, ellipsoid organs, whitish in color, located on the kidneys in positions similar to that of the ovary in the female. They show asymmetry, however, one, generally the left, being larger than the other. During the breeding season the testes increase greatly in circumference, more than doubling their size. The medial border of each testis is rather concave, with a minute projection, barely visible, called the **epididymis.** From this springs the **vas deferens** (plural, **vasa deferentia**), a small convoluted duct that passes directly to the cloaca lateral to the ureters. It gradually widens into a **seminal vesicle** as it approaches the cloaca. In life the vas deferens bears the sex cells or spermatozoa from the testis, and the seminal vesicle stores them.

The cloaca is practically the same as in the female save that it is smaller, while the lips of the vent are thicker and tend to protrude in a more conspicuous manner. No penis occurs in the pigeon. (See Appendix A, page 376, for a description in several groups of birds.) The spermatozoa are passed to the female when the lips of the cloacae of the two sexes meet during copulation. The urodeum receives the two vasa deferentia, instead of an oviduct; these ducts enter the chamber laterally, just above the ureters.

If a specimen of the Common Pigeon or any other bird is to be dissected for the sole purpose of examining the **gonads** (ovary or testes), a longitudinal incision should be made on the left side of the abdominal cavity, halfway between the mid-ventral and mid-dorsal lines and through the posteriormost ribs. Then the viscera ventral to the left kidney should be lifted up and pressed aside. This will bring into view either the ovary or the larger testis. Should the specimen be immature and not otherwise in breeding condition, the gonad will be very small. Consequently, care must be taken not to confuse it with the adrenal gland, which may be more prominent than the gonad and may, in some species, be shaped like an ovary. Avoid confusion by noting color. Gonads are always whitish; adrenal glands are more highly colored—usually orange-yellow, though occasionally light yellow and in a few instances pink or red.

Nervous System

Remove the remaining organs from the body cavity and note the ventral branches of the spinal nerves passing over the dorsal wall. Trace them dorsally and note the ganglia of the sympathetic nervous system where they converge to go between the vertebrae to the spinal cord.

Sense Organs

A knowledge of three organs of special sense is necessary in understanding the behavior of a bird.

Organs of Smell Cut into one of the nasal cavities by making an incision in the palate slightly to the side of the midline. Find three **conchae** extending into the nasal cavity from its lateral wall. (See Figure 37a.) They are in line from front to back and are consequently referred to as the anterior, median, and posterior. The last is the most superior in position and is also the smallest, being quite rounded. It supports the **olfactory membrane,** which is connected to the brain by the olfactory nerve. In man there are three conchae, which occupy a similar position in the nasal cavity; the olfactory membrane is supported by the nasal septum for a short distance as well as by the posterior concha.

Organs of Sight The eyes of birds are highly specialized and deserve careful attention. Cut away the tissues surrounding one of the eyes and chip away the roof of the orbit. (Refer to Figure 38a.) Note the peculiar turnip-shaped eyeball: the whitish tapering sclerotic part and the transparent, more curved **cornea.** Move it, and notice that there are six eye muscles attached to it. Find the optic nerve entering on the inner side. Ventral to one of the eye muscles on the anterior part of the eyeball is a whitish mass, the **Harderian gland.** Dissect away from the eyeball the tissues below and find a small **lacrimal gland.**

Remove the eye from the head by severing the muscles and optic nerve and (holding the eyeball in one hand) cut off the dorsal wall. Observe the large **lens.** Its shape is characteristic of birds, being rather flat externally and convex internally. As in mammals, it is held in place by, and focused with the aid of, the **ciliary body.** The chambers of the eye and the two humors are similar to those in mammals. The three layers in the eye's wall, the outer **sclera,** the middle dark **choroid,** and the inner **retina,** together with the thin **conjunctiva** which passes over the cornea, are also similar to those in mammals. Note, however, a brown vascular fringe that projects from the lower medial wall toward the lens. This structure is the **pecten.** It probably plays a nutritional role for the retina. Scrape away a bit of the tissue of the eyeball near the point where the cornea joins the sclera. Here a bony ring encircles the eye. It is called the **sclerotic ring** and strengthens the eyeball. Such a ring is also found in reptiles.

Organs of Hearing The ears of the bird are without external appendages. In the pigeon the opening of the ear is rounded and covered by a fringe of feathers, the **auriculars;** it leads directly into a passage, the **external auditory meatus,** situated below and behind the eye. (See Figure 37b.) Cut into the meatus and locate the transparent **tympanum** or

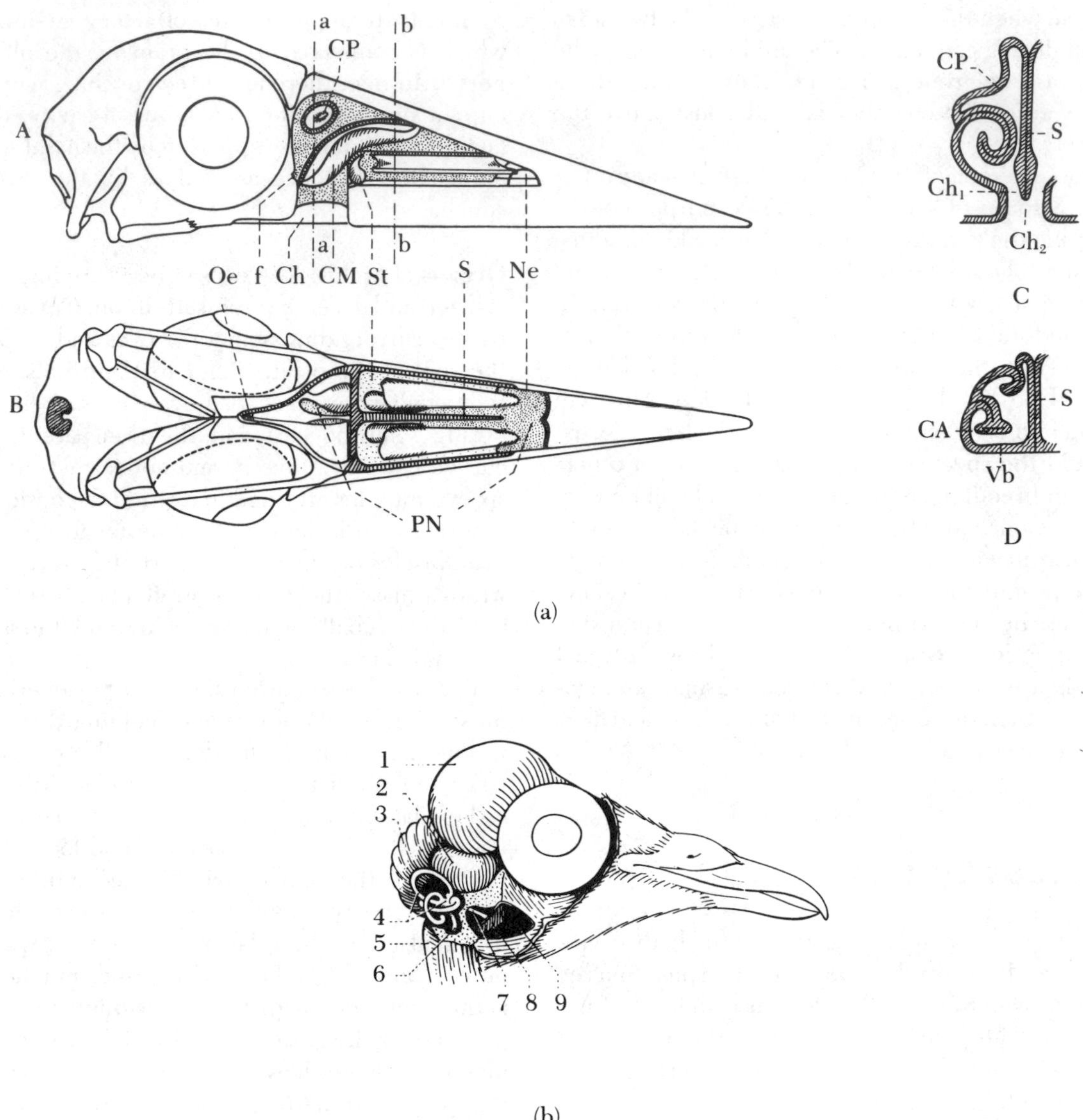

Figure 37 **Nasal Cavity of the Bird and Ear of the Common Pigeon**
(a) A, lateral view; B, ventral view; C, cross section of the cavity (a-a of A); D, cross section of the cavity (b-b of A); CA, anterior concha; Ch, choana; Ch 1, primary choana; Ch 2, secondary choana; CM, median concha; CP, posterior concha; f, connection between the nasal cavity and the infraocular air space; Ne, nostril (anterior naris); Oe, opening of the Eustachian tube; PN floor of the nasal cavity; S, nasal septum; St, transverse fold in the floor of the nasal cavity; Vb, floor of anterior nasal cavity. (From Portmann, 1950, in *Traité de Zoologie,* edited by Grassé, Vol. 15, Figure 148, Masson et Cie; after Technau.) (b) Lateral view of the head of a Common Pigeon showing the position and principal parts of the ear. 1, cerebral hemisphere; 2, optic lobe; 3, cerebellum; 4, semicircular canals; 5, neck; 6, blind apex of the cochlear duct; 7, columella in middle ear; 8, tympanum; 9, external auditory meatus. (From Portmann, 1950, in *Traité de Zoologie,* edited by Grassé, Vol. 15, Figure 153, Masson et Cie; after Krause.)

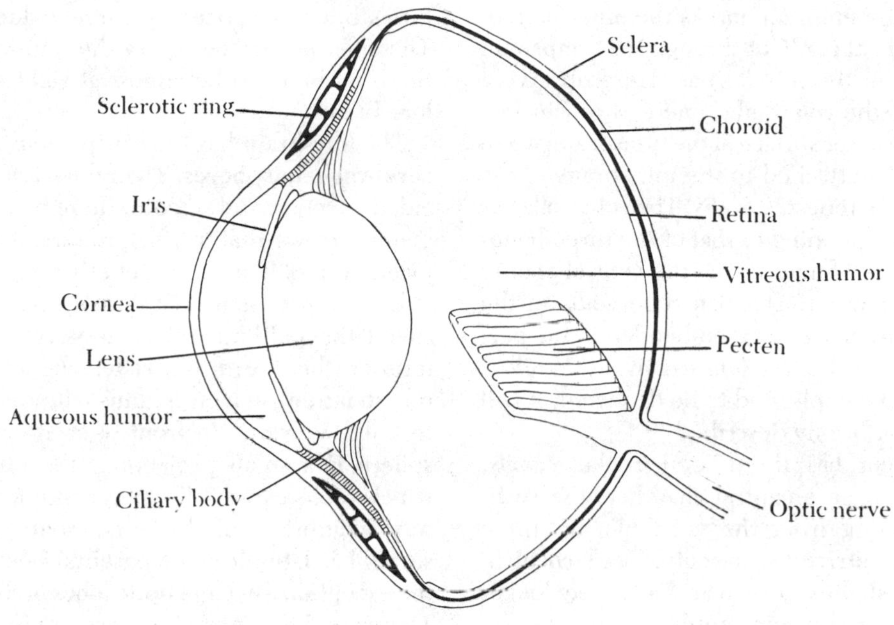

(a)

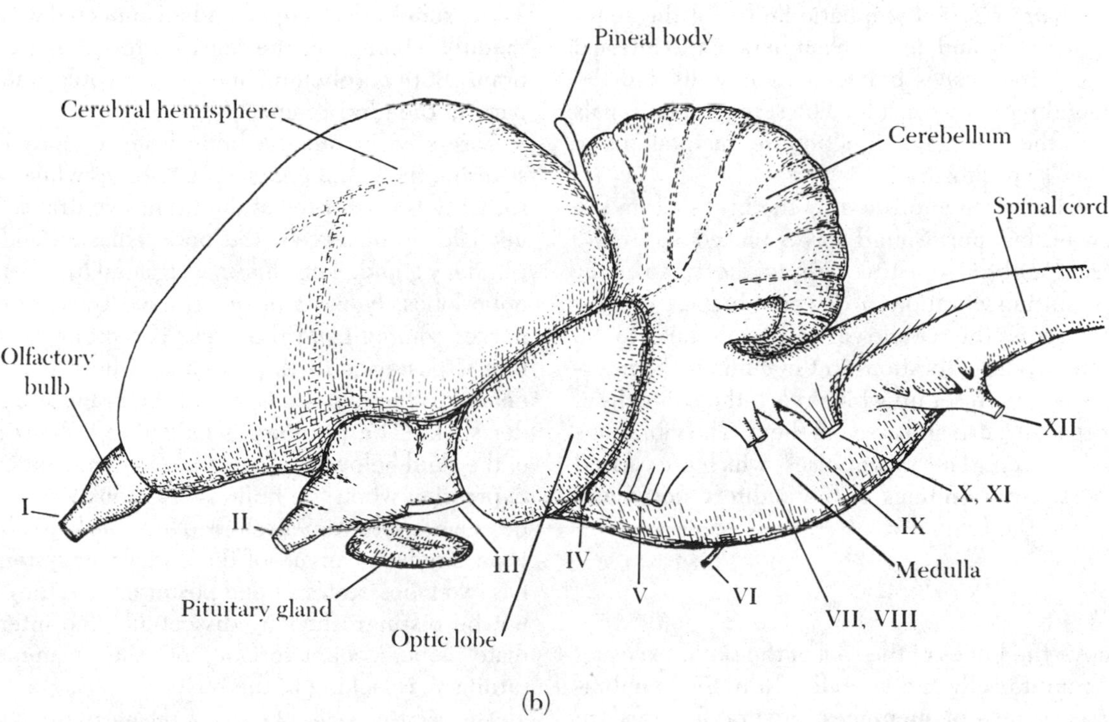

(b)

Figure 38 **Eye and Brain of Common Pigeon**
(a) Eye. (b) Brain.

eardrum. This membrane marks the inner boundary of the **external ear.** Cut through the tympanum and thus expose the **middle ear.** Here observe a rod-like bone, the **columella,** one end of which is attached to the inner surface of the tympanum while the other end is attached to the membrane of the oval window, the **fenestra ovalis.** The columella has a function corresponding to that of the three bones of the middle ear of man. From the ventral floor of the middle ear the Eustachian tube leads to the pharynx. The **internal ear** is embedded in the periotic capsule medial to the fenestra ovalis. While it is too small and complicated to be dissected, it will nevertheless be briefly described.

The inner ear has three **semicircular canals,** which emerge from a central chamber, the **vestibule.** Also arising from the vestibule are three chambers, the **utriculus, sacculus,** and **cochlea.** The cochlea is slightly curved and relatively longer than the same structure in reptiles but not so long as the spiraled cochlea of mammals, to which it is homologous. In the cochlea is the important **organ of Corti** with its basement membrane of ciliated cells ("hair cells"). Lymphatic fluids fill the semicircular canals and all the chambers of the internal ear, also the spaces between their walls and the surrounding bony capsule. The semicircular canals have as their principal function the maintaining of the bird's equilibrium.

Hearing is accomplished by the bird's ear in the following manner. Sound waves picked up by the external auditory meatus pass to the tympanum. The resulting vibrations of this membrane are transported across the middle ear of the columella to the fenestra ovalis. Vibrations in the fluids of the internal ear are then set up which reach the hair cells of the organ of Corti and activate them. The vibrations are converted to nerve impulses, which are carried to the sensory endings of the auditory nerve and thence to the brain.

Brain

Remove the bones of the roof of the skull, exposing the brain dorsally and laterally. Note the membranous coverings or **meninges.** Several features are worthy of attention. Refer to Figure 38b.

The brain is divided, as in other vertebrates, into three main parts: the **fore-brain, mid-brain,** and **hind-brain.** Unlike the quadrupeds, however, the bird's brain is noticeably curved, due to flexures. The chief flexure occurs in the mid-brain, causing the hind-brain to be almost at right angles to the fore-brain.

The fore-brain has the conspicuous right and left **cerebral hemispheres.** Their dorsal surfaces are decidedly convex and smooth, there being none of the many furrows that are so prominent in the mammalian brain. They are, nonetheless, separated by a deep dorsal fissure. Each hemisphere has a roof, called the **pallium,** which is very thin; a hollow interior (the **lateral ventricle**); and a floor (the **corpus striatum**), which is unusually thick and tends to bulge upward. In front of each cerebral hemisphere is a small projection, the **olfactory lobe,** which receives the olfactory nerve from the olfactory membrane of the nasal cavity. Ventral and somewhat lateral to the cerebral hemispheres are the exceptionally large **optic lobes** of the mid-brain. There are the principal centers for visual reception. Back of the cerebral hemispheres is the rounded **cerebellum** with its very noticeable middle portion, the **vermis,** marked by many transverse fissures. The cerebellum lies upon and is connected with the **medulla oblongata,** the most posterior part of the brain. Both cerebellum and medulla oblongata are parts of the hind-brain.

Carefully lift out the brain from its bony base, severing the connecting cranial nerves while doing so. Only two features of the brain's ventral surface need to be observed: the **optic chiasma** and the **pituitary gland.** The chiasma, situated between the optic lobes, consists of the crossed tracts of optic nerves coming from the eyes. The pituitary gland is just posterior to the chiasma, where it is connected to the **hypothalamus** of the brain by a slender stalk. (The pituitary is lodged in a depression of the skull below the brain, and may remain in the depression when the brain is lifted out.) The pituitary gland, sometimes referred to as the **hypophysis,** is an important organ of the endocrine system. It has two lobes, anterior and posterior, but they cannot be distinguished by dissection. The intermediate lobe, characteristic of the mammalian pituitary, is lacking in the bird.

The relative sizes of parts of the avian brain show several significant correlations with the bird's sensory capacities and motor activities. The olfactory lobes are small, the sense of smell rather poorly developed; the optic lobes, directly associated with enormous eyes that have numerous, highly organized sensory elements, are comparatively enormous. The cerebellum, as in all vertebrates, coordinates the muscular activities concerned with locomotion and bodily equilibrium; it is particularly large in the bird to accommodate the extremely complicated neuromuscular mechanisms for flight. Parts of the avian brain, when compared with those of the mammalian, exhibit striking differences that are closely correlated with differences in behavior. For example, the bird's pallium is a thin, smooth wall of the cerebral hemispheres and lacks any notable aggregation of nerve cells. But the mammal's pallium is a thick, usually furrowed, cortex and contains countless nerve cells that form the dominant association center of the brain, with marked capacities for learning. In the bird the dominate association center is the large corpus striatum wherein lie nerve elements to regulate mechanisms for instinctive behavior (i.e., behavior that is innate and thus unlearned). Neither the corpus striatum, nor any other part of the bird's brain, has developed the capacities for learning to the level attained by the mammal's—a fact that becomes readily apparent in any comparative behavior study.

Some Anatomical and Physiological Considerations

Literature

There is no one treatise on avian anatomy, either gross or microscopic, that readily provides all the desired details on all the organ systems. The most extensive works on anatomy relate to domestic birds and are intended for use in poultry husbandry, aviculture, and veterinary medicine. However, the following general works are useful.

Beddard, F.E.

1898 *The structure and classification of birds*. London: Longmans, Green, and Company. (Though somewhat out of date, this is still a good book on anatomy.)

Berger, A.J.

1961 *Bird study*. New York: John Wiley & Sons. (The chapter "Structure and Function" provides the most concise review of the subject available; recommended for introductory reading.)

Coues, E.

1903 *Key to North American birds*. 5th ed. 2 vols. Boston: Dana Estes and Company. (Volume 1, pp. 139–233, has one of the best general summaries on the organ systems in the English language.)

Evans, H.E.

1969 Anatomy of the Budgerigar. In *Diseases of cage and aviary birds*, ed. M.L. Petrak. Philadelphia: Lea & Febiger. (An excellent, illustrated discussion of all anatomical systems.)

Farner, D.S.; King, J.R.; and Parkes, K.C., eds.

1971–1983 *Avian biology*. 7 vols. New York: Academic Press. (Detailed chapters on anatomy and physiology of birds.)

Harvey, E.B.; Kaiser, H.E.; and Rosenberg, L.E.

1968 *An atlas of the Domestic Turkey (Meleagris gallopavo): myology and osteology*. U.S. Atomic Energy Commission; Division of Biology and Medicine. (A well-illustrated and extensively labeled work, usable as a guide to the dissection of many species.)

Horton-Smith, C., and Amoroso, E.C., eds.

1966 *Physiology of the Domestic Fowl*. Edinburgh: Oliver and Boyd. (Contains 34 papers, grouped in five sections: "Reproductive Physiology and Endocrinology," "Metabolism and Nutrition," "Calcium Matabolism and Eggshell Formation," "Environmental Physiology," "Pharmacology and General Physiology.")

King. A.S., and McLelland, J., eds.

1980–1981 *Form and function in birds*. 2 vols. New York: Academic Press. (Clear discussions of anatomy and physiology by different authors.)

Lucas, A.M., and Stettenheim, P.

1965 Avian Anatomy. In *Diseases of poultry*. 5th ed., eds. H.E. Biester and L.H. Schwarte. Ames: Iowa State University Press. (A review of the subject; highly recommended.)

Marshall, A.J., ed.

1960–1961 *Biology and comparative physiology of birds*. 2 vols. New York: Academic Press. (Organ systems reviewed by different authors in separate chapters.)

Newton, A.

1896 *A dictionary of birds*. London: Adam & Charles Black. (Still an indispensable work, briefly covering many parts of organ systems.)

Romanoff, A.L.

1960 *The avian embryo*. New York: Macmillan.

Sturkie, P.D.

1976 *Avian physiology*. 3rd ed. New York: Springer-Verlag.

Thomson, A.L., ed.

1964 *A new dictionary of birds*. New York: McGraw-Hill. (A good reference for general information.)

Skeleton

The skeleton in birds differs between taxonomic groups chiefly in the number of vertebrae; in the loss of certain bones or processes of the skull, pectoral appendage, or foot; in the configuration of the palatal complex, sternum, humerus, and pelvis; and in the degree of ossification of nasal cartilages and of tendons. Such differences may reflect different evolutionary origins; they may, along with other differences in the shape and proportions of skeletal elements within orders or families of birds, also reflect adaptations to different ways of life. The skeleton is thus a rich source both for systematic studies and for the study of functional anatomy and structural adaptation.

Although the literature on descriptive and functional osteology of birds is voluminous, there is no one source for a comprehensive survey of the avian skeletal system. The accounts of the skeleton in Newton (1896), Coues (1903), and Marshall (1960) provide an adequate background for understanding the more specialized papers listed below.

Ashley, J.F.
1941 A study of the structure of the humerus in the Corvidae. *Condor* 43:184–195. (A comparative study of humeral configuration in crows, jays, and magpies. Shows osteological differences at the generic level. Includes a drawing of the brachial plexus and a description of the muscles at the head of the humerus in the American Crow, *Corvus brachyrhynchos*.)

Beecher, W.J.
1962 The Bio-mechanics of the bird skull. *Bull. Chicago Acad. Sci.* 11(2):10–33. (A good introduction to the subject of avian kinesis.)

Berger, A.J.
1952 The comparative functional morphology of the pelvic appendage of three genera of Cuculidae. *Amer. Midland Nat.* 47:513–605. (Includes a discussion of variation and functional aspects of the postcranial skeleton and shows positions of muscle attachments on bones of the hindlimb.)
1955 Suggestions regarding alcoholic specimens and skeletons of birds. *Auk* 72:300–303. (Discusses aspects of the skeleton that need special care in preparation; gives many important points for the production of a scientifically useful bird skeleton.)

Bock. W.J.
1962 The pneumatic fossa of the humerus in the Passeres. *Auk* 79:425–443.

Bock, W.J., and McEvey, A.
1969 The radius and relationship of owls. *Wilson Bull.* 81:55–68.

Curtis, E.L., and Miller, R.C.
1938 The sclerotic ring in North American birds. *Auk* 55:225–243.

Feduccia, A.
1972 Variation in the posterior border of the sternum in tree-trunk foraging birds. *Wilson Bull.* 84:315–328.

Hargrave, L.L.
1972 *Comparative osteology of the chicken and American grouse*. Arizona: Prescott College Press.

Heimerdinger, M.A., and Ames, P.L.
1967 Variation in the sternal notches of suboscine passeriform birds. *Peabody Mus. Postilla* No. 105:1–44.

Howard, H.
1929 The avifauna of Emeryville Shellmound. *Univ. Calif. Publ. in Zool.* 32:301–394. (Contains an excellent series of labeled drawings of the skeletons of the Golden Eagle, *Aquila chrysaetos*, and Snow Goose, *Chen caerulescens*. A standard reference for the names of bones and parts of bones.)

Huxley, T.H.
1867 On the classification of birds; and on the taxonomic value of the modifications of certain of the cranial bones observable in that class. *Proc. Zool. Soc. London* 1867:415–472. (The classical work on palatal structure.)

Jollie, M.T.
1957 The head skeleton of the Chicken and remarks on the anatomy of this region in other birds. *Jour. Morph.* 100:389–436. (Describes and illustrates all bones of the skull as they appear in the young bird.)

Nero, R.W.
1951 Pattern and rate of cranial "ossification" in the House Sparrow. *Wilson Bull.* 63:84–88. (A useful paper for determining age of bird specimens by skull condition. Other important papers on the subject are cited.)

Owre, O.T.
1967 Adaptations for locomotion and feeding in the Anhinga and the Double-crested Cormorant. *Amer. Ornith. Union, Ornith. Monogr.* No. 6. (Analyzes the proportions of the skeleton and muscles in relation to the skeleton, thus showing adaptations for two different kinds of feeding under water.)

Simpson, G.G.
1946 Fossil penguins. *Bull. Amer. Mus. Nat. Hist.*

87:1–99. (The relationships of fossil forms are very convincingly worked out through the osteology of Recent species.)

Tiemeier, O.W.
1950 The os opticus of birds. *Jour. Morph.* 86:25–46.

Tordoff, H.B.
1954 A systematic study of the avian family Fringillidae. Based on the structure of the skull. *Univ. Michigan Mus. Zool. Misc. Publ.* No. 81:1–41. (An important approach to an understanding of the relationships of the former Fringillidae and closely allied families.)

Woolfenden, G.E.
1961 Postcranial osteology of the waterfowl. *Bull. Florida St. Mus.* 6(1):1–129.

Musculature

In spite of a number of significant studies on the anatomy and function of avian muscles, many more comparative studies are necessary to determine how much variation occurs in the muscle pattern and in the relative development of muscles within a family, and, therefore, the taxonomic value of muscle differences. More needs to be learned about wing muscles and about Garrod's leg-muscle formulas as expanded by Hudson (1937) and Van Tyne and Berger (1976). Some of the following papers demonstrate how research on muscle adaptations and functions leads to a better understanding of both phylogenetic relationships and locomotor patterns.

Ames, P.L.
1971 The morphology of the syrinx in passerine birds. *Peabody Mus. Nat. Hist. Bull.* No. 37.

Baumel, J.J.; King, A.S.; Lucas, A.M.; Breazile, J.E.; and Evans, H.E., eds.
1979 *Nomina Anatomic avium: An annotated anatomical dictionary of birds*. London: Academic Press. (Attempts to standardize the nomenclature for birds; many illustrations.)

Berger, A.J.
1956a The expansor secundariorum muscle, with special reference to passerine birds. *Jour. Morph.* 99:137–168.
1956b Anatomical variation and avian anatomy. *Condor* 58:433–441.
1969 Appendicular myology of passerine birds. *Wilson Bull.* 81:220–223.

Cracraft, L.J.
1971 The functional morphology of the hind limb of the Domestic Pigeon, *Columba livia*. *Bull. Amer. Mus. Nat. Hist.* 144:171–268.

Evans, H.E.
1969 Anatomy of the Budgerigar. In *Diseases of cage and aviary birds,* ed. M.L. Petrak. Philadelphia: Lea & Febiger.

Fisher, H.I.
1966 Hatching and the hatching muscle in some North American ducks. *Trans. Illinois St. Acad. Sci.* 59:305–325.

Fisher, H.I., and Goodman, D.C.
1955 The myology of the Whooping Crane, *Grus americana*. *Illinois Biol. Monogr.* 24, No. 2. (An excellent study of all the muscles of this species.)

Forbes, W.A.
1885 The collected scientific papers of the late William Alexander Forbes, ed. F.E. Beddard. London: R.H. Porter. (Contains all the works of this British anatomist.)

Garrod, A.H.
1873–1874 On certain muscles of the thigh of birds, and their value in classification. I and II. *Proc. Zool. Soc. London* 1873:626–664; 1874:111–123. (The classic work on thigh muscle formulas.)
1881 *The collected scientific papers of the late Alfred Henry Garrod*, ed. W.A. Forbes. London: R.H. Porter. (A valuable collection of the works of this pioneering British anatomist who first proposed the use of the leg-muscle formulas.)

Hudson, G.E.
1937 Studies on the muscles of the pelvic appendage in birds. *Amer. Midland Nat.* 18:1–108. (An important work, based on the dissections of muscles in 16 bird orders.)

Hudson. G.E.; Parker, R.A.; Vanden Berge, J.; and Lanzillotti, P.J.
1966 A numerical analysis of the modifications of the appendicular muscles in various genera of gallinaceous birds. *Amer. Midland Nat.* 76:1–73.

Klemm, R.D.
1969 Comparative myology of the hind limb of procellariiform birds. *So. Illinois Univ. Monogr., Sci. Ser.*, 2.

Mitchell, P.C.
1913 The peroneal muscles in birds. *Proc. Zool. Soc. London* 1913:1039–1072. (A good treatise on the peroneus longus and p. brevis in different groups of birds.)

Raikow, R.
1977 Pectoral appendage myology of the Hawaiian honeycreepers (Drepanididae). *Auk* 94:331–342.
1979 The appendicular myology and phylogenetic relationships of the Ploceidae and Estrildidae

(Aves: Passeriformes). *Bull. Carnegie Mus. Nat. Hist.* No. 15.

Vanden Berge, J.C.
1970 A comparative study of the appendicular musculature of the order Ciconiiformes. *Amer. Midland Nat.* 84:289–364.

Van Tyne, J., and Berger, A.J.
1976 *Fundamentals of ornithology.* 2nd ed. New York: John Wiley & Sons. (Detailed discussion of anatomical characters that have been used in the classification of birds.)

Weymouth, R.D.; Lasiewski, R.C.; and Berger, A.J.
1964 The tongue apparatus in hummingbirds. *Acta Anat.* 58:252–270.

Zusi, R.L.
1962 Structural adaptations of the head and neck in the Black Skimmer, *Rynchops nigra* Linnaeus. *Publ. Nuttall Ornith. Club* [*Monogr.*] No. 3. Cambridge, Massachusetts.

Zusi, R.L., and Storer, R.W.
1969 Osteology and myology of the head and neck of the Pied-billed Grebes *(Podilymbus)*. *Univ. Michigan Mus. Zool. Misc. Publ.* No. 139:1–49.

Zweers, G.A.
1974 Structure, movement and myology of the feeding apparatus of the Mallard *(Anas platyrhynchos* L.). *Neth. Jour. Zool.* 24:323–467.

The Oil Gland

In birds, as in reptiles, skin glands of any kind are rare; sweat and sebaceous glands, so common in mammals, are completely lacking. The only integumentary gland of any prominence in birds is the **oil gland,** sometimes referred to as the **uropygial** or **preen gland,** situated dorsally at the base of the tail and concealed by contour feathers. Most of the following discussion is based on a summary of the literature by Elder (1954).

The oil gland secretes a substance containing much fatty acid plus some fat and wax. This the bird smears on its bill and head plumage; then rubs it off on the various feathers over the body and on the wings. Secretion in the gland is probably stimulated, through a reflex mechanism, by the act of preening.

Generally the gland is a relatively large structure, terminating in a nipple-like opening. Depending on the species of bird, there may be one to eight outlets (Grassé, 1950) and there may be a tiny cluster of slender feathers around the nipple that lengthen the nipple into a brush for dispensing the secretion. The gland reaches its largest size in aquatic birds. Among parrots and doves, it may be absent in some species and in others show various stages in development, from the rudimentary to the fully functional. It is entirely wanting in ostriches, rheas, emus, and cassowaries and in certain species of the orders Galliformes, Gruiformes, Caprimulgiformes, and Apodiformes.

Elder (1954) conducted experiments on the function of the oil gland in ducks and arrived at a number of conclusions. Several are given below more or less verbatim. The secretion maintains the water-repellent quality of feathers either directly or by preserving their physical structure. Without this secretion the feathers lose much of their efficiency in their normal functions as a flight mechanism and as a heat-insulating medium. It seems unlikely that a bird rendered glandless could survive in the wild. Degenerative plumage changes following removal of the glands are more pronounced in waterfowl than in chickens (Domestic Fowl) and more pronounced in chickens than in pigeons. This seems reasonable in view of relative gland size and probable need for "waterproofing." The secretion is used to anoint the bill and maintain its surface structure and gloss: without the secretion the bill becomes dry and shows some sloughing.

Elder, W.H.
1954 The oil gland of birds. *Wilson Bull.* 66:6–31. (Includes an extensive bibliography on the subject.)

Grassé, P.P.
1950 La glande uropygienne. In *Traité de zoologie.* Vol. 15. Paris: Masson et Cie.

Digestive System

Mouth Cavity Birds may have as many as seven different pairs of **salivary glands** (see Farner, 1960). In general, such glands are well developed in birds that eat seeds and other vegetable food that require moistening, and are reduced in aquatic birds whose food is pre-moistened; they are sometimes completely absent in the Pelecaniformes (Portmann, 1950). Besides lubricating the mouth with mucous and perhaps secreting a digestive enzyme (ptyalin), they may occasionally have special functions. In woodpeckers a pair of large salivary glands, whose ducts enter the floor of the mouth, coats the tongue with a sticky fluid to assist the organ in retrieving insects. Gray Jays *(Perisoreus canadensis)* have similarly large salivary glands in the same position to coat the tongue (Bock, 1961). The copious mucous in this case is presumably used by the bird in making a bolus of food so sticky that it will adhere to any surface where the bird may choose to store it (Dow, 1965). Some species of swifts have salivary glands that become especially enlarged in the breeding season, producing mucous for nest construction (see page 284). The Black Swift *(Cypseloides niger)*, however, shows no such modifications since it nests in situations where saliva is unnecessary (Johnston, 1961).

Most birds have no soft palate, but pigeons at least show a softer condition of the posterior palate, enabling them to swallow successive boluses of water, as hoofed mammals do, without lifting the head.

The Rosy Finch *(Leucosticte arctoa)* has an extension of the mouth cavity in the form of two well-developed sacs or pouches, each with a separate opening on either side of the tongue and glottis (Miller, 1941). During the nesting season they use the pouches for carrying large quantities of food to the young, which are usually in high barren places, such as lofty cliffs, far from the food supply. The Pine Grosbeak *(Pinicola enucleator)* has pouches apparently identical with those of the Rosy Finch (French, 1954). Such structures are unusual; most birds bear food to their young either in the mouth (by distending the floor), in the esophagus, or in the crop.

Modifications of the Esophagus The esophagus is a distensible tube, lined with mucous epithelium. Though unmodified along its course in most birds, it nevertheless serves in all birds as a temporary reservoir for food. Geese may fill it so greatly from pharynx to stomach as to make the neck bulge perceptibly. Occasionally the esophagus will have a simple dilation, as in cormorants, or be considerably dilated ventrally, as in diurnal birds of prey. The Common Redpoll *(Carduelis flammea)* has a single enlargement that extends "laterally around the right of the vertebral column and forms a lesser enlargement on the left side of the column" (Fisher and Dater, 1961). Similarly, enlargements occur in other seed-eating birds. Only in a few groups of birds—e.g., gallinaceous birds, pigeons, and parakeets—does the esophagus show a true crop, an outpocketing of the ventral wall with a more or less constricted connection. The crop of pigeons is characteristically bi-lobed.

Any dilation of the esophagus, or any crop, is primarily for food storage. While ptyalin from the mouth may act upon food in the crop slightly and mucous and the muscular action of the crop's wall may soften the food partially, the crop actually plays a very minor part in digestion.

In pigeons during the breeding season the crop lining of both sexes produces, by the proliferation and sloughing of its epithelial cells, a white substance ("pigeon's milk") rich in fats and proteins. After being mixed with food received by the crop, the substance is regurgitated into the mouths of the young birds.

The Greater Prairie-Chicken *(Tympanuchus cupido)* has the anterior end of the esophagus modified as a vocal sac or resonating chamber. In the Heath Hen, the recently extinct eastern subspecies of the Greater Prairie-Chicken, Gross (1928) found that air is forced into the sac directly from the trachea by way of the pharynx. As the sac is inflated, sound waves from the syrinx strike the tense walls, which act as resonators.

Stomach The proventriculus provides the gastric juice (mainly mucous, pepsin, and hydrochloric acid); in this respect it corresponds closely with the fundus region of the mammalian stomach. In point of development, however, the proventriculus is believed to be a differentiation peculiar to birds, whereas the other division of the stomach, the gizzard, is a reptilian legacy. The waxy, musky oil that petrels eject when disturbed is a secretion from glands of the proventriculus (Matthews, 1949).

The gizzard tends to be very strongly muscular in birds that eat seeds and herbage, much less so in birds that subsist chiefly on fruits and animal flesh. This difference in muscularity depends on the amount of mechanical action required to macerate the particular type of food. Mucous glands are present in the gizzard walls. In the highly muscular gizzard of seed-eaters these glands secrete a horny lining with ridges to facilitate the process of maceration. In the less muscular gizzard of fruit-eaters the glands secrete only mucous and the gizzard itself has the appearance inside and out of being simply a large adjunct to the proventriculus.

In addition to preparing food for digestion, the gizzard serves as the principal barrier to indigestible materials such as feathers, fur, bones, animal shells, and so on. These are usually ejected via the esophagus and mouth; certain groups of birds, notably hawks and owls, eliminate them in spindle-shaped wads called pellets. Pebbles, normally present in most gizzards as an aid to maceration, may eventually be discharged through the esophagus and mouth or, if worn down sufficiently, voided by way of the intestinal tract.

Liver The two-lobed liver, the largest visceral organ of the body, is larger in birds than in mammals of equivalent body size. Its main contribution to the digestive process is the secretion of bile into the small intestine by way of two bile ducts. In most birds a reservoir for bile, the gall bladder, is present along the course of the duct from the larger (right) lobe. The gall bladder is particularly large in penguins and plantain-eaters (Musophagidae). Only in pigeons and relatively few other groups of birds—ostriches, rheas, parakeets, and hummingbirds—is it absent.

Pancreas This organ is relatively large in birds; as a rule, it is larger per unit of body weight in small birds than in large (Sturkie, 1976). Apparently the avian pancreas once consisted of three lobes, one dorsal and two ventral; though they are now united to form one glandular mass, their ducts still persist separately. Through these ducts important digestive enzymes (e.g., amylase, trypsin, and lipase) are secreted by the pancreas into the small intestine. In what was probably the dorsal lobe are the islets of Langerhans, which secrete the hormones insulin and glucagon.

Small Intestine This is the principal organ for the digestion and absorption of food. Here food is received from the stomach, where it has been mixed with, and perhaps slightly changed by, gastric juices and has been extensively macerated. Once in the intestine it undergoes complete chemical alteration by the action of enzymes accumulated from the proventriculus and pancreas, by bile from the liver, and by hydrochloric acid from the proventriculus. Possibly the intestine itself contributes enzymes, but they have not yet been satisfactorily determined (Sturkie, 1976). Subjected to the peristaltic movements of the intestinal wall, the intestinal contents are mixed thoroughly and moved along. Gradually the end-products of this digestive process are absorbed by the intestinal mucosa and reach the blood stream. The process of absorption is practically completed by the time the remainder of the intestinal contents reaches the large intestine.

The intestine is generally longer in seed- and herbage-eaters than in flesh-eaters, for the food received is more bulky in proportion to its nutritive content; the intestine thus needs more space to accommodate it and more inner surface for absorbing the end-products of digestion. In some groups of birds—ostriches, tinamous, waterfowl (swans, geese, and ducks), and gallinaceous birds, all primarily vegetarians—the intestinal ceca are greatly elongated. Not only do they serve to increase absorptive surface, but they assist digestion by lodging high concentrations of bacteria which reduce cellulose.

Rate of Digestion The physiology of avian digestion, which includes both physical processes (taking and swallowing of food, storing food in crop and esophagus, and macerating food in the gizzard) and chemical processes (alteration of food by the action of juices from the mouth, proventriculus, liver, pancreas, and intestine, and by the action of bacteria), has yet to be thoroughly studied. Basically it is the same as in mammals, though the rate of digestion is probably more rapid. In the Domestic Fowl, grain has been found to pass through the digestive tract in 2.5 to 12 hours, depending on the type and amount of food and the physiological state of the individual (Sturkie, 1976). Ordinarily a bird digests animal food more rapidly. For example, a magpie *(Pica)* may digest a mouse in three hours (Hewitt, 1948).

Bock, W.J.
1961 Salivary glands in the Gray Jays *(Perisoreus)*. *Auk* 78:355–365.

Dow, D.D.
1965 The role of saliva in food storage by the Gray Jay. *Auk* 82:139–154.

Farner, D.S.
1960 Digestion and the digestive system. In *Biology and comparative physiology of birds*. Vol. 1, ed. A.J. Marshall. New York: Academic Press.

Fisher, H.I., and Dater, E.E.
1961 Esophageal diverticula in the Redpoll, *Acanthis flammea*. *Auk* 78:528–531.

French, N.R.
1954 Notes on breeding activities and on gular sacs in the Pine Grosbeak. *Condor* 56:83–85.

Gross, A.O.
1928 The Heath Hen. *Mem. Boston Soc. Nat. Hist.* 6:487–588.

Hewitt, E.A.
1948 Digestion. In *Diseases of poultry*, eds. H.E. Biester and L.H. Schwarte. 2nd ed. Ames: Iowa St. Univ.

Hickey, J.J., and Elias, H.
1954 The structure of the liver of birds. *Auk* 71:458–462.

Johnston, D.W.
1961 Salivary glands in the Black Swift. *Condor* 63:338. (The salivary glands are smaller than in some other swifts.)

Leopold, A.S.
1953 Intestinal morphology of gallinaceous birds in relation to food habits. *Jour. Wildlife Mgmt.* 17:197–203.

Lucas, F.A.
1897 The tongues of birds. *Rept. U.S. Natl. Mus. for 1895*, pp. 1003–1020.

Matthews, L.H.
1949 The origin of stomach oil in the petrels, with comparative observations on the avian proventriculus. *Ibis* 91:373–392.

Miller, A.H.
1941 The buccal food-carrying pouches of the Rosy Finch. *Condor* 43:72–73.

Portmann, A.
1950 Le tube digestif. In *Traité de zoologie*. ed. P.P. Grassé. Vol. 15. Paris: Masson et Cie.

Stevenson J.
1933 Experiments on the digestion of food by birds. *Wilson Bull.* 45:155–167.

Sturkie, P.D.
1976 *Avian physiology*. 3rd ed. New York: Springer-Verlag.

Walsberg, G.E.
1975 Digestive adaptations of *Phainopepla nitens* associated with the eating of mistletoe berries. *Condor* 77:169–174.

Respiratory System

The Syrinx and Its Function The syrinx, or voice organ, is usually situated at the tracheo-bronchial junction. In the oscines, or songbirds, the form of the syrinx is reasonably constant (see Fig. 33, p. 75). However, it shows wide variation among the nonpasserine taxa, even at the level of species. The syringeal musculature within the songbirds is complex, with as many as five pairs of intrinsic muscles as well as two pairs of extrinsic muscles. All other birds have fewer, with many having only the two pairs of extrinsic muscles. Complexity of the syrinx seems to have little to do with the kinds of modulations it can make but is related to the variety of sounds a species may use. The ability to learn all or part of the repertoire appears to be confined to species possessing at least one pair of intrinsic muscles.

Differences in syringeal structure may occur between the sexes of some species. In some cases—e.g., the Anatidae, in which males have a large bulla on one side—the differences can be spectacular. In species in which only the male sings the syrinx and portions of the brain controlling it are less well developed in the female.

Sound production depends on two things, an increase in air flow through the syrinx and a reconfiguration of the syrinx so that flexible membranes move into the lumen and interact with the air flow. Sound is produced either by direct vibration of the membranes or by the formation of a series of vortices as the air stream is forced through the constricted lumen (Gaunt et al., 1982). The first mechanism produces complex sounds rich in overtones; the latter elicits pure tones (whistles). The two mechanisms need not be mutually exclusive and may interact. Some avian phonations appear to be harmonic—i.e., the overtones are at whole number multiples of the fundamental frequency. This rather common phenomenon is not well understood. Of the several mechanisms suggested for the production of harmonics, none seems well suited to the avian situation.

In one of the most thorough and important studies of avian voice, Greenewalt (1968) showed that the avian vocal system differs from that of humans in two important ways. First, modulations of the sound are not brought about by the resonance properties of the vocal passage but rather are source generated. In those syringes for which membrane vibration is the sound source, frequency (related to pitch) is determined by the tension of the membranes, which is controlled by action of the syringeal muscles. Frequency modulation of whistles is less well understood, but it is probably related to a combination of air velocity and syringeal shape. In both systems amplitude (related to loudness) is modulated primarily by two factors, the rate of air flow through the source and presence or absence of oscillators that can augment or suppress the sound. A good example of the latter is the labia, which can muffle sound by occluding the lumen.

Beat frequencies can also cause apparent changes in loudness. Beats are possible because of the second major difference, the "two-voice" phenomenon. Most birds can produce tones from both sides, and the tones need not be the same. Indeed, in some species—e.g., owls—the two sides are quite different and must produce different tones. Greenewalt suggested that birds with the appropriate musculature could control the two sides independently to produce quite different tones simultaneously, in effect duetting with themselves. The independence of the two sides was confirmed by Nottebohm (1971, 1977). In an elegant series of experiments he showed that control of the two sides is learned and lateralized—i.e., birds have a preferred side. Further, lateralization continues into the brain where one hemisphere is dominant. In this latter regard songbirds show a characteristic previously thought to be unique to humans.

Modifications of the Trachea Among different groups of birds are cases where the trachea is elaborated. A particularly unusual case occurs in two diverse groups of birds, the cranes and swans, represented by the Whooping Crane *(Grus americana)* and the Trumpeter Swan *(Cygnus buccinator)*. Through parallel evolution, each has developed an enormously long trachea that is coiled and fitted into the bony mass of the sternum. A somewhat similar case appears in the bird-of-paradise, *Phonygammus keraudrenii*, in which an exceedingly elongated trachea is coiled in the breast region between the skin and the big flight muscles. The males of many ducks have the trachea dilated to form a bulbous area, either along its course or where it bifurcates to form the bronchi.

Most ornithologists have long accepted the assumption that the trachea in all birds modifies the sound produced in the syrinx and that its elaborations, such as those just described, further modify the sound, either by increasing the volume of the sound by resonance or by altering the quality of the tone. After studying the syringeotracheal system of sound production in geese, which shows no special elaborations, Sutherland and McChesney (1965) likened the effect of the trachea to that of an open pipe as in a trumpet. Other ornithologists, from their investigations of systems without elaborations, have likened its effect to that of a closed or reed pipe. Greenewalt (1968), however, has presented strong evidence that the trachea plays no role in sound modulation and rejects the function attributed to any of its elaborations. Why then the elaborations? Their function or the cause of their development calls for renewed investigation.

The Function of the Air Sacs Air sacs play a multifarious role in the lives of birds. Among the many functions credited to them, the following seem the most acceptable:

1. Air sacs moisten air, a function that the bird's greatly reduced nasal cavities cannot effectively perform.

2. Air sacs provide complete ventilation, an advantage to the bird's flying mechanisms. With each inspiration and each expiration, air sweeps over the lung's respiratory surfaces, providing a "double tide" of air (see pages 73–74 for the anatomical arrangement in the lung permitting this), whereas in the mammal there is always incomplete ventilation, there being ever present in the lung a considerable residuum of vitiated air.

3. Air sacs play a role in the regulation of body temperature by permitting loss of heat through radiation and vaporization. Thus the bird, with its feather-covered, dry skin—without the sweat glands of the mammal—is able to cool itself by internal radiation and "prespiration."

4. Air sacs serve in an excretory capacity by being the principal outlets for all waste body fluids that are eliminated by vaporization.

5. The interclavicular air sac functions in sound production by effecting air pressure on the syrinx.

6. The cervical air sacs in certain species are elaborated for display purposes (e.g., in a frigate-

bird, *Fregata*) or sound production (in the Brown Jay, *Cyanocorax morio;* see Sutton and Gilbert, 1942).

7. In pelicans (e.g., Brown Pelican, *Pelecanus occidentalis*), and possibly in a few other water-plunging species, the interclavicular sac is connected to specially developed subcutaneous air sacs that act as a cushion to reduce the force of impact with the water when the bird is diving from a height.

The concept that air sacs lessen specific gravity of a bird by harboring a great amount of warm air and making it more buoyant seems to have no basis in fact. There is actually not enough warm air in a bird's air-sac-system to cause any material effect.

Mechanics of Respiration Certain muscles that act in breathing also move the wings. For a long time there was a controversy as to whether or not the movements are synchronized, performing the two functions simultaneously. It now is known that respiration, synchronous with wing beats, does occur in some birds (e.g., pigeons and crows) but that other species may exhibit one of 11 other types of coordination. Birds that have relatively high wing-beat frequencies (e.g., quails, pheasants, ducks) have a 5:1 coordination between wing beats and respiratory movements (Berger et al., 1970). Moreover, the type of coordination sometimes varies during flight, and the intake of air may occur on either the upstroke or the downstroke of the wing (Tomlinson, 1963).

Ames, P.L.
1971 The morphology of the syrinx in passerine birds. *Peabody Mus. Nat. Hist. Bull.* No. 37.

Berger, M.; Hart, J.S.; and Roy, O.Z.
1970 Respiration, oxygen consumption and heart rate in some birds during rest and in flight. *Z. Vergl. Physiol.* 66:201–204.

Bock, W.J.
1978 Morphology of the larynx of *Corvus brachyrhynchos* (Passeriformes: Corvidae). *Wilson Bull.* 90:553–565.

Brackenbury, J.
1980 Respiration and production of sound by birds. *Biol. Rev.* 55:363–378.

Bretz, W.L., and Schmidt-Nielsen, K.
1971 Bird respiration: flow patterns in the duck lung. *Jour. Exp. Biol.* 54:103–118.

Calder, W.A.
1968 Respiratory and heart rates of birds at rest. *Condor* 70:358–365.

Duncker, H.R.
1971 The lung air sac system of birds. *Ergeb. Anat. Entwickl. Gesch.* Vol. 45, 171 pp.

Gaunt, A.S.; Gaunt, S.L.L.; and Casey, R.M.
1982 Syringeal mechanics reassessed: evidence from *Streptopelia*. *Auk* 99:474–494.

Greenewalt, C.H.
1968 *Bird song: acoustics and physiology.* Washington, D.C: Smithsonian Institution Press.

King, A.S., and Molony, V.
1971 The anatomy of respiration. In *Physiology and biochemistry of the Domestic Fowl.* Vol. 1. New York: Academic Press.

Lasiewski, R.C.
1972 Respiratory function in birds. In *Avian biology.* Vol. 2, eds. D.S. Farner and J.R. King. New York: Academic Press.

Lord, R.D., Jr.; Bellrose, F.C.; and Cochran, W.W.
1962 Radiotelemetry of the respiration of a flying duck. *Science* 137(3523):39–40.

Nottebohm, F.
1971 Neural lateralization of vocal control in a passerine bird. I. Song. *Jour. Exp. Zool.* 177:229–261.
1977 Asymmetries in neural control of vocalization in the Canary. In *Lateralization in the nervous system*, eds. S. Harnard, R.W. Doty, L. Goldstein, J. Jaynes, and G. Krauthamer. New York: Academic Press.

Richardson, F.
1939 Functional aspects of the pneumatic system of the California Brown Pelican. *Condor* 41:13–17.

Sutherland, C.A., and McChesney, D.S.
1965 Sound production in two species of geese. *Living Bird* 4:99–106.

Sutton, G.M., and Gilbert, P.W.
1942 The Brown Jay's Furcular Pouch. *Condor* 44:160–165.

Tomlinson, J.T.
1963 Breathing of birds in flight. *Condor* 65:514–516.

Tucker, V.A.
1968 Respiratory exchange and evaporative water loss in the flying Budgerigar. *Jour. Exp. Biol.* 48:67–87.

Circulatory System

Blood Cells and Their Numbers Birds share with mammals the distinction of having the "richest" blood—i.e., blood with the highest number of erythrocytes, or red blood cells, per unit of measure. According to Stresemann (1927–34), the figures among different species of birds range from 1,500,000 to 5,500,000 in a cubic millimeter, in mammals from 2,000,000 to 18,000,000. In man the count is 4,500,000 for adult females, 5,000,000 for adult males. The number of erythrocytes tends to be higher in small birds than in large. Thus, Nice, et al. (1935) found that the number in 15 species of passerine birds averaged from 4,200,000 to 6,055,000; the lowest figure, 3,930,000, was for a Tufted Titmouse *(Parus bicolor)*, the highest, 7,645,000, for a Dark-eyed Junco *(Junco hyemalis)*. The figures published by these authorities actually extend Stresemann's range, previously mentioned. In contrast to the figures for small birds are the following for large birds, given by Ponder (1924): 1,620,000, Ostrich *(Struthio camelus)*; 2,189,000, White Stork *(Ciconia ciconia)*; 2,547,000, Peregrine Falcon *(Falco peregrinus)*. As a rule, when the erythrocytes are higher in number, they are smaller and have more hemoglobin (the oxygen-carrying pigment). Such a condition, characteristic of smaller birds, indicates greater efficiency with respect to the oxygen carrying capacity of the blood.

Within a bird species the number of erythrocytes is usually higher in the male. Domm et al. (1943) found the normal male Domestic Fowl to have 3,600,000 and the female 2,700,000—a difference of about a million. They concluded that gonadal hormones were responsible for this difference. Their experimental evidence showed that the higher erythrocyte number in the male is due to the presence of androgen and the lower number in the female is caused either by a lack in androgen or by the presence of estrogen. Within a species there may also be seasonal variation in the number of erythrocytes. Counts reported by Riddle and Braucher (1934) in the Common Pigeon and by Young (1937) in the Common Canary *(Serinus)* show that the higher figures appear in the cooler months of the year.

The Heart Rate The rate of the heart beat, long considered an indicator of the physiological activity in animals, is higher in most birds than in mammals of equivalent size; moreover, it is higher in small birds than in larger ones (Sturkie, 1976). The rate in birds varies widely, being affected by such features as muscular movements, mental activity, feeding, and air temperature. Captive birds show a lower rate than wild birds (Odum, 1941), and probably the same condition occurs in domesticated birds, Woodbury and Hamilton (1937) found the average rate (i.e., heart beats per minute) of the adult Common Pigeon at rest to be 221. Much higher rates in smaller birds were found by Odum, who captured wild specimens and made measurements while the birds were kept undisturbed and in darkness. Some of his figures on adult male specimens were: 614, Ruby-throated Hummingbird *(Archilochus colubris)*; 522, Black-capped Chickadee *(Parus atricapillus)*; 455, House Wren *(Troglodytes aedon)*; 480, Yellow Warbler *(Dendrocia petechia)*; 305, House Sparrow; 391, Northern Cardinal *(Cardinalis cardinalis)*. For comparison, about 72 times per minute is the rate of heart beat of the normal human adult at rest. While a bird is incubating, the rate of heart beat is apparently higher than when it is resting on a perch. In the House Wren, for example. Odum determined the rate to be from 550 to 650 on the nest during the day and as high as 701 on the nest late at night. This difference between day and night rates is correlated with changes in air temperature, the higher rate occurring late at night as the air temperature dips to its lowest point.

Heart Size Birds as a group exhibit relatively larger hearts than mammals, while among different bird species the size of the heart varies inversely with the size of the body. Thus small birds have proportionately larger hearts. Hummingbirds, the smallest and probably the most active of all birds, have by far the largest hearts (Hartman, 1954). Species of birds that fly less frequently than others have correspondingly smaller hearts (Johnston, 1963). Heart size may therefore be just as important an indicator of physiological activity in birds as heart rate (see Brush, 1966).

Heart size may vary within bird genera and species. Thus species and subspecies in northern regions appear to have proportionately larger hearts than their counterparts in more southerly latitudes (Johnston, 1963; Hartman, 1955); and there is evidence that populations of a species living at high altitudes show an increase in heart size relative to body size (Norris and Williamson, 1955).

Akester, A.R.
1971 The blood vascular system. In *Physiology and biochemistry of the Domestic Fowl.* Vol. 2. New York: Academic Press.

Baumel, J.J., and Gerchman, L.
1968 The avian intercarotid anastomosis and its homologue in other vertebrates. *Amer. Jour. Anat.* 122:1–18.

Brush, A.H.
1966 Avian heart size and cardiovascular performance. *Auk* 83:266–273.

Domm, L.V.; Taber, E.; and Davis, D.E.
1943 Comparison of Erythrocyte numbers in normal and hormone-treated Brown Leghorn Fowl. *Proc. Soc. Exp. Biol. and Med.* 52:49–50.

Glenny, F.H.
1955 Modifications of pattern in the aortic arch system of birds and their phylogenetic significance. *Proc. U.S. Natl. Mus.* 104 (3346):525–621.

Hartman, F.A.
1954 Cardiac and pectoral muscles of trochilids. *Auk* 71:467–469.
1955 Heart weight in birds. *Condor* 57:221–238.

Hartman, F.A., and Lessler, M.A.
1963 Erythrocyte measurements in birds. *Auk* 80:467–473. (Measurements in 124 species among 46 families in Panama and the United States.)

Johnston, D.W.
1963 Heart weights of some Alaskan birds. *Wilson Bull.* 75:435–446. (Weights from 563 individuals representing 77 species.)

Kanwisher, J.W.; Williams, T.C.; Teal, J.M.; and Lawson, Jr., K.O.
1978 Radiotelemetry of heart rates from free-ranging gulls. *Auk* 95:288–293.

Lucas, A.M., and Jamroz, C.
1961 Atlas of avian hematology. *U.S. Dept. Agric., Agric. Monogr.* 25. (Concerns mainly the Domestic Fowl.)

Nice, L.B.; Nice, M.M.; and Kraft, R.M.
1935 Erythrocytes and hemoglobin in the blood of some American birds. *Wilson Bull.* 47:120–124.

Norris, R.A.
1963 A preliminary study of avian blood groups with special reference to the Passeriformes. *Bull. Tall Timbers Res. Sta.* No. 4.

Norris, R.A., and Williamson, F.S.L.
1955 Variation in relative heart size of certain passerines with increase in altitude. *Wilson Bull.* 67:78–83.

Odum, E.P.
1941 Variations in the heart rate of birds: A study in physiological ecology. *Ecol. Monogr.* 3:299–326.

Ponder, E.
1924 *Erythrocytes and the action of simple haemoloysis.* Edinburgh: Oliver and Boyd. (Cited by Nice, Nice, and Kraft, 1935.)

Riddle, O., and Braucher, P.F.
1934 Hemoglobin and erythrocyte differences according to sex and season in doves and pigeons. *Amer. Jour. Physiol.* 108:554–566.

Simmons, J.R.
1960 The blood-vascular system. In *Biology and comparative physiology of birds.* Vol. 1, Ed. A.J. Marshall, New York: Academic Press.

Stresemann, E.
1927–1934 Aves. In *Kükenthal u. Krumbach, Handbuch der Zoologie.* Vol. 7, Part 2. Berlin: De Gruyter.

Sturkie, P.D.
1976 *Avian physiology.* 3rd ed. New York: Springer-Verlag.

Williamson, F.S.L. and Norris, R.A.
1958 Data on relative heart size of the Warbling Vireo and other passerines from high altitudes. *Wilson Bull.* 70:90–91. (Data on 73 specimens representing 24 species.)

Woodbury, R.A., and Hamilton, W.F.
1973 Blood pressure studies in small animals. *Amer. Jour. Physiol.* 119:663. (Cited by Sturkie, 1976)

Young, M.D.
1937 Erythrocyte counts and hemoglobin concentration in normal female canaries. *Jour. Parasitology* 23:421–426.

Body Temperature

The body temperatures of birds range from about 98.6 degrees F (37.0 degrees C) to 112.3 degrees F (44.6 degrees C) (Dawson and Hudson, 1970). The average resting temperature of 311 passerine species was found to be 105.1 degrees F (40.6 degrees C) and that of 90 charadriiform species was about 104.2 degrees F (40.1 degrees C). In nocturnal species—e.g., kiwis (*Apteryx* spp.), some sea birds, owls, and nighthawks—the body temperature is higher at night than during the daytime when the birds are less active. The primitive kiwis have both a lower and a greater fluctuation in body temperature than do other birds. (Farner et al., 1956).

In their study of the body temperature in passerine birds Baldwin and Kendeigh (1932) concluded that the normal temperature is highly variable. Such factors as emotional excitement, muscular activity, high air temperature, and the digestion of food cause a bird's temperature to rise, while low air temperature and lack of food bring about a decrease in temperature. The same two investigators found a daily rhythm in body temperature. Briefly, the temperature rises gradually during the morning from the beginning of the day's activities until the middle of the day; it decreases during the late afternoon. When the bird settles on the nest for the night, the temperature first falls rapidly, then decreases gradually until about midnight. Thereafter the body temperature fluctuates more or less until the short period, just before the bird leaves the nest for the first time in the morning, when there is a rapid rise in body temperature.

Adult birds, being warm-blooded (homeothermal), are able to remain active under varying environmental conditions. By temperature-control mechanisms centered in the brain, they can maintain their normal body temperature even when the surrounding air temperature undergoes marked fluctuations. If the weather becomes extremely cold, they can conserve their body heat by lifting up or "fluffing" the feathers, thereby widening the insulating layer of air between feathers and skin; they can increase their body heat by stimulating muscle activity through shivering. On the other hand, if it gets extremely warm, they can press the feathers against the skin to eliminate the insulating layer of air and thus facilitate reduction of their body heat by radiation; they can pant—i.e., increase respiratory movements and consequently increase the amount of internal radiation of heat and vaporization ("perspiration") through the lungs and air sacs. Some birds such as pelicans, boobies, and nighthawks lower body heat by "gular flutter"—fluttering their gular area—instead of panting.

In the course of their early development, all birds pass through a cold-blooded (poikilothermal) stage, in which the temperature-control mechanisms are lacking. In precocial birds the stage is completed in the egg; by hatching time, temperature control is already partially established. A day-old chick of the Domestic Fowl has an average body temperature of 103.4 degrees F (39.7 degrees C), which is about 3 degrees F (1.7 degrees C) below that of the adult; at 20 days of age its temperature averages the same as the adult's, and temperature control is fully established. Altricial young at hatching time are still distinctly cold-blooded. In the House Wren *(Troglodytes aedon)* temperature control is gradually established after hatching; when nine days old the young bird obtains full temperature control (Baldwin and Kendeigh, 1928); thereafter it is able to maintain an average temperature of 104.7 degrees F (40.4 degrees C) (Baldwin and Kendeigh, 1932), the same as that of the adult. When young birds, whether precocial or altricial, have acquired temperature control, they are able to tolerate moderate air temperatures, but for a few days they may still require brooding or other form of protection if air temperatures are extreme.

Torpidity in Birds A number of different kinds of birds—e.g., goatsuckers, swifts, and hummingbirds—enter a torpid or hibernating state during which the body temperature is lowered to a level near that of the air, breathing is sometimes indiscernible, and reactions to handling are slow, if they occur at all. Their general physiological condition often closely resembles hibernation in certain mammals, especially bats. Jaeger (1949) made observations on a Common Poorwill *(Phalaenoptilus nuttallii)* in profound torpidity for a known winter period of 85 days. Its body temperature, taken on five different days, ranged from 64.4 degrees F to 67.6 degrees F (18 degrees C to 19.8 degrees C), while the surrounding air temperature on the same days ranged from 63.5 degrees F to 75.3 degrees F (17.5 degrees C to 24.1 degrees C). During this winter period insects—the bird's usual food—were scarce, thus suggesting the possibility that torpidity, at least in the case of this species, is an adaptation for tiding the bird over the period of the year when food is unobtainable.

Baldwin, S.P., and Kendeigh, S.C.

1928 Development of temperature control in nestling House Wrens. *Amer. Nat.* 62:249–278.

1932 Physiology of the temperature of birds. *Sci. Publ. Cleveland. Mus. Nat. Hist.* 3:i–x; 1–196.

Bartholomew, G.A., and Dawson, W.R.

1958 Body temperatures in California and Gambel's Quail. *Auk* 75:150–156. (A conspicuous diurnal cycle of body temperature correlates with activity in both juveniles and adults at moderate environmental temperatures.)

Bartholomew, G.A.; Hudson, J.W.; and Howell, T.R.

1962 Body temperature, oxygen consumption, evaporative water loss, and heart rate in the Poorwill. *Condor* 64:117–125.

Dawson, W.R., and Hudson, J.W.
1970 Birds. In *Comparative physiology of thermo-regulation*. Vol. 1. Ed. G.C. Whittow. New York: Academic Press.

Farner, D.S.; Chivers, N.; and Riney, T
1956 The body temperatures of North Island Kiwis. *Emu* 56:199–206.

Heath, J.E.
1962 Temperature fluctuation in the Turkey Vulture. *Condor* 64:234–235. (An example of lowered body temperature normally occurring in a large bird.)

Howell, T.R., and Bartholomew, G.A.
1961 Temperature regulation in Laysan and Black-footed Albatrosses. *Condor* 63:185–197. (Young birds effect heat loss by exposing vascularized foot webbing to the air.)
1962a Temperature regulation in the Red-tailed Tropic Bird and the Red-footed Booby. *Condor* 64:6–18. (The importance of panting in the tropicbird and gular flutter in the booby is emphasized.)
1962b Temperature regulation in the Sooty Tern *Sterna fuscata*. *Ibis* 104:98–105. (Chicks able to lose heat by vigorous panting and seeking shade.)

Jaeger, E.C.
1949 Further observations on the hibernation of the Poor-will. *Condor* 51:105–109.

Kendeigh, S.C.
1961 Energy of birds conserved by roosting in cavities. *Wilson Bull.* 73:140–147.

Lasiewski, R.C.
1963 Oxygen consumption of torpid, resting, active, and flying hummingbirds. *Physiol. Zool.* 36:122–140.

Marshall, J.T., Jr.
1955 Hibernation in captive goatsuckers. Condor 57:129–134.

Morrison, P.
1962 Modification of body temperature by activity in Brazilian hummingbirds. *Condor* 64:315–323. (Tropical hummingbirds appear to be more sensitive to the lowering of the body temperature than do temperate species.)

Ohmart, R.D., and Lasiewski, R.C.
1971 Roadrunners: energy conservation by hypothermia and absorption of sunlight. *Science* 172:67–69.

Pearson, O.P.
1960 Torpidity in birds. *Bull. Mus. Comp. Zool.* 124:93–103.

Sladen, W.J.L.; Boyd, J.C.; and Pedersen, J.M.
1966 Biotelemetry studies on penguin body temperatures. *Antartic Jour. U.S.* 1:142–143.

Urogenital System

Kidneys and Urine Elimination

The avian kidneys are relatively larger than those of reptiles and mammals, ranging from one to two percent of the body weight, and are more voluminous in aquatic birds (Benoit, 1950). Structurally they are more reptilian, since they show no demarcation between cortex and medulla as in mammals. The glomeruli, located in the cortical area, range in number from 30,000 in small passerine birds to 200,000 in the Domestic Fowl; in general, they are small and more numerous than those in mammals. The functional unit of the kidneys, the uriniferous tubule or nephron, closely resembles the mammal's.

Besides regulating the salts and liquids of the body, the kidneys eliminate the end products of protein metabolism. The urine that the nephron produces by filtration consists chiefly of water but also contains excess salts in solution and nitrogenous wastes from protein metabolism. In bird urine, as in reptile urine, the nitrogenous wastes are in the form of uric acid, a practically insoluble substance of high nitrogen concentration, whereas in mammal urine the wastes comprise urea, which has half the concentration.

Once the urine is in the cloaca, some of its water content is absorbed by the cloaca. The residue is distinctly whitish, semi-solid, and usually encapsulated with a mucous material. Eventually it is voided, along with the fecal material, but nevertheless remains recognizable by its color.

About 98 percent of the water filtered by the kidney is reabsorbed either in the kidney or the cloaca (Sturkie 1976). In this way the bird conserves water for further use. Thus the bird's urine when voided lacks the watery consistency of the mammal's. Actually the bird's body loses more water through vaporization, probably through the lungs and air sacs, than through the urine.

Reproductive System

Many studies have been made on the reproductive system of domestic birds, comparatively few on wild birds. Most of the data in the ensuing paragraphs, unless otherwise indicated, are drawn from studies on the Domestic Fowl as summarized by Sturkie (1976).

Ovary and Ova The avian ovary consists of two parts: outer cortex and inner medulla. The cortex contains the ova, enormous in number and only a few of which ever reach maturity. In one ovary as many as 1,906 ova have been counted with the unaided eye and an additional 12,000 under the microscope. Each ovum develops within a follicle of its own. At the time of ovulation (i.e., the release of the ovum from the follicle) a follicle may have a diameter of 40 mm. Following ovulation the ovum begins its passage through the oviduct, taking about 4½ hours to reach the uterus and remaining in the uterus 18 to 20 hours. During the last 5 hours of its stay in the uterus it receives its coloration (pigment). From the uterus the egg is forced by contraction through the vagina, cloaca, and vent; it is not held for any length of time in either the vagina or cloaca.

Testes and Spermatozoa Birds' testes have a regular shape, which is usually oval or ellipsoid. Within each testis are the seminiferous tubules, lined with epithelium in which the spermatozoa are produced. In birds the testis is not divided into lobes as it is in mammals; instead, the interior of the testis is one chamber in which the tubules form a complicated network.

In the majority of birds the seminal vesicles, which store spermatozoa, are in the body cavity, but in some passerine birds during the breeding season they take up a position in a cloacal protuberance (see Appendix A, page 377) outside the body cavity. Here the body temperature is lower, as in a mammalian scrotum. Possibly the spermatozoa ultimately mature in the seminal vesicles of the protuberance, in which case the lower temperature might assist normal spermatogenesis (Wolfson, 1954).

The spermatozoa in different birds vary considerably in form; unlike those of mammals, most have elongated heads and extremely long, whip-like tails with undulating membranes. There are generally said to be two types of spermatozoa in birds: sauropsidian and passeriform (Benoit, 1950). The first, common to both reptiles and all birds except passerine species, has a more or less curved head; the second, found in passerine species only, has a spiraled head.

The number of spermatozoa in bird semen from a single ejaculation during copulation has been estimated to range from 1,700,000,000 to 3,500,000,000. In man the number from one ejaculation is usually estimated at 300,000,000. By their own motility, the spermatozoa pass from the cloaca up the oviduct to the magnum or isthmus where, presumably, the ovum is fertilized. On the average, the time between copulation and fertilization is about 72 hours, but it may be as short as 19.5 hours.

Asymmetry in the Reproductive System

Asymmetry in the reproductive system of the bird has long been an intriguing subject for comment and investigation, and its probable cause has aroused much speculation. Though the male often shows asymmetries with respect to the size and location of the gonads, it is the female that has the more marked variations, ranging from almost perfect bilateral symmetry in a few birds (e.g., hawks) to a unilateral condition in most birds in which the right ovary is missing and the right oviduct reduced to a mere vestige. Witschi (1935a), who studied the embryonic development of gonadic asymmetry in the Domestic Fowl, House Sparrow, and Red-winged Blackbird *(Agelaius phoeniceus),* found the germ cells evenly distributed on both the right and left sides up to the end of the third day of incubation, but by the end of the fourth day asymmetry was established, the germ cells from the right side having migrated to the left. He concluded that asymmetry in the bird is due to a genetically inherited deficiency or inability on the part of the ovarian cortex to attract and keep the germ cells. The same author also put forward the provocative idea that "asymmetry in the female oviducts might have evolved as an adaptation to the aviatic life habits of the birds, while the more or less complete reduction of the right ovary followed later, in order that eggs may more safely reach the one remaining fallopian tube (oviduct)."

Benoit, J.
1950 Organes uro-génitaux. In *Traité de zoologie,* ed. P.-P. Grassé. Vol. 15 Paris: Masson et Cie.

Marshall, A.J.
1961 Reproduction. In *Biology and comparative physiology of birds.* Vol. 2, ed. A.J. Marshall. New York: Academic Press.

Middleton, A.L.A.
1972 The Structure and possible function of the avian seminal sac. *Condor* 74:185–190.
Romanoff, A.L., and Romanoff, A.J.
1949 *The avian egg*. New York: John Wiley & Sons. (Chapter 4 has an extensive description of the female reproductive system of the Domestic Fowl.)
Sperber, I.
1960 Excretion. *In Biology and comparative physiology of birds*. Vol. 1, ed. A.J. Marshall. New York: Academic Press.
Stanley, A.J.
1937 Sexual dimorphism in North American hawks. I. Sex Organs. *Jour. Morph.* 61:321–339.
Sturkie, P.D.
1976 *Avian physiology*. 3rd ed. New York: Springer-Verlag.
Witschi, E.
1935a Origin of asymmetry in the reproductive system of birds. *Amer. Jour. Anat.* 56:119–141.
1935b Seasonal sex characters in birds and their hormonal control. *Wilson Bull.* 47:177–188. (Has illustrations of the House Sparrow's urogenital system in active and quiescent stages.)
Wolfson, A.
1954 Sperm storage at lower-than-body temperature outside the body cavity in some passerine birds. *Science* 120:68–71.
1960 The ejaculate and the nature of coition in some passerine birds. *Ibis* 102:124–125.

Salt Glands

Many aquatic birds have a large pair of multilobular glands on the skull above the orbits; each lies in a crescentric depression and sends a duct into the nasal cavity. Called the supraorbital or nasal glands, their function remained undetermined until 1957–1958 when Knut Schmidt-Nielsen and his co-workers made a surprising discovery. In the Double-crested Cormorant *(Phalacrocorax auritus)*, these glands excrete salt (sodium chloride) in concentrated solution (Schmidt-Nielsen, Jorgensen, and Osaki, 1958); in the Humboldt Penguin *(Spheniscus humboldti)*, they could perform the same function (Schmidt-Nielsen and Sladen, 1958). Since these initial observations it has become apparent that the glands—now more properly called salt glands—have a true excretory function in all marine birds as well as in other species that normally acquire large amounts of salt in their food or drinking water. Excess salt, picked up by the blood stream and transferred to the salt glands, is conveyed to the nasal cavity and out through the nostrils or (particularly in the case of pelecaniform birds with occluded nostrils) into the mouth cavity through the choana and thence to the tip of the bill. In 10 to 15 minutes after a bird feeds on a salty substance, a clear fluid appears at the nostrils or bill tip in large droplets, and the bird dispatches them by vigorously shaking its head.

Salt glands are subject to variation within a species depending on the amount of salt particular individuals or populations normally acquire. Domestic ducks *(Anas platyrhynchos)* drinking only salt water developed larger glands than control birds that were allowed to drink only fresh water (Schmidt-Nielsen and Kim, 1964). The size of the gland therefore appears to be determined by individual adaptation rather than genetically.

Cooch, F.G.
1964 A preliminary study of the survival value of a functional salt gland in prairie Anatidae. *Auk* 81:380–393. (Ducks and other water birds inhabiting the Great Plains have a salt gland that definitely has survival value for birds living in an alkaline environment.)
Peaker, M., and Linzell, J.L.
1975 Salt glands in birds and reptiles. Cambridge: Cambridge Univ. Press.
Schmidt-Nielsen, K., and Fange, R.
1958 The function of the salt gland in the Brown Pelican. *Auk* 75:282–289.
Schmidt-Nielsen, K.; Jorgensen, C.B.; and Osaki, H.
1958 Extrarenal salt excretion in birds. *Amer. Jour. Physiol.* 193:101–107.
Schmidt-Nielsen, K., and Kim, Y.T.
1964 The effect of salt intake on the size and function of the salt gland of ducks. *Auk* 81:160–172.
Schmidt-Nielsen, K., and Sladen, W.J.L.
1958 Nasal salt secretion in the Humboldt Penguin. *Nature* 181:1217–1218.

Endocrine Organs

The endocrine organs are small, ductless glands that pass their secretions (called hormones) into the blood. Once in the circulatory system, the hormones are taken to various parts of the body where they effect changes in the cells or tissues. The organs form no real "system" in the anatomical sense, as they are widely scattered over the body and sometimes far removed from structures on which they nonetheless exercise profound influence. Briefly outlined below are the principal endocrine organs of the bird, their better-known hormones, and some of the more important functions attributed to them.

Pituitary Gland Anterior Lobe, Gonadotrophic Hormones: Stimulate growth of ovarian follicles, growth of seminiferous tubules, production of spermatozoa, and production of gonadal hormones. Adrenocorticotrophic Hormone: Stimulates development of adrenal gland and controls function of adrenal cortex. Thyrotrophic Hormone: Stimulates secretion of the thyroid gland. Lactogenic Hormone or Prolactin: Stimulates production of "milk" in the pigeon's crop, possibly induces broodiness, and inhibits gonadal activity. Posterior Lobe, Vasopressin: Regulates blood pressure. Oxytocin: Causes premature expulsion of eggs from oviduct. Antidiuretic: Retards flow of urine.

Adrenal Gland Each gland consists of two functionally unrelated parts, named because of the structure of the human adrenal: cortex (shell) and medulla (pith). In birds, however, two kinds of tissues are intermingled. The *interrenal tissue* (cortical) secretes hormones that regulate carbohydrate and salt metabolism and activities of the kidneys and liver as well as both androgens and estrogens. The *chromaffin tissue* (medullary) secretes adrenalin and noradrenalin; these regulate blood pressure and heart rate and inhibit digestion.

Thyroid Gland Thyroxin: Stimulates body growth (by increasing metabolic rate); induces molting; increases growth rate of feathers; influences size and shape of feathers; affects coloration (small doses increase melanins, while heavy doses may inhibit coloration or cause discoloration).

Parathyroid Gland Parathormone: Regulates calcium and phosphorus metabolism; may, in the laying female, synergize with estrogen (see below) in the mobilization of calcium for egg shells.

Ultimobranchial Glands Calcitonin. The glands lie posterior to the parathyroids.

Pancreas Insulin and glucagon: Regulate carbohydrate metabolism.

Gonads Androgen (from testis): Stimulates development of accessory reproductive organs, secondary sex characters (e.g., bright coloration, tarsal spurs, plumage for display purposes), and courtship behavior patterns; influences number of erythrocytes and amount of hemoglobin in blood; stimulates general body growth. Estrogen (from ovary): Stimulates development of accessory reproductive organs and incubation patch, secondary sex characters, and breeding behavior patterns; increases blood calcium, phosphorus, and lipids; depresses secretion of prolactin from pituitary gland.

For more than 50 years a significant line of research has been conducted on the role of light in stimulating the activity of avian gonads. William Rowan (1925) was primarily responsible for initiation of the work; he found that the inactive gonads of male Dark-eyed Juncos *(Junco hyemalis)* could be stimulated in winter to produce spermatozoa by exposure of the birds to an additional several hours per day of (artificial) light. His work was later confirmed by others, notably by Bissonnette (1937) in experiments on the European Starling *(Sturnus vulgaris)*. From these investigations has come the generalization that increased day length induces sexual activity prior to the nesting season. The gonadotrophic hormones from the pituitary gland are the factors that stimulate gonadal activity, and the pituitary gland is stimulated by secretions from the hypothalamus in the brain. Attention has also been given to the factors that bring about, after the nesting season, a so-called refractory period during which light cannot induce gonadal activity. In some way, apparently, gonadotrophic hormones from the pituitary are blocked. Bailey (1950) demonstrated that prolactin is capable of inhibiting light-induced gonadal activity and is probably one of the hormones responsible for the refractory period. Estrogen is probably another. The stimulus that causes the secretion of prolactin is yet to be found. Students interested in the work that has been done on photoperiodism in birds will find the reviews by

Burger (1949), Wolfson (1958), Höhn (1961), Farner and Lewis (1971), and Farner (1973) particularly instructive.

Bailey, R.E.
1950 Inhibition with prolactin of light-induced gonad increase in White-Crowned Sparrows. *Condor* 52:247–251.
1952 The incubation patch of passerine birds. *Condor* 54:121–136.

Benoit, J.
1950 Les glandes endocrines. In *Traité de zoologie*, ed. P.P. Grassé. Vol. 15. Paris: Masson et Cie.

Bissonnette, T.H.
1937 Photoperiodicity in birds. *Wilson Bull.* 49:241–270.

Burger, J.W.
1949 A review of experimental investigations on seasonal reproduction in birds. *Wilson Bull.* 61:211–230.

Epple, A., and Stetson, M.H., eds.
1980 *Avian endocrinology.* New York: Academic Press.

Farner, D.S., ed.
1973 *Breeding biology of birds: Proceedings of a symposium on breeding behavior and reproductive physiology in birds.* Washington, D.C.: National Academy of Science.

Farner, D.S., and Lewis, R.A.
1971 Photoperiodism and reproductive cycles in birds. In *Photophysiology.* Vol. 6. New York: Academic Press.

Freeman, B.M.
1971 The endocrine status of the bursa of Fabricius and the thymus gland. In *Physiology and biochemistry of the Domestic Fowl.* Vol. 1, ed. B.M. Freeman. New York: Academic Press.

Hartman, F.A., and Brownell, K.A.
1961 Adrenal and thyroid weights in birds. *Auk* 78:397–422.

Höhn, E.O.
1961 Endocrine glands, thymus and pineal body. In *Biology and comparative physiology of birds.* Vol. 2, ed. by A.J. Marshall. New York: Academic Press.

Lofts, B.
1962 Photoperiod and the refractory period of reproduction in an equatorial bird, *Quelea quelea. Ibis* 104:407–414.

Miller, A.H.
1959 Response to experimental light increments by Andean sparrows from an equatorial area. *Condor* 61:344–347.

Riddle, O.; Bates, R.W.; and Lahr, E.L.
1935 Prolactin induces broodiness in fowl. *Amer. Jour. Physiol.* 111:352–360.

Rowan, W.
1925 Relation of light to bird migration and developmental changes. *Nature* 115:494–495.
1929 Experiments in bird migration. I. Manipulation of the reproductive cycle: seasonal histological changes in the gonads. *Proc. Boston Soc. Nat. Hist.* 39:151–208.
1938 Light and seasonal reproduction in animals. *Biol. Rev.* 13:374–402.

Wallin, H.E., and Kendeigh, S.C.
1966 Seasonal and taxonomic differences in the size and activity of the thyroid glands in birds. *Ohio Jour. Sci.* 66:369–379.

Wingstrand, K.G.
1951 *The structure and development of the avian pituitary from a comparative and functional viewpoint.* Lund: C.W.K. Gleerup.

Wolfson, A.
1945 The role of the pituitary, fat deposition, and body weight in bird migration. *Condor* 47:95–127.
1958 Role of light in the photoperiodic responses of migratory birds. *Science* 129 (3360):1425–1426.

Special Senses and Their Organs

Taste

Taste buds in birds, owing to the extensive horny covering of the tongue and palate, are restricted to the soft tissue at the base and sides of the tongue and to the softer region of the palate. Probably the number of taste buds seldom exceeds 100 in any species and averages about 40 in most species. Compared to man, whose taste buds approximate 9,000, birds are ill equipped to detect food by taste.

Birds vary widely in their response to different substances. Nectar-feeding species such as hummingbirds and many fruit-eating species show a strong preference for sugary substances, while insect- and seed-eating species are indifferent to anything particularly sweet. Birds that drink or obtain their food in salt or brackish water show considerable tolerance for salty food, and certain finches actually seek salt as part of their diet. By contrast, many species reject food or water with a strong salty taste. Some birds tolerate food with a sour or bitter taste while others avoid it.

The broad generalization can be made that all birds are able to detect certain substances by taste, but in selecting their food they depend more on its visual properties.

Smell

Practically all birds have olfactory organs that appear functional in detecting odors. However, some species are credited with a keen sense of smell, others with none whatever.

Kiwis (*Apteryx* spp.) no doubt have a sense of smell far superior to most birds. With nostrils (anterior nares) at the tip of the bill, large nasal cavities, and greatly elaborated posterior conchae (Portmann, 1950), they have an olfactory apparatus remarkably like a mammal's. Supposedly, this condition enables kiwis, nocturnal birds with notoriously poor vision, to orient themselves by odors.

Vultures (Cathartidae) are almost certainly assisted at times in finding carrion by a sense of smell. Chapman (1929, 1938), testing the ability of the Turkey Vulture *(Cathartes aura)* to locate animal matter by smell, hid dead mammals completely from view at Barro Colorado Island (Panama) where this species occurs. As soon as the carcasses produced odors through advanced decay, numerous vultures arrived at the spot of concealment. When Chapman similarly hid decaying fish with strong odors, no birds showed up. Chapman concluded that the Turkey Vulture is dependent for its existence on a discriminating use of its senses of smell and sight. Years later Owre and Northington (1961) proved from tests that Turkey Vultures do have an olfactory ability. There is also ample evidence, according to Bang (1960) and Stager (1964), that Turkey Vultures have a well-developed sense of smell.

Other birds have been tested for their sense of smell. Strong (1911) reported the Common Pigeon able to recognize the oil of bergamot by smell. Calvin et al. (1957) found that pigeons could not be trained to odor alone as a cue in conditioned experiments, but Michelsen (1959) reported that pigeons could be stimulated by an odor. Frings and Boyd (1952) noted the ability of the Northern Bobwhite *(Colinus virginianus)* in captivity to distinguish between two feeders by odor and consequently develop a preference for one. In such birds as cormorants, whose nostrils are normally occluded on approaching adulthood, one might expect the loss of an olfactory sense, but observations indicate that the nasal cavities of these birds may receive odors by way of the mouth and choana (Portmann, 1950).

Vision

Vision reaches such a high state of development in birds that it is worth more than passing attention. Most of the data and interpretations in the paragraphs that follow are drawn from Walls' *The Vertebrate Eye,* a thorough treatise with further details and additional information.

Eye Glands Lacrimal and Harderian glands vary in their size and amount of secretion in accordance with the habits of the bird. Aquatic birds have very small lacrimals, the reason being perhaps that their secretions for moistening the surface of the eyes are not greatly needed in a watery environment. On the other hand, Harderian glands are apt to be very large in marine birds (e.g., cormorants), for their thick oily secretions serve to protect the exposed surface of the eyes from the harmful effects of salt water.

Size and Shape of Birds' Eyes The eyes of birds are notoriously large; they are the biggest structures of the head, and they often weigh more than the brain. Large hawks and owls have eyes whose size in diameter rivals a full-grown man's; ostriches, with an eye diameter of about two inches, have the largest eyes among birds and, reputedly, among all land vertebrates.

Avian eyes may be classified on the basis of shape as flat, globose, and tubular. Flat eyes, characteristic of pigeons and the majority of birds, have relatively slight convexities on their corneal surfaces, while the distance through the eye from cornea to retina is shorter than from dorsal to ventral walls—in other words, flat eyes are more broad than deep. Globose and tubular eyes, found in birds of prey, have much greater corneal convexitites; in globose eyes either distance through them is about the same; in tubular eyes the distance from cornea to retina tends to be longer and the sides of the eyes are noticeably concave. Both globose and tubular eyes, with greater distances between the lens and retina, broaden and sharpen the image thrown on the retina. This affords better vision at longer distances—an adaptation with obvious advantages for a bird, such as an eagle, that often seeks food from great heights.

Accommodation In the eye of any land vertebrate light waves from a given object are bent or refracted as they pass through the cornea and lens, and to a slight extent as they pass through the two humors. In order for the light waves to strike the retina in proper focus and thus form a satisfactory image of the object, the eye must make certain adjustments. This is called **accommodation.** In most birds, as well as in reptiles and mammals, accommodation is effected by changing the curvature of the lens—flatter for distant objects, more rounded for near objects; in a few birds accommodation is assisted by changing the curvature of the cornea. The amount of range of accommodation of which an eye is capable is expressed in diopters, one diopter being equal to the reciprocal of the focal length in meters. A lens of one diopter will focus on an object a meter away, a lens of two diopters one-half meter away, a lens of four diopters a quarter of a meter away, and so on. Land birds vary considerably in their range of accommodation. Generally they have an accommodation of only a few diopters: from 8 to 12 in the Common Pigeon and Domestic Fowl; from 2 to 4 in owls and most other nocturnal birds. (Adult man has an accommodation of about 10.) On the other hand, aquatic diving birds (e.g., cormorants) have 40 to 50 diopters, a remarkably extensive range. Vision in the majority of birds is probably emmetropic—i.e., when the eye is resting, it is focused on distant objects. A few birds are myopic or nearsighted under certain conditions. Penguins, for example, cannot see far when out of water though their eyes are very sensitive to light, enabling them to detect movements quite readily. The kiwis, nocturnal birds, are perhaps the most myopic of all avian species, especially during the day when they can scarcely see anything unless it is very near them.

The structure chiefly responsible for accommodation in birds is the ciliary body, whose muscles bring pressure to bear on the lens via the suspensory ligament, which holds the lens in place. The ciliary body is very much like that of reptiles but has a few minor modifications. Its muscles are strongly striated, capable of very quick action; moreover, they are completely separated into two parts. Brücke's muscle, the posterior part, acts on the lens—usually soft in birds—increasing its curvature, particularly that of its outer surface, by pulling it forward, for near vision. Crampton's muscle, the anterior part, aids further in accommodation by changing the shape of the cornea. This double action is most fully developed in birds, such as hawks, whose eyes must adjust rapidly from very far to very near vision during pursuit of prey. In aquatic diving birds, whose corneas are thick and inflexible, Crampton's muscle is all but absent, whereas Brücke's muscle is notably large, able to effect, in conjunction with a strong iris muscle, and extreme lens curvature for very near vision under water.

The Birds' Visual Field The eyes of birds are more or less fixed in their sockets and so can be moved only to a very limited degree. Change in direction of vision is accomplished mainly by turning the head and neck. An owl can rotate its head through 270 degrees or more. The bird's visual field—i.e., the area which the eyes cover when the head is stationary—varies among different species, depending on the shape of the eyes and on their position in the head. Flat eyes allow for the coverage of wide areas; globose and tubular eyes have more restricted views. When the eyes are in the sides of the head so as to be directed laterally and give what is called **monocular vision,** the visual field of the two eyes is very wide, but when the eyes are farther forward in the head so as to be aimed anteriorly and give **binocular vision**—both eyes taking in part of the same area—the visual field is consequently more limited.

Monocularity and Binocularity Probably very few birds have strictly monocular vision. Though the majority of birds have eyes in the sides of the head and quite commonly cock their heads to examine an object on the ground, looking with one eye while vision in their other eye is suppressed, they can nevertheless view binocularly by peering straight ahead at an object with both eyes. The Common Pigeon, a good example of a bird with eyes in the sides of the head, enjoys a total visual field of 340 to 342 degrees and a binocular field—

the field on which both eyes may be aimed—of 24 degrees. Those birds with eyes situated more frontally have a greater binocular field, which aids in the pursuit of prey by allowing better distance-determination. The insect-catching goatsuckers and swallows have a wider binocular field than birds that normally pick food from the ground. Hawks and owls have a still wider field—35 to 50 degrees in hawks, 60 to 70 degrees in owls. Usually binocularity is more or less limited to objects in front of the head, but there are some species of birds whose head structure and eye positions give them binocularity for objects in other directions. An American Woodcock *(Scolopax minor)*, with eyes far back and far up in the head, no doubt can watch the sky or overhanging tree canopy for enemies while probing the ground for food with its long bill. The eyes of bitterns, which are long-billed birds, are directed somewhat "under their chins," so that they can see food on the ground when the bill is pointed ahead, and look ahead when the bill is pointed upward.

Binocular vision, in addition to providing better distance-determination, allows a keener perception of depth and solidity. In other words, binocular vision enhances **stereopsis**—gives a stronger third-dimensional effect. Many birds with limited binocularity and wide monocularity have developed actions that serve as substitutes for stereopsis. For example, the Common Pigeon in walking, or the American Coot *(Fulica americana)* in swimming, thrusts the head forward, then jerks it backward, with brief pauses between each motion; meanwhile the body moves forward steadily, and thus, presumably, the eyes do not go backward through space. With each pause, the bird gets a good lateral view; with each head-motion, vision is temporarily suppressed. The result is a quick succession of views at rapidly changing angles and distances that gives the bird a better estimation of relief or spatial relationships in its immediate environment. The bobbing of the foreparts, including the head, of the Killdeer *(Charadrius vociferus)*, and similar motions in other shorebirds may produce the same effect. Another substitute for stereopsis is rapid peering (Grinnell, 1921). Thus, when birds are about to pick up motionless food, such as seeds, they frequently tilt or cock their heads at different angles to get a better notion of the shape of the food—perhaps necessary for identification—before seizing it.

Retinal Cells, Their Function and Distribution The retina, which is the bird's all-important cellular layer for receiving visual stimuli and transferring them to the central nervous system, contains the receptor cells called **rods** and **cones.** The rods are effective in dull light; cones function in bright light, give sharp visual details, and play a major role in color vision. In birds both rods and cones have oil-droplets and the rods have the pigment rhodopsin.

Diurnal birds have an abundance of cones and relatively few rods, whereas nocturnal birds have many more rods. The cones, in both diurnal and nocturnal birds, are sparingly distributed over most of the retina except in certain spots, called **foveas,** where the cones are highly concentrated to form points of sharpest vision. Some mammals have but one, the fovea centralis, which is centrally located in the retina; most birds have a central fovea and some have another, the temporal fovea, in the posterior part or temporal quadrant of the retina. The central fovea serves in monocular vision of birds by registering the lateral field, while the temporal assists in binocular vision by covering the field that lies ahead. This "visual trident" (three simultaneous views, two lateral and one forward) is well developed in hawks and other fast-flying, predaceous birds and allows wide views of the field as a whole and concentrated, distance-discriminating views of the field into which the birds are flying. Paradoxically, owls, which would seemingly need the trident, have only the temporal fovea for forward, binocular vision. The lack of the central fovea is, however, satisfactorily compensated for by the ability of the birds to sweep an extraordinarily wide area with their eyes by turning their heads.

Visual Acuity Bird eyes have a remarkable visual acuity or resolving power—the capacity to form distinguishable, unblurred images of objects as they become smaller and come closer together. The large size of the eyes and the big images they cast partly account for this characteristic; other factors are the high concentration of visual cells and the high ratio of nerve fibers to the cells. In the human fovea there are only 200,000 visual cells per square millimeter, but in the House Sparrow *(Passer domesticus)* the number is presumably 400,000. The hawks have the highest known concentrations. In the Buzzard *(Buteo buteo)*, whose eye is about the size of a man's, but whose visual acuity seems to be many times greater than man's, the fovea has about a million per square millimeter.

Nocturnal Vision The abundance of rods in eyes of nocturnal birds is not unexpected; however, there may also be considerable numbers of cones, even in the eyes of owls. The importance of rods in nocturnal vision lies in their particular sensitivity to dim light through the presence in their outer parts of a purple-red pigment called **rhodopsin** ("visual purple"). Rhodopsin, containing a protein plus a derivative of vitamin A, breaks down (bleaches out) in the presence of bright light into its two components, reducing or destroying the responsiveness of the retina to light waves; but in the presence of dim light or darkness the components recombine by synthesis, increasing retinal sensitivity. If a bird moves suddenly from a bright light to a dim light, there is a brief period of blindness while the rhodopsin resynthesizes. It hardly need be said that the greater the amount of rhodopsin in the rods, the greater their sensitiveness to light.

Nocturnality in birds is possible not only because of the presence of rhodopsin in the retina but also because of the dilation of the pupils, giving the retina access to as much light as possible.

Color Vision Careful work by Watson (1915) and later by Lashley (1916) revealed that color vision in the Domestic Fowl is, as in man, **trichromatic**—i.e., the retina is sensitive to mixtures of light waves or wave lengths from the middle to both sides of the spectrum. The longest and slowest light waves give the sensation of red, the shorter and faster waves give the sensations of orange, yellow, green, blue, and violet. Probably most diurnal birds have trichromatic vision, but subject to adaptive modifications in different species with respect to the distribution of oil-droplets in the retina. Oil-droplets (they are not present in man and the other higher mammals) are of two types, the colorless and the colored. The colored are red, orange, and yellow; all act as color filters by cutting out violet and blue light, thereby reducing chromatic aberration and glare, but the red droplets are more effective filters than orange, and the orange more than yellow. About 60 percent of the droplets in a kingfisher are red, a presumed adaptation for reducing glare on the water when looking for food below the surface. Red droplets are known to make red more distinct. On the basis of this fact, it is tempting to assume that an abundance of red droplets occurs in the retina of those species of hummingbirds that are particularly attracted to red flowers. Passerine birds, with an average of 20 percent red droplets, and hawks, with about 10 percent, probably have a color perception that is more like man's. Strongly nocturnal birds, such as owls, are believed to have achromatic vision in which the retina possesses merely colorless or faintly pigmented droplets and is sensitive only to black, gray, and white.

The Pecten This conspicuous structure of the avian eye, projecting into the lower half of the posterior chamber from the head of the optic nerve, has long aroused much speculation as to its function. It is now assumed to have a role in the nutrition of the retina and in maintaining intraocular pressure (Seaman and Himelfarb, 1963; Brachi, 1977). It is pigmented, highly vascular (consists mostly of minute blood vessels), and undoubtedly a derivative of the retina. In reptiles it is a simple cone.

Hearing

The avian ear, like the mammalian ear, is sensitive to a wide range of sounds, or sound waves—i.e, vibratory disturbances in the air. The number of vibrations—very often called cycles—per second is referred to as the frequency and establishes the pitch. A sound of high frequency, or a great many vibrations per second, produces a high pitch, just as a sound of low frequency causes a low pitch.

The hearing of birds, as shown by data gathered by Schwartzkopff (1955), ranges from 40 cycles per second (c.p.s.) in the Budgerigar *(Melopsittacus undulatus)* to 29,000 c.p.s. in the Chaffinch *(Fringilla coelebs)*. The upper limit, however, seldom exceeds 20,000 c.p.s. Brand and Kellogg (1939) found that the European Starling *(Sturnus vulgaris)* has a hearing range of 700 to 15,000 c.p.s.; the House Sparrow 675 to 11,500; and the Common Pigeon 200 to 7,500. All the birds became gradually less sensitive to sounds as the extremes in frequencies were approached. Comparing the hearing of these birds to that of man, who has a hearing range of about 16 to 17,000 c.p.s. (roughly nine octaves), man can hear about four octaves lower than the

Common Pigeon and five lower than the European Starling and House Sparrow; and he can hear as high as the European Starling and House Sparrow and about an octave higher than the Common Pigeon. Owls show an exceptionally wide hearing range. The Great Horned Owl *(Bubo virginianus)* can respond to sounds as low as 60 c.p.s. (Edwards, 1943), while the Common Barn-Owl *(Tyto alba)*, which catches mice in total darkness by detecting their highly pitched squeaks, is sensitive to sounds above 8,500 c.p.s. (Payne, 1962).

For a critical review of the findings on the hearing of birds by different authors, see Schwartzkopff, 1968.

Bang, B.G.
1960 Anatomical evidence for olfactory function in some species of birds. *Nature* 188:547–549. (In the Turkey Vulture, *Cathartes aura*: Oilbird, *Steatornis caripensis*; Laysan and Black-footed Albatrosses, *Diomedea immutabilis* and D. *nigripes*.)

Bang, B.G., and Cobb, S.
1968 The size of the olfactory bulb in 108 species of birds. *Auk* 85:55–61.

Bartholomew, G.A., and MacMillen, R.E.
1961 Water economy of the California Quail and its use of sea water. *Auk* 78:505–514. (Change in salinity shows no significant change in fluid consumption. When sufficiently dehydrated, the species will drink even 70 percent sea water.)

Brachi, V.
1977 The functional significance of the avian pecten: A review. *Condor* 79:321–327.

Brand, A.R., and Kellogg, P.P.
1939 Auditory responses of Starlings, English Sparrows, and Domestic Pigeons. *Wilson Bull.* 51:38–41.

Calvin, A.D.; Williams, C.M.; and Westmoreland, N.
1957 Olfactory sensitivity in the Domestic Pigeon. *Amer. Jour. Physiol.* 188:255–256. (The species cannot learn to detect odor alone.)

Chapman, F.M.
1929 *My tropical air castle: Nature studies in Panama*. New York: D. Appleton.
1938 *Life in an air castle: Nature studies in the tropics*. New York: D. Appleton-Century.

Edwards, E.P.
1943 Hearing ranges of four species of birds. *Auk* 60:239–241.

Frings, H., and Boyd, W.A.
1952 Evidence for olfactory discrimination by the Bobwhite Quail. *Amer. Midland Nat.* 48:181–184.

Goldsmith, K.M., and Goldsmith, T.M.
1982 Sense of smell in the Black-chinned Hummingbird. *Condor* 84:237–238.

Grinnell, J.
1921 The principle of rapid peering in birds. *Univ. Calif. Chron*. 23:392–396.

Grubb, T.C., Jr.
1972 Smell and foraging in shearwaters and petrels. *Nature* 237:404–405.

Hamrum, C.L.
1953 Experiments on the senses of taste and smell in the Bob-white Quail *(Colinus virginianus virginianus)*. *Amer. Midland Nat.* 49:872–877.

Hutchison, L.V., and Wenzel, B.M.
1980 Olfactory guidance in foraging by Procellariiforms. *Condor* 82:314–319.

Lashely, K.S.
1916 Color vision in chickens: The spectrum of the Domestic Fowl. *Jour. Animal Psychol.* 6:1–26.

Michelsen, W.J.
1959 Procedure for studying olfactory discrimination in pigeons. *Science* 130 (3376):630–631. (Pigeons can discriminate by an olfactory sense.)

Moore, C.A., and Elliot, R.
1946 Numerical and regional distribution of taste buds on the tongue of the bird. *Jour. Comp. Neurol*. 84:119–131.

Owre, O.T., and Northington, P.O.
1961 Indication of the sense of smell in the Turkey Vulture, *Cathartes aura* (Linnaeus), from feeding tests. *Amer. Midland Nat.* 66:200–205.

Payne, R.S.
1962 How the Barn Owl locates prey by hearing. *Living Bird* 1:151–159. (The owl orients the head in such a way as to hear all frequencies, audible to it in a complex sound, at maximum intensity in both ears. When it has achieved such an orientation, it will automatically be facing the source of the sound with a theoretical accuracy of less than one degree.)

Portmann, A.
1950 Les organes des sens. In *Traité de zoologie,* ed. by P.-P. Grassé. Vol. 15. Paris: Masson et Cie.

Pumphrey, R.J.
1961 Sensory organs: Vision and hearing. In *Biology and comparative physiology of birds*. Vol. 2, ed. by A.J. Marshall. New York: Academic Press.

Schwartzkopff, J.
1955 On the hearing of birds. *Auk* 72:340–347.
1963 Morphological and physiological properties of

the auditory system in birds. *Proc. XIIIth Internatl. Ornith. Congr.* pp. 1059–1068.
1968 Structure and function of the ear and of the auditory brain areas in birds. In *Hearing mechanisms in vertebrates,* ed. by A.V.S. de Reuck and J. Knight. Boston: Little, Brown.

Seaman, A.R., and Himelfarb, T.M.
1963 Correlated ultrafine structural changes in avian pecten oculi and ciliary body of *Gallus domesticus*. *Amer. Jour. Ophthalmol.* 56:278–296.

Stager, K.E.
1964 The role of olfaction in food location by the Turkey Vulture *(Cathartes aura). Los Angeles County Mus., Contrib. in Sci.*, No. 81. (From the evidence it is concluded that, among the cathartine vultures, the Turkey Vulture "possesses and utilizes a well-developed olfactory food-locating mechanism.")

Strong, R.M.
1911 On the olfactory organs and the sense of smell in birds. *Jour. Morph.* 22:619–658.

Walls, G.L.
1942 The vertebrate eye and its adaptive radiation. *Cranbrook Inst. Sci. Bull.* No. 19. Bloomfield Hills, Michigan.

Watson, J.B.
1915 Studies on the spectral sensitivity of birds. *Papers Dept. Marine Biol., Carnegie Inst. Washington* 7:87–104.

Wood, C.A.
1917 *The fundus oculi of birds, especially as viewed by the ophthalmoscope*. Chicago: Lakeside Press. (A classic work on the subject.)

The Brain

Relatively few studies have been made on the avian brain. However, the following works introduce the available literature.

Goodman, I.J., and Schein, M.W., eds.
1974 *Birds: Brain and behavior.* New York: Academic Press. (A collection of papers on avian neurobehavioral research.)

Pearson, R.
1972 The avian brain. New York: Academic Press.

Portmann, A.
1946–1947 Études sur la cérébralisation chez les oiseaux. I-III. *Alauda* 14:2–20; 15:2–15, 161–171. (Research on the relative degree of development of the brain in birds; a rather unusual approach to the study of avian evolution.)

Portmann, A., and Stingelin, W.
1961 The central nervous system. In *Biology and comparative physiology of birds*. Vol. 2, ed. A.J. Marshall. New York: Academic Press.

Sutter, E.
1951 Growth and differentiation of the brain in nidifugous and nidicolous birds. *Proc. Xth Internatl. Ornith. Congr.* pp. 636–644. (Concerned chiefly with postembryonic development.)

Systematics and Taxonomy

Systematics and taxonomy have often been used interchangeably, but some authors such as Mayr (1969) differentiate them as follows: *systematics* is the scientific study of the diversity of organisms and their relationships; *taxonomy* is the theory and practice of classifying organisms into *taxa* (singular *taxon*). Taxa are organized into nested groups. Thus the class Aves (birds) include a number of orders, each of which is characterized by attributes held in common; orders include one or more families; families include one or more genera; genera include one or more species. The inventory of the world's species of birds is probably more nearly complete than that of any other group of animals.

There are recognized today about 9,000 species of birds, and it is estimated that less than one percent of the total number of species still remain unknown. It was long believed that the classification of birds was nearly as definitive as the inventory. In recent years, however, new evidence from a variety of sources—fossils, anatomical studies, comparative behavior, and biochemistry—has shown that some of the long-held ideas of bird relationships must be reexamined. The classification of birds in this book, or any other classification of birds, must be considered tentative.

The Species

The **species** (plural also **species**) is one of the several taxa into which all organisms are classified. In ornithology—and probably all other branches of biology—it is by far the most important taxon, because student and practicing scientist alike work at one time or another with species. No person can hope to be an ornithologist without knowing different species of birds, though one may succeed without a knowledge of other categories, such as genera and subspecies. In ornithology an understanding of the nature of species is a prerequisite.

A species may be defined as a population, or populations, of mutually fertile individuals, reproductively isolated from individuals of other populations and possessing in common certain characters that distinguish them from any other similar population, or populations. If cross-breeding of two species occurs, the offspring are often sterile.

Species, like other categories into which organisms are classified, are distinguished by combinations of inherent peculiarities called **taxonomic characters.** In avian systematics some of the commonly used taxonomic characters for species are morphological, such as those having to do with minor details of size, shape, and color; other characters are ecological requirements (e.g., type of niche), reproductive traits (kind of song, type of nest) and general behavior patterns. When learning to identify different species of birds, be inclined at first to pay special attention to morphological characters, particularly size, form, and color. Gradually it will become evident, however, that morphological characters alone are not always helpful, especially in the field where species are sometimes better recognized by the way they sing or the way they behave than by their appearance. Therefore be prepared to identify species by characters other than morphological.

Subspecies

Any species shows variation among the individuals comprising it. When a species is spread over a wide area, the variations seldom, if ever, occur uniformly, but tend to be grouped in local populations. A widespread species, then, is actually comprised of numerous local populations, each one with a variable combination of characters by which it differs from all other populations in the species. A local population with a combination of characters making it sufficiently distinct from other populations is called a **subspecies** (sometimes called a **race**). The species comprising these subspecific divisions is **polytypic.** On the other hand, local populations, not sufficiently distinct, have no categorical designation and are considered **monotypic.** Some examples of monotypic species are the American Woodcock *(Scolopax minor),* Ruby-throated Hummingbird *(Archilochus colubris),* Pinyon Jay *(Gymnorhinus cyanocephalus),* and Bobolink *(Dolichonyx oryzivorus).*

A subspecies may be defined as a geographically limited population whose members possess in common certain taxonomic characters that distinguish them from all other populations in the species. All the subspecies of a species are mutually fertile; hence interbreeding occurs where two or more subspecies meet. Any one subspecies usually consists of a group of local populations differing slightly from one another unless it is confined to a small, isolated area, such as an island, in which case the characters may be rather uniform among all the individuals.

Subspecies normally replace one another geographically; their ranges adjoin and frequently produce a zone where individuals show marked intergradation through interbreeding. If the subspecies replace one another over a wide area, a progressive change in certain taxonomic characters may be evident, often correlated with a progressive change in one or more environmental factors, such as climate. This kind of character gradient is referred to as a **cline.**

Clines, occurring as they do with a gradual change in environment, probably represent adaptive change. Sometimes the succession of changes producing a cline conforms so closely in different species that systematists have established several so-called rules, examples of which follow:

Bergmann's Rule Body size tends to be larger in cooler climates, smaller in warmer climates.

Gloger's Rule Coloration tends to be darker in humid climates, lighter in arid climates. An increase in melanins produces the darker coloration.

Allen's Rule Bills, tails, and other extensions of the body tend to be longer in warmer climates, shorter in cooler climates.

The differences between subspecies are often average differences, and single individuals may not always be identifiable to subspecies. To an experienced taxonomist, the morphological differences between one subspecies population and another are soon apparent, but to the unpracticed eye or to an ornithologist not engaged in systematics, most subspecies "look alike." This is to be expected. Unfortunately, because the subspecies is an established category in systematic ornithology, many students and ornithologists feel that, in order to be accurate, they must try to recognize subspecies, rather than just species, in their laboratory and field work. If they fail, as is often the case, they "bluff" their identifications by using the available information on the ranges that the subspecies are supposed to occupy. Such a procedure is wholly unscientific and is in no sense a contribution to knowledge. In the foregoing paragraphs the statement was made that the species is the most important category. It should be emphasized here that the subspecies is a category of special interest to systematists and those scientists engaged in the study of speciation (discussion following). The beginning student and most ornithologists should not expect to identify subspecies and should never pretend to except by proper scientific analysis.

Speciation

As indicated above, a widespread species is usually divided into many local populations, and each one has its own variable combination of characters. Ordinarily there is little outbreeding in any one population. With the passage of time a population may become split by some geographical factor into two wholly or partially isolated populations, and each of these populations may build up distinct characters. Eventually the distinctions may become sharp enough to designate each population as a subspecies. If the isolating factor continues to operate, further distinctions may accumulate and establish a physiological or behavioral barrier that eliminates any further possibility of interbreeding. The two populations are then reproductively isolated from one another and each may be termed a species.

If the two species continue to be separate geographically, even though their ranges may be contiguous, they are **allopatric.** If, however, the two species eventually come together so that their breeding ranges merge and even overlap entirely, they are **sympatric.** But they will not ordinarily interbreed since they are now so distinctive in one or more ways—morphologically, physiologically, behaviorally (including voice), or ecologically (including choice of habitat)—as to be reproductively isolated.

In southeastern Canada and northeastern United States sympatry is illustrated among four species of small flycatchers of the genus *Empidonax*—the Least Flycatcher *(E. minimus)*, Willow Flycatcher *(E. traillii)*, Alder Flycatcher *(E. alnorum)*, and Yellow-bellied Flycatcher *(E. flaviventris)*. All four appear very much alike, yet they are prevented from interbreeding by virtue of their different vocalizations as well as by their habitats—the Least preferring park-like woodlands; the Willow and Alder, habitats suggested by their names; and the Yellow-bellied, deeply shaded boreal forests of conifers. The Eastern and Western Meadowlarks. *(Sturnella magna* and *S. neglecta)* exhibit sympatry in the grasslands of midwestern North America. Here their ranges meet and overlap broadly, yet the two species do not normally interbreed since not only their songs and calls differ markedly, but also their habitats— the Eastern Meadowlark generally occupying the more moist lowlands and the Western the drier uplands.

Occasionally, when two closely related species become sympatric with their ranges partly overlapping, they are apt to differ more sharply from each other where their ranges overlap than where their ranges are still separate. Called **character displacement,** this phenomenon in which two species may so rapidly diverge possibly reduces the chances of the two species interbreeding and/or minimizes competition when the two species are together. Parkes (1965) explained character displacement at some length and demonstrated it in two species of cuckoos inhabiting the Philippine archipelago.

Species formation, or **speciation,** is the result of splitting of an evolutionary (phyletic) line over a prolonged period of time. According to this concept, local populations are incipient subspecies, and isolated (not clinal) subsepecies are incipient species. For information on some of the geographical and ecological factors affecting isolation, see "Distribution," pages 180 and 174.

One should bear in mind that speciation is a continuous evolutionary process going on today as in the past—and much more rapidly than once supposed. A good example is the rate at which adaptive differentiation has occurred in the House Sparrow *(Passer domesticus)* since its introduction to North America from England and Germany in 1851. The species breeds now across the continent in extremes of environment to which the original introduced stock was never accustomed; and, like many American polytypic species, it has adapted itself clinally in size, color, and length of body extensions in accordance, respectively, with Bergmann's, Gloger's, and Allen's rules (Johnston and Selander, 1964; Packard, 1967). These adaptations have evolved in no more and probably fewer than 90 generations or—assuming one generation a year—90 years (Packard, 1967).

Ornithologists have usually assumed that geographical differences in plumage and in size and shape have a genetic basis and are the result of natural selection for types that are best adapted to their respective environments. But recent experiments with Red-winged Blackbirds *(Agelaius phoeniceus)* suggest that a substantial amount of clinal variation in shape may be induced directly by the environment. James (1983) and her students transplanted eggs of blackbirds between nests in northern and southern Florida and from nests in Colorado to nests in Minnesota. In each case the shape of nestlings in the foster nests showed similarities to the shape of unmanipulated nestlings in the foster population. The characters that changed most—bill shape, wing length/tarsus length—were the characters that best distinguished the normal populations of both nestlings and adults.

In order to determine the taxonomic rank of

closely allied populations of birds and their mode of evolution, the systematist must analyze the morphological characters of many hundreds of representative specimens and make detailed observations in the field on ecological preferences, reproductive traits, and other habits. Any study of speciation involves many complex problems.

One of the problems is that of deciding whether certain populations deserve subspecific rank. Criteria vary. For example, if a systematist finds that 75 percent of the specimens examined in one population, or an assemblage of populations, of a species are distinctly separate from all specimens of other populations of the species, the systematist may consider it a subspecies. This is the application of the **75 percent rule.** See Amadon (1949) for a discussion of the rule and an example of its application.

Another problem is that of deciding whether certain populations are subspecies of a polytypic species and therefore **conspecific** or are distinct allopatric species. The answer lies largely in what the systematist can learn about their origin and their degree of isolation. If the systematist finds evidence that certain allopatric species were once subspecies of a single species but have since achieved sufficient differentiation and isolation to be accorded the status of species, they may be designated **allospecies** and grouped under the heading **superspecies.** Consult Amadon (1966) for details. The superspecies is not an established category in the system of classification. It is merely a convenient term to express the concept that a group of species are very closely related, yet not closely enough to be given subspecific rank. Some examples of superspecies are the Great-tailed and Boat-tailed Grackles *(Quiscalus mexicanus* and *Q. major),* the Golden-fronted and Red-bellied Woodpeckers *(Melanerpes aurifrons* and *M. carolinus),* the Barred and Spotted Owls *(Strix varia* and *S. occidentalis),* and the Snowy and Little Egrets *(Egretta thula* and *E. garzetta).* Perhaps as many as a third of the bird species in the world may be components of superspecies.

Still another example of speciation problems is that of evaluating the occurrence of **hybrids** and **hybridization.** Hybrids result from the crossing of two taxonomically unlike individuals. **Interspecific hybrids** are produced by the crossing of individuals of two different species. Such hybrids are usually designated by writing the names of the two parent species one after the other with a cross between. For example: *Passerina cyanea* × *Passerina amoena*. Hybridization commonly means, in systematics, the crossing between individuals of two different species.

Various isolating mechanisms ordinarily prevent extensive hybridization between species though sometimes two species may hybridize to a limited extent. In Saskatchewan, Pettingill (Sibley and Pettingill, 1955) collected a hybrid between the Chestnut-collared and McCown's Longspurs *(Calcarius ornatus* and *C. mccownii),* two sympatric species normally separated by their habitat, the Chestnut-collared preferring the long-grass prairie, the McCown's the short-grass. Hybridization of this sort between sympatric species is unusual. Much more common is hybridization between allopatric species in areas where their ranges adjoin. A well-known case is that of two essentially allopatric species, the Blue-winged and Golden-winged Warblers *(Vermivora pinus* and *V. chrysoptera).* The Blue-winged Warbler has a more southern breeding range, and the Golden-winged a more northern, but their ranges overlap extensively (see Parkes, 1951)—hence they are not typical allopatric species. In the zone of overlap the hybrids sort genetically into two distinct types, the so-called Brewster's Warbler and the Lawrence's Warbler. The hybrids, however, are relatively few, considering the size of the parent populations, and there is no combining of characters of the two species to form a true hybrid population.

In the western Great Plains of North America the ranges of several eastern species—e.g., the Indigo Bunting *(Passerina cyanea)* and Rose-breasted Grosbeak *(Pheucticus ludovicianus)*—meet the ranges of their western allopatric counterparts, the Lazuli Bunting *(Passerina amoena)* and Black-headed Grosbeak *(Pheucticus melanocephalus).* These species pairs interbreed rather regularly where their ranges are in contact. Some systematists believe that such hybridization indicates that the parent populations are subspecies of a polytypic species rather than separate species. However, in these examples the hybridization has not led to extensive hybrid zones where hybrid individuals are at least as common as the parental types; there appears to be at least a partial barrier to full genetic interchange. In some other eastern and western pairs that were formerly considered separate species—e.g., the Yellow-shafted Flicker *(Colaptes auratus)* and Red-shafted Flicker *(C. cafer)*, and the Baltimore Oriole *(Icterus galbula)* and Bullock's Oriole *(I. bullockii)*—the hybridization is so extensive that in some areas it is almost impossible to find an individual of a pure parental type. These pairs are now considered to represent polytypic species, the Northern Flicker and Northern Oriole, which were given new English names because it would be inappropriate to designate the entire polytypic species by a name formerly used for only one of its distinctive components. The presence of an occasional hybrid, as in the case of the Chestnut-collared Longspur × McCown's Longspur, or a few hybrids as in the case of the Blue-winged Warbler × Golden-winged Warbler, is not sufficient evidence to prove that the parent stocks are subspecies of a single species. In some groups, such as the waterfowl (Anatidae) and the wood warblers (Parulinae), even members of distinctive genera may occasionally pick the wrong mate and hybridize. Examples are known of hybrids between the Common Goldeneye *(Bucephala clangula)* and Hooded Merganser *(Lophodytes cucullatus)*, and the Blue-winged Warbler *(Vermivora pinus)* and Kentucky Warbler *(Oporornis formosus)*. Offspring of such "reproductive accidents" are almost certainly infertile.

Many papers deal with hybridization. Examples include description of single hybrid specimens (Sprunt, 1954); field observations of a single hybrid (Wells and Baptista, 1979); detailed analyses of a hybrid zone (Sibley and West, 1959); and analyses of the significance of hybridization (Sibley, 1957). Other recommended papers are Gill (1980), Barrowclough (1980), Parkes (1961, 1978), Short (1963, 1965, 1969), Sibley (1950, 1954), Sibley and Short (1959), and West (1962).

Some Examples of Bird Speciation

Speciation, as already explained, is dependent on the isolation of populations long enough for them to build up distinctive characters that will insure a reproductive barrier. Oceanic archipelagos, remote from continents with many small islands separated by channels, often provide adequate isolation for terrestrial animals and consequently make ideal laboratories for the study of speciation. This is the case with the Galapagos and Hawaiian Islands.

Rising from the Pacific Ocean some 600 miles west of Ecuador, the Galapagos Islands are of volcanic origin with no history of a connection to South America. How long they remained unexploited by plants and animals, nobody knows. In due course—and quite by chance—terrestrial life arrived and, stranded, proceeded to adapt to the peculiarities of an unused environment. The few bird species existing on the Galapagos, when discovered by man, included Darwin's finches, so called because they were discovered by Charles Darwin during the historic voyage of "The Beagle."

Darwin's finches, numbering 14 species (Lack, 1947), probably stemmed from a finch-like form that arrived from South America and established a population on one of the islands. From this population, emigrants reached other islands where they in turn established populations. In time, as these populations adapted themselves to the new and unoccupied environments, they evolved distinctions as subspecies and eventually species. This diverging or branching out from one species into several to suit different ecological niches that may include different types of vegetation is a good example of **adaptive radiation,** a common evolutionary process. But the story does not end here. Having evolved on certain islands, these new finch species reached other islands where they entered into sympatry with other new finch species and proceeded to compete with them for habitat and food. This sort of rivalry between sympatric species intensified or reinforced their distinctions. Today, each of the principal Galapagos Islands has at least three finch species and some have as many as ten. In summary, the formation of Darwin's finches is the result, first, of initial isolation of populations on separate islands and, second, of competition between species when sympatry occurred.

Measuring four to eight inches (10.16–20.32 cm) in length, Darwin's finches are generally drab in appearance; both sexes are colored alike in browns and grays and occasionally black. Their primary distinctions are in the bill, which varies for different kinds of foraging (Grant, 1981; Bowman, 1961; Lack, 1945, 1947). One type of bill persists finch-like for seed-eating; other types range from long and down-curved for obtaining nectar from cactus flowers and parrot-like for fruit-eating to small and somewhat chickadee-like or even warbler-like for insect-eating. One species, the Woodpecker Finch *(Cactospiza pallida),* has a bill suited to excavating in wood and to manipulating a cactus spine as a probe for extracting insects from holes and crevices (see "Behavior," p. 217).

Like the Galapagos, the Hawaiian Islands in the north-central Pacific are of volcanic origin and even more remote from continental life. Here a comparable development by adaptive radiation took place among the honeycreepers, an endemic subfamily (Drepanidinae). See Berger (1981) and Raikow (1976). The colonizing form is now thought to have been a cardueline finch, either from North America or from northern Asia (Sibley and Ahlquist, 1982). From this evolved as many as 43 species, of which nine became extinct in prehistoric times and 15 are known from prehistoric fossils (Olson and James, 1982). Although similar in size to Darwin's finches, adaptive radiation proceeded further, producing not only more species but greater distinctions in color and more pronounced extremes in bill shapes. Some species are predominantly green or yellow; others are bright red or black; one species is mainly black, with orange and white streaks, and features a crest. The sexes may or may not be alike. In some species the bill is long, thin, and remarkably down-curved for extracting nectar and taking insects from flowers; in other species it is grosbeak-like for seed- and fruit-crushing, or woodpecker-like for obtaining insects from bark and wood.

The necessary isolation for speciation on the continents is not as obvious as in oceanic archipelagos. Nevertheless, widely separated physiographic features such as high mountains or mountain ranges or widely spaced areas with sharp climatic differences can be veritable islands, providing adequate isolation.

A good example of speciation in a wide-ranging continental group of birds is among the juncos (genus *Junco*) of North America. After studying many hundreds of representative specimens and observing many forms—species and subspecies—in their natural environment, Miller (1941) found in Mexico and Central America that some forms resulted from populations restricted to different mountains where there was no opportunity for interbreeding. Farther north, where the range of juncos is less interrupted by physiographic features, he noted that different climatic conditions, correlated with particular geographical areas, served as isolating factors. If different forms of juncos came in contact, as was sometimes the case, each tended to remain attached to its own habitat. When interbreeding occurred, hybridization rather than intergradation resulted, provided the distinctive characters had become stabilized. The hybrids were different from their parents and formed a distinctive and self-perpetuating population.

Speciation in the large subfamily of American wood warblers (Parulinae) offers a fruitful field for speculation. Probably the family originated in the North American tropics during the Miocene (see the Geologic Time Scale, p. 352) and by the early Pliocene early forms were well established in the temperate ancestral deciduous forest of eastern North America. Mengel (1964) has postulated that speciation leading to the vast array of present-day parulines began with the advent of the Pleistocene or Ice Age. His hypothesis follows briefly. Paruline forms in the temperate ancestral deciduous forest in eastern North America developed adaptations to the northern coniferous (boreal) forest when it was forced deep into the southeast by the first glacial advance. Upon glacial recession a transcontinental coniferous forest formed in the wake of the retreating ice and was soon occupied by the newly formed parulines. When the next glacial advance separated the continent-spanning coniferous forest into eastern and western parts, the parulines separated into corresponding eastern and western populations, each developing its own distinctions. In the west the process of further separation, isolation, and adaptive radiation continued among the parulines through the warm interglacial period for at this time the birds were forced into the "islands" of coniferous forest still remaining high on mountain slopes. Repetition of the process in subsequent glacial cycles— there were four altogether in the Pleistocene—very likely completed the formation of the present-day western parulines. Most of the present-day eastern species seem to have descended directly from the original paruline stock in the eastern deciduous forest.

The Higher Categories

All species are classified into higher taxa on the basis of presumed "blood" relationships. This is a "natural" system because it relies on **phylogeny**—the evolution of related groups of organisms. For the beginning student, as well as the biologist, it brings order into an otherwise confusing array of species; at the same time, it provides a means of showing species relationships.

Frequently, some relationships are obscured by **divergence,** others by **convergence.** Two closely related groups of species may in outward appearances seem distantly related owing to their having diverged greatly in their structural and behavioral adaptations; on the other hand, two distantly related groups may have developed very similar adaptations, thereby converging in their outward appearances. Swifts, for example, seem more closely related to swallows than to hummingbirds. Yet such is not the case. Swifts and hummingbirds presumably diverged from common ancestry. Swifts and swallows, which have very different ancestry, have converged in form and habits essential for the aerial pursuit of insects.

No single study can provide a sufficient basis for establishing relationships. Various studies are necessary to meet and solve the problems posed by divergence and convergence.

Of paramount importance are studies of anatomy, particularly the more stable features of the skeleton and other organ systems; basic reproductive traits and behavior patterns; and geographic ranges and habitat preferences. See Brodkorb (1968) for many of the taxonomic characters denoting the higher categories of birds.

Useful in determining taxonomic relationships of birds is a study of their external parasites, most notably lice of the insect order Mallophaga (see Clay, 1951). As a rule, mallophagans are host-specific, each species being restricted and adapted to a single bird species or groups of closely related bird species. With little opportunity to parasitize other bird species, since different species of birds rarely contact one another enough to allow any significant interchange of their ectoparasites, mallophagans have evolved along with their hosts. Thus the mallophagans are themselves as closely related as their hosts and consequently provide clues to the phylogeny of their hosts.

More recently a study of bird proteins, which are genetically controlled and therefore specific to a given group of birds, has proven highly useful in determining taxonomic relationships. By a fractionating process known as electrophoresis, carried out in a special laboratory apparatus, one may analyze and compare, for example, the egg–white proteins of different birds by observing their electrophoretic "profiles." Frequently the pictures corroborate morphological evidence of phylogenetic relationships and in some instances show relationships not previously detected. For further information, see Barrowclough and Corbin (1978); Peakall (1962); Sibley (1960), (1967); Sibley and Ahlquist (1982); and Zink (1982).

The three higher taxa most commonly used in the classification of birds warrant discussion here; they are, in their line of ascending rank or hierarchy, the genus, the family, and the order.

Genus Embraces one or more species exhibiting a combination of taxonomic characters shared with no other taxon of the same rank. Some of the characters are morphological (e.g., the general color and details of shape and structure), some are ecological (the type of breeding habitat such as a forest, grassland, or shore), and some are behavioral (the type of displays and nesting habits). The identification of a polytypic genus (i.e., a genus containing two or more species) is a matter of determining the peculiarities common to different species, whereas the identification of a species is a problem of denoting distinctiveness between it and related species. In a few cases of monotypic genera the identification of the single species of each genus in turn identifies the genus. The genus is the only higher category whose name is part of a species' technical name (discussion following).

Family Embraces one genus, or two or more genera, exhibiting a combination of obvious morphological peculiarities such as the shape of the bill, presence of notches on the bill, the nature of the tarsal covering, and the number of primaries. See Storer (1960) for a review of world families that include some distinguishing external characteristics. Relatively few families contain only one genus. Usually a family has a large number of genera that show a variety of habitat preferences. The fact that peculiarities of a family are obvious is an indication that the families have diverged widely in the evolution from a common type. All family names end in **-idae,** making the categorical designation easy to recognize.

Order Embraces one or more families exhibiting peculiarities of the skeleton and other parts of the anatomy that have been least modified by adaptive change and are consequently more basic, or stable. Whereas a family is often confined in its distribution to a continent or neighboring continents, an order is often world-wide in its distribution. All names of bird orders end in **-iformes.**

When any category contains a large number of taxonomic groups from the category below, systematists have sometimes found it desirable to express the relationships of these groups in a more precise manner by using intermediate categories whose names bear super- or sub- as a prefix. If, for example, a particular family includes a large number of genera that seem to show a natural division into two or more groups, then the groups are considered subfamilies. Among the various intermediate categories, two have standard endings for their names, **-oidea** for superfamily and **-inae** for subfamily. In some large subfamilies the more closely allied genera are grouped into tribes, the name for each tribe with a standard ending, **-ini.** Below, in hierarchic order, are all the categories generally used in systematic ornithology. To their right is the complete classification of the common subspecies of Mallard *(Anas platyrhynchos platyrhynchos),* the eastern subspecies of Belted Kingfisher *(Ceryle alcyon alcyon),* the western subspecies of American Robin *(Turdus migratorius propinquus),* and the Wood Thrush *(Hylocichla mustelina),* a monotypic species. Note that certain intermediate categories are not always used.

A classification of the living (not fossil) birds of the world is given in the following pages. Categories are shown down to subfamilies. For the sake of simplification, most all the intermediate categories are omitted, as are all the families of birds not regularly represented in North America north of Mexico. (For coverage of all families of birds in the world, see Austin, 1971, and Harrison, 1978.)

Become familiar with this classification. Birds, whenever listed in modern guides and checklists, are grouped under these categories and, in American works, presented in sequence starting with the oldest—presumably most like the ancestral form—and ending with the most advanced—the least like the ancestral form. By learning the sequence of orders and the sequence of families and subfamilies within the orders, one can readily find a category without scanning an entire list or referring to an index.

CLASS				
Subclass	Neornithes	Neornithes	Neornithes	Neornithes
Superorder	Neognathae	Neognathae	Neognathae	Neognathae
Order	Anseriformes	Coraciiformes	Passeriformes	Passeriformes
Suborder	Anseres	Alcedines	Passeres	Passeres
Superfamily		Alcedinoidea		
Family	Anatidae	Alcedinidae	Muscicapidae	Muscicapidae
Subfamily	Anatinae	Cerylinae	Turdinae	Turdinae
Tribe	Anatini			
Genus	*Anas*	*Ceryle*	*Turdus*	*Hylocichla*
Subgenus	*Anas*	*Megaceryle*		
Species	*platyrhynchos*	*alcyon*	*migratorius*	*mustelina*
Subspecies	*platyrhynchos*	*alcyon*	*propinquus*	

- Class Aves, Birds
  - Subclass Neornithes, True Birds
    - Superorder Paleognathae, Tinamous and Ratites
      - Order TINAMIFORMES, Tinamous
      - RHEIFORMES, Rheas
      - STRUTHIONIFORMES, Ostriches
      - CASUARIIFORMES, Emus and Cassowaries
      - AEPYORNITHIFORMES, Elephant Birds (extinct)
      - DINORNITHIFORMES, Moas (extinct)
      - APTERYGIFORMES, Kiwis
    - Superorder Neognathae, Typical Birds
      - Order GAVIIFORMES, Loons
        - Family Gaviidae, Loons
      - Order PODICIPEDIFORMES, Grebes
        - Family Podicipedidae, Grebes
      - Order PROCELLARIIFORMES, Tube-nosed Swimmers
        - Family Diomedeidae, Albatrosses
        - Procellariidae, Petrels, Shearwaters, and Fulmars
        - Hydrobatidae, Storm-Petrels
      - Order SPHENISCIFORMES, Penguins
        - Family Spheniscidae, Penguins
      - Order PELECANIFORMES, Totipalmate Swimmers
        - Family Phaethontidae, Tropicbirds
        - Sulidae, Boobies and Gannets
        - Pelecanidae, Pelicans
        - Phalacrocoracidae, Cormorants
        - Anhingidae, Anhingas
        - Fregatidae, Frigatebirds
      - Order CICONIIFORMES, Herons, Ibises, Storks, and allies
        - Family Ardeidae, Bitterns and Herons
        - Threskiornithidae, Ibises and Spoonbills
        - Ciconiidae, Storks
      - Order PHOENICOPTERIFORMES, Flamingos
        - Family Phoenicopteridae, Flamingos
      - Order ANSERIFORMES, Screamers and Waterfowl
        - Family Anatidae, Waterfowl
          - Subfamily Anserinae, Whistling-Ducks, Geese, and Swans
          - Anatinae, Ducks
      - Order FALCONIFORMES
        - Family Cathartidae, New World Vultures
        - Accipitridae, Kites, Eagles, Hawks, and allies
          - Subfamily Pandioninae, Osprey
          - Accipitrinae, Kites, Eagles, Hawks, and allies
        - Family Falconidae, Caracaras and Falcons
      - Order GALLIFORMES, Gallinaceous Birds
        - Family Cracidae, Chachalacas, Guans, and Curassows
        - Family Phasianidae, Partridges, Pheasants, Grouse, Quail, and Turkeys
          - Subfamily Phasianinae, Partridges and Pheasants
          - Tetraoninae, Grouse
          - Meleagridinae, Turkeys
          - Odontophorinae, New World Quail
      - Order GRUIFORMES, Rails, Cranes, and allies
        - Family Rallidae, Rails, Gallinules, and Coots
        - Aramidae, Limpkins
        - Gruidae, Cranes
      - Order CHARADRIIFORMES, Shorebirds, Gulls, Auks, and allies
        - Family Charadriidae, Plovers and Lapwings
        - Haematopodidae, Oystercatchers
        - Recurvirostridae, Stilts and Avocets
        - Jacanidae, Jacanas
        - Scolopacidae, Sandpipers, Phalaropes, and allies
          - Subfamily Scolopacinae, Sandpipers and allies
          - Phalaropodinae, Phalaropes
        - Family Laridae, Skuas, Gulls, Terns, and Skimmers

Subfamily Stercorariinae, Jaegers and Skuas
Larinae, Gulls
Sterninae, Terns
Rynchopinae, Skimmers
Family Alcidae, Auks, Murres, Puffins, and allies

Order COLUMBIFORMES, Pigeons, Doves, and allies
Family Columbidae, Pigeons and Doves

Order PSITTACIFORMES, Parrots
Family Psittacidae, Parrots

Order CUCULIFORMES, Cuckoos and Turacos
Family Cuculidae, Cuckoos
Subfamily Coccyzinae, New World Cuckoos
Neomorphinae, Roadrunners and allies
Crotophaginae, Anis and allies

Order STRIGIFORMES, Owls
Family Tytonidae, Barn-Owls
Strigidae, Typical Owls

Order CAPRIMULGIFORMES, Goatsuckers, Oilbirds, and allies
Family Caprimulgidae, Goatsuckers
Subfamily Chordeilinae, Nighthawks
Caprimulginae, Nightjars

Order APODIFORMES, Swifts and Hummingbirds
Family Apodidae, Swifts
Subfamily Cypseloidinae, Collared Swifts and allies
Chaeturinae, Needletails and allies
Apodinae, Typical Swifts and allies
Family Trochilidae, Hummingbirds

Order COLIIFORMES, Colies

Order TROGONIFORMES, Trogons
Family Trogonidae, Trogons

Order CORACIIFORMES, Kingfishers, Rollers, Hornbills, and allies
Family Alcedinidae, Kingfishers

Order PICIFORMES, Woodpeckers, Barbets, Toucans, and allies
Family Picidae, Woodpeckers

Order PASSERIFORMES, Perching Birds
Family Tyrannidae, New World or Tyrant Flycatchers
Subfamily Elaeniinae, Elaenias, Tyrannulets, and allies
Fluvicolinae, Pewees, Phoebes, and allies
Tyranninae, Crested Flycatchers, Kingbirds, and allies
Tityrinae, Tityras and Becards
Family Alaudidae, Larks
Hirundinidae, Swallows
Corvidae, Jays, Magpies, and Crows
Paridae, Titmice
Remizidae, Verdin and allies
Aegithalidae, Bushtit and allies
Sittidae, Nuthatches
Certhiidae, Creepers
Pycnonotidae, Bulbuls (Introduced)
Troglodytidae, Wrens
Cinclidae, Dippers
Muscicapidae, "Old World Insecteaters"
Subfamily Sylviinae, Kinglets and Old World Warblers
Turdinae, Thrushes
Timaliinae, Babblers, Wrentit, and allies
Family Mimidae, Thrashers and Mockingbirds
Motacillidae, Wagtails and Pipits
Bombycillidae, Waxwings
Ptilogonatidae, Silky Flycatchers
Laniidae, Shrikes
Sturnidae, Starlings and Mynas (Introduced)
Vireonidae, Vireos
Emberizidae, New World Nine-primaried Songbirds

Subfamily Parulinae, Wood Warblers
Coerebinae, Bananaquit
Thraupinae, Tanagers and Honeycreepers
Cardinalinae, Cardinals, Grosbeaks, and allies
Emberizinae, Buntings, New World Sparrows, and allies
Icterinae, American Orioles and Blackbirds

Family Fringillidae, Cardueline Finches, Hawaiian Honeycreepers, and allies
Passeridae, Old World Sparrows and allies (Introduced)

The above classification, or any such classification, though relying on phylogeny, does not attempt to show the phylogeny of birds. That tinamous and ratite birds head the list in no sense implies that all the birds following evolved from them. The phylogeny of birds is best demonstrated by a tree. The trunk from base to top represents the lineage from the oldest birds to the most advanced and its major branches the sequence of orders as they are believed to have emerged.

Nomenclature

The system of giving technical names to birds and other animals in the United States and Canada conforms closely to the International Rules of Zoological Nomenclature, more familiarly known as the **International Code.** Power to act on the Rules is within the province of the International Commission on Zoological Nomenclature, which obtains its authority from the International Congresses of Zoology.

The technical (i.e., "scientific") name is a combination of two, sometimes three, Latin or latinized words. The first is the name of the genus in which the bird is placed and is always written with a capital letter. The second and third names are the names of the species and subspecies, respectively, and are never capitalized. The name of the author who first described and named a species or subspecies, is placed directly after the technical name with no intervening punctuation. Parentheses are placed around the author's name if the genus is now different from the one in which the author placed the species originally. When the technical name is written in longhand, or is typewritten, it must always be underscored; when printed, it must be italicized. The author's name, however, is never underscored or italicized. Following are several complete technical names and their authors, selected to illustrate certain nomenclatural procedures.

Anas platyrhynchos platyrhynchos Linnaeus
The common subspecies of Mallard in Europe, Asia, and North America is the ***nominate subspecies*** because it shares the species name given by Linnaeus. That subspecies of every polytypic species that has the earliest valid name is called nominate.

Ceryle alcyon alcyon (Linnaeus)
The subspecies of Belted Kingfisher in eastern North America is another nominate subspecies. In this case Linnaeus' name is in parentheses because he named and described the species originally in *Alcedo,* a different genus.

Turdus migratorius propinquus Ridgway
The subspecies of American Robin in western United States and parts of Mexico. Originally described and named by Ridgway as a species, *Turdus propinquus,* this form was later considered conspecific with *Turdus migratorius;* thus it was relegated to a subspecific status. Since the form still remains in the same genus, Ridgway's name is not placed in parentheses.

Hylocichla mustelina (Gmelin)
The Wood Thrush is a monotypic species. Gmelin originally assigned *mustelina* to the genus *Turdus* but later systematists considered it a species of *Hylocichla*. Consequently Gmelin's name belongs in parentheses.

Dendroica pensylvanica (Linnaeus)
The Chestnut-sided Warbler is another monotypic species. The name *pensylvanica* exemplifies two nomenclatural procedures:

1. The species name must remain uncapitalized even when based on a proper name.

2. The spelling of a species name must be preserved as given by the author even when the spelling of the name is erroneous or does not otherwise conform to standard usage.

The established classification and nomenclature of birds in North America is based on the *Check-list of North American Birds,* prepared by the Committee on Classification and Nomenclature of the American Ornithologists' Union. This is the standard work used by American ornithologists. The fifth edition (1957) included only the area north of Mexico, but the sixth edition (1983) expanded its coverage to include Middle America (from Mexico through Panama), the West Indies, and Hawaii. The sixth edition is a species list only. For subspecies, consult Peters (1931–79) and others.

Become fully acquainted with *"The A.O.U. Check-list,"* as it is familiarly called, and habitually rely upon it for the proper presentation of all technical and English (vernacular) names. Moreover, rely upon it when seeking the following information:

1. The categories into which birds are classified.

2. The sequence of categories.

3. The original authors and references to their original descriptions.

4. The **type localities**—the places of collection of the **type specimens** upon which the original descriptions were based.

5. The known ranges of species and subspecies.

For further information on nomenclatural procedures refer to the book by Mayr (1969) that includes detailed treatment of the various rules governing the naming of birds and other animals, the significance of type specimens, the methods in naming new species and subspecies, and related matters. Several lists of the species of birds of the world have been published, and all vary in amount of information for each species. Among the lists are those by Edwards (1974, 1982); Morony, Bock, and Farrand (1975); Howard and Moore (1980); Walters (1980); and Clements (1981).

References

Amadon, D.
1949 The seventy-five per cent rule for subspecies. *Condor* 51:250–258.
1966 The superspecies concept. *Systematic Zool.* 15:245–249.

American Ornithologists' Union
1957 *Check-list of North American birds.* 5th ed. Published by the Union.
1983 *Check-list of North American birds.* 6th ed. Published by the Union; available from Allen Press, P.O. Box 368, Lawrence, Kansas 66044.

Austin, O.L., Jr.
1971 *Families of birds.* New York: Golden Press. (Thumbnail sketches.)

Barrowclough, G.F.
1980 Genetic and phenotypic differentiation in a wood warbler (genus *Dendroica*) hybrid zone. *Auk* 97:655–668.

Barrowclough, G.F., and Corbin, K.W.
1978 Genetic variation and differentiation in the Parulidae. *Auk* 95:691–702.

Behle, W.H.
1950 Clines in the Yellow-throats of western North America. *Condor* 52:193–219. (A model study of clines in one species.)

Berger, A.J.
1981 *Hawaiian birdlife.* 2nd ed. Honolulu: Univ. Press of Hawaii.

Bowman, R.I.
1961 Morphological differentiation and adaptations in the Galapagos finches. *Univ. Calif. Publ. in Zool.* 58:i–viii; 1–326.

Brodkorb, P.
1968 Birds. In *Vertebrates of the United States.* By W.F. Blair, A.P. Blair, P. Brodkorb, F.R. Cagle, and G.A. Moore. 2nd ed. New York: McGraw-Hill.

Brown, W.L., Jr., and Wilson, E.O.
1956 Character displacement. *Systematic Zool.* 5:49–64. (The authors propose the term for the phenomenon and illustrate it.)

Clay, T.
1951 The Mallophaga as an aid to the classification of birds, with special reference to the structure of feathers. *Proc. Xth Internatl. Ornith. Congr.* pp. 207–215.

Clements, J.
1981 *Birds of the world: A checklist.* 3rd ed. New York: Facts on File, Inc.

Cockrum, E.L.
1952 A check-list and bibliography of hybrid birds in North America north of Mexico. *Wilson Bull.* 64:140–159.

Cory, C.B.; Hellmayr, C.E.; and Conover, B.
1918–1949 Catalogue of birds of the Americas. *Field Mus. Nat. Hist. Zool.* Ser. 13.

Delacour, J., and Mayr, E.
1945 The family Anatidae. *Wilson Bull.* 57:3–55. (A revision based on both morphological and behavioral characters.)

1946 Supplementary notes on the family Anatidae. *Wilson Bull.* 58:104–110.

Edwards, E.P.

1974 *A coded list of birds of the world.* Sweet Briar, Virginia: Privately published.

1982 *A coded workbook of birds of the world. Non-passerines,* Vol. 1. 2nd ed. Sweet Briar, Virginia: Privately published.

Ficken, M.S.

1965 Mouth color of nestling passerines and its use in taxonomy. *Wilson Bull.* 77:71–75. (With a few exceptions usually a good family character.)

Gill, F.B.

1980 Historical aspects of hybridization between Blue-winged and Golden-winged Warblers. *Auk* 97:1–18.

Grant, P.R.

1981 Speciation and the adaptive radiation of Darwin's finches. *Amer. Scientist* 69(6):653–663.

Harrison, C.J.O., ed.

1978 *Bird families of the world.* New York: Harry N. Abrams.

Howard, R., and Moore, A.

1980 *A complete checklist of the birds of the world.* New York: Oxford Univ. Press.

James F.C.

1983 Environmental component of morphological differentiation in birds. *Science* 221:184–186.

Johnsgard, P.A.

1961 The taxonomy of the Anatidae—a behavioural analysis. *Ibis* 103a:71–85.

Johnson, N.K.

1963 Biosystematics of sibling species of flycatchers in the *Empidonax hammondii-oberholseri-wrightii* complex. *Univ. Calif. Publ. in Zool.* 66:79–237. (In these three sympatric species, reproductive isolation is maintained "solely through ecologic means." Individuals of all three species come into contact at the time of pair formation and then "segregate by behavioral means into conspecific pairs for reproduction.")

Johnston, R.F., and Selander, R.K.

1964 House Sparrows: Rapid evolution of races in North America. *Science* 144:548–550.

1971 Evolution in the House Sparrow. II. Adaptive differentiation in North American populations. *Evolution* 25:1–28.

1973 Evolution in the House Sparrow. III. Variation in size and sexual dimorphism in Europe and North and South America. *Amer. Nat.* 107:373–390.

Lack, D.

1945 The Galapagos finches (Geospizinae): A study in variation. *Calif. Acad. Sci. Occas. Papers* No. 21.

1947 *Darwin's finches.* Cambridge, England: Univ. Press.

Lanyon, W.E.

1957 The comparative biology of the meadowlarks *(Sturnella)* in Wisconsin. *Publ. Nuttall Ornith. Club* No. 1.

1962 Specific limits and distribution of meadowlarks of the desert grassland. *Auk* 79:183–207.

Mayr, E.

1963 *Animal species and evolution.* Cambridge, Massachusetts: Harvard Univ. Press.

1969 *Principles of systematic zoology.* New York: McGraw-Hill.

Mayr, E., and Amadon, D.

1951 A classification of recent birds. *Amer. Mus. Novitates* No. 1496:1–42.

Mengel, R.M.

1964 The probable history of species formation in some northern wood warblers (Parulidae). *Living Bird* 3:9–43.

Miller, A.H.

1941 Speciation in the avian genus *Junco. Univ. Calif. Publ. in Zool.* 44:173–434.

1955 Concepts and problems of avian systematics in relation to evolutionary processes. In *Recent Studies in Avian Biology,* ed. A. Wolfson. Urbana: Univ. of Illinois Press. (Highly recommended reading for any student interested in modern systematics.)

Morony, J.J., Jr.; Bock, W.J.; and Farrand, J., Jr.

1975 *Reference list of birds of the world.* (Looseleaf) New York: Dept. Ornith., Amer. Mus. Nat. Hist.

Norris, R.A.

1958 Comparative biosystematics and life history of the nuthatches *Sitta pygmaea* and *Sitta pusilla. Univ. Calif. Publ. in Zool.* 56:119–300.

Olson, S.L., and James, H.F.

1982 Fossil birds from the Hawaiian Islands: Evidence for wholesale extinction by man before western contact. *Science* 271:633–635.

Packard, G.C.

1967 House Sparrows: Evolution of populations from the Great Plains and Colorado Rockies. *Systematic Zool.* 16:73–89.

Parkes, K.C.

1951 The genetics of the Golden-winged × Blue-winged Warbler complex. *Wilson Bull.* 63:5–15.

1961 Intergeneric hybrids in the family Pipridae. *Condor* 63:633–635.

1965 Character displacement in some Philippine cuckos. *Living Bird* 4:89–98.

1978 Still another parulid intergeneric hybrid (*Mniotilta* × *Dendroica*) and its taxonomic and evolutionary implications. *Auk* 95:682–690.

Peakall, D.B.
1962 Electrophoresis of egg-white proteins as a taxonomic tool: A critical note. *Ibis* 104:567–568.

Peters, J.L.
1931–1979 *Check-list of birds of the world*. Cambridge, Massachusetts: Harvard Univ. Press. (Vols. 1–7, 1931–51, by Peters. Vol. 1 revised. Vols. 8 and 10–15 by other authors. Vol. 9 not yet published.)

Pitelka, F.A.
1950 Geographic variation and the species problem in the shore-bird genus *Limnodromus*. *Univ. Calif. Publ. in Zool.* 50:1–108.
1951 Speciation and ecologic distribution in American jays of the genus *Aphelocoma*. *Univ. Calif. Publ. in Zool.* 50:195–464.

Raikow, R.J.
1976 The origin and evolution of the Hawaiian honeycreepers Drepanididae. *Living Bird* 15:95–117.

Rand, A.L.
1959 Tarsal scutellation of song birds as a taxonomic character. *Wilson Bull.* 71:274–277. (An instance in classifying shrikes where tarsal scutellation fails as a key taxonomic character.)

Ridgway, R., and Friedmann, H.
1901–1950 The birds of North and Middle America: A descriptive catalogue of the higher groups, genera, species, and subspecies of birds known to occur in North America, from the Arctic lands to the Isthmus of Panama, the West Indies and other islands of the Caribbean Sea, and the Galapagos Archipelago. *Bull. U.S. Natl. Mus.* No. 50.

Salomonsen, F.
1965 The geographical variation of the Fulmar (*Fulmarus glacialis*) and the zones of marine environment in the North Atlantic. *Auk* 82:327–355.

Selander, R.K., and Giller, D.R.
1959 Interspecific relations of woodpeckers in Texas. *Wilson Bull.* 71:107–124.
1961 Analysis of sympatry of Great-tailed and Boat-tailed Grackles. *Condor* 63:29–86.

Short, L.L., Jr.
1963 Hybridization in the wood warblers *Vermivora pinus* and *V. chrysoptera*. *Proc. XIIIth Internatl. Ornith. Congr.* pp.147–160.
1965 Hybridization in the flickers (*Colaptes*) of North America. *Bull. Amer. Mus. Nat. His.* 129:307–428.
1969 Taxonomic aspects of avian hybridization. *Auk* 86:84–105.

Sibley, C.G.
1950 Species formation in the Red-eyed Towhees of Mexico. *Univ. Calif. Publ. in Zool.* 50:109–194. (The author demonstrates that the Eastern Towhee, *Pipilo erythrophthalmus*, Swainson's Towhee, *P. macronyx*, and Spotted Towhee, *P. maculatus*, are conspecific and, therefore, subspecies.)
1954 Hybridization in the Red-eyed Towhees of Mexico. *Evolution* 8:252–290.
1957 The evolutionary and taxonomic significance of sexual dimorphism and hybridization in birds. *Condor* 59:166–191.
1960 The electrophoretic patterns of avian egg-white proteins as taxonomic characters. *Ibis* 102:215–284.
1967 Proteins: History books of evolution. *Discovery* 3:5–20.

Sibley, C.G., and Ahlquist, J.E.
1982 The relationships of the Hawaiian honeycreepers (Drepaninini) as indicated by DNA-DNA hybridization. Auk 99:130–140.

Sibley, C.G., and Pettingill, O.S., Jr.
1955 A hybrid longspur from Saskatchewan. *Auk* 72:423–425.

Sibley, C.G., and Short, L.L., Jr.
1959 Hybridization in the buntings (*Passerina*) of the Great Plains. *Auk* 76:443–463.

Sibley, C.G., and West, D.A.
1959 Hybridization in the Rufous-sided Towhees of the Great Plains. *Auk* 76:326–338.

Sprunt, A., Jr.
1954 A hybrid between the Little Blue Heron and the Snowy Egret. *Auk* 71:314.

Stein, R.C.
1963 Isolating mechanisms between populations of Traill's Flycatchers. *Proc. Amer. Phil. Soc.* 107:21–50. (Song is the main isolating mechanism.)

Storer, R.W.
1952 A comparison of variation, behavior and evolution in the sea bird genera *Uria* and *Cepphus*. *Univ. Calif. Publ. in Zool.* 52:121–222.
1960 The classification of birds. In *Biology and comparative physiology of birds*. Vol. 1, ed. by A.J. Marshall. New York: Academic Press.

Tordoff, H.B.
1954a Relationships in the New World nine-primaried Oscines. *Auk* 71:273–284.
1954b A systematic study of the avian family Fringillidae based on the structure of the skull. *Univ. Michigan Mus. Zool. Misc. Publ.* No. 81.

Walters, M.
1980 *The complete birds of the world*. London, England: David & Charles.

Wells, S., and Baptista, L.F.
1979 Displays and morphology of an Allen × Anna Hummingbird hybrid. *Wilson Bull.* 91:524–532.

West, D.A.
1962 Hybridization in grosbeaks *(Pheucticus)* of the Great Plains. *Auk* 79:399–424.

Wetmore, A.
1960 A classification for the birds of the world. *Smithsonian Misc. Coll.* 139(11):1–37.

Woods, R.S.
1944 *The naturalist's lexicon. A list of classical Greek and Latin words used or suitable for use in biological nomenclature.* Pasadena, California: Abbey Garden Press.

Zink, R.M.
1982 Patterns of genic and morphologic variation among sparrows in the genera *Zonotrichia*, *Melospiza*, *Junco*, and *Passerella*. *Auk* 99:632–649.

External Structural Characters

The preceding section of this book indicated the methods used in classifying birds into categories and pointed out that combinations of taxonomic characters determined these categories.

The laboratory identification of bird specimens to orders, families, genera, and sometimes species is commonly based on those taxonomic characters that are morphological and at the same time visible externally. They are usually called **external structural characters.** All keys and synopses employ them. Therefore, before undertaking laboratory identification, know the external structural characters in order to be able to use keys and synopses effectively.

In the following pages are outlined the common external structural characters. Definitions of the terms used in describing them are included, together with the names of certain birds in which the characters are exemplified. Study these characters from actual specimens, and, when spaces are available on the left-hand side of the pages, make a series of sketches similar to those already drawn.

Although "popular" manuals, keys, and synopses do not always employ technical terms in describing characters, a thorough student should be familiar with such terms and be able to use them when the occasion demands.

Characters Of the Bill

Long: the bill is decidedly longer than the head, as in a bittern.

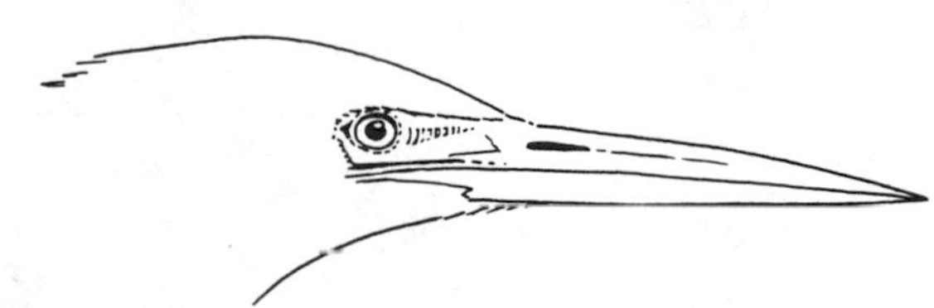

Short: the bill is decidedly shorter than the head, as in a redpoll.

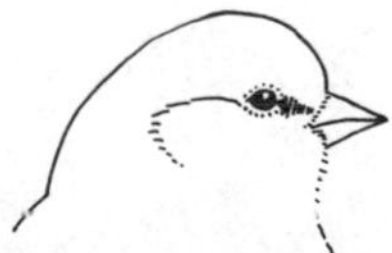

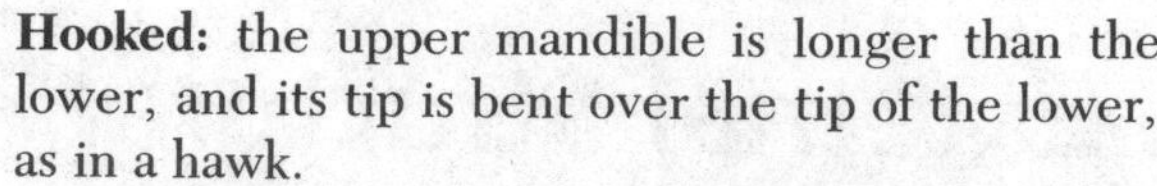

Hooked: the upper mandible is longer than the lower, and its tip is bent over the tip of the lower, as in a hawk.

Crossed: the tips of the mandibles cross each other, as in a crossbill.

Compressed: the bill for a good part of its length is higher than wide, as in a puffin or a kingfisher. (Show both a lateral and a frontal view.)

Depressed: the bill is wider than high, as in a duck. (Show both a lateral and a frontal view.)

Stout: the bill is conspicuously high and wide, as in a grouse. (Show both a lateral and a dorsal view.)

Terete: the bill is generally circular either in cross section, or when viewed anteriorly, as in a hummingbird. (Show both a lateral and a frontal view.)

Straight: the line along which the mandibles close (i.e., the commissure) is in line with the axis of the head, as in a bittern.

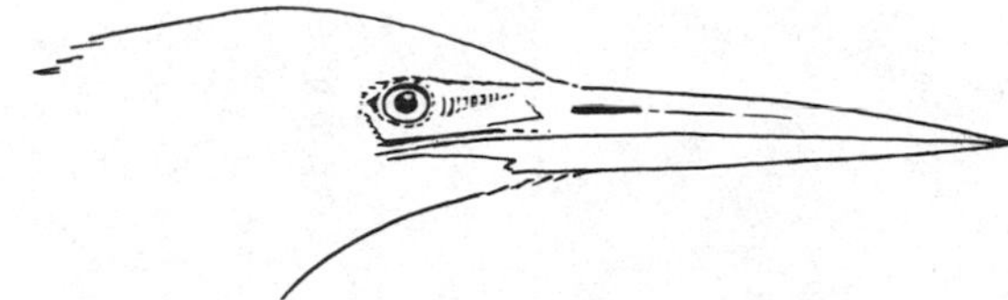

Recurved: the bill curves upward, as in a godwit.

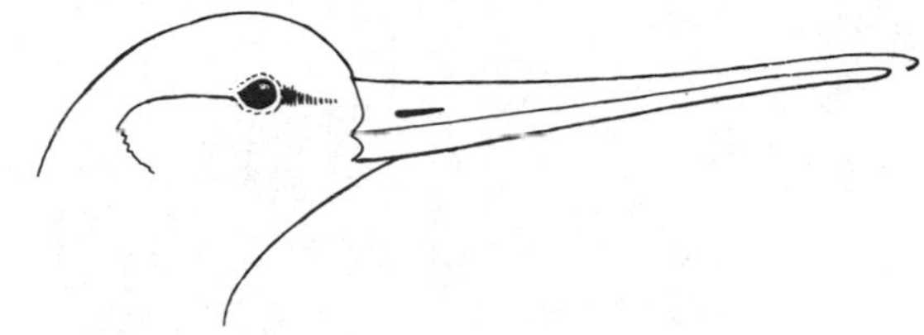

Decurved: the bill curves downward, as in the Brown Creeper *(Certhia americana)*, or a curlew.

Bent: the bill is deflected at an angle (usually deflected downward at the middle), as in a flamingo.

Swollen: the sides of the mandibles are convex, as in a tanager *(Piranga)*. (Show a dorsal view.)

Acute: the bill tapers to a sharp point, as in the Yellow Warbler *(Dendroica petechia)*.

Chisel-like: the tip of the bill is beveled, as in the Hairy Woodpecker *(Picoides villosus)*. (Show both a lateral and a dorsal view.)

Toothed: the upper mandibular tomium has a "tooth," as in a falcon, or several "teeth," as in a trogon.

Serrate: the bill has saw-like tomia, as in a merganser.

Gibbous: the bill has a pronounced hump, as in a scoter.

Spatulate, or **spoon-shaped:** the bill is much widened, or depressed, toward its tip, as in the Northern Shoveler *(Anas clypeata)*.

Notched: the bill has a slight nick in the tomia of one or both mandibles. Most frequently the notch occurs near the tip of the upper mandible, as in a thrush.

Conical: the bill has the shape of a cone, as in a redpoll.

Lamellate, or **sieve-billed:** the mandibles have just within their tomia a series of transverse tooth-like ridges, as in swans, geese, and ducks.

With **angulated commissure:** the commissure forms an angle at the point where the tomium proper meets the rictus, as in a grosbeak, finch, sparrow, or bunting. (Show the mouth closed. See Figure 3a for the character in the House Sparrow with mouth open.)

With **gular sac:** the chin, gular region, and jugulum are distended. In the pelican the gular sac is conspicuous, outwardly membranous, and featherless; in the cormorant it is inconspicuous and partially feathered.

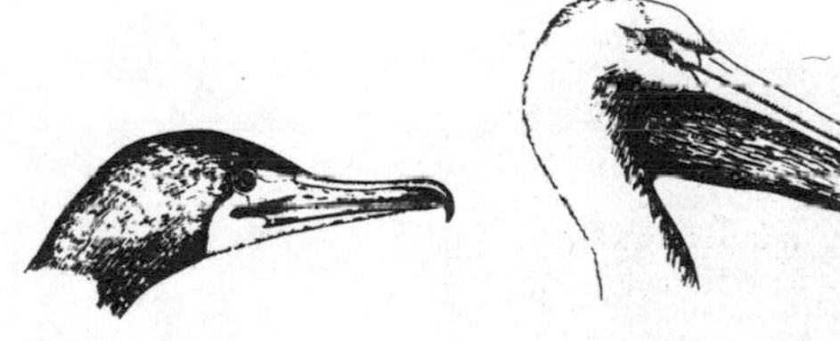

Cormorant Pelican

Nostrils

The nostrils are generally separated from each other by a complete wall, or septum; they are, therefore, **imperforate.** A few groups of birds, such as the vultures, have nostrils without a medial septum; they communicate with each other and are, therefore, **perforate.** (See Figure 39, p. 138.) Nostrils show other characters:

Tubular: the nostrils are in the ends of short prolongations of the base of the upper mandible, as in an albatross, a shearwater, or petrel.

Operculate: nostril openings are partly covered by an **operculum**—membranous, as in the Barn Swallow (*Hirundo rustica*), fleshy, as in the pigeon.

Linear, oval, or **circular:** the nostril openings are thus shaped, as in a gull, an accipitrid hawk, and a falcon, respectively. The nostrils in the falcon possess **bony tubercles.**

Covering

The covering of the bill is generally horny throughout and may be divided into distinct sections, as in petrels and gannets. Sometimes, as in shorebirds, it is soft throughout. The covering may show other modifications that constitute important characters.

Cere: the distal end of the upper mandible may be horny, and the proximal portion may be thick and soft, as in a hawk.

Nail: the tip of the upper or of both mandibles may be conspicuously harder and set off in grooves, as in ducks.

Characters of the Tail

A tail is said to be **long** when it is decidedly longer than the trunk, as in a pheasant or a cuckoo, and **short,** when it is either approximately the length of, or shorter than, the trunk, as in the shorebirds.

Due to the different relative lengths of the rectrices, the posterior margin of the tail assumes various shapes that are distinguishing characters.

Square: the rectrices are all of the same length, as in the Sharp-shinned Hawk *(Accipiter striatus).*

Rounded: the rectrices shorten successively from the inside to the outside, in slight gradations, as in a crow.

Graduated: the rectrices shorten successively from the inside to the outside, in abrupt gradations, as in a cuckoo.

Pointed, or **acute:** the middle rectrices are much longer than the others, as in a Ring-necked Pheasant *(Phasianus colchicus).*

Emarginate: the rectrices increase in length successively from the middle to the outermost pair, in slight gradations, as in a finch.

Forked: the rectrices increase in length successively from the middle to the outermost pair, in abrupt gradations, as in a tern.

Characters of the Wings

A wing is said to be **long** when the distance from the bend to the tip is decidedly longer than the trunk, as in a tern, and **short** when the distance is either approximately the length of, or shorter than, the trunk, as in a grebe.

Spurred: the bend of the wing has a peculiar horny structure in the shape of a spur, as in a jacana.

The varying length of the primaries in different species causes the wing to assume different shapes.

Rounded: the middle primaries are the longest, and the remaining primaries are graduated, as in the Sharp-shinned Hawk *(Accipiter striatus)*.

Pointed: the outermost primaries are the longest, as in a gull.

The varying length of both primaries and secondaries in different species causes wings to show differences in width. A wing is **narrow** when the primaries, and particularly the secondaries, are relatively short throughout, as in a gull. A wing is **broad** when both the primaries and the secondaries are very long throughout, as in the Sharp-shinned Hawk. (See drawings.)

The surface of the spread wing may vary in curvature. Although it is somewhat convex above and concave below, the curvature may sometimes be extreme, or it may sometimes be very slight. If the curvature is extreme, the wing is said to be **concave,** as in a grouse. If it is very slight, the wing is said to be **flat,** as in a swift or a hummingbird. (See drawings.)

Wing of Grouse

Wing of Swift

Characters Of the Feet

The tibia, when featherless, and the tarsus and toes usually have a horny investment. In different birds this investment is variously cut up.

Tarsus

Scutellate: the investment is cut up into more or less imbricated (overlapping) scales, as the tarsus of a grosbeak, finch, sparrow, or bunting.

Reticulate: the investment is cut up into small irregular plates, as the tarsus of a plover.

Serrate: the investment plates have serrations, as on the posterior edge of the tarsus in a grebe.

Scutellate-reticulate: the investment is scutellate in front and reticulate behind, as in a pigeon.

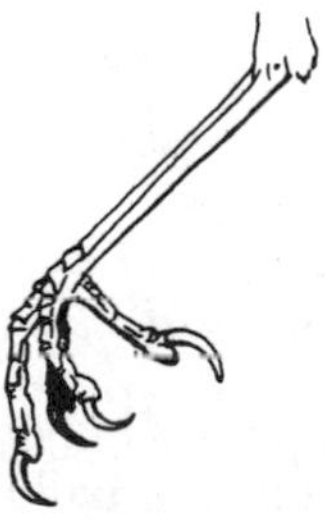

Booted: the investment of the tarsus is continuously horny without scales or plates, as in a thrush.

Scutellate-booted: the tarsus is scutellate in front and booted behind, as in the Gray Catbird *(Dumetella carolinensis)*.

Spurred: the posterior investment of the tarsus is peculiarly modified to form a spur, as in the Ring-necked Pheasant *(Phasianus colchicus)*.

The tarsus may assume several shapes in cross section.

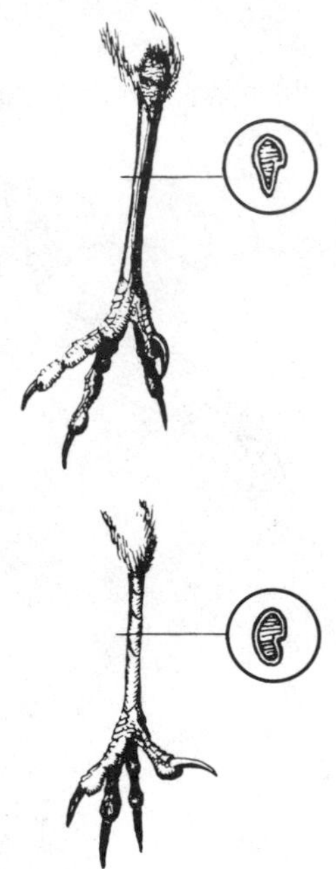

Rounded in front: this is the most common case, with somewhat flattened sides converging to a rather sharp ridge behind.

Rounded in front and behind: Occasionally, the tarsus is rounded on both sides, as in a tyrannid flycatcher and in the Horned Lark *(Eremophila alpestris)*.

Compressed: the tarsus is very flat from side to side with rather sharp edges in front and behind, as in loons, grebes, and a few other aquatic birds.

Toes

The position of the toes is important. In all birds the front toes are inserted on the metatarsus at the same level. But the hind toe, or **hallux,** varies in position.

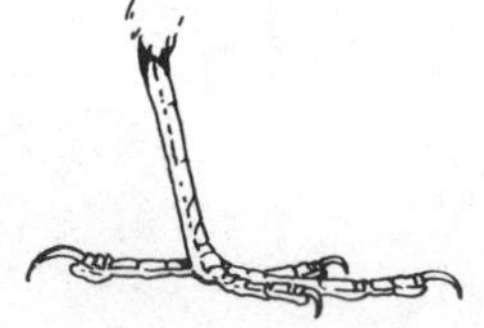

Incumbent: the hallux is inserted on the metatarsus at the level of the other toes, as in a meadowlark.

Elevated: the hallux is inserted so high on the metatarsus that its tip does not reach the ground, as in a rail.

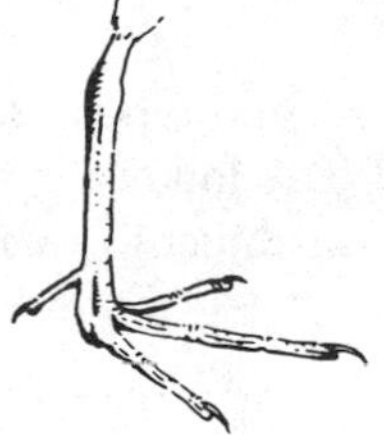

Nails

The nails of birds are generally curved and sharp-pointed. They are rounded above, flattened from side to side, and somewhat concave below. In certain birds these nails vary from the ordinary.

Acute: the nails are extremely curved and sharp-pointed, as in a woodpecker.

Obtuse: the nails are less curved and have rather blunt points, as in a grouse.

Lengthened: the nails are rather straight and elongated but sharp-pointed, as in the hallux nail of the Horned Lark *(Eremophila alpestris)*.

Pectinate: the nails have serrated edges, as the middle nail of a heron.

Flattened: the nails are so extremely flattened and broadened as to resemble a human finger nail, as in a grebe.

Feet

Birds' feet are of several types, depending on the arrangement of the toes and/or the particular functions the feet perform. The following ten types are commonly used as characters for distinguishing groups of birds. Each type is subject to variation.

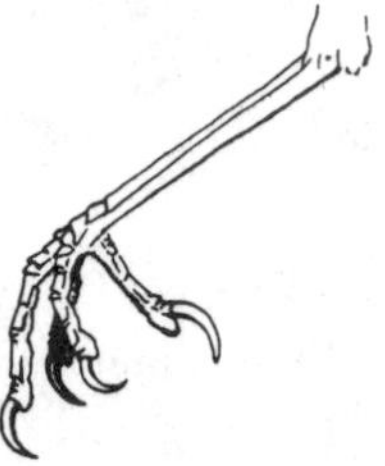

Anisodactyl: the hallux is behind and the other three toes are in front, as in a thrush.

Syndactyl: the third and fourth toes (outer and middle) are united for most of their length and have a broad sole in common, as in the Belted Kingfisher *(Ceryle alcyon)*.

Zygodactyl: the toes are arranged in pairs, the second and third toes in front, the fourth and hallux behind, as in a woodpecker.

Heterodactyl: the toes are arranged in pairs, in this case, the third and fourth toes in front, the second and hallux behind, as in a trogon.

Pamprodactyl: all four toes are in front, the hallux being turned forward, as in a swift.

Raptorial: the toes are deeply cleft, with large, strong, sharply curved nails (talons), as in a hawk.

Semipalmate, or half-webbed: the anterior toes are joined part way by a small webbing, as in the Semipalmated Plover *(Charadrius semipalmatus),* Willet *(Catoptrophorus semipalmatus),* or Semipalmated Sandpiper *(Calidris pusilla).*

Totipalmate, or fully webbed: all four toes are united by ample webs, as in a cormorant.

Palmate: or webbed: the front toes are united by ample webs, as in ducks and gulls.

Lobate: or lobed: a swimming foot with a series of lateral lobes on the toes, as in a grebe. Sometimes the foot may be palmate, but the hallux may bear a lobe, as in a diving duck.

Characters of the Plumage

For examples of the following characters, refer to Figure 39.

Distribution

Certain parts usually covered by feathers may be without well-developed feathers, or **bare,** as the lores of herons and the entire heads and upper necks of New World vultures (for example, the Turkey Vulture, *Cathartes aura).* Certain parts usually uncovered may be **feathered,** as the tarsi of some owls.

Texture

The feathers of the goatsuckers and owls are generally **soft;** the rectrices of woodpeckers are **stiffened;** the tufts of feathers covering the nostrils of crows are **tough** and **bristle-like;** the plumage of wrentits is **lax;** certain rectrices of Anhingas, *Anhinga anhinga,* possess ripple-like **flutings;** and the barbs of the outer

Figure 39 **Characters of the Plumage**

vanes of the outermost primaries of Northern Rough-winged Swallows, *Stelgidopteryx serripennis*, have stiffly hooked tips, which give the bird its name.

Shape and Structure

Some of the contour feathers may be modified to form **"horns,"** as in the Horned Lark, *Eremophila alpestris;* **crests,** as in the Cedar Waxwing, *Bombycilla cedrorum;* **ruffs,** as in the Ruffed Grouse, *Bonasa umbellus;* and **pinnae,** as in the Greater Prairie-Chicken, *Tympanuchus cupido;* **"ears"** and **facial discs,** as in the Great Horned Owl, *Bubo virginianus;* the highly colored area, the **speculum,** on the secondaries of several ducks; and the **rictal bristles,** as in the Brown Thrasher, *Toxostoma rufum,* and Whip-poor-will, *Caprimulgus vociferus.*

Contour feathers may be peculiarly modified in shape and structure.

Notched: a vane of the contour feather is incised toward the end, as the proximal vanes of the outer primaries of the Broad-winged Hawk *(Buteo platypterus).*

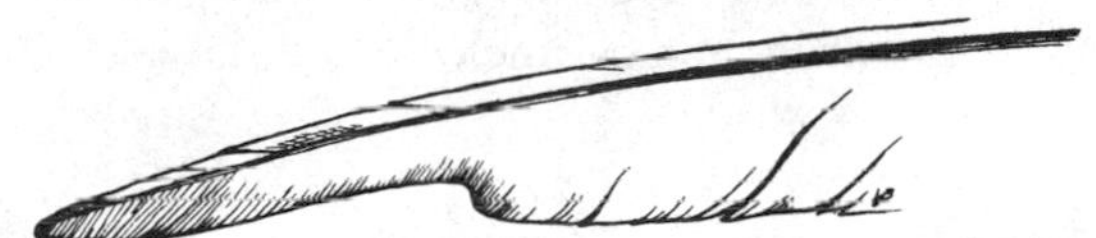

Spinose: the shaft of the contour feather is prolonged distally without barbs, as in the rectrices of the Chimney Swift *(Chaetura pelagica).*

Acuminate: the contour feather ends in a sharp point, as the rectrices of woodpeckers.

Attenuate: the contour feather is long and extremely narrow, as the outer rectrices of the Barn Swallow *(Hirundo rustica),* or the outermost primaries of the American Woodcock *(Scolopax minor).*

Broad: the contour feather is extremely wide, as the rectrices of a trogon.

Figure 40 **Miscellaneous Characters**
In the head region: (a) eye scales in the Atlantic Puffin; (b) frontal shield in a gallinule; (c) wattles and caruncles in a turkey.

Miscellaneous Characters

Numerous integumentary outgrowths may occur elsewhere than on the bill, wings, and feet, thus providing distinguishing characters. In the head region there may be small **eye scales** above and below the eyes, as in the Atlantic Puffins *(Fratercula arctica);* a **frontal shield** at the base of the upper mandible, as in gallinules and moorhens; and **wattles** and **caruncles,** as in turkeys. (See Figure 40.)

Synopsis of the External Structural Characters of North American Orders and Families of Birds

The orders and families of birds native in North America and north of Mexico are presented in the following synopsis, together with their distinguishing external structural characters. Classification and nomenclature follow the sixth edition (1983) of *The A.O.U. Check-list of North American Birds*.

The synopsis requires brief explanation. Characters are limited to those of the bill, tail, wings, feet (including legs), and plumage, as outlined in the preceding pages. "Negative" characters—i.e., the absence of characters—are usually not mentioned. Unless otherwise indicated, characters apply not only to the North American representatives of an order, or a family, but also to the entire group. Measurements given are the extremes in length among the species in each group. They are taken from bird skins and are at best only approximate.

Use the synopsis as a reference as well as a direct means of learning the distinguishing characters of each bird group. This can be done by first selecting conspicuous characters, such as hooked bill, forked tail, or palmate feet, then listing after each character the bird groups possessing it. Later, as time permits, select the less conspicuous characters and proceed in the same manner. By learning these lists one will gain a knowledge that will greatly facilitate the laboratory identification of birds, which is discussed in the next chapter.

For characters distinguishing all the bird families of the world, consult Chapter 13, "The Classification of World Birds by Families," in *Fundamentals or Ornithology* by J. Van Tyne and A.J. Berger (John Wiley & Sons, New York, 1976).

Order GAVIIFORMES

Family Gaviidae. Loons. (Length: 24–38 in.; 60.96–96.52 cm)

Bill: straight; acute; compressed. *Tail:* short; rectrices stiff. *Wings:* well-developed but short; somewhat pointed. *Feet:* tarsi compressed and reticulate; toes four, and palmate.

Order PODICIPEDIFORMES

Family Podicipedidae. Grebes. (Length: 7.5–26 in.; 19.05–66.04 cm)

Bill: straight; acute (one exception); compressed. *Tail:* rudimentary. *Wings:* poorly developed and short; somewhat pointed. *Feet:* tarsi compressed and scutellate with serrate posterior edges; toes four, and lobed; nails flattened.

Order PROCELLARIIFORMES. Tube-nosed Swimmers.

Bill: hooked; nostrils tubular. *Tail:* short to moderately long. *Wings:* long, narrow. *Feet:* palmate; hallux rudimentary or absent. *Plumage:* predominantly black and gray.

Family Diomedeidae. Albatrosses. (Length: 28–40 in.; 71.12–101.6 cm)

Bill: nostril tubes lateral, separated by culmen.

Family Procellariidae. Shearwaters and Fulmars. (Length: 10–34 in.; 25.4–86.36 cm)

Bill: nostril tubes on culmen; nostrils imperforate.

Family Hydrobatidae, Storm-Petrels. (Length: 6–10 in.; 15.24–25.4 cm)

Bill: nostrils on culmen, united in one tube.

Order PELECANIFORMES. Totipalmate Swimmers.

Bill: with gular sac. *Feet:* toes four, and totipalmate.

Family Phaethontidae. Tropicbirds. (Length: 24–40 in.; 60.96–101.6 cm)

Bill: as long as head; straight; compressed; acute; gular sac rudimentary; nostrils small and linear. *Tail:* pointed; middle rectrices filamentous. *Wings:* long; pointed. *Feet:* relatively small; hallux more elevated than in following families of order.

Family Sulidae. Gannets. (Length: 28–36 in.; 71.12–91.44 cm)

Bill: slightly longer than head; straight, slightly decurved at tip; bluntly acute; exterior nostrils absent; gular sac very small. *Tail:* long; pointed. *Wings:* long; pointed. *Feet:* legs short and stout.

Family Pelecanidae. Pelicans. (Length: 52–72 in.; 132.08–182.88 cm)

Bill: very long; straight; hooked; nostrils absent; gular sac large. *Tail:* very short. *Wings:* very long; rounded. *Feet:* legs short and stout; tarsi compressed and reticulate.

Family Phalacrocoracidae. Cormorants. (Length: 22–36 in.; 55.88–91.44 cm)

Bill: long as head; straight; hooked; exterior nostrils absent; gular sac very small. *Tail:* long; rounded. *Wings:* short; rounded.

Family Anhingidae. Anhingas. (Length: 28–36 in.; 71.12–91.44 cm)

Bill: long; straight; slender; acute; nostrils minute; gular sac moderate. *Tail:* long: rounded; middle pair of rectrices with flutings. *Wings:* moderately long.

Family Fregatidae. Frigatebirds. (Length: 30–40 in.; 76.2–101.6 cm)

Bill: long; straight; hooked; nostrils small and linear. *Tail:* long; deeply forked. *Wings:* long; pointed. *Feet:* legs short; feet small; tarsi partly feathered; middle nail pectinate.

Order CICONIIFORMES. Deep-water Waders.

Bill: long. *Tail:* short. *Wings:* long; broad; rounded. *Feet:* legs very long; toes four. *Plumage:* lores usually bare.

Family Ardeidae. Herons and Bitterns. (Length: 10–56 in.; 25.4–142.24 cm)

Bill: straight; acute. *Feet:* tarsi usually scutellate in front; toes long and on same level; middle nail pectinate. *Plumage:* lax; frequently with modified plumes; powder-down tracts.

Family Threskiornithidae. Ibises and Spoonbills. (Length: 19–38 in.; 48.26–96.52 cm)
Bill: either decurved, slender throughout, and somewhat terete; or straight, broad, and spatulate. *Feet:* tarsi usually reticulate; hallux slightly elevated.

Family Ciconiidae. Storks and Wood Ibises. (Length: 28–60 in.; 71.12–152.4 cm)
Bill: straight; stout at base; sometimes recurved at tip; sometimes decurved at tip (American form). *Feet:* tarsi reticulate; toes long; hallux slightly elevated.

Order PHOENICOPTERIFORMES

Family Phoenicopteridae. Flamingos. (Length: 36–53 in.; 91.44–134.62 cm)
Bill: bent in middle; lamellate. *Feet:* tarsi scutellate; hallux elevated; palmate.

Order ANSERIFORMES

Family Anatidae. Lamellate-billed Swimmers. (Length: 12–72 in.; 30.48–182.88 cm)
Bill: either lamellate or serrate; narrowly compressed or broadly depressed; with nail-like hook. *Tail:* short (except one tribe); usually rounded. *Wings:* either pointed, or rounded (one tribe). *Feet:* toes four, and palmate; hallux elevated and either lobate or not lobate. *Plumage:* lores either feathered or bare.

Subfamily Anserinae. Whistling-Ducks, Swans, and Geese. (Length 13–72 in.; 33.02–182.88 cm)
Feet: tarsi completely reticulate.
Tribe Dendrocygnini. Whistling-Ducks. (Length: 13–30 in.; 33.02–76.2 cm)
Wings: rounded. *Feet:* tarsi longer than middle toe with nail; legs, feet, and nails exceptionally long.
Tribe Cygnini. Swans. (Length: 47–72 in.; 119.38–182.88 cm)
Feet: tarsi shorter than middle toe with nail. *Plumage:* lores bare.
Tribe Anserini. Geese. (Length: 21–39 in.; 53.34–99.06 cm)
Bill: compressed at base, narrowing toward tip. *Feet:* tarsi longer than middle toe with nail.

Subfamily Anatinae. Ducks. (Length: 12–29 in.; 30.48–73.66 cm)
Feet: tarsi scutellate in front.
Tribe Cairinini. (Genus *Aix* only.) Wood Ducks. (Length 17–20 in.; 43.18–50.8 cm)
Bill: somewhat compressed; colorful in males. *Feet:* tarsi shorter than middle toe without nail. *Plumage:* head with prominent crest.
Tribe Anatini. Dabbling Ducks. (Length: 12–29 in.; 30.48–73.66 cm)
Wings: usually with iridescent speculum. *Feet:* tarsi shorter than middle toe without nail.
Tribe Aythyini. Pochards and Scaups. (Length: 15–24 in.; 38.1–60.96 cm)
Wings: with noniridescent speculum. *Feet:* tarsi shorter than middle toe without nail; hallux lobate.
Tribe Mergini. Mergansers, Eiders, and other sea ducks. (Length: 12–29 in.; 30.48–73.66 cm)
Bill: somewhat laterally compressed, terete in mergansers; lamellate or serrate. *Feet:* tarsi shorter than middle toe with nail; hallux lobate.
Tribe Oxyurini. Ruddy and Masked Ducks. (Length: 13–16 in.; 33.02–40.64 cm)
Bill: broad and depressed. *Tail:* long; rectrices narrow and stiffened; coverts extremely short. *Feet:* tarsi shorter than middle toe with nail; hallux lobate.

Order FALCONIFORMES. New World Vultures and Diurnal Birds of Prey.
Bill: hooked; with cere in which the nostrils open centrally. *Feet:* anisodactyl and raptorial. *Plumage:* lores with bristle-like feathers only.

Family Cathartidae. New World Vultures. (Length: 24–54 in.; 60.96–137.16 cm)
Bill: moderately hooked; nostrils large, oval, and perforate. *Wings:* long, broad, and rounded. *Feet:* weakly raptorial; hallux less than half the length of the middle toe; tarsi reticulate. *Plumage:* head in adults bare.

Family Accipitridae. Ospreys, Kites, Eagles, and Hawks. (Length: 10–48 in.; 25.4–121.92 cm)
Bill: strongly hooked; nostrils oval or slit-like and imperforate. *Wings:* long and usually broad. *Feet:* strongly raptorial.

Subfamily Pandioninae. Ospreys. (Length: 21–24.5 in.; 53.34–62.23 cm)
Bill: nostrils obliquely oval, with anterior end uppermost. *Wings:* long and pointed. *Feet:* hallux about the same length as the middle toe; hallux nail approximately the same size as other nails; under surfaces of nails rounded; under surfaces of toes with spiny scales; tarsi reticulate.

Subfamily Accipitrinae. Kites, Eagles and Hawks. (Length: 10–48 in.; 25.4–121.92 cm)
Bill: nostrils small, usually oval, or slit-like, with anterior end uppermost. *Wings:* long, broad, and rounded (except in kites, in which they are long, narrow and pointed). *Feet:* hallux usually the same length as, or slightly longer than, the shortest front toe; hallux nail larger than other nails; under surfaces of all nails grooved; tarsi more often scutellate than reticulate; feathered or booted.

Family Falconidae. Caracaras and Falcons. (Length: 10–25 in.; 25.4–63.5 cm)
Bill: strongly hooked; upper mandible toothed near tip (faintly so in caracaras); nostrils either circular with prominent central bony tubercle, or (as in caracaras) slit-like, with posterior end uppermost, and imperforate. *Wings:* long, narrow, and pointed (except in caracaras, in which they are long, broad, and rounded). *Feet:* strongly raptorial; hallux usually the same length as, or slightly longer than, the shortest front toe; under surfaces of nails grooved; tarsi reticulate (front distinctly scutellate in caracaras).

Order GALLIFORMES. Gallinaceous Birds.
Bill: short, stout, culmen decurved; tip of upper mandible bent slightly over tip of lower mandible. *Wings:* short; concave; rounded; primaries stiff and usually curved. *Feet:* strong, tarsi scutellate when not feathered; nails obtuse.

Family Cracidae. (Genus *Ortalis* only.) Chachalacas. (Length: 20–24 in.; 50.8–60.96 cm)
Bill: nostrils exposed. *Tail:* long; rounded. *Feet:* hallux incumbent. *Plumage:* area around the eyes and sides of the throat bare.

Family Phasianidae. Pheasants, Grouse, Turkeys, and Quail. (Length: 8–40 in.; 20.32–101.6 cm)
Bill: nostrils partly feathered or exposed, usually covered by an operculum. *Feet:* hallux elevated and sometimes spurred. *Plumage:* head often bare and wattled (except one subfamily).

Subfamily Phasianinae. Pheasants. (Length 20–36 in.; 50.8–91.44 cm)
Bill: nostrils exposed. *Tail:* long and usually pointed. *Feet:* spurred in males. *Plumage:* areas on head often bare and wattled.

Subfamily Tetraoninae. Grouse and Ptarmigan. (Length: 20–24 in.; 50.8–60.96 cm)
Bill: nostrils feathered. *Tail:* variable in shape. *Feet:* tarsi feathered wholly or in part; toes somewhat pectinate, especially in winter. *Plumage:* sides of neck often with inflatable air sacs and/or modified, erectile feathers.

Subfamily Meleagridinae. Turkeys. (Length: 30–40 in.; 76.2–101.6 cm)
Bill: nostrils exposed. *Tail:* wide and rounded. *Feet:* tarsus spurred in males. *Plumage:* with evident luster; head and neck bare, wattled, and carunculated; individual body feathers and rectrices wide and somewhat square at ends.

Subfamily Odontophorinae. Quail. (Length: 7–10 in.; 17.78–25.4 cm)
Bill: nostrils exposed; mandible may be finely serrated. *Tail:* short and rounded.

Order GRUIFORMES. Suborder Grues. Marsh Birds.
Bill: variable in shape; nostrils perforate. *Tail:* short. *Wings:* rounded; tertiaries often as long as the primaries. *Plumage:* lores feathered, or with bristles..

Family Rallidae. Rails, Gallinules, Moorhens, and Coots. (Length: 5–20 in.; 12.7–50.8 cm)
Bill: variable in length and shape. *Wings:* short. *Feet:* legs moderately long, hallux elevated and longer than nail of middle toe.

Family Aramidae. Limpkins. (Length: 24–28 in.; 60.96–71.12 cm)
Bill: long; compressed; decurved at tip. *Wings:* short; outer primary stiff, attenuate, and incurved. *Feet:* legs moderately long; hallux elevated and twice the length of nail of middle toe. *Plumage:* predominately brownish.

Family Gruidae. Cranes. (Length: 33–54 in.; 83.82–137.16 cm)
Bill: as long as, or slightly longer than, the head; straight; compressed. *Wings:* broad and long. *Feet:* legs very long; hallux elevated and short, being equal to the length of nail of middle toe. *Plumage:* lores and crown with bristles.

Order CHARADRIIFORMES. Suborder Charadrii. Plovers, Oystercatchers, Avocets, and Stilts.
Bill: short or long; straight or recurved. *Tail:* short. *Wings:* long and pointed; tertiaries greatly lengthened. *Feet:* legs usually long and near center of body; tarsi reticulate; hallux absent or rudimentary.

Family Charadriidae. Plovers. (Length: 5–15 in.; 12.7–38.1 cm)
Bill: moderate in length, usually under one in.; lateral profile constricted in middle and swollen toward tip. *Feet:* toes three (four in Black-bellied Plover, *Pluvialis squatarola*).

Family Haematopodidae. Oystercatchers. (Length: 16–21 in.; 40.64–53.34 cm)
Bill: twice the length of the head; straight; compressed; constricted near base; tip chisel-like. *Tail:* square to slightly rounded. *Feet:* legs and toes stout; tarsi shorter than bill; hallux absent; small webbing between toes. *Plumage:* black and white.

Family Recurvirostridae. Avocets and Stilts. (Length: 13–18 in.; 33.02–45.72 cm)
Bill: long; slender; recurved. *Feet:* legs extremely long; hallux either absent (stilts), or rudimentary (avocets); toes either palmate (avocets), or with webbing cleft nearly to base of toes (stilts).

Order CHARADRIIFORMES. Suborder Scolopaci. Jacanas and Sandpipers.
Bill: variable in length; straight or curved; usually slender, pliable in life. *Tail:* short. *Wings:* long and pointed (except jacanas); tertiaries greatly lengthened. *Feet:* legs usually long and near center of body; tarsi partially or completely scutellate; hallux usually present and elevated.

Family Jacanidae. Jacanas. (Length: 6–12 in.; 15.24–30.48 cm)
Bill: as long as head; straight; compressed; frontal shield on bill above forehead. *Wings:* somewhat rounded; spurred. *Feet:* toes very long; nails long, slender, and acute; hallux nail much longer than hallux and somewhat recurved.

Family Scolopacidae. Woodcock, Snipe, Sandpipers, and Phalaropes. (Length: 5–24 in.; 12.7–60.96 cm)
Bill: variable in length; usually slender, pliable in life, and soft, with tip somewhat depressed, or tapering to acute tip. *Feet:* tarsi scutellate in front; toes four (three in Sanderling, *Calidris alba*); toes with lateral membranes, developed into lobes (phalaropes only).

Order CHARADRIIFROMES. Suborder Lari. Gull-like Birds.

Family Laridae. Skuas, Gulls, Terns, and Skimmers. (Length: 9–30 in.; 22.86–76.2 cm)
Bill: relatively short and stout; variable in shape; nostrils perforate. *Tail:* variable in length and shape. *Wings:* very long and relatively narrow; pointed. *Feet:* legs relatively short and stout and attached

near center of body; tibiae partly bare; tarsi scutellate in front and reticulate elsewhere; hallux small and elevated (though sometimes rudimentary) or absent; anterior toes palmate. *Plumage:* compact on under parts; predominant coloration black, white, and gray.

Subfamily Stercorariinae. Skuas and Jaegers. (Length: 20–24 in.; 50.8–60.96 cm)
Bill: with nail-like hook; cere present. *Tail:* shorter than wing, except when middle rectrices are elongated; rounded to pointed. *Feet:* strong; tarsi longer than middle toe without nail; nails long, strongly hooked, and acute. *Plumage:* brown.

Subfamily Larinae. Gulls. (Length: 10–30 in.; 25.4–76.2 cm)
Bill: hooked but without nail-like hook. *Tail:* square to slightly rounded or wedge-shaped. *Feet:* of moderate size and length, the tarsus being more than one-tenth as long as the wings.

Subfamily Sterninae. Terns. (Length: 8–20 in.; 20.32–50.8 cm)
Bill: straight and sometimes acute. *Tail:* forked. *Feet:* extremely small and short, tarsus being less than one-tenth as long as the wings.

Subfamily Rynchopinae. Skimmers. (Length: 16–20 in.; 40.64–50.8 cm)
Bill: straight, compressed to thinness of knife blade; lower mandible notably longer than the upper and blunt at tip; upper mandible less blunt at tip. *Tail:* forked. *Feet:* small; tarsus short, approximately one-eleventh as long as the wing.

Order CHARADRIIFORMES. Suborder Alcae. Auks.

Family Alcidae. Auks, Murres, and Puffins. (Length: 8–18 in.; 20.32–45.72 cm)
Bill: variable in shape and length; nostrils imperforate. *Tail:* short. *Wings:* moderately long to short; pointed. *Feet:* legs of moderate length and size and attached far back; tibiae bare near heel; tarsi compressed and either wholly reticulate or partly scutellate; hallux absent; palmate; nails curved and acute but not large. *Plumage:* compact throughout; head sometimes crested; predominant coloration black and white.

Order COLUMBIFORMES

Family Columbidae. Doves and Pigeons. (Length: 6–17 in.; 15.24–43.18 cm)
Bill: relatively small and slender; basal part soft; terminal part horny with decurved culmen; middle part constricted; nostrils usually slit-like and overhung by operculum. *Tail:* long; in North American forms may be either square, rounded, or pointed. *Wings:* long and flat but variable in shape. *Feet:* tarsi scutellate in front (sometimes feathered on proximal part) and reticulate elsewhere; toes cleft to base; sometimes with slight webbing; hallux incumbent to slightly elevated. *Plumage:* dense; region in vicinity of eyes often quite bare.

Order PSITTACIFORMES

Family Psittacidae. Parrots, Parakeets, and Macaws. (Length: 5–36 in.; 12.7–91.44 cm)
Bill: short and stout; culmen greatly decurved; strongly hooked; cere in which nostrils open. *Tail:* variable in form. *Feet:* tarsi shorter than longest toe and reticulate (sometimes peculiarly granulated); zygodactyl; fourth toe reversible.

Order CUCULIFORMES

Family Cuculidae. Cuckoos, Roadrunners, and Anis. (Length: 5–24 in.; 12.7–60.96 cm)
Bill: variable in size and shape; usually compressed; more or less decurved. *Tail:* usually long and graduated. *Wings:* variable in form. *Feet:* variable in size; zygodactyl; fourth toe permanently reversed. *Plumage:* predominantly brown and gray but occasionally black.

Order STRIGIFORMES. Nocturnal Birds of Prey.

Bill: hooked; culmen strongly decurved; cere present; nostrils opening at edge of cere except in genus *Athene*. *Tail:* variable in form. *Wings:* variable in form; inner webs of certain primaries may be either smooth or notched. *Feet:* strongly raptorial; tibiae and tarsi usually feathered; toes frequently feathered; zygodactyl, the fourth toe reversible. *Plumage:* soft and lax; facial disc present; "ears" in many species; lores with dense feathers that cover base of bill and hide nostrils; sexes alike in coloration.

Family Tytonidae. (Genus *Tyto* only.) Barn-Owls. (Length: 15–21 in.; 38.1–53.34 cm)

Tail: short; square to emarginate. *Wings:* long; pointed; inner webs of all primaries without notching. *Feet:* tarsi twice as long as middle toe without nail; feathers on back of tarsi pointed upward; middle toe as long as inner; middle nail with inner edge pectinate. *Plumage:* facial disc triangular.

Family Strigidae. Typical Owls. (Length: 6–30 in.; 15.24–76.2 cm)

Tail: variable in length; usually somewhat rounded, rarely square. *Wings:* variable in form; from one to six of the outer primaries with notched edges. *Feet:* feathers on back of tarsi pointed downward; middle toe shorter than the inner. *Plumage:* facial disc circular; "ears" commonly present.

Order CAPRIMULGIFORMES

Family Caprimulgidae. Goatsuckers. (Length: 7–12 in.; 17.78–30.48 cm)

Bill: short (small); weak; depressed; slightly hooked; gape very wide with commissural point below eyes; nostrils circular (sometimes tubular) and exposed. *Tail:* variable in form. *Wings:* long; pointed. *Feet:* small; weak; tarsi partly feathered and twice the length of hallux; hallux short and elevated; outer toe noticeably shorter than middle toe, having four phalanges only; nail of middle toe pectinate. *Plumage:* soft; lax; feathers with aftershafts; rictal bristles usually evident; coloration dull and usually streaked, mottled, or barred.

Order APODIFORMES. Suborder Apodi. Swifts.

Family Apodidae. Swifts. Length: 4–9 in.; 10.16–22.86 cm)

Bill: very short (small); culmen decurved; depressed; gape extremely wide; commissural point below eyes. *Tail:* variable in form, usually forked or emarginate; rectrices in genus *Chaetura* stiffened and spinose. *Wings:* long; flat; pointed; secondaries extremely short. *Feet:* small; weak; tarsi unfeathered or feathered; pamprodactyl, the small hallux frequently directed inward, but reversible; three anterior toes about equal in length; nails strongly curved and acute. *Plumage:* compact; feathers of forehead may partially conceal nostrils; coloration of sexes alike, uniformly dull but sometimes with white areas.

Order APODIFORMES. Suborder Trochili. Hummingbirds.

Family Trochilidae. Hummingbirds. (Length: 2–9 in.; 5.08–22.86 cm)

Bill: variable in length; slender; straight (sometimes decurved; rarely recurved); terete (sometimes compressed). *Tail:* variable in form. *Wings:* long; flat; pointed; secondaries extremely short. *Feet:* small; weak; tarsi unfeathered or feathered and not longer than middle toe with claw; hallux large and incumbent; three anterior toes of different lengths; nails strongly curved and acute. *Plumage:* compact; usual coloration brilliantly metallic in both sexes but less so in females.

Order TROGONIFORMES

Family Trogonidae. (Genus *Trogon* only.) Trogons. (Length: 11–12 in.; 27.94–30.48 cm)

Bill: short; wide at base; culmen decurved; somewhat hooked; upper mandible with several "teeth." *Tail:* long, with broad rectrices; shape variable, usually graduated. *Wings:* short; concave; rounded. *Feet:* small and weak; tarsi shorter than longest toe and usually feathered; heterodactyl, the inner toe reversed; approximately half of anterior toes united. *Plumage:* soft; lax; feathers with afterfeathers; bristle-like feathers covering base of bill; coloration of males brilliantly metallic, of females less so.

Order CORACIIFORMES

Family Alcedinidae. Kingfishers. (Length: 5–17 in.; 12.7–43.18 cm)
Bill: long; straight; compressed; acute; nostrils linear. *Tail:* moderately long, being one-half to two-thirds as long as wing; slightly rounded. *Wings:* moderately long; pointed. *Feet:* small and weak; tibiae partly bare; tarsi extremely short and irregularly scutellate in front; syndactyl; hallux notably shorter than inner toe and partly connected with it; nails very acute, the middle one somewhat flattened. *Plumage:* head frequently crested.

Order PICIFORMES

Family Picidae. Woodpeckers. (Length: 5–21 in.; 12.7–53.34 cm)
Bill: strong; usually straight; usually chisel-like but sometimes acute. *Tail:* either pointed, rounded, or graduated; rectrices acuminate and with stiffened tips. *Wings:* moderately long; more or less pointed but with outermost primary very short or rudimentary. *Feet:* strong; tarsi scutellate in front and reticulate behind; zygodactyl; fourth toe permanently reversed; hallux occasionally absent; nails strong, decurved, and very acute. *Plumage:* nostrils concealed by bristle-like feathers.

Order PASSERIFORMES. Perching or Passerine Birds.
Bill: variable in form; covering horny; nostrils imperforate. *Tail:* variable in form; with 12 rectrices. *Wings:* variable in form; with 10 primaries, the outermost frequently rudimentary; with 6 or more secondaries. *Feet:* anisodactyl; hallux incumbent and as long as middle toe; hallux nail often as long as, or longer than, nail of middle toe.

Family Tyrannidae. New World Flycatchers. (Length: 2.5–10 in.; 6.35–25.4 cm)
Bill: variable in size; straight; wide and depressed at base (triangular in outline when viewed from above); slightly hooked; culmen somewhat decurved toward tip; commissural point almost below nasal canthus of eye; nostrils circular. *Tail:* usually square; sometimes forked. *Wings:* obvious primaries 10, the outermost usually longer than the secondaries. *Feet:* small and weak; tarsi short (seldom longer than middle toe), irregularly scutellate, and rounded behind. *Plumage:* rictal bristles usually evident; coloration of sexes usually similar.

Family Alaudidae. (Genus *Eremophila* only.) Horned Larks. (Length 5–9.5 in.; 12.7–24.13 cm)
Bill: short; conical; acute. *Tail:* shorter than wing; nearly square. *Wings:* long; pointed; nine primaries. *Feet:* moderate size; tarsi longer than middle toe, rather stout, rounded behind and scutellate; hallux nail very long, equaling hallux in length. *Plumage:* nostrils concealed by feather tufts; head frequently crested or "horned"; coloration predominantly brown, white, and black; conspicuous black patches on head and neck.

Family Hirundinidae. Swallows. (Length: 4–8 in.; 10.16–20.32 cm)
Bill: short; wide at base (triangular in outline when viewed from above); depressed; slightly hooked; culmen somewhat decurved toward tip; gape wide and twice the length of culmen; commissural point below nasal canthus of eye. *Tail:* not longer than wing; emarginate or forked; lateral rectrices sometimes filamentous. *Wings:* long; pointed; nine obvious primaries; secondaries generally very short. *Feet:* small and weak; tarsi usually shorter than middle toe with nail, and scutellate. *Plumage:* compact; coloration often partly metallic.

Family Corvidae. (North American species only.) Crows, Magpies, and Jays. (Length: 8–25 in.; 20.32–63.5 cm)

Bill: usually long; stout; culmen decurved toward tip; somewhat acute. *Tail:* rounded, sometimes graduated. *Wings:* either long and pointed (crows and ravens), or short and rounded (magpies and jays); 10 primaries. *Feet:* large and strong; tarsi longer than middle toe with nail, and scutellate. *Plumage:* nostrils usually concealed by dense tufts of stiff feathers; rictal bristles also evident.

Family Paridae. Titmice. (Length: 4–8 in.; 10.16–20.32 cm)

Bill: short; straight; stout; compressed. *Tail:* either as long as or shorter than wings; slightly rounded. *Wings:* rounded; 10 primaries. *Feet:* strong; tarsi longer than middle toe with nail, and scutellate. *Plumage:* nostrils concealed by dense tufts of stiff feathers; rictal bristles sometimes evident; coloration without bars, streaks, or spots.

Family Remizidae. (Genus *Auriparus* only.) Verdins. (Length: approximately 3.5 in.; 8.89 cm)

Bill: short; acute. *Tail:* rounded; shorter than wing. *Wings:* rounded; 10 primaries. *Feet:* tarsi longer than middle toe with nail, and scutellate. *Plumage:* coloration grayish with yellow head, chestnut at bend of wing (concealed); nostrils concealed by feathers.

Family Aegithalidae. (Genus *Psaltriparus* only.) Bushtits. (Length: approximately 3.5 in.; 8.89 cm)

Bill: very short; compressed. *Tail:* longer than wing; much rounded and graduated. *Wings:* rounded; 10 primaries. *Feet:* tarsi longer than middle toe with nail, and scutellate. *Plumage:* nostrils concealed by feathers.

Family Sittidae. Nuthatches. (Length: 4–7 in.; 10.16–17.78 cm)

Bill: as long as head; straight; slender; compressed; acute; gonys somewhat recurved toward tip. *Tail:* much shorter than wings; nearly square; rectrices broad with rounded tips. *Wings:* long; pointed; 10 primaries. *Feet:* strong; tarsi usually as long as middle toe with nail and scutellate in front; nails very curved and compressed. *Plumage:* compact; nostrils more or less covered with stiff feathers; rictal bristles evident; coloration in American species plain blue or gray above.

Family Certhiidae. (Genus *Certhia* only.) Creepers. (Length: 5–7 in.; 12.7–17.78 cm)

Bill: variable in length, sometimes much shorter or longer than head; decurved; slender; compressed; nostrils entirely exposed. *Tail:* about as long as, or slightly longer than, the wings; rounded; rectrices stiff and acuminate. *Wings:* somewhat long; rounded; 10 primaries. *Feet:* strong tarsi shorter than middle toe with nail, and scutellate; nails long and very curved. *Plumage:* coloration of upper parts brownish and streaked.

Family Troglodytidae. Wrens. (Length: 3–9 in.; 7.62–22.86 cm)

Bill: varying in length from half as long to about as long as head; usually decurved; slender; compressed. *Tail:* Varying in length from slightly longer than wings to two-thirds as long; rounded; rectrices soft, with rounded tips. *Wings:* short; concave; rounded; 10 primaries. *Feet:* strong; tarsi longer than middle toe with nail, and scutellate in front as well as (sometimes) behind. *Plumage:* brown coloration predominating; wings and tail usually barred.

Family Cinclidae. Dippers. (Length: 5–8 in.; 12.7–20.32 cm)

Bill: short; straight; slender; compressed; culmen decurved toward tip; gonys recurved toward tip; upper mandible notched near tip. *Tail:* short, more than half as long as wings; square to slightly rounded; rectrices broad with rounded tips. *Wings:* short; concave; rounded; 10 primaries. *Feet:* strong; tarsi longer than middle toe with nail, and booted; nails very curved. *Plumage:* soft; compact; coloration with brown, or gray, predominating.

Family Muscicapidae. (Subfamilies Sylviinae, Turdinae, and Timaliinae only.) "Old World" Insecteaters. (Length: 3.5–9.5 in.; 8.89–24.13 cm)

Bill: variable in length, usually short, straight, and slender; nostrils exposed. *Tail:* variable in length and shape. *Wings:* variable in length and shape; 10 primaries. *Feet:* usually strong; tarsus variable in length, booted (except in gnatcatchers, Genus *Polioptila*). *Plumage:* rictal bristles present.

Subfamily Sylviinae. (Tribes Sylviini and Polioptilini only.) Kinglets and Gnatcatchers. (Length 3–6 in.; 9.14 cm)

Bill: short; straight; slender; somewhat depressed at base; culmen decurved toward tip; upper mandible notched near tip. *Tail:* shorter than wing (except in gnatcatchers, in which it is longer than wings); variable in shape, usually square to rounded (except in kinglets, in which it is emarginate); rectrices broad and either acuminate at tips (kinglets) or rounded at tips (gnatcatchers). *Wings:* long, rounded. *Feet:* moderately strong; tarsi longer than middle toe with nail, and either booted (kinglets) or scutellate (gnatcatchers). *Plumage:* nostrils either exposed or partly covered by bristle-like feathers.

Subfamily Turdinae. Thrushes. (Length: 5–12 in.; 12.7–30.48 cm)

Bill: variable in length, usually short, straight; slender; compressed; culmen decurved toward tip; upper mandible notched near tip. *Tail:* usually shorter than wings; square to slightly rounded. *Wings:* long; pointed. *Feet:* strong; tarsi usually longer than middle toe with nail, and booted. *Plumage:* juvenal plumage always spotted above and below.

Subfamily Timaliinae. (Genus *Chamaea* only). Wrentits. (Length: approximately 6 in.; 15.24 cm)

Bill: short; straight; stout; compressed; culmen very decurved; nostrils entirely exposed. *Tail:* much longer than wings; graduated; rectrices narrow, but rather broad toward rounded tips. *Wings:* short; rounded. *Feet:* strong; tarsi much longer than middle toe with nail. *Plumage:* soft; lax; slightly crested; lores with bristly feathers, coloration plain olive-brown above.

Family Mimidae. Thrashers and Mockingbirds. (Length: 8–12 in.; 20.32–30.48 cm)

Bill: variable in length; usually rather slender, terete, and decurved toward tip; upper mandible notched toward tip (except in genus *Toxostoma*). *Tail:* variable in length, usually somewhat longer than wings; rounded, sometimes graduated. *Wings:* variable in length, usually short; rounded; 10 primaries. *Feet:* strong; tarsi distinctly longer than middle toe without nail and scutellate in front (often booted behind). *Plumage:* rictal bristles evident.

Family Motacillidae. Wagtails and Pipits. (Length: 4.5–8 in.; 11.43–20.32 cm)

Bill: short; straight; slender; acute; culmen decurved toward tip; upper mandible notched near tip. *Tail:* variable in length, never shorter than wings; variable in shape; rectrices narrow and acuminate (except possibly middle pair). *Wings:* long; pointed; tertiaries elongated, nearly equaling primaries in length; nine primaries. *Feet:* strong; tarsi usually longer than middle toe with nail, and scutellate; hallux nail elongated, equaling hallux in length. *Plumage:* rictal bristles present but not evident.

Family Bombycillidae. Waxwings. (Length: 6–7.5 in.; 15.24–19.05 cm)

Bill: short; stout; straight; upper mandible slightly hooked, with notch near tip; culmen decurved; gape deeply cleft and wide, nearly equal to length of exposed culmen. *Tail:* shorter than wing; square to slightly rounded; upper tail coverts greatly elongated. *Wings:* long; pointed; 10 primaries, the outermost very short, being less than half as long as the primary coverts. *Feet:* strong; tarsi shorter than middle toe without nail, and scutellate. *Plumage:* soft and dense; nostrils clearly concealed by small dense feathers; head crested; coloration predominantly brownish, with black band from bill through eye, and with tail tipped yellow, orange, or red.

Family Ptilogonatidae. Silky Flycatchers. (Length: 7–9 in.; 17.78–22.86 cm)

Bill: short; stout (less so than in Bombycillidae); straight; upper mandible slightly hooked, with notch near tip; culmen decurved; gape deeply cleft and wide, much less than length of exposed culmen. *Tail:* usually equal to, or longer than, the wings; variable in shape. *Wings:* short; rounded; 10 primaries, the outermost much longer than the primary coverts. *Feet:* strong; tarsi usually shorter than middle toe with nail, and scutellate; hallux very short. *Plumage:* soft rictal bristles evident; head usually crested; coloration plain without markings.

Family Laniidae. (Subfamily Laniinae only.) American Shrikes. (Length: 8–10 in.; 20.32–25.4 cm)

Bill: short; compressed; hooked; upper mandible toothed near tip. *Tail:* variable in length, being nearly as long as, and sometimes much longer than, the wings; variable in shape, being either square, rounded, or graduated. *Wings:* short, rounded, 10 primaries. *Feet:* strong; tarsi longer than middle toe with nail, and scutellate; nails very curved and acute. *Plumage:* soft; rictal bristles evident; nostrils fringed by bristle-like feathers; coloration of sexes alike; with plain gray and brown, mixed with black and white, predominating.

Family Sturnidae. (Genus *Sturnus* only.) Starlings. (Length: 8–9 in.; 20.32–22.86 cm)
Bill: as long as head; straight; slightly depressed toward tip; commissure somewhat angulated; feathers of forehead partially divided by culmen. *Tail:* short, being half the length of the wings; nearly square to slightly emarginate. *Wings:* long; pointed; 10 primaries, the outermost rudimentary and acuminate. *Feet:* strong; tarsi longer than middle toe without nail, and scutellate. *Plumage:* feathers of head, neck, and breast long and narrow; coloration metallic and somewhat iridescent.

Family Vireonidae. Vireos. (Length: 5–7 in.; 12.7–17.78 cm)
Bill: usually short; rather straight; somewhat compressed at base; hooked; upper mandible notched near tip. *Tail:* usually much shorter than wings; slightly rounded or emarginate; rectrices narrow. *Wings:* long; variable in shape, usually somewhat rounded; 10 primaries but the outermost rudimentary. *Feet:* strong; tarsi longer than middle toe with nail, and scutellate. *Plumage:* rictal bristles present but not evident; nostrils somewhat concealed by bristle-like feathers; coloration of sexes alike, with plain olive, olive-green, or gray above.

Family Emberizidae. Wood Warblers, Tanagers, Cardinal-Grosbeaks, Sparrows, and Blackbirds. (Length: 4–19 in.; 10.16–48.26 cm)
Bill: extremely variable in length and shape; nostrils exposed. *Tail:* extremely variable in length and shape; *Wings:* variable in length and shape; nine primaries. *Feet:* strong; tarsi variable in length, and scutellate. *Plumage:* rictal bristles usually present (except Icterinae; see below).

Subfamily Parulinae. Wood Warblers. (Length: 4–7.5 in.; 10.16–19.05 cm.)
Bill: variable in length, usually short; usually slender, straight, compressed, and acute, with variations too numerous to itemize. *Tail:* generally shorter than wings; varying from square to slightly rounded. *Wings:* variable in length and shape but long and somewhat pointed. *Feet:* tarsi usually less than twice as long as middle toe without nail. *Plumage:* rictal bristles present (sometimes conspicuous, sometimes not evident) or absent; males usually brightly colored.

Subfamily Thraupinae. (Genus *Piranga* only.) Tanagers. (Length: 6.5–8 in.; 16.51–20.32 cm)
Bill: usually as long as head; somewhat conical, stout, and swollen; slightly hooked; upper mandible notched near tip and its tomium toothed near middle. *Tail:* shorter than wings; square to slightly rounded, sometimes emarginate. *Wings:* moderately long; more or less pointed. *Feet:* tarsi longer than middle toe with nail. *Plumage:* rictal bristles present but not conspicuous; coloration in adult males more or less red, replaced by olive-green in females.

Subfamily Cardinalinae. Buntings, Cardinal–Grosbeaks, and Dickcissels. (Length: 4–8 in.; 10.16–20.32 cm)
Bill: short; stout; conical; culmen slightly decurved; commissure sharply angulated; lower mandibular tomia flat, not rolled inward; nasal fossa oval. *Tail:* variable in length and shape. *Wings:* variable in length and shape. *Feet:* tarsi variable in length; hallux nail equal to middle toe nail. *Plumage:* rictal bristles usually present; males often brightly colored.

Subfamily Emberizinae. Towhees, Sparrows, and Longspurs. (Length: 4–8 in.; 10.16–20.32 cm)
Bill: short; stout; conical; commissure angulated; lower mandibular tomia rolled inward; nasal fossa somewhat triangular. *Tail:* extremely variable in length and shape. *Wings:* variable in length and shape but generally rounded. *Feet:* tarsi variable in length; hallux nail distinctly longer than middle toe nail. *Plumage:* rictal bristles usually present; sexes in sparrows usually alike.

Subfamily Icterinae. Blackbirds, Orioles, and Meadowlarks. (Length: 7–19 in.; 17.78–48.26 cm)
Bill: variable in length, seldom conspicuously longer than head; somewhat conical; acute; culmen sometimes slightly decurved, elevated toward base, and extending far back to part the feathers of the forehead; commissure somewhat angulated. *Tail:* variable in length and shape, usually rather short (always more than half as long as wings, never conspicuously longer) and rounded. *Wings:* variable in length and shape, usually long and pointed. *Feet:* very strong; tarsi usually equal to, or slightly longer than, middle toe with nail.

Family Fringillidae. (Subfamily Carduelinae only.) Finches, Grosbeaks, and Crossbills. (Length: 3.5–8 in.; 8.89–20.32 cm)
Bill: short; stout; conical (mandibles crossed in crossbills); culmen decurved; commissure abruptly angulated. *Tail:* variable in length and shape but generally square or slightly forked. *Wings:* variable in length and shape; nine primaries. *Feet:* tarsi variable in length, but always scutellate. *Plumage:* nostrils concealed by tufts of bristle-like feathers; rictal bristles usually present.

Family Passeridae. (Genus *Passer* only.) Old World Sparrows. (Length: 6–7 in.; 15.24–17.78 cm)
Bill: short; conical; culmen slightly decurved; commissure angulated. *Tail:* shorter than wings; somewhat square. *Wings:* long; pointed; 10 primaries, the outermost rudimentary. *Feet:* strong; tarsi shorter than middle toe without nail, and scutellate. *Plumage:* rictal bristles present but not conspicuous.

Laboratory Identification

Having learned the methods used in classifying and naming birds, and having also acquired a preliminary knowledge of certain characters used in distinguishing birds, the next step is to determine the categories to which given birds belong. This is most quickly and satisfactorily accomplished by the use of **keys** based on external structural characters.

Considerable practice is needed in order to use the keys effectively. Therefore, with the guidance of the instructor, identify a series of specimens representing a diversity of orders, families, and species.

The instructor will make available a series of unidentified specimens of birds native to North America north of Mexico. Each specimen will be numbered. On ruled sheets, number the lines to correspond to the numbers in the series of specimens. Also, on the same ruled sheets, mark off three vertical columns, the first headed "Order," the second "Family," and the third "Species." Then proceed with identification, starting with any number in the series.

Identification of Orders and Families

Using the following keys to the orders and families, "run down" the specimens to their proper categories. On the numbered lines, which correspond to the numbers on the specimens, and under the appropriate column headings, write the technical names of the orders and families as identified. After the specimens have been run down, have the identifications checked by the instructor.

Key to the Orders of Birds of North America North of Mexico

(Roman numerals after the names of orders refer to the position of the same orders in the key to families. See page 154.)

A. Feet webbed; neither semipalmated nor lobed.
  1. All four toes joined in web (totipalmate) .... PELECANIFORMES (IV)
  2. Only front toes joined in web (palmate).
    a) Nostrils tubular .... PROCELLARIIFORMES (III)
    b) Nostrils not tubular.
      (1) Bill lamellate or serrate.
        (*a*) Bill decurved
          i) Bill slender and decurved abruptly .... CICONIIFORMES (V)
          ii) Bill stout and bent in middle .... PHOENICOPTERIFORMES (VI)

(*b*) Bill more or less straight with nail-like hook at tip of upper mandible.......... ANSERIFORMES (VII)

(2) Bill not lamellate or serrate.

(*a*) Legs inserted far behind middle of body; tarsi compressed.

i) Hallux present.......... GAVIIFORMES (I)

ii) Hallux absent.......... CHARADRIIFORMES (XI)

(*b*) Legs not inserted far behind middle of body; tarsi rounded.......... CHARADRIIFORMES (XI)

B. Feet not webbed; may be semipalmated or lobed.

1. Toes four.

a) Toes three in front and one behind.

(1) Toes with flattened nails.......... PODICIPEDIFORMES (II)

(2) Toes without flattened nails.

(*a*) Feet syndactyl.......... CORACIIFORMES (XIX)

(*b*) Feet not syndactyl.

i) Feet raptorial.......... FALCONIFORMES (VIII)

ii) Feet not raptorial.

1) Middle toe pectinate.

a. Bill long and acute; legs long; lores bare; plumage ordinary.......... CICONIIFORMES (V)

b. Bill very short and slightly hooked; feet small and weak; lores feathered; plumage soft.......... CAPRIMULGIFORMES (XVI)

2) Middle toe not pectinate.

a. Small birds abundantly colored metallic green above; or uniformly sooty brown throughout; or sooty brown except for white on chin, throat, or rump; tarsus without obvious scales; feet small and weak.......... APODIFORMES (XVII)

b. Birds without the above combination of characteristics.

(i) Lores entirely bare; rest of head feathered entirely or in part.......... CICONIIFORMES (V)

(ii) Lores feathered; or bare when rest of head is bare.

a) Entire head bare.

i. Hallux incumbent.......... CICONIIFORMES (V)

ii. Hallux elevated.

(a) Nostrils perforate.......... FALCONIFORMES (VIII)

(b) Nostrils imperforate.......... GALLIFORMES (IX)

b) Entire head not bare.

i. Hallux incumbent.

(a) Nails acute.......... PASSERIFORMES (XXI)

(b) Nails obtuse.......... GALLIFORMES (IX)

ii. Hallux elevated.

(a) Bill with operculum.......... COLUMBIFORMES (XII)

(b) Bill without operculum.

i) Bill short and stout; tip of upper mandible noticeably curved over lower one; all primaries stiff and curved.......... GALLIFORMES (IX)

ii) Bill may or may not be short and stout; tip of upper mandible not noticeably curved over lower one; all primaries not stiff and curved.
- (*i*) Wings pointed (in one species the three outermost primaries are attentuate and short giving rounded effect .... CHARADRIIFORMES (XI)
- (*ii*) Wings rounded .... GRUIFORMES (X)

b) Toes two in front and two behind.
- (1) Eyes directed forward in facial discs; plumage soft .... STRIGIFORMES (XV)
- (2) Eyes not in facial discs; plumage ordinary.
  - (*a*) Bill with several "teeth" .... TROGONIFORMES (XVIII)
  - (*b*) Bill without "teeth."
    - i) Bill with cere and conspicuously hooked .... PSITTACIFORMES (XIII)
    - ii) Bill without cere and not hooked.
      - 1) Rectrices acuminate, with stiffened tips .... PICIFORMES (XX)
      - 2) Rectrices not acuminate, without stiffened tips .... CUCULIFORMES (XIV)

2. Toes three.
   - *a*) Two toes in front and one behind .... PICIFORMES (XX)
   - *b*) Three toes in front .... CHARADRIIFORMES (XI)

Key to the Families of Birds of North America North of Mexico

I. Order GAVIIFORMES. Family Gaviidae. Loons

II. Order PODICIPEDIFORMES. Family Podicipedidae. Grebes

III. Order PROCELLARIIFORMES. Tube-nosed Swimmers

Key to Families

A. Large birds 28–40 in. long (71.12–101.6 cm); nostrils opening on each side of the culmen in independent tubes .... Diomedeidae: Albatrosses

B. Medium-sized to small birds usually less than 28 in. long (71.12 cm); nostrils not opening in independent tubes.
1. Nostrils opening on top of the culmen in tubes separated only by a septum .... Procellariidae: Shearwaters and Fulmars
2. Nostrils opening on top of the culmen in one tube .... Hydrobatidae: Storm-Petrels

IV. Order PELECANIFORMES. Totipalmate Swimmers

Key to Families

A. Bill hooked.
1. Tarsus partly feathered, tail deeply forked .... Fregatidae: Frigatebirds
2. Tarsus not feathered, tail not forked.
   - *a*) Bill over 10 in. long (25.4 cm) with huge gular sac, bare and suspended from the lower mandible and extending its entire length .... Pelecanidae: Pelicans

b) Bill less than 4 in. long (10.16 cm) with inconspicuous gular sac .... Phalacrocoracidae: Cormorants

B. Bill not hooked.
  1. Middle rectrices filamentous; gular sac rudimentary and feathered .... Phaethontidae: Tropicbirds
  2. Middle rectrices not filamentous; gular sac small and bare.
    a) Tail rounded; middle rectrices with flutings .... Anhingidae: Anhingas
    b) Tail pointed; rectrices without flutings .... Sulidae: Gannets

V. Order CICONIIFORMES. Deep-water Waders

Key to Families

A. Middle toe nail pectinate; bill long, straight, and acute .... Ardeidae: Herons and Bitterns

B. Middle toe nail not pectinate; bill not acute.
  1. Large birds, never under 28 in. long (71.12 cm); bill stout at base, tapered to tip (bill decurved toward tip) .... Ciconiidae: Storks
  2. Smaller birds, never over 28 in. long (71.12 cm); bill either slender, decurved, and somewhat terete; or straight, very depressed, and spatulate .... Threskiornithidae: Ibises and Spoonbills

VI. Order PHOENICOPTERIFORMES. Family Phoenicopteridae. Flamingos

VII. Order ANSERIFORMES. Lamellate-billed Swimmers

Key to Subfamilies of Anatidae

A. Tarsus completely reticulate .... Anserinae

B. Tarsus scutellate in front .... Anatinae

Key to Tribes of Anserinae

A. Lores bare; neck longer than body .... Cygnini: Swans

B. Lores feathered; neck not longer than body.
  1. Medium-sized birds not longer than 20 in. (50.8 cm); wings rounded .... Dendrocygnini: Whistling-Ducks
  2. Large birds never shorter than 21 in. (53.34 cm); wings pointed .... Anserini: Geese

Key to Tribes of Anatinae

A. Rectrices narrow and stiff, with coverts extremely short .... Oxyurini: Ruddy and Masked Ducks

B. Rectrices of ordinary shape and texture, with coverts of ordinary length.
  1. Bill not broadly depressed (terete and serrate in Mergansers) .... Mergini: Mergansers, Eiders, and other sea ducks
  2. Bill broadly depressed and lamellate.
    a) Hallux broadly lobed; no iridescent speculum .... Aythyini: Pochards and Scaups
    b) Hallux slightly or not lobed; iridescent speculum usually present.
      (1) Head with prominent crest .... Cairinini: Wood Duck
      (2) Head not crested .... Anatini: Dabbling Ducks

VIII. Order FALCONIFORMES. Vultures and Diurnal Birds of Prey

Key to Families

A. Head bare, nostrils perforate; feet weakly raptorial .... Cathartidae: New World Vultures

B. Head feathered; nostrils imperforate; feet strongly raptorial.
   1. Nostrils circular, with central bony tubercle, or (as in caracaras) slit-like with posterior end uppermost.......... Falconidae: Falcons and Caracaras
   2. Nostrils oval or slit-like, with anterior end uppermost.......... Accipitridae

Key to Subfamilies of Accipitridae

A. Under surfaces of toes spiny; hallux approximately the same size of the other nails; under surfaces of nails rounded .... Pandioninae: Osprey

B. Under surfaces of toes smooth; hallux nail larger than other nails; under surfaces of nails grooved.......... Accipitrinae: Kites, Eagles, and Hawks

IX. Order GALLIFORMES. Gallinaceous Birds

Key to Families

A. Hallux incumbent.......... Cracidae: Chachalacas

B. Hallux elevated.......... Phasianidae

Key to Subfamilies of Phasianidae

A. Tarsus either partly or wholly feathered; nostrils feathered.......... Tetraoninae: Grouse and Ptarmigan

B. Tarsus and nostrils not feathered.
   1. Head bare; rectrices somewhat square at ends.......... Meleagridinae: Turkeys
   2. Head mostly feathered; rectrices ordinary, not square at ends.
      a) Small birds, less than 10 in. (25.4 cm); tail short and rounded.......... Odontophorinae: Quail
      b) Large birds, more than 20 in. (50.8 cm); tail long and pointed.......... Phasianinae: Pheasants

X. Order GRUIFORMES. Marsh Birds

Key to Families

A. Crown more or less bare .... Gruidae: Cranes

B. Crown feathered.
   1. Length of bird approximately 26 in. (66.04 cm); outer primary stiff, attenuate, and incurved .... Aramidae: Limpkins
   2. Length of bird not over 20 in. (50.8 cm); outer primary of ordinary shape .... Rallidae: Rails, Gallinules, Moorhens, and Coots

XI. Order CHARADRIIFORMES. Shorebirds, Gulls, and Auks

Key to Families

A. Wings spurred.......... Jacanidae: Jacanas

B. Wings not spurred.
  1. Toes not completely webbed.
    a) Toes four, but hallux sometimes inconspicuous.
      (1) Bill slender, pliable in life, and slightly depressed toward tip; bill frequently straight but sometimes either quite recurved or decurved; (toes lobed or with lateral membranes in Phalaropes only) .... Scolopacidae: Snipe, Woodcock, Sandpipers, and Phalaropes
      (2) Bill in lateral profile either constricted in middle or swollen toward tip, or constricted near base and tapering to an acute tip (Turnstones) .... Black-bellied Plover of Charadriidae and Turnstones of Scolopacidae

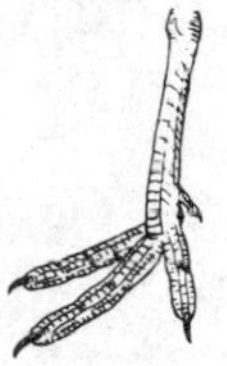

Toes with lateral membranes

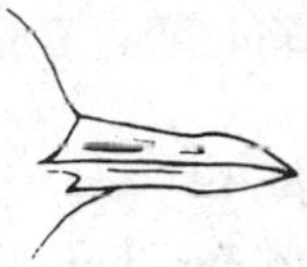

Bill of plover

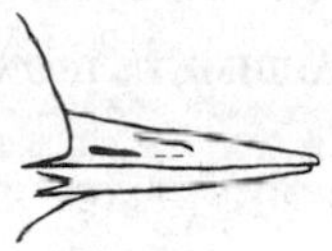

Bill of turnstone

    b) Toes three.
      (1) Tarsus reticulate.
        (*a*) Bill longer than head.
          i) Bill slender; legs extremely long and slender... Stilts of Recurvirostridae
          ii) Bill compressed, constricted toward base and chisel-like at tip; legs short and stout .... Haematopodidae: Oystercatchers
        (*b*) Bill never longer, usually shorter than head.. Charadriidae: Plovers
      (2) Tarsus scutellate.. Sanderlings of Scolopacidae
  2. Toes more or less completely webbed (palmate).
    a) Toes three .... Alcidae: Auks
    b) Toes four.
      (1) Bill strongly recurved .... Avocets of Recurvirostridae
      (2) Bill not recurved.. Laridae

Key to Subfamilies of Laridae

A. Lower mandible blade-like, compressed and decidedly longer than upper mandible.. Rynchopinae: Skimmers
B. Lower mandible not as above.
  1. Bill with a cere and a nail-like hook; tail rounded to pointed (middle rectrices often greatly elongated)... Stercorariinae: Skuas and Jaegers

2. Bill without a cere and without a nail-like hook.
   a) Bill plainly hooked .... Larinae: Gulls
   b) Bill straight, sharply pointed .... Sterninae: Terns

XII. Order COLUMBIFORMES. Family Columbidae. Pigeons and Doves

XIII. Order PSITTACIFORMES. Family Psittacidae. Parrots

XIV. Order CUCULIFORMES. Family Cuculidae. Cuckoos, Roadrunners, and Anis

XV. Order STRIGIFORMES. Nocturnal Birds of Prey

Key to Families

A. Middle toe nail pectinate .... Tytonidae: Barn-Owls
B. Middle toe nail not pectinate .... Strigidae: Typical Owls

XVI. Order CAPRIMULGIFORMES. Family Caprimulgidae. Goatsuckers

XVII. Order APODIFORMES. Swifts and Hummingbirds

Key to Families

A. Bill very short and depressed; gape extremely wide .... Apodidae: Swifts
B. Bill variable in length (usually long) and terete; gape narrow .... Trochilidae: Hummingbirds

XVIII. Order TROGONIFORMES. Family Trogonidae. Trogons

XIX. Order CORACIIFORMES. Family Alcedinidae. Kingfishers

XX. Order PICIFORMES. Family Picidae. Woodpeckers

XXI. Order PASSERIFORMES.

Key to Families

A. Tarsus rounded in front and behind.
   1. Bill wider than high; bill also hooked; rictal bristles present; hallux nail ordinary length .... Tyrannidae: Flycatchers
   2. Bill not wider than high; bill not hooked; rictal bristles not present; hallux nail extraordinarily long .... Alaudidae: Larks
B. Tarsus rounded in front and ridged behind.
   1. Tarsus booted.
      a) Rictal bristles present .... Muscicapidae (part)

Key to Subfamilies of Muscicapidae

(1) Small birds, not longer than 6 in. (15.24 cm); wings rounded
   (*a*) Wings shorter than tail; tail graduated; plumage lax .... Timaliinae: Wrentit
   (*b*) Wings longer than tail; tail emarginate; plumage ordinary .... Sylviinae (Tribe Sylviini only): Kinglets
(2) Birds of moderate size, not longer than 12 in. (30.48 cm); wings pointed .... Turdinae: Thrushes

b) Rictal bristles not present .... Cinclidae: Dippers

2. Tarsus not booted.
   a) Brownish birds with a combination of yellow-tipped tail and black band from bill through eye; head crested .... Bombycillidae: Waxwings
   b) Birds without the above color combination; head may or may not be crested.
      (1) Bill strongly hooked and notched.
         (*a*) Small birds, between 5 and 7 in. (12.7–17.78 cm) in length, with olive-green (occasionally olive or gray) upper parts .... Vireonidae: Vireos
         (*b*) Birds of moderate size, between 8 and 10 in. (20.32–25.4 cm) in length, with black, white, and gray plumage .... Laniidae: Shrikes
      (2) Bill not strongly hooked, though it may or may not be slightly notched.
         (*a*) Bill slender and decidedly decurved; rectrices stiff and acuminate .... Certhiidae: Creepers
         (*b*) Bill and rectrices without the above combination of characters.
            i) Nostrils entirely covered by feathers (or bristles) or birds mostly blue and greater than 8 in. (20.32 cm) in length.
               1) Wings with ten primaries; bill not conical or crossed.
                  a. Small birds between 3.5 and 6 in. (8.89–15.24 cm) in length.
                     (i) Yellow on head and throat .... Remizidae: Verdin
                     (ii) No yellow on head and throat.
                        a) Tail longer than wing; tail much rounded or graduated .... Aegithalidae: Bushtit
                        b) Tail shorter than or equal to wing; tail slightly rounded or nearly even .... Paridae: Chickadees and Titmice
                  b. Large birds, between 8 and 25 in. (20.32–63.5 cm) in length .... Corvidae: Crows, Jays, and Magpies
               2) Wings with nine primaries; bill conical or crossed .... Fringillidae: Finches, Grosbeaks, and Crossbills
            ii) Nostrils uncovered.
               1) Wings with ten primaries (outermost may be rudimentary).
                  a. Bill conical .... Passeridae: Old World Sparrows
                  b. Bill not conical.
                     (i) Bill somewhat depressed toward tip (when viewed dorsally); in adult plumage, feathers of head, neck, and breast long and narrow with metallic coloration; in immature plumage, upper parts grayish brown, under parts streaked with grayish white .... Sturnidae: Starlings
                     (ii) Combination of characters not as above.
                        a) Wings rounded.

i. Rictal bristles evident.
 (a) Head crested.............................. Ptilogonatidae: Phainopepla
 (b) Head not crested.
  i) Small birds, between 4 and 5 in. (10.16–12.7 cm) in length; upper parts bluish gray; under parts white; outer rectrices white............................ Muscicapidae (Tribe Polioptilini only): Gnatcatchers
  ii) Birds of moderate size, between 8 and 12 in. (20.32–30.48 cm) in length; coloration not always as above .... Mimidae: Thrashers and Mockingbirds

ii. Rictal bristles not evident; small brownish birds under 9 in. (22.86 cm) in length with bill slender and usually decurved; wings and tail indistinctly barred .... Troglodytidae: Wrens

b) Wings pointed.
 i. Legs relatively short and weak; hallux nail shorter than hallux itself.. Hirundinidae: Swallows
 ii. Legs of ordinary size and strength; hallux nail as long as hallux itself .... Motacillidae: Pipits and Wagtails

2) Wings with nine primaries.. Emberizidae

Key to Subfamilies of Emberizidae

A. Bill slender and straight; small variously colored birds under 6 in. (15.24 cm) in length except in one group (Yellow-breasted Chat), which is approximately 7 in. (17.78 cm) in length with bill stout and decurved......................... Parulinae: Wood Warblers

B. Bill conical.
 1. Commisure of bill not angulated; upper mandibular tomia toothed near middle... Thraupinae (Genus *Piranga*): Tanagers
 2. Commisure of bill angulated.
  a) Bill conical; rictal bristles usually present.
   (1) Hallux nail distinctly longer than middle claw nail; lower mandibular tomia usually rolled inward; plumage with no large areas of bright colors.......................... Emberizinae: Towhees, Sparrows, and Longspurs
   (2) Hallux nail approximately equal to middle claw nail; lower mandibular tomia usually flat; not rolled inward; males frequently with brightly colored plumage................. Cardinalinae: Buntings, Cardinal-Grosbeaks, and Dickcissel
  b) Bill less conical; commisure less abruptly angulated; rictal bristles absent; in some species culmen extends far backward and parts the feathers of forehead..................... Icterinae: Blackbirds, Orioles, and Meadowlarks

Identification of Species

Having identified the orders and families of the specimens, run them down to species. Enter their English names on the ruled sheets in the designated column. Use the keys to species in Part 5 by P. Brodkorb in *Vertebrates of the United States* (New York: McGraw-Hill. 1968); or obtain keys to species on the birds of the region where the study is undertaken. If the region is in North America east of the Mississippi River, use the keys in F.M. Chapman's *Handbook of Birds of Eastern North America* (D. Appleton and Company, 1932; Dover reprint available). If the region is in north-central North America, including the northern prairie states and prairie provinces, use T.S. Roberts' *A Manual for the Identification of Birds of Minnesota and Neighboring States,* 2nd ed. (Minneapolis: Univ. of Minnesota Press. 1955). Before using any of these keys, read the introduction in which the methods of measurements are explained and all terms are defined. A clear understanding of measurements and terms is necessary because the keys are based largely upon them.

After having identified the series of specimens to species, one is ready to identify specimens representing all the species found in the region. To save time, the instructor will group the various specimens by orders, or families. The label on each specimen may, or may not, give the name of the species, depending on the judgment of the instructor, but it should give the sex of the specimen and the locality and date where the specimen was collected. The keys already used will serve as adequate guides. If the specimens are already identified by species, use the keys anyway to gain a knowledge of distinctions. There is no better way than by keys to learn the structural characteristics of different species.

Plumages and Plumage Coloration

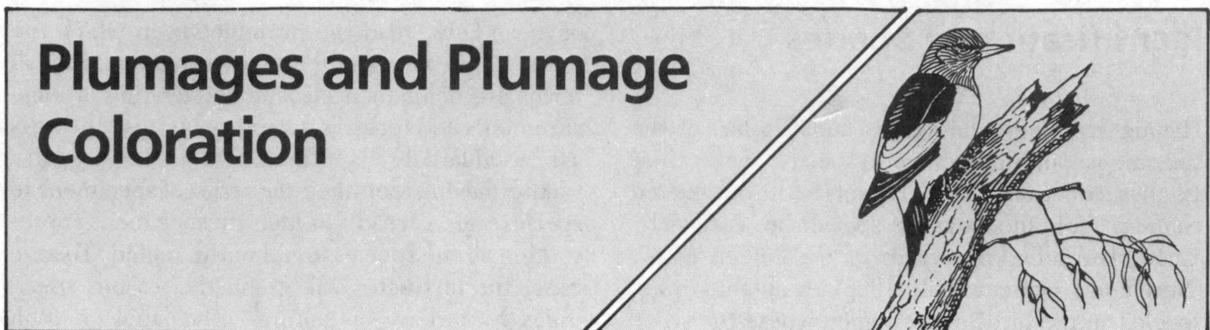

Birds periodically shed and renew their **plumage** or feather covering by a process known as **molting.** Every species normally exhibits a definite sequence of plumages and molts.

Sequence of Plumages and Molts

The sequence of plumages and molts in each species proceeds in a relatively consistent manner, so consistent that the successive plumages and intervening molts can be identified and named. Although there is much variation among birds in the extent and the timing of molts, the following summary is typical of many North American species.

Natal Down This plumage consists of down feathers. In birds belonging to such orders as the Anseriformes, Galliformes, and Charadriiformes, natal down covers the entire body; in birds belonging to the Passeriformes down occurs only on the pterylae of the upper part of the body; in birds belonging to such orders as the Piciformes the body is naked at hatching, entirely without feathers of any sort. In a few orders, such as the Gaviiformes and Procellariiformes, there are two successive, structurally continuous generations of natal downs. The natal down (or second natal down) is lost by the **postnatal molt,** which is always complete. The postnatal molt differs from later molts in that the feathers—the natal down feathers—are pushed out from their follicles by the tips of the feathers comprising the next generation—the juvenal feathers. In due course the down feathers, which are structurally continuous with the juvenal feathers, become dislodged, drop off, and disappear.

Juvenal Plumage Consisting of juvenal feathers, which are the first true contour feathers. In passerine species and some others the juvenal feathers are noticeably loose-textured, giving the plumage a soft, fluffy appearance. Species that remain in the nest long after hatching acquire the juvenal plumage before they leave; species that leave the nest soon after hatching acquire it later and often more slowly. The juvenal plumage is lost during the late summer and fall of the year by the **postjuvenal molt,** which may or may not be complete.

First Winter Plumage Retained at least throughout the first winter. It has the same texture as the adult plumage but is frequently different in coloration. In species that normally have, as adults, two molts a year, it is lost in late winter or early spring by a **first prenuptial molt** involving varying amounts of body feathers but—with very rare exceptions, such as the Franklin's Gull *(Larus pipixcan)*—not the remiges. In species that lack this molt, the first winter plumage is retained until the following summer or fall.

First Nuptial Plumage Retained during the first breeding (i.e., nesting) season. In certain species the males are less brilliantly colored in their first nuptial plumages than they are in later years. The nuptial plumage is lost immediately after the nesting season by the **first postnuptial molt,** which is always complete.

Second Winter Plumage Retained throughout the second winter of the bird's life and usually indistinguishable from winter plumages of later years. Depending on the species, it may be lost by the **second prenuptial molt.**

Second Nuptial Plumage Retained throughout the second breeding season and only rarely distinguishable from later nuptial plumages. It is lost immediately after the nesting season by the **second postnuptial molt.**

Third and Fourth Winter and Nuptial Plumages In a few species the plumages acquired in the third and fourth year of life are distinguishable from later plumages. The plumages and molts are, therefore, designated accordingly.

Adult Winter and Adult Nuptial Plumages When plumages become indistinguishable from those of adults, the numerical designation (first, second, etc.) is dropped before the names of the plumages and the term "adult" is substituted. In the majority of species, plumages become indistinguishable from adult plumages after the first year. In those species the plumages following the First Winter Plumage are commonly referred to as the Adult Nuptial Plumage and Adult Winter Plumage. The names of the intervening molts are changed correspondingly.

Eclipse Plumage This plumage is acquired by only certain birds, notably most male ducks, after the nesting season is under way. It is retained for only about two months, during which time the males very closely resemble the females. Actually the eclipse plumage is homologous to the winter plumage of other birds, being acquired by a postnuptial molt and lost by a hastening of the prenuptial molt. As a result, male ducks wear their bright nuptial plumage during the fall, winter, and spring. Female ducks have corresponding molts but usually offset chronologically from those of males. The two seasonal plumages of females differ only inconspicuously from one another.

The names of the plumages and their intervening

Activity

In the following chart, indicate, in the left column, the sequence of plumages and, in the adjacent column, the molts by which they are lost. Thoroughly memorize this chart because there will be many occasions in later studies when a knowledge of this sequence will be essential.

molts apply satisfactorily to most birds in the temperate regions and have been used extensively, with some variation (e.g., summer plumage instead of nuptial plumage, nonnuptial plumage instead of winter plumage), in most works on North American and European birds. They are used in this book. The names, however, are not entirely suitable for tropical, oceanic, and antarctic birds. The difficulty arises from a nomenclature based on seasonal and annual cycles. In some tropical species the molts—as well as the reproductive cycles—are correlated with rainy or dry seasons, rendering "winter plumage" meaningless. Other tropical species, in areas with little or no seasonality, may have long, protracted molting periods. Oceanic and antarctic species do not always breed annually; some breed less, others more often, than once a year. And some birds do not breed in their first "nuptial" plumage.

In an effort to derive a uniform series of terms for plumages and molts that express homologies between different groups of birds anywhere in the world, Humphrey and Parkes (1959) proposed a new terminology, which is being adopted in many works on birds. A comparison of the Humphrey-Parkes terminology is given below.

Traditional Terminology	Humphrey-Parkes Terminology
Natal Down	Natal Down
Postnatal Molt	Prejuvenal Molt
Juvenal Plumage	Juvenal Plumage
Postjuvenal Molt	First Prebasic Molt
First Winter Plumage	First Basic Plumage
First Prenuptial Molt	First Prealternate Molt
First Nuptial Plumage	First Alternate Plumage
First Postnuptial Molt	Second Prebasic Molt
Second Winter Plumage	Second Basic Plumage
Second Prenuptial Molt	Second Prealternate Molt
Second Nuptial Plumage	Second Alternate Plumage
Second Postnuptial Molt	Third Prebasic Molt
Etc.	Etc.

Humphrey and Parkes name the molt from the incoming plumage and confine the word "plumage" to a single generation of feathers acquired by a specific molt. In place of the "plumage" in the traditional sense they use the term **feather coat,** which constitutes all the feathers that a bird is wearing at a given time and may include not just one but two or more plumages or generations of feathers. Humphrey and Parkes choose the word "basic" in place of winter plumage because in adult birds, which have only one plumage per cycle, this is the one almost invariably lost and renewed by a complete molt; and they consider "alternate" the logical name for the nuptial plumage because it is the second or alternate plumage in birds that have two plumages per cycle. Humphrey and Parkes also suggest a useful term, **definitive,** for any plumage that will not change further with age. Thus, the plumage of the Bald Eagle *(Haliaeetus leucocephalus)* is definitive when, after four or more years, the head and tail feathers are completely white.

Molting in Adults

Almost all fully adult birds renew their feathers at least once a year by the postnuptial molt, and many birds renew twice a year by the additional prenuptial molt. The postnuptial molt is complete in most all species. The prenuptial molt, on the other hand, is complete in very few species. A third molt of certain pterylae, giving rise to a so-called **supplementary plumage,** occurs in the Oldsquaw *(Clangula hyemalis)* during the winter, the Ruff *(Philomachus pugnax)* and possibly some other scolopacids during the late winter/early spring, and ptarmigan (*Lagopus* spp.) during the summer.

In general, the prenuptial molt immediately precedes the nesting season, and the postnuptial molt immediately follows. Thus the times of the year during which a species molts depends on when its nesting season begins and when it ends. Adults whose breeding attempts failed, and nonbreeding individuals (often yearlings), usually undergo the postnuptial molt earlier than successful breeders. In most migratory species the prenuptial molt is completed before they arrive on the nesting grounds, and the postnuptial molt is completed before they start southern migration. In the northern United States and Canada the majority of species undergo the postnuptial molt from late July through August to early September, although a few

Activity

Examine the wings and tails of a series of specimens of a passerine bird arranged in a sequence to show the progress of molt.

species begin as early as June and a few others are still molting as late as October. Swallows and some tyrannid flycatchers do not even begin their postnuptial molt until they are on their wintering grounds.

Molting is a gradual process; the feathers of a particular pteryla are normally not shed all at once, and thus the bird is assured continuance of adequate plumage protection. Among groups of birds there are marked differences with respect to the molting of the flight feathers (remiges and rectrices).

In the majority of flying birds the remiges are shed and replaced in an orderly sequence so that flight is still possible during molt. In a passerine bird, for instance, the primaries, secondaries, and tertiaries are shed one by one. The innermost primary is usually the first remex to drop out. As soon as the feather replacing it is partly grown, the adjoining feather drops out. The remaining primaries are molted successively until all are replaced. Coincident with the dropping out of the fifth or sixth primary is the molt of the outermost tertiary, followed by the adjoining tertiary, and then the third tertiary, which in most passerine birds is the last or innermost. The molt of the secondaries, beginning with the outermost, follows the molt of the last tertiary and proceeds in series, the innermost secondary being the last secondary (and also the last remex) to drop out. The greater primary coverts are shed one at a time; each one is molted at about the same time as the primary it overlies. The greater secondary coverts, though molted successively and in the same direction as the secondaries, are dropped out and replaced much more rapidly than the secondaries; as a consequence, a new set of greater secondary coverts is fully developed before the molt of the secondaries is well under way.

In a few groups of aquatic birds—notably, the loons, grebes, swans, geese, and ducks—all the remiges are shed simultaneously, with the result that the birds are temporarily incapable of flight. This occurs usually before fall migration, except in loons, which delay the molt until they have reached southern coastal waters.

As a rule, the rectrices of flying birds are molted and replaced beginning in most passerine species and many others with the innermost pair, followed by the second innermost pair, and so on, until the outermost pair is shed. Thus the birds, even during molt, have ample tail surface for flight. There is more variation in the sequence of the rectrix molt than in that of the remiges, and in some species it is entirely irregular, following no fixed pattern. In tree-climbing birds, such as woodpeckers and creepers, the tail molt is similar, except that it begins with the second innermost pair, leaving the innermost pair to be shed last. In this way the birds during molt retain a tail that gives adequate support in climbing. Swans, geese, and ducks molt the rectrices simultaneously, as they do the remiges. The smaller owls also molt the rectrices simultaneously, but this peculiarity probably does not impair their flight, as they use their tails very little in flying.

Plumage-Change Without Molt

A number of species change the appearance of their winter plumage before the breeding season without molting. This is accomplished in two ways: (1) by **wear**, or abrasion—i.e., contrastingly colored tips of feathers freshly acquired by the postnuptial molt wear away during the winter, revealing, before the breeding season, the basic color of the feathers; (2) by **fading**—i.e., the color of the winter plumage gradually fades until, by the time of the breeding season, it has become noticeably lighter.

Both of these processes continue through the spring and summer; the feathers that are replaced at the time of the postnuptial molt may be decidedly worn, especially in grass inhabiting birds (e.g., sparrows and meadowlarks), and faded to the point of bleaching in birds exposed to much sunlight as in desert and arctic regions.

Activity

1. Examine bird skins of one or more species arranged in sequence to show plumage-change by wear and by fading.

2. Study a series of male specimens of six species, the Mallard *(Anas platyrhynchos)*, Ring-billed Gull *(Larus delawarensis)*, Black-capped Chickadee *(Parus atricapillus)*, American Redstart *(Setophaga ruticilla)*, Bobolink *(Dolichonyx orizivorus)*, and Rose-breasted Grosbeak *(Pheucticus ludovicianus)*, that will show the sequence of plumages and molts in their first two years while at the same time demonstrating some of the differences among species as to the relative completeness of certain molts and the modification of molts by age. Except in some of the waterfowl (Anatidae), the sequence and timing of molts in females are essentially like those in males. Note the date when each specimen was collected. During the study consult the chart, "Sequences of Plumages in the Males of Six Species," on the inside of the back cover of this book. The Humphrey-Parkes names for plumages and molts are in parentheses. Not shown in the chart is the fully adult plumage of the Ring-billed Gull with entirely white tail, acquired in its third year by the Second Postnuptial (Third Prebasic) Molt.

Plumage Coloration

Abnormal Plumage Coloration

Birds sometimes vary from their normal coloring because of the lack or excess of pigments in one or more of the feathers. There are four variations.

Albinism This is a variation caused by the reduction or absence of melanin pigments in the feathers. In the complete absence of these pigments the feathers' structure refracts light, giving white. The pigments may also be absent from the irises and from the normally featherless, colored skin of the lores, forehead, crown, and other topographical parts of certain species of birds. The colors of these areas will then have a light red or pinkish hue, owing to the blood showing through from the capillaries. There are four degrees of albinism in birds.

Total albinism When all melanin pigments are completely absent from the plumage, irises, and skin. Total albinos of species that have carotenoid pigments in parts of the plumage normally retain these; thus an albino male Rose-breasted Grosbeak *(Pheucticus ludovicianus)* will be pure white except for the pink of the breast and under wing coverts, which will retain their color.

Incomplete albinism When the pigments are completely absent from the plumage, or irises, or skin, but not from all three.

Imperfect albinism When all the pigments are reduced ("diluted"), or at least one of the pigments is absent, in any or all three areas.

Partial Albinism When the pigments are reduced, or one or more is absent, from *the parts* of any or all three areas. Of the four degrees of albinism, partial albinism is the commonest. It frequently involves certain feathers only, and it is often symmetrical, each side of the bird being affected in the same way.

Albinism may have a genetic basis and be inherited. It may also be spontaneous, developing in an individual as a result of some physiological disturbance. An instance is reported (Frazier, 1952) of an American Robin *(Turdus migratorius)* that was in normal plumage when banded, but was partially albinistic when recovered two years later.

Melanism, Erythrism, and Xanthochroism These are variations in plumage color caused by either excessive amounts of pigments or the absence of certain pigments. (See "The Coloration of Feathers") pages 42–47, for information on the different pigments.) Melanism results from large amounts of melanin, giving colors ranging from brown to black; erythrism and xanthochroism usually result from the absence of melanin and the retention of red or yellow carotenoid pigments. Xanthochroism may develop in captive birds, due to dietary deficiency. Thus green or blue feathers of parrots and red feathers of tanagers and finches are sometimes replaced by yellow feathers.

Color Phases

In certain wild species there are normally two or more color phases that presumably have a genetic basis. When a species has only two color phases, the condition is known as **dichromatism.** In Atlantic populations of the Common Murre *(Uria aalge)* many individuals show a "ringed" or "bridled" phase distinguished by a white line encircling the eye and then running backward and downward for a short distance. The Eastern Screech-Owl *(Otus asio)* and Ruffed Grouse *(Bonasa umbellus)* have "red" (erythristic) and "gray" phases, although they are not completely clearcut because intermediate individuals exist. When a species has two or more phases often intergrading, the condition is **polychromatism.** The Swainson's Hawk *(Buteo swainsoni)* and Rough-legged Hawk *(B. lagopus)* have "light" and "dark" (melanistic) phases; the Ferruginous Hawk *(B. regalis)* has "normal" and "rufous" phases. The Gyrfalcon *(Falco rusticolus)* features three phases designated as "white," "gray," and "black." Color phases are sometimes clearly correlated with geographical location. The Snow Goose *(Chen caerulescens)* has a pigmented phase, formerly considered to be a separate species ("Blue Goose"), that breeds only in the center of its range near Hudson Bay but not in the westernmost or easternmost parts of its range.

Sexual, Age, and Seasonal Differences in Plumage Coloration

Within a species, plumage coloration is often subject to differences, sometimes extreme, dependent on sex, age, and season. The usual differences within many familiar species may be summarized as follows:

1. The nuptial plumage of the adult male is more colorful, or more strikingly marked, than the plumage of the adult female which is characteristically dull and inconspicuously marked the year-round.

2. The colorful, or strikingly marked, nuptial plumage of the male is replaced, following the breeding season, by a winter plumage closely resembling the plumage of the adult female.

3. The juvenal plumage of the immature bird, regardless of sex, is characteristically dull, or inconspicuously marked, and closely resembles the plumage of the adult female.

Activity

Examine several bird skins that show some of the color abnormalities and phases described in text. In the following chart, list the names of the species and state the kinds of abnormalities or phases observed.

Species	Color Abnormality or Phase

Activity

Table 2 on page 169 gives the usual differences and several well-known exceptions in sexual, age, and seasonal differences in plumage coloration. Complete the chart in the following manner: (1) Fill in the spaces across the top of the chart with the names of species selected to illustrate the differences and exceptions. (2) Examine skins of these species. Then, in the columns under the species names, put a check in the space opposite the usual difference or exception that each species represents. *Note:* when attempting to describe quality of coloration, use these criteria: Plumage is *colorful* when possessing one or more brilliant colors such as red, yellow, green, blue, etc.; *strikingly marked* when possessing conspicuous, contrasting dark or white markings with or without brilliant colors; *dull* when possessing colors predominantly brown or gray; and *inconspicuously marked* when the markings are not sharply contrasted and are not bright in color.

While many species conform to the above statements, a great number of species do not. For example, the sexes are alike in color and show no seasonal change in petrels, geese, most hawks and owls, swifts, crows and jays, mimids, and many others. The sexes are alike in color, but both molt into duller winter plumages in loons, grebes, most shorebirds, auks, and terns. Juveniles look essentially like the adults, although slightly duller in color, in petrels, swifts, crows, most jays, and titmice.

Uses of Plumage Coloration

Plumage coloration has numerous uses, or functions, some of which are roughly classified as cryptic, deflective, and epigamic.

Cryptic Coloration When coloration serves to conceal the bird in its environment, it is cryptic. Cryptic coloration functions in two ways: (1) It may conceal by **mimicking** the normal background of the bird. A bird thus colored is frequently **countershaded,** having light under parts that counteract dark shadows cast by its body, making it appear flat against the background. The bird has a tendency to remain motionless when approached by an enemy and to flush only as a last resort. (2) It may conceal by having **ruptive patterns** — designs with irregular, sharply defined areas of contrasting colors that "break up" the general contour of the bird into apparently unrelated, shapeless parts. When a bird with ruptive pattern is seen at a distance, these separate areas are more conspicuous than the bird itself.

Deflective Coloration Certain birds have feathered areas colored in such a way as to become conspicuous when the birds are in flight. Examples are the white areas on the wings, rump, and tail, which are sometimes called **banner-marks.** These areas are considered to have a protective function in that they deflect a pursuing enemy's attack from a vital to a less vital part of the body. These areas are sometimes considered to have another protective function in that they serve to confuse an enemy and, in a sense, deflect its attention. When, for instance, a bird suddenly flushes before an enemy, the conspicuous areas unexpectedly appear and momentarily startle it, thus delaying (deflecting) the attack.

Epigamic Coloration When coloration is used to bring the sexes together in any manner during the breeding season, it is epigamic. Plumage colors found in males, but not in females, are a means of intraspecific sex recognition. Facial markings, masks, and throat patches are examples. Brilliant colors in the feathers of many males enhance the role of plumage display prior to and during mating.

Bear in mind that plumage coloration serves many uses other than those classified and described

Table 2 **Sexual, Age, and Seasonal Differences in Plumage Coloration**

Adult Plumage Usual: *♂ in breeding season only: more colorful, or more strikingly marked, than ♀*													
Exceptions: ♂ the year-round: more colorful, or more strikingly marked, than ♀													
♂ and ♀ similar: both colorful, or strikingly marked, in breeding season only													
♂ and ♀ similar: both colorful, or strikingly marked, the year-round													
♂ and ♀ similar: neither colorful nor strikingly marked the year-round													
♀ in breeding season only: more colorful, or more strikingly marked, than ♂													
♀ the year-round: more colorful, or more strikingly marked, than ♂													
Juvenal Plumage Usual: *Immature birds dull, or inconspicuously marked, and similar to adult ♀; unlike adult ♂*													
Exceptions: Immature birds dull, or inconspicuously marked, but different from both adult ♂ and ♀													
Immature birds similar to, and as colorful as, both adult ♂ and ♀													

***Table* 3 Uses of Plumage Coloration**

	Cryptic Coloration			Deflective Coloration			Epigamic Coloration	
	Mimicking	Countershading	Ruptive Patterns	Banner-marks on Tail	Banner-marks on Wings	Banner-marks on Rump	Sex-recognition Marks	Display-colors

Activity

Examine a group of bird skins to identify cryptic, deflective, and epigamic coloration. On Table 3, list the species examined in the left-hand column. Check the spaces opposite the species that illustrate the types of coloration indicated across the top.

above. Certain plumage colors in the same species may have more than one function. Banner-marks, while serving as a protective feature in one way or another, may act in intraspecific recognition as signals to keep members of a flock together. Epigamic coloration, besides bringing the opposite sexes together, may assist in threat displays against rivals of the same sex.

References

Bailey, S.F.
1978 Latitudinal gradients in colors and patterns of passerine birds. *Condor* 80:372–381. (Study indicates that "tropical passerines in general are no more colorful than temperate zone passerines in general, with respect to number and area of bright colors.")

Behle, W.H., and Selander, R.K.
1953 The plumage cycle of the California Gull *(Larus californicus)* with notes on color changes of soft parts. *Auk* 70:239–260.

Clay, W.M.
1953 Protective coloration in the American Sparrow Hawk. *Wilson Bull*. 65:129–134. (Shows how countershading, ruptive patterns, and deflective coloration in *Falco sparverius* assist the species in deceiving prey and/or enemies.)

Cott, H.B.
1964 Coloration, adaptive. In *A new dictionary of birds,* ed, A.L. Thompson. New York: McGraw-Hill.

Dwight, J., Jr.
1900a The sequence of plumages and moults of the passerine birds of New York. *Annals of New York Acad. Sci*. 13:73–360. (Reprinted in 1975 by Dover Publications, New York, with an added index to scientific and English names.)
1900b The moult of the North American shore birds (Limicolae). *Auk* 17:368–385.
1920 The plumages of gulls in relation to age as illustrated by the Herring Gull *(Larus argentatus)* and other species. *Auk* 37:262–268.

Foster, M.S.
1975 Temporal patterns of resource allocations and life history phenomena. *Florida Scientist* 38:129–139. (Discusses timing of molting in relation to breeding.)

Frazier, F.P.
1952 Depigmentation of a Robin. *Bird-Banding* 23:114.

Friedmann, H.
1944 The natural-history background of camouflage. *Smithsonian Rept. for 1943,* pp. 259–274.

Graul, W.D.
1973 Possible functions of head and breast markings in Charadriinae. *Wilson Bull*. 85:60–70. (Study of 37 species; markings may have a disruptive function, may also enhance sex recognition.)

Gross, A.O.
1965a The incidence of albinism in North American birds. *Bird-Banding* 36:67–71.
1965b Melanism in North American Birds. *Bird-Banding* 36:240–242.

Gullion, G.W., and Marshall, W.H.
1968 Survival of Ruffed Grouse in a boreal forest. *Living Bird* 7:117–167. (In Minnesota the gray color phase predominates in grouse populations occupying northern coniferous forests, the red phase in southern hardwoods. There appears to be a correlation between color phase and rate of survival.)

Holmes, R.T.
1966 Molt cycle of the Red-backed Sandpiper *(Calidris alpina)* in western North America. *Auk* 83:517–533.

Humphrey, P.S., and Clark, G.A., Jr.
1964 The anatomy of waterfowl. In *The waterfowl of the world,* vol. 4, by J. Delacour. London: Country Life. ("Plumage Succession," pp. 175–180.)

Humphrey, P.S., and Parkes, K.C.
1959 An approach to the study of molts and plumages. *Auk* 76:1–31. (The authors propose a new terminology for plumages and molts.)
1963 Comments on the study of plumage succession. *Auk* 80:496–503. (The authors' response to the criticism by Stresemann, 1963, of the new terminology in their previously listed paper.)

Huxley, J.S.
1938 Threat and warning coloration in birds with a

general discussion on the biological functions of colour. *Proc. VIIIth Internatl. Ornith. Congr.* pp. 430–455.

Johnson, N.K.
1963 Comparative molt cycles in the tyrannid genus *Empidonax*. *Proc. XIIIth Internatl. Ornith. Congr.* pp. 870–883. (A great interspecific variation in the timing of molts, suggesting that it is "a reflection of adaptation at the species level.")

Lesher, S.W., and Kendeigh, S.C.
1941 Effect of photoperiod on molting of feathers. *Wilson Bull.* 53:169–180. (Experimental procedures show that captive birds, *Zonotrichia albicollis* and *Colinus virginianus,* which normally go through a prenuptial molt in spring as the days are increasing in length, will molt out of season when exposed to light-periods artificially increased to 15 hours.)

Mayr, E., and Mayr, M.
1954 The tail molt of small owls. *Auk* 71:172–178. (The authors suggest that simultaneous tail molt may be due to a relaxation of selection pressure.)

Middleton, A.L.A.
1977 The molt of the American Goldfinch. *Condor* 79:440–444. (Species unique among carduelines in acquiring its breeding, or alternate, plumage by a prenuptial body molt.)

Miller, A.H.
1928 The molts of the Loggerhead Shrike, *Lanius ludovicianus* Linnaeus. *Univ. Calif. Publ. in Zool.* 30:393–417.

Mueller, C.D., and Hutt, F.B.
1941 Genetics of the fowl. 12. Sex-linked, imperfect albinism. *Jour. Heredity* 32:71–80. (Contains the classification of albinism used, with slight modification, in this book.)

Nero, R.W.
1954 Plumage aberrations of the Redwing *(Agelaius phoeniceus). Auk* 71:137–155. (The author found a surprisingly high percentage of aberrations, classified as total albinism; imperfect albinism; partial albinism—random and specific. Practically all 219 males collected near Madison, Wisconsin, showed some deviation from the wholly black plumage.)

1960 Additional notes on the plumage of the Red-winged Blackbird. *Auk* 77:298–305. (The author found a similarly high percentage of aberrations among 100 males near Regina, Saskatchewan, leading him to suggest that such aberrations are characteristic of the species.)

Niles, D.M.
1972 Molt cycles of Purple Martins *(Progne subis). Condor* 74:61–71. (Describes certain aspects of the postnuptial and postjuvenal molts, with emphasis on age-related or sex-related differences in timing and extent of molts. Also discusses molts relative to other major events of the species' annual cycle.)

Noble, G.K.
1936 Courtship and sexual selection of the Flicker *(Colaptes auratus luteus). Auk* 53:269–282.

Noble, G.K., and Vogt, W.
1935 An experimental study of sex recognition in birds. *Auk* 52:278–286.

Oring, L.W.
1968 Growth, molts, and plumages of the Gadwall. *Auk* 85:355–380. (An excellent study of an individual species.)

Palmer, R.S.
1972 Patterns of molting. In *Avian Biology,* Vol. 2, ed. D.S. Farner and J.R. King. New York: Academic Press.

Payne, R.B.
1972 Mechanisms and control of molt. In *Avian Biology,* Vol. 2, ed. D.S. Farner and J.R. King. New York: Academic Press.

Petrides, G.A.
1945 First-winter plumages in the Galliformes. *Auk* 62:223–227. (Age may be determined by winter feathers retained after postjuvenal molt in grouse, quail, and partridges but not in pheasants.)

Pitelka, F.A.
1945 Pterylography, molt, and age determination of American jays of the genus *Aphelocoma*. *Condor* 47:229–260.

Rohwer, S.
1978 Passerine subadult plumages and the deceptive acquisition of resources: Test of a critical assumption. *Condor* 80:173–179. (Subadult plumages of year-old males of certain temperate passerines have "a female mimicry feature which deceives older males and, thus, enables young males to settle in good habitat.")

Salomonsen, F.
1939 Moults and sequence of plumages in the Rock Ptarmigan *(Lagopus mutus* (Montin)). *Vid. Medd. Dansk Naturh. Foren.* 103:1–491. Copenhagen: Reprinted by P. Haase and Son. (A thorough study based on an enormous amount of data.)

Stresemann, E.
1963 The nomenclature of plumages and molts. *Auk* 80:1–8. (A strongly critical review of the new nomenclature proposed by Humphrey and Parkes, 1959.)

Stresemann, E., and Stresemann, V.
1966 *Die Mauser der Vögel*. Jour. f. Ornith. 107 Sonderheft. (Although titled "The Molting of Birds," this monograph of 445 pages is concerned almost exclusively with the molting of the remiges and rectrices in selected species.)

Sutton, G. M.
1935 The juvenal plumage and postjuvenal molt in several species of Michigan sparrows. *Cranbrook Inst. Sci. Bull.* No. 3, Bloomfield Hills, Michigan.

Test, F. H.
1945 Molt in flight feathers of flickers. *Condor* 47:63–72.

Thayer, G. H.
1909 *Concealing-coloration in the animal kingdom: An exposition of the laws of disguise through color and pattern*. New York: Macmillan. (Although much criticized, the work is nonetheless an important contribution to the subject.)

Tuck, L. M.
1960 *The Murres: Their distribution, populations and biology*. Canadian Wildlife Series, 1. Ottawa: Canadian Wildlife Service. (Genetic Variations, pp. 32–33.)

Verbeek, N. A. M.
1973 Pterylosis and timing of molt of the Water Pipit. *Condor* 75:287–292. (Breeding and molt partly overlap.)

Watson, G. E.
1963 Feather replacement in birds. *Science* 139:50–51. (Report of "passive" pushing out of old generation of pennaceous feathers; supports the Humphrey-Parkes system for naming plumages.)

Woolfenden, G. E.
1967 Selection for a delayed simultaneous wing molt in loons (Gaviidae). *Wilson Bull.* 79:416–420. (The author suggests selective factors that may account for the simultaneous molt following fall migration.)

Distribution

Despite their remarkable flying abilities, most birds are limited, as are other animals, to specific areas called **ranges.** Only a few bird species have ranges that are cosmopolitan. The Osprey *(Pandion haliaetus)* that resides in all the continents except Antarctica is one example of a cosmopolitan species. Some birds represent the opposite extreme by having highly restricted ranges. The Kirtland's Warbler *(Dendroica kirtlandii),* for instance, nests solely in an area of Michigan, 100 by 60 miles (160.9 by 96.54 km), and winters in the Bahamas. Other species, confined the year round to oceanic islands or archipelagos, may have still smaller ranges.

Most birds occupy only parts of the ranges to which they are particularly suited and therefore are not evenly distributed. In some species the ranges are clearly **disjunct** or **discontinuous** with great distances between segments. The Scrub Jay *(Aphelocoma coerulescens)* has a small population in peninsular Florida, separated by hundreds of miles from the rest of its population in the far western contiguous states and Mexico.

Numerous factors account for the present ranges of species. Some are past or historical factors, such as the shaping, separating, and rejoining of continents; the invasion and recession of glaciers; the shifts in climate; and the changes in dominant vegetation. Others are present-day factors. They include the following:

Geographical Barriers Stretches of open water may limit the ranges of land species; large land masses, the ranges of marine species. Vast areas of continuous forest, grassland, or desert may be barriers to birds that are in no way adapted to them. High mountain ranges are often effective barriers. No general rule can explain why a barrier limits the range of one species and not another, and no correlation exists between the size or performance of the species and the limiting effect of a barrier.

Air Temperature High and low temperatures, sometimes in conjunction with humidity and precipitation, may be a factor in restricting a bird's range. In the Northern Hemisphere low minimum temperature, usually reached at night, may limit a species' range northward; or a high maximum temperature, usually attained in the late afternoon, may prevent its expansion southward.

Sunlight Seasonal variations in periods of light may be a limiting factor. For example, certain species cannot withstand the long cool nights and the loss of feeding time resulting from the long absence of light at northern latitudes in winter.

Winds Some oceanic birds, such as the larger albatrosses, require winds for sustaining flight and consequently range only in areas where winds blow constantly.

Geographical Distribution

The arrangement of organisms with relation to areas is **geographical distribution.** The aforementioned ranges of the Osprey and Kirtland's Warbler are geographical—the general areas where the species reside. When a bird species is migratory, its range is customarily divided into **breeding range,** where the species nests, and **winter range,** where the species resides between fall and spring migrations. For authoritative descriptions of the breeding and winter ranges of North American birds, consult the A.O.U. *Check-list of North American Birds* (American Ornithologists' Union, 1983). For abbreviated but nonetheless useful descriptions, consult Reilly (1968). The range maps in several field guides (see "Field Identification," p. 199) are helpful for general reference.

The ranges of birds, except for species confined to oceanic islands, are seldom static; indeed, they change so constantly, one should think of them as fluid.

Birds have an inherent tendency to disperse from their place of birth and invade new areas. If the areas harbor favorable conditions, birds can soon establish themselves and thereby expand their ranges.

Various factors aid in the constant dispersal of birds and may lead to the expansion of their ranges. Prevailing winds and cyclonic storms move individuals over seas, even to different continents where, provided the environment is suitable, they survive and reproduce. The Cattle Egret *(Bubulcus ibis)* may have arrived in South America from its native Africa by way of a storm. A failure in food supply may cause an **irruption, or invasion,** of a population of a somewhat sedentary species and hasten its expansion. In certain years and for some reason, perhaps in search of food, a number of northern species, such as Northern Hawk-Owls, *(Surnia ulula)* and Boreal Chickadees *(Parus hudsonicus),* suddenly move, in fall and winter, far south of their normal ranges. Although they usually retreat northward with the coming of spring, the possibility prevails that some individuals will find suitable breeding areas and establish themselves south of their present ranges.

Yet to be explained is why wading birds such as the Snowy Egret *(Egretta thula),* Little Blue Heron *(E. caerulea),* Tricolored Heron *(E. tricolor),* and Glossy Ibis *(Plegadis falcinellus)* have gradually extended their breeding range northward along the Atlantic coast to Maine. Similarly unexplained is why the Laughing Gull *(Larus atricilla)* has extended its breeding range north to the coast of Maine, while the Herring Gull *(L. argentatus)* and Great Black-backed Gull *(L. marinus)* have extended their breeding range south to the coasts of Virginia and North Carolina.

Civilization has changed the ranges of impressive numbers of North American birds. Some species, their populations reduced and natural environments depleted by man, have retreated to very limited ranges where favorable conditions still hold. The Whooping Crane *(Grus americana),* which was once widespread through the interior of the continent, steadily declined in numbers through excessive shooting and loss of suitable marshes for feeding and nesting. Today its unmanaged breeding in the wild is confined to a small, remote area in northwestern Canada and its winter range to a peninsula on the coast of Texas.

On the other hand, some species seem to have benefited so greatly by man's alteration of environments that they have extended their ranges and probably increased in numbers. Over the years the Northern Cardinal *(Cardinalis cardinalis)* has moved northwestward along the bottomlands of the Mississippi and Missouri Rivers into Minnesota and the Dakotas and northeastward into New York State and New England, where it is now well established. Other examples of birds breeding farther north are the Northern Mockingbird *(Mimus polyglottos)* and Tufted Titmouse *(Parus bicolor).* The American Robin *(Turdus migratorius),* House Wren *(Troglodytes aedon),* and Song Sparrow *(Melospiza melodia)* have extended their breeding ranges southward into Alabama and Georgia. All these species show a marked attraction to open places — e.g., dooryards and shrubby places—following the depletion of forests. Apparently their earlier ranges were restricted less by climate than by available places for feeding and nesting. In the mid-continent, forest and forest-edge birds such as the Rose-breasted Grosbeak *(Pheucticus ludovicianus)* and Indigo Bunting *(Passerina cyanea)* have extended their breeding ranges steadily across the Great Plains into western North and South Dakota, Nebraska, and Kansas, encouraged, no doubt, by the addition of tree and shrub plantations ("shelter belts") westward across the prairie where heretofore the only natural woodlands were along the rivers. At the same time, the Scissor-tailed Flycatcher *(Muscivora forficata)* has moved eastward into western

Missouri and the Brewer's Blackbird *(Euphagus cyanocephalus)* into Michigan, both finding, in the place of formerly extensive forests, the open country to which they are adapted.

The rate at which a species may expand its range is sometimes astonishing. The most dramatic example is perhaps the movement of the Cattle Egret northward following its arrival in South America about 1930. In 1941 or 1942 it made its first appearance in the United States at Clewiston, Florida. By 1952 it was numerous in Florida and noted in New Jersey, Massachusetts, and northern Illinois. By 1954 it was breeding in Florida, where it now numbers in the thousands. Since that time it has reached all the contiguous states from coast to coast and has nested in Canada (see Crosby, 1972).

The extension of the range of one species may spell the decline of another species, into whose range it intrudes, by intensely and successfully competing with it. The spread of the European Starling *(Sturnus vulgaris)* westward through conterminous United States very likely reduced the population of the Red-headed Woodpecker *(Melanerpes erythrocephalus)*, whose nesting sites the starling usurps. The Cattle Egret is said to have harmful effects in southern wading bird colonies by displacing Snowy Egrets and Little Blue Herons.

In any area, large or small, the ranges of different species overlap, forming an aggregation of organisms. All the animals in the aggregation are collectively called the **fauna.** Part of the study of geographical distribution concerns the species composition and development of faunas.

Continental Distribution

Each continent has its own fauna with certain forms that are **endemic,** or **indigenous,** meaning that they are native to, or restricted to, the area and reside nowhere else.

It has long been the custom to divide the land areas of the world into **zoogeographical regions** based on the presence or absence of distinctive forms of animals. The exact number of these major units varies, depending on whether certain regions are considered as full regions or subdivisions, but six is the usually accepted number, as follows:

Nearctic Greenland and North America south to and including the Mexican highlands.

Palaearctic Europe and Africa south to and including most of the Sahara Desert; Asia north from the Himalayas and the Yangtze River.

Neotropical South America, Central America, the Mexican lowlands, and the West Indies.

Ethiopian Africa, south of the Sahara Desert, and southern Arabia; Madagascar.

Oriental Asia south of the Himalayas and the Yangtze River; Sri Lanka, Sumatra, Java, Borneo, Celebes, Taiwan, and the Philippines.

Australian Australia, New Zealand, New Guinea, all the islands of the southwest Pacific, and the Hawaiian Islands.

Some summations of the bird **faunas,** or **avifaunas,** of these regions follow. Figures for the numbers of families and breeding species are approximate since authorities differ in awarding full familial rank to certain groups of genera and full specific rank to many of the breeding populations. Marine birds are not included in the figures. For further data, see Darlington (1957), Serventy (1960), and Vuilleumier (1975).

The Nearctic and Palaearctic regions, sometimes combined as the Holarctic region because of their faunal similarities, have 650 and 750 species, respectively. Although these two regions have fewer species than any of the other regions, the number of individuals in each species is generally greater because of less interspecific competition and more extensive areas suitable for breeding. The Nearctic has 52 families, the Palaearctic 69; many are shared. No family is endemic to the Nearctic; only one—the Prunellidae (accentors)—to the Palaearctic. Read Udvardy (1958) for an illuminating and detailed comparison of the Nearctic avifauna with the Palaearctic.

The Neotropical region that embraces all of South America ("The Bird Continent") is by far the richest for birds, with about 2,900 species classified into 97 families, 32 of which are endemic. A peculiarity of the avifauna, besides its great degree of endemism, is the absence of numerous forms widespread elsewhere in the world. Many of the species have exceedingly restricted ranges, some amounting to little more than a lake, valley, or mountain slope. Mountainous Colombia, with some 1,600 species within its borders, boasts more different kinds of birds than any other country in the world.

The Ethiopian region is next to the Neotropical in variety of birds with 1,900 species, representing 67 families, eight of which are endemic to Africa and five to Madagascar.

The Oriental region has perhaps 1,500 species among some 83 families, only one of which is endemic. All the other families are shared with one or more of the neighboring regions.

The Australian region has about 1,200 species among 83 families and is next to the Neotropical region in the number of endemic families—14 altogether.

History of a Continental Avifauna

The zoogeographical regions, as just demonstrated, indicate that forms of land life are not uniformly distributed over the earth but tend to be grouped on different continents or parts of continents, and that between one such group and another there are at least a few recognizable differences. The most serious criticism of this procedure is that it implies a fixed grouping of organisms within geographical boundaries, as if all the animals of one region developed independently of the faunas of neighboring regions and became a sharply demarked assemblage. The truth is that the fauna of a region, continent, or smaller area has been changing and is still changing, ever modified by the disappearance of some forms and the invasion of forms from elsewhere.

In any continent as a whole the avifauna constitutes a complex of forms from various sources. Some forms originated within the area; others arrived from neighboring areas at different periods of geologic time, including recent centuries. In order to determine how any avifaunal complex has evolved, one must consider the geological history of the area, investigate the fossil record of bird forms, and analyze its present-day avifauna for clues as to where different forms came from and how they reached their present ranges. By combing these three approaches one can, possibly, derive a working hypothesis.

Mayr (1946) carefully analyzed the bird fauna of the North American continent — the Nearctic region plus Central America and West Indian parts of the Neotropical region. Early in his report he pointed out two well-established facts relating to the continent's history during the Tertiary (see the Geologic Time Scale, p. 352): (1) North America was separated from South America when the isthmus between Colombia and central Mexico was broken up into islands with wide, intervening oceanic channels; (2) meanwhile North America was repeatedly connected to Asia by a land "bridge" across Bering Strait. Mayr then went on to show that the bird fauna of North America contains six presently applicable elements as follows:

Panboreal Element Forms equally well represented in the northern latitudes of North America and the Old World, but possibly originating in the Old World. Examples: loons (Gaviidae), phalaropes (Phalaropodinae), and auks (Alcidae).

Old World Element Three groups of forms originating in the Old World and arriving in North America by way of the Bering Strait "bridge": (1) Very recent arrivals that have undergone no change. Examples: *Motacilla flava tschutschensis*, a subspecies of the Yellow Wagtail, and *Oenanthe oenanthe oenanthe*, a subspecies of Northern Wheatear, which occur in both Alaska and the Eurasian continent to the west. (2) Fairly recent arrivals that have had time to attain subspecific distinction from Old World forms. Examples: *Pica pica hudsonia*, the North American race of the Black-billed Magpie, and *Lanius excubitor borealis* and *L. e. invictus*, the North American races of the Northern Shrike. (3) Much earlier arrivals that have had time to evolve into species, genera, or even subfamilies distinct from their Old World relatives. Examples: a variety of well-known North American birds, such as pigeons (Columbidae), cuckoos (Cuculidae), typical owls (Strigidae), kingfishers (Alcedinidae), crows and jays (Corvidae), titmice (Paridae), and nuthatches (Sittidae).

North American Element Forms, developed in North America during the Tertiary, while the continent was partially isolated and much more of its southern area had a tropical or subtropical climate. Examples: New World vultures (Cathartidae), turkeys (Meleagridinae), limpkins (Aramidae), dippers (Cinclidae), wrens (Troglodytidae), thrashers and mockingbirds (Mimidae), waxwings (Bombycillidae), silky flycatchers (Ptilogonatidae), vireos (Vireonidae), and wood warblers (Parulinae).

Pan-American Element Forms of South American origin that "island hopped," during the Tertiary, from South to North America and evolved endemic genera either on the islands, now Central America, or in southern North America. Examples: hummingbirds (Trochilidae), New World flycatchers (Tyrannidae), and blackbirds (Icterinae).

Pantropical Element Forms common to both the New and Old World tropics. Examples: parrots (Psittacidae) and trogons (Trogonidae), which differ at the generic level. Just where the parent stock originated and how the New and Old World forms came to be so widely separated geograpically provide fascinating lines of speculation (see Mayr's paper for a review).

Unanalyzed Element Forms so cosmopolitan in their distribution as to make any positive determination of their origin difficult, if not impossible. Examples: most oceanic birds, fresh-water birds, shorebirds, and such land birds as the Osprey, hawks, eagles, goatsuckers, woodpeckers, and swallows.

Mayr concluded that most North American families and subfamilies of birds are clearly either of Old World origin, of South American origin, or members of an indigenous North American element that developed while the continent was partially isolated during the Tertiary.

Insular Distribution

Sea islands hold breeding populations of land birds and provide nesting sites for the great majority of marine birds. Since the main concern with distribution in this book is with land birds, including fresh-water birds and shorebirds, the distribution of insular land birds is discussed first.

Practically all sea islands that are habitable—i.e., adequate in size with suitable environment—support, or have supported in recent times, land birds originating from the continents. In general, the more remote these islands are from the mainland, the fewer kinds of birds that colonize them, since the chances of stragglers, or "pioneers," reaching them in sufficient numbers to found colonies decrease with increased distance. Nevertheless, lists of birds reported over the years from sea islands include stragglers, representing a great array of species, far more than ever established residence.

Why have some birds successfully colonized remote islands and others have not? The size of the bird is not a determining factor. Certain passerines, despite their small size, have reached and founded populations on many of the distant islands. And flying ability is not necessarily a factor. Swallows, surely among the strongest flyers, have never colonized distant islands, while rails and gallinules, seemingly among the weakest flyers, have been some of the most successful colonists. The answer seems to lie not so much in the physical peculiarities of the birds as in their ability to adapt to the limited resources of an insular environment and to compete with other birds already established there.

The nonmigratory forms among the populations on sea islands will probably show taxonomic distinctions, and the farther they are from their parent populations, the greater will be these distinctions.

Forms on islands close to mainlands, such as those on the continental shelf, may be expected to develop no more than minor peculiarities. The avifauna on the Queen Charlotte Islands, separated from the coast of British Columbia by a channel only 25 miles wide, contains essentially the same birds as that of the nearby mainland, except for the population of several permanent-resident species—e.g., the Northern Saw-whet Owl *(Aegolius acadicus)* and Hairy Woodpecker *(Picoides villosus)*—whose Queen Charlotte Island forms are peculiar enough to be considered endemic subspecies. Presumably, the bird fauna has not been isolated long enough or completely enough to evolve more than subspecific distinctions.

By contrast, the avifaunas of both the Hawaiian Islands in the north-central Pacific and the islands of Tristan da Cunha in the middle of the South Atlantic have developed marked peculiarities.

Besides endemic subspecies of a gallinule, coot, stilt, and an owl, the avifauna of the Hawaiian Islands includes 44 endemic forms ranked as full species: a goose, two ducks, two rails, hawk, crow, two thrushes, one Old World warbler (Sylviinae), one Old World flycatcher (Muscicapinae), five honeyeaters (Meliphagidae), and 28 honeycreepers grouped together as an endemic subfamily, Drepanidinae. Four of the honeyeaters and nine of the honeycreepers are now extinct. See "Systematics and Taxonomy," page 115, for a discussion of the origin of, and speciation in, the Drepanidinae. Refer to Berger (1981) for the species of birds in the Hawaiian Islands.

On the three main islands of the Tristan da Cunhas, all very small with grass, ferns, shrubs, and (formerly) trees as the principal cover, the avifauna is remarkably distinguished by a monotypic genus of flightless rail on one island; a monotypic genus of thrush whose single species has developed three subspecies, one on each island; and two species of sparrows that constitute one genus and are divided into subspecies, each occupying one or two but not all three islands. Also included in the avifauna is a monotypic genus of flightless gallinule whose single species developed two subspecies, one (now extinct) on one of the Tristans and the other on Gough Island, some 217 miles (349.15 km) to the southeast. Consult Rand (1955) for further details. Most of the evidence suggests that these endemic forms are of American origin, descended from stragglers that reached the islands accidentally, aided perhaps by the prevailing west winds.

The flightless condition of the Tristan rails is in keeping with rallid fauna endemic to the islands of the Pacific and Indian Oceans—and in line with the condition in a few other island forms of birds as well. Carlquist (1965) lists altogether 20 species of island Rallidae, all flightless. Nine are extinct and six are in danger of extinction—undoubtedly because man and the dogs, cats, rats, and other mammals that accompany him can kill them with ease.

In a sense, forested mountains that rise from deserts or grasslands far from large mountains are "islands," attracting an avifauna distinct from the surroundings. Always of interest in these isolated situations is the composition of bird species and the factors that attracted the species and how they survive. See Thompson (1978) for a study of forested buttes rising from the Montana plains.

Marine Distribution

True marine birds—also called oceanic birds or sea birds—are those species regularly inhabiting the sea and consistently obtaining their food from the sea the year round. All nest in colonies, usually on islands, or, more rarely, in continental situations adjacent to the sea. All feature salt glands (see "Anatomy and Physiology," p. 101).

Marine birds vary in their habitual distance from land and general feeding habits. Albatrosses, shearwaters, and petrels—indeed, all members of the Procellariiformes—are pelagic, occurring on the open sea, ordinarily out of sight of land, and feeding largely on plankton. Most all the other marine species are confined to the shallow waters above the continental shelf, or waters adjacent to islands rising from the deep sea, and feed primarily on fish. Penguins, tropicbirds, boobies and gannets, frigatebirds, and the auks, auklets, murres, and puffins commonly stay offshore; cormorants, skuas and jaegers, skimmers, and the marine species of pelicans, gulls, and terns stay inshore, seldom straying beyond sight of land. But there are exceptions. For example, small gulls, called kittiwakes, are at times pelagic as are some of the terns during their migrations. And some species of inshore birds partly forsake the marine environment altogether and move inland.

The seas of the world, like the land areas, may be divided into regions, distinguished by the presence or absence of animal forms. Ornithologists (e.g., Serventy, 1960) recognize three major regions as follows:

Northern Marine Region The frigid waters of the Arctic south to the Subtropical Convergence of the Northern Hemisphere where, between Latitudes 35 and 40 degrees, the water temperature rises rapidly.

Southern Marine Region The frigid waters around Antarctica north to the Subtropical Convergence of the Southern Hemisphere where, between the same latitudes, the water temperature rises similarly.

Tropical Marine Region The warm equatorial waters between the Subtropical Convergences of both hemispheres.

The Northern Marine region is distinguished by the Alcidae (auks, etc., totaling 22 species), the only family wholly confined to the northern seas. More alcid species, probably with a higher average of individuals per species, occur in the North Pacific than in the North Atlantic.

The Southern Marine region is the richest in its variety of bird species if not in numbers of individuals. Its principal distinctions are 16 of the 17 species of penguins, the other species reaching north to the Galapagos Islands at the equator; all but three of the dozen or more species of albatrosses; and perhaps the majority of the 75 or more species of shearwaters and petrels. The penguins seem to fill the position in the Southern region that the alcids occupy in the Northern. The high density of plankton and the strong, almost constant winds account for the presence of so many albatrosses; the plankton alone attracts the shearwaters and petrels.

The Tropical Marine region is distinguished particularly for its wide variety of pelecaniform birds—tropicbirds, boobies, and frigatebirds—and terns. Compared to the open seas of the Southern region, those of the Tropical seem devoid of bird life. This is due partly to a low supply of plankton, the food of pelagic birds, and to the habit of other avian residents of staying close to land where their food, small fish, is more abundant.

No doubt the Tropical Marine region, with its warm water and low food content in the open sea, is an effective barrier to Northern Marine birds entering into the Southern Marine region and vice versa. Almost certainly it has kept the poor-flying alcids and the flightless penguins in their respective cool, food-rich seas. But it seems to have in no way impeded the dispersal of many marine species. One, the Great Skua *(Catharacta skua)*, is comprised of two widely separate populations, one residing in the Northern region, the other in the Southern. While it is true that there are more species of shearwaters and petrels in the Southern region than elsewhere, some species are nevertheless cosmopolitan, frequenting nearly all the seas in one season or another.

Seasonal Distribution

The natural occurrence of animals in an area with relation to the seasons is **seasonal distribution.** In any part of North America the bird population is subject to changes during the course of the year as a result of seasonal migrations. In one season of the year certain species appear that are not ordinarily present at another time. Accordingly, in any area species may be grouped into seasonal categories as follows:

Summer Residents Species in an area during the summer, coming from the South in the spring to breed and returning in the fall.

Transients Species stopping temporarily in an area during their northward migration in the spring and during their southward migration in the fall.

Winter Visitants Species in an area during the winter, having come from their northern nesting grounds to pass the winter in less rigorous climate and departing north in the spring.

Permanent Residents Species not undergoing a regular periodical migration and consequently staying in one area the year round.

Ecological Distribution

The natural arrangement of organisms with relation to environment is called **ecological distribution.** Whereas geographical distribution deals with organisms in areas, ecological distribution has to do with organisms in environments.

The environment that a species normally occupies in its geographical range is its **habitat.** For example, the deciduous forest is the habitat of a number of bird species whose breeding ranges cover wide areas in eastern North America. Generally, a number of different organisms have the same habitat and together constitute a biotic community.

A **biotic community** is an aggregation of organisms, both plant and animal, living in a given habitat. Any such community is characterized by the physical features of its habitat, by the complex of organisms living in it, and by the relationships of the organisms to the physical features of the habitat and to one another. Among the three principal biotic communities of the world—marine, fresh-water, and terrestrial—birds occupy essentially the terrestrial, even though many species have become secondarily adapted to either one of the others. It is beyond the scope of this section to consider birds with relation to the marine and fresh-water communities.

Major Biotic Communities

In North America north of Mexico, the bottomlands of the lower Rio Grande, and southern Florida, there are nine Major Biotic Communities (some are termed **Biomes** by a number of authorities). Each Major Community is made up of a series of communities, any one of which may be recognized by certain plant forms, or types of vegetation, that dominate the aggregation of organisms and certain other organisms (plants and animals) that influence the aggregation in varying degrees, depending on their abundance, size, and mode of life. The communities are usually named for their dominant plant forms.

One community in the Major Biotic Community is the **climax community;** the other communities in the Major Biotic Community are successive stages of development, or a **sere,** leading toward the climax community and are termed **seral communities.** The climax community is the final stage of development over a prolonged period of time and represents the highest possible development (from an ecological viewpoint) in an area under the prevailing conditions of soil and climate. A beech-maple forest is an example of a climax community in certain areas. The seral communities are transitory communities and eventually will be replaced. Thus a shrub community in certain areas will be succeeded by a beech-maple forest because the environmental conditions favor the development of the beech-maple forest as the climax community. Occasionally, the term **subclimax community** is used to designate a seral community that lasts for a very long time.

The meaning of "dominant plant forms" requires explanation. The controlling factor in plants for birds, as well as for other animals, is the form or type of plant, rather than the species of plant. Thus certain birds are limited in habitat by their adaptations to coniferous trees, regardless of whether the trees are red spruces, eastern hemlocks, or alpine firs.

Following are the nine Major Biotic Communities of North America. Nearly all are named for their dominant plant forms:

Tundra
Coniferous Forest
Deciduous Forest
Grassland
Southwestern Oak Woodland
Pinyon-Juniper Woodland
Chaparral
Sagebrush
Scrub Desert

The map on page 182 (Fig. 41) entitled "Major Biotic Communities of North America," shows the general extent of the climax communities in eight of the nine Major Biotic Communities. (The Southwestern Oak Woodland is not shown.) It also shows certain ecotones, to be defined and discussed later. Bear in mind that this map, or any attempt to depict distribution of communities over a wide area, has definite limitations. This map does not reveal seral communities because they are usually too small. It does not give the full extent of climax communities and ecotones but only those parts broad enough to be mapped; in the western part of the conterminous United States many parts of climax communities and ecotones are merely narrow belts on steep mountain slopes and consequently cannot be indicated.

In the following pages each Major Biotic Community is characterized by a condensed description of the climax community (see Kendeigh, 1961, for further details) and a list of some of the bird species either confined to it, or showing a marked preference for it, during the breeding season. An asterisk before the species name indicates that there is one or more subspecies with an affinity in the breeding season for that particular climax community. Following each list are useful references pertaining to the community.

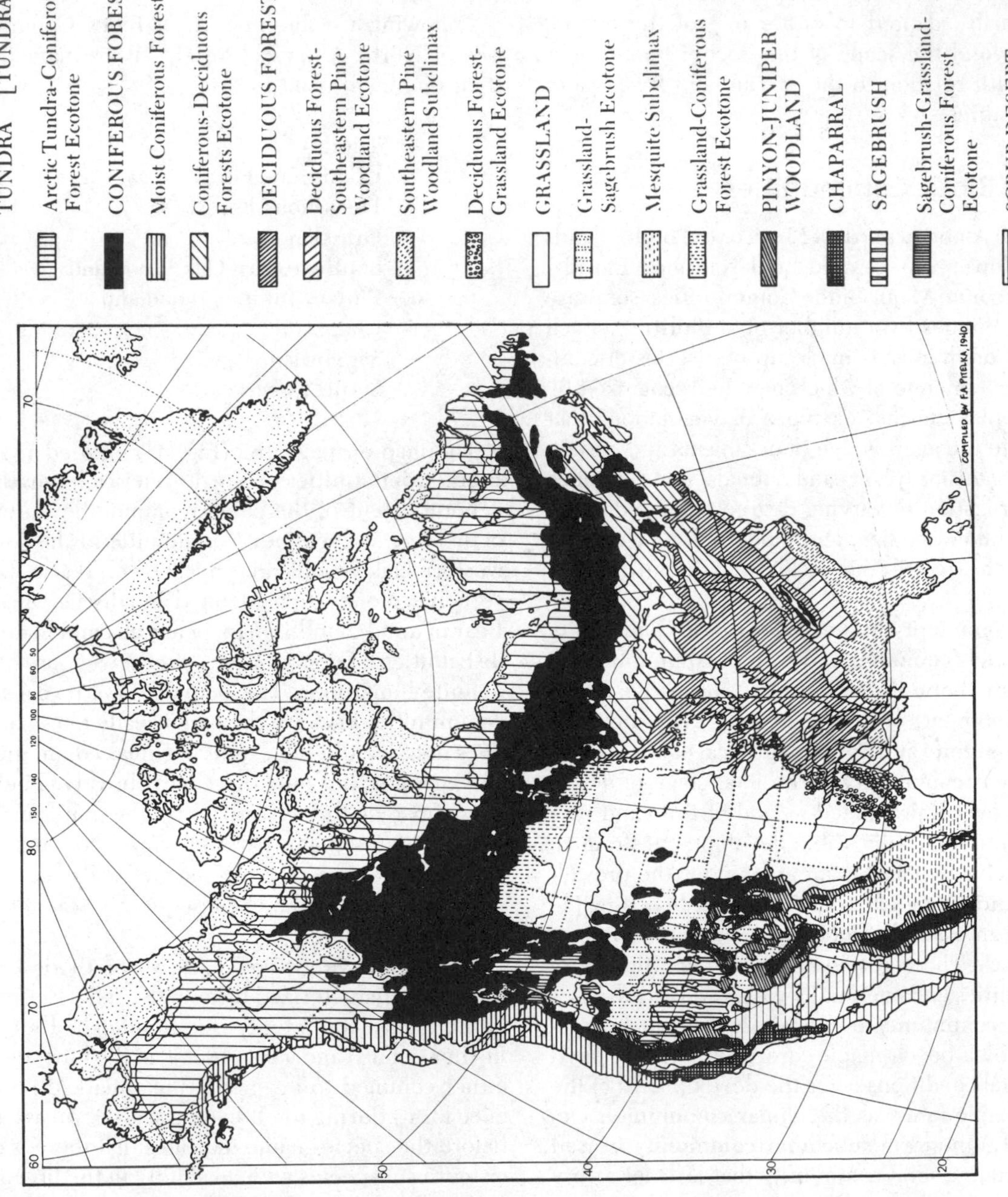

Figure 41 **Major Biotic Communities of North America**

Tundra

Arctic Tundra Northern Alaska and northern Canada, south along the coasts of Hudson Bay and through most of Labrador to Newfoundland.
Alpine Tundra In discontinuous areas above 10,000 feet on the Rocky Mountains and the Sierra Nevada-Cascade system. Winter temperatures extremely cold; rainfall moderate to heavy. General aspect: treeless terrain (level to undulating in the Arctic Tundra). Dominant plants: lichens, grasses, sedges, and occasionally dwarf willows *(Salix)* and other small woody plants.

Arctic Tundra

Willow Ptarmigan *(Lagopus lagopus)*
Rock Ptarmigan *(Lagopus mutus)*
Snowy Owl *(Nyctea scandiaca)*
*Horned Lark *(Eremophila alpestris)*
*Water Pipit *(Anthus spinoletta)*
Lapland Longspur *(Calcarius lapponicus)*
Smith's Longspur *(Calcarius pictus)*
Snow Bunting *(Plectrophenax nivalis)*

Alpine Tundra

White-tailed Ptarmigan *(Lagopus leucurus)*
*Water Pipit *(Anthus spinoletta)*
*Rosy Finch *(Leucosticte arctoa)*

References: Cooke et al. (1975); French and French (1974); Hayward (1952); Irving (1960); Shelford and Twomey (1941); Verbeek (1967).

Coniferous Forest

Transcontinental Coniferous Forest (also called *Taiga*), Southern Canada, paralleling the Arctic Tundra on the north.
Eastern Montane Coniferous Forest Northeastern United States southward at higher elevations of the Appalachians to North Carolina.
Western Montane Coniferous Forest Southward on the Rocky Mountains to New Mexico and on the Sierra Nevada-Cascade system to California, below the Alpine Tundra. Climates subject to seasonal changes, summers always cool and winters very cold; rainfall moderate to heavy. General aspect: dense forest with deeply shaded floors supporting few small plants to open forest supporting shrubs and herbaceous plants. Dominant plants: coniferous, evergreen trees belonging to several genera—pines *(Pinus)*, spruces *(Picea)*, hemlocks *(Tsuga)*, firs *(Abies)*, cedars *(Thuja)*, and in the Western Montane division only, Douglas fir *(Pseudotsuga)*. See Table 4, page 184.

References: Franzreb (1978); Kendeigh (1947,1948); N.D. Martin (1960); Rasmussen (1941); Richards (1974); Shelford and Olson (1935); Snyder (1950).

Deciduous Forest

Eastern United States, from southern New England and the Appalachians, at elevations below the coniferous forests, west to and including the bottomlands and bluffs along the Mississippi River and its tributaries. Climate moderate, with summers warm and winters cool; rainfall moderate to heavy. General aspect: forest of varying density, usually with partially shaded floors supporting small shrubs and herbaceous plants. Dominant plants: broad-leaved deciduous trees of several genera—beech *(Fagus)*, maples *(Acer)*, basswood *(Tilia)*, oaks *(Quercus)*, hickory *(Carya)*, walnut *(Juglans)*, and others.

*Red-shouldered Hawk *(Buteo lineatus)*
*Broad-winged Hawk *(Buteo platypterus)*
*Barred Owl *(Strix varia)*
*Whip-poor-will *(Caprimulgus vociferus)*
Red-bellied Woodpecker *(Melanerpes carolinus)*
*Downy Woodpecker *(Picoides pubescens)*
*Hairy Woodpecker *(Picoides villosus)*
Eastern Wood-Pewee *(Contopus virens)*
Acadian Flycatcher *(Empidonax virescens)*
*Great Crested Flycatcher *(Myiarchus crinitus)*
*Carolina Chickadee *(Parus carolinensis)*
Tufted Titmouse *(Parus bicolor)*
*White-breasted Nuthatch *(Sitta carolinensis)*
*Blue-gray Gnatcatcher *(Polioptila caerulea)*
Wood Thrush *(Hylocichla mustelina)*
Yellow-throated Vireo *(Vireo flavifrons)*
Cerulean Warbler *(Dendroica cerulea)*
Worm-eating Warbler *(Helmitheros vermivorus)*
Kentucky Warbler *(Oporornis formosus)*
Hooded Warbler *(Wilsonia citrina)*

References: Bond (1957); D.W. Johnston and Odum (1956); Kendeigh (1946, 1948); Williams (1936).

Table 4 **Birds of the Coniferous Forest**

	Eastern Montane Coniferous Forest	Trans-continental Coniferous Forest	Western Montane Coniferous Forest
Northern Goshawk *(Accipiter gentilis)*	X	X	X
Spruce Grouse *(Dendragapus canadensis)*		X	
Blue Grouse *(Dendragapus obscurus)*			X
Northern Hawk-Owl *(Surnia ulula)*		X	
Williamson's Sapsucker *(Sphyrapicus thyroideus)*			X
Three-toed Woodpecker *(Picoides tridactylus)*		X	X
Black-backed Woodpecker *(Picoides arcticus)*		X	X
Olive-sided Flycatcher *(Contopus borealis)*	X	X	X
Yellow-bellied Flycatcher *(Empidonax flaviventris)*		X	
Hammond's Flycatcher *(Empidonax hammondi)*			X
Gray Jay *(Perisoreus canadensis)*		X	X
Clark's Nutcracker *(Nucifraga columbiana)*			X
Mountain Chickadee *(Parus gambeli)*			X
Boreal Chickadee *(Parus hudsonicus)*		X	
Red-breasted Nuthatch *(Sitta canadensis)*	X	X	X
Brown Creeper *(Certhia americana)*	X	X	X
Winter Wren *(Troglodytes troglodytes)*	X	X	X
Golden-crowned Kinglet *(Regulus satrapa)*	X	X	X
Ruby-crowned Kinglet *(Regulus calendula)*		X	X
Townsend's Solitaire *(Myadestes townsendi)*			X
Swainson's Thrush *(Catharus ustulata)*		X	X
Hermit Thrush *(Catharus guttata)*	X	X	X
Tennessee Warbler *(Vermivora peregrina)*		X	
Magnolia Warbler *(Dendroica magnolia)*	X	X	
Cape May Warbler *(Dendroica tigrina)*		X	
Yellow-rumped Warbler *(Dendroica coronata)*		X	X
Townsend's Warbler *(Dendroica townsendi)*			X
Hermit Warbler *(Dendroica occidentalis)*			X
Blackburnian Warbler *(Dendroica fusca)*	X	X	
Palm Warbler *(Dendroica palmarum)*		X	
Bay-breasted Warbler *(Dendroica castanea)*		X	
Canada Warbler *(Wilsonia canadensis)*	X	X	
White-throated Sparrow *(Zonotrichia albicollis)*	X	X	
Dark-eyed Junco *(Junco hyemalis)*	X	X	X
Pine Grosbeak *(Pinicola enucleator)*		X	X
Purple Finch *(Carpodacus purpureus)*	X	X	
Cassin's Finch *(Carpodacus cassinii)*			X
Red Crossbill *(Loxia curvirostra)*	X	X	X
White-winged Crossbill *(Loxia leucoptera)*		X	
Pine Siskin *(Carduelis pinus)*	X	X	X

Grassland

Primarily the interior plains ("prairies") from the forested bottomlands and bluffs along the Mississippi River and its tributaries west to the Rocky Mountains, and from south-central Texas north into Canada. Climate subject to abrupt seasonal changes, winters being cold (southern division) to very cold (northern division); rainfall moderate to low. General aspect: flat to rolling, treeless, grass-covered terrain. Dominant grasses in the eastern section, where elevation is lower and rainfall moderate, of the "tall" type—bluestem *(Andropogon)*, Indian grass *(Sorghastrum)*, switch grass *(Panicum virgatum)*, and others; dominant grasses in the western section, on the Great Plains, where the elevation is higher and the rainfall low, of the "short" type—chiefly buffalo *(Buchloë)* and grama *(Bouteloua)*. Bird species marked by N breed only in the far northern sections.

Swainson's Hawk *(Buteo swainsoni)*
Ferruginous Hawk *(Buteo regalis)*
Greater Prairie-Chicken *(Tympanuchus cupido)*
Sharp-tailed Grouse *(Tympanuchus phasianellus)*
Long-billed Curlew *(Numenius americanus)*
Burrowing Owl *(Athene cunicularia)*
N Sprague's Pipit *(Anthus spragueii)*
Lark Bunting *(Calamospiza melanocorys)*
N Baird's Sparrow *(Ammodramus bairdii)*
Grasshopper Sparrow *(Ammodramus savannarum)*
Western Meadowlark *(Sturnella neglecta)*
N McCown's Longspur *(Calcarius mccownii)*
N Chestnut-collared Longspur *(Calcarius ornatus)*

References: Breckenridge (1974); Carpenter (1940); Gammell (1974); Hurley and Franks (1976); Kendeigh (1941); Mitchell (1961); Tester and Marshall (1961); Wiens (1969, 1974).

Southwestern Oak Woodland (not shown on map)

Southwestern United States—mainly Utah, Nevada, California, New Mexico, Arizona, and parts of Colorado and Oregon; usually on hills and mountain slopes. Summers warm, contrasting sharply with cool winters; rainfall low to moderate. General aspect: partially open woodland of oaks, 20 to 50 feet high (6.09 to 15.24 m); trees ordinarily close enough for branches to touch, but there may be wide spaces between trees that are covered with grasses and shrubs. Dominant plants: practically all oaks of one genus *(Quercus)*.

Nuttall's Woodpecker *(Picoides nuttallii)*
Bridled Titmouse *(Parus wollweberi)*
Hutton's Vireo *(Vireo huttoni)*
Virginia's Warbler *(Vermivora virginiae)*
Black-throated Gray Warbler *(Dendroica nigrescens)*

References: Anderson (1970); Marshall (1957); Miller (1951).

Pinyon-Juniper Woodland

Primarily the Great Basin and the Colorado River region in Colorado, Utah, Nevada, Arizona, New Mexico, and the east side of the Sierra Nevada-Cascade system in California; situated on hills and mountain slopes at elevations above deserts or grasslands and below the coniferous forests. Summers warm, contrasting sharply with cool winters; low rainfall. General aspect: open, park-like woodland of small trees ("pigmy conifers"), 15 to 35 feet high (4.57 to 9.14 m). Dominant plants: pinyon pines *(Pinus edulis* or *P. monophylla)* and several species of juniper *(Juniperus)*; yuccas often prevalent.

Gray Flycatcher *(Empidonax wrightii)*
Pinyon Jay *(Gymnorhinus cyanocephalus)*
Plain Titmouse *(Parus inornatus)*
Bushtit *(Psaltriparus minimus)*
Bewick's Wren *(Thryomanes bewickii)*

References: Hardy (1945); Laudenslayer and Balda (1976); Rasmussen (1941); Woodbury (1947); Woodin and Lindsey (1954).

Chaparral

California, chiefly on hills and mountains of southwestern part of state, also on inner Coast Ranges and hills bordering the Great Valley. Mild climate; summers very dry and winters with abundant rain fall. General aspect: extensive tracts of densely growing, heavily branched shrubs and stunted trees, 2 to 8 feet high (0.6 to 2.44m), interrupted now and then by grassy areas and rocky outcrops. Kinds of plants vary according to elevation above sea level and available moisture, but growth forms similar. Among dominant plants in one situation or another: snowbush *(Ceanothus)*, chamise *(Adenstoma)*, scrub oak *(Quercus)*, mountain mahogany *(Cercocarpus)*, coffeeberry *(Rhamnus)*, manzanita *(Arctostaphylos)*, poison oak *(Rhus)*, and baccharis *(Baccharis)*.

*Wrentit *(Chamaea fasciata)*
California Thrasher *(Toxostoma redivivum)*
Gray Vireo *(Vireo vicinior)*
*Orange-crowned Warbler *(Vermivora celata)*
Black-chinned Sparrow *(Spizella atrogularis)*
*Sage Sparrow *(Amphispiza belli)*
*White-crowned Sparrow *(Zonotrichia leucophrys)*

References: Cogswell (1974); Hayward (1948); Miller (1951).

Sagebrush

At elevations above lower deserts and valley floors of Great Basin Plateau between the Rocky Mountains and the Sierra Nevada-Cascade system. Relatively dry climate with slight rainfall; summer temperatures high; winter temperatures cool. General aspect: densely foliaged shrubs, 2 to 5 feet high (0.6 to 1.52m); plants not so widely spaced as in Scrub Desert, not so close together as in Chaparral. Dominant plants: sagebrush *(Artemisia)*, shadscale *(Atriplex)*, rabbitbrush *(Chrysothamnus)*, greasewood *(Sarcobatus)*, winterfat *(Eurotia)*, and occasional clumps of grasses.

Sage Grouse *(Centrocercus urophasianus)*
Sage Thrasher *(Oreoscoptes montanus)*
Brewer's Sparrow *(Spizella breweri)*
*Sage Sparrow *(Amphispiza belli)*

References: Fautin (1946); Miller (1951).

Scrub Desert

Lowlands and valley floors from western Texas to southwestern California. Semiarid climate with scant rainfall; summer temperatures very high. General aspect: bare ground with widely spaced plants, 3 to 6 feet high (0.91 to 1.83m). The dominant plant is creosote bush *(Larrea)*; prominent plants include mesquite *(Prosopis)*, paloverde *(Cercidium)*, catclaw *(Acacia)*, ironwood *(Olneya)*, ocotillo *(Fouquieria)*, agaves, cactuses, yuccas, and (occasionally) grasses.

Gambel's Quail *(Callipepla gambelii)*
Greater Roadrunner *(Geococcyx californianus)*
Elf Owl *(Micrathene whitneyi)*
Lesser Nighthawk *(Chordeiles acutipennis)*
Costa's Hummingbird *(Calypte costae)*
Gila Woodpecker *(Melanerpes uropygialis)*
Vermilion Flycatcher *(Pyrocephalus rubinus)*
Verdin *(Auriparus flaviceps)*
Cactus Wren *(Campylorhynchus brunneicapillus)*
Black-tailed Gnatcatcher *(Polioptila melanura)*
Bendire's Thrasher *Toxostoma bendirei)*
Crissal Thrasher *(Toxostomaa dorsale)*
Le Conte's Thrasher *Toxostoma lecontei)*
Phainopepla *(Phainopepla nitens)*

References: Behle (1976); Dixon (1959); Hensley (1954); Jaeger (1957); Miller (1963); Miller and Stebbins (1964); Monson (1974); Riatt (1976); Schmidt-Nielson (1964); Tomoff (1974).

Subclimaxes

Several communities within the Major Biotic Communities, besides those with climax rating already treated, are great enough in extent and importance to warrant special mention.

Southeastern Pine Woodland Subclimax On the Coastal Plain of the southeastern states (see map), it is characterized by the presence of open pine forests which, because of the poor soils and many fires, have never been succeeded by deciduous stands. Perhaps the most typical breeding birds confined to the Pine Woodland are the Red-cockaded Woodpecker *(Picoides borealis)*, Brown-headed Nuthatch *(Sitta pusilla)*, and a subspecies of the Bachman's Sparrow *(Aimophila aestivalis)*. Reference: D.W. Johnston and Odum (1956).

Mesquite Subclimax The wide areas of mesquite *(Prosopis)* in southwestern Texas and southern New Mexico (see map). In this community one finds such birds as the Golden-fronted Woodpecker *(Melanerpes aurifrons)* and the Tufted (Black-crested)

Titmouse *(Parus bicolor atricristatus).* Reference: Hamilton (1962).

Moist Coniferous Forest Sometimes called the **Coast Forest,** it extends as a narrow belt along the western coast of North America from Alaska south into California to approximately the San Francisco area (see map). It is regarded by most ecologists as a division of the Coniferous Forest Climax. Owing to the great humidity in the region, the community has a distinctive vegetational aspect and, in turn, a few species of birds—e.g., the Chestnut-backed Chickadee *(Parus rufescens)* and Varied Thrush *(Ixoreus naevius)*—that are generally limited to this particular biotic situation. References: Macnab (1958); Miller (1951).

Ecotones

When one community gives way to another, there is an area of overlap, or **ecotone,** where their respective plant and animal associations are intermixed or blended. In some of the broader ecotones, resulting from the overlapping of two Major Biotic Communities, one may find not only bird species from both communities but occasionally also species restricted more or less to the ecotone.

(Arctic) Tundra-Coniferous Forest Ecotone An exceptionally good example of the broader ecotone, it is characterized by the interdigitation of Arctic Tundra and Coniferous Forest extensions (see map). This area, loosely referred to as "timberline," has numerous low tundra shrubs and stands of stunted conifers. Here may be found such breeding birds as the Gray-cheeked Thrush *(Catharus minimus),* Northern Shrike *(Lanius excubitor),* Blackpoll Warbler *(Dendroica striata),* American Tree Sparrow *(Spizella arborea),* Harris' Sparrow *(Zonotrichia querula),* and Common Redpoll *(Carduelis flammea),* none of them characteristic of the open, uninterrupted tundra or the deep spruce forest. Reference: Jehl and Smith (1970).

Coniferous-Deciduous Forests Ecotone Another good example, it is where conifers (especially hemlocks) and maples, beech, and other hardwoods intermingle (see map). The eastern subspecies of the Solitary Vireo *(Vireo solitarius)* and the Black-throated Blue Warbler *(Dendroica caerulescens)* are two of the several birds with affinities for this kind of forest during the breeding season. References: Kendeigh (1946, 1948).

None of the broader ecotones represent a perfect blending of neighboring communities; one finds numerous "islands" of these communities intermixed. Thus in a Coniferous-Deciduous Forests ecotone there are isolated, pure stands of conifers and groves of hardwoods where the conditions of soil and moisture are especially suitable to one plant form. If such an island is large enough, it will attract the bird usually associated with its particular plant form.

Seral Communities

Within the Major Biotic Community are the seral communities showing stages of development toward the climax communities. Some of the more common, natural communities are listed here. For extensive information about them, consult Farb (1963), Kendeigh (1961), and Smith (1966).

Beaches Bare areas of rock and sand, with little or no vegetation, bordering large bodies of water.

Lakes Large bodies of water of varying depth from a few to many feet, with little or no rooted vegetation.

Ponds Small, shallow bodies of water with plants, floating and submerged, rooted in most of the bottom.

Marshes Innundated land with emergent, herbaceous plants—bulrushes, cattails, sedges, grasses, etc. References: Allen (1914); Beecher (1942); Errington (1974); Hochbaum (1974); Provost (1947); Robertson (1974); Weller and Spatcher (1965).

Swamps and Bogs Undrained wetlands with wooded vegetation. Swamps are common in warm southern regions and feature deciduous growth. Bogs, more common in the cooler northern regions, feature coniferous growth and may have at their margins a semifloating mat of sphagnum moss and other cushiony plants. References: Aldrich (1943); Brewer (1967); Cottrille (1974); Murray (1974); Robertson (1974).

Shrublands Areas with low, thickly growing woody plants predominating. Reference: D.W. Johnston and Odum (1956).

Riparian Woodlands Mostly of deciduous trees, near streams, or bottomlands where there is an adequate supply of subsurface water; especially prominent in the more arid sections of western conterminous United States. References: Austin (1970); Ingles (1950).

Any one of the above communities may be a stage in a succession leading to a particular climax community. For example, in eastern United States a pond may be succeeded by a marsh, a marsh by a shrubland, and a shrubland by a climax community. During the course of many years the seral communities will disappear as the climax community takes over. While the succession is in progress, many birds are associated with the seral stages. Pied-billed Grebes *(Podilymbus podiceps)* and Black Terns *(Chlidonias niger)* feed in the open water of the pond and nest in the adjacent fringe of the marsh. American and Least Bitterns *(Botaurus lentiginosus* and *Ixobrychus exilis)*, Common Moorhens *(Gallinula chloropus)*, Virginia Rails *(Rallus limicola)* and Soras *(Porzana carolina)*, Sedge and Marsh Wrens *(Cistothorus platensis* and *C. palustris)*, and Red-winged Blackbirds *(Agelaius phoeniceus)* nest in the bulrushes, cattails, or sedges in accordance with their preference. In the shrublands are nesting Common Yellowthroats *(Geothlypis trichas)* as well as many other species typical of shrublands. See Beecher (1942) for a classic study of marsh birds with relation to vegetation. For other studies of succession, see Beckwith (1954), Evans (1978), D.W. Johnston and Odum (1956), and Shugart and James (1973).

Primary and Secondary Communities

Seral communities are stages of primary succession, or **primary communities.** Any primary community is presumably the most highly developed association that has ever existed in a given area during the present geological age.

In addition to primary communities there are stages in a secondary succession called **secondary communities.** A secondary community is any biotic association that has developed in an area where a natural, primary community has been removed. The most common secondary communities are "artificial" in that man's activities have produced or influenced them directly or indirectly. Secondary communities include croplands, fields, pastures, shrub and tree plantations around dwellings, marshes and swamps resulting from dammed waterways, and deserts caused by overgrazing. Most woodlands in conterminous United States and southern Canada are secondary communities, replacing the primeval forests that have long since been lumbered or burned.

All secondary communities are more transitory than primary communities because areas previously occupied by primary communities are more receptive to replacement, and many plants, already present in neighboring areas, rapidly spread into them. As an illustration, a plot of cleared ground in a Deciduous Forest may, in the course of a man's lifetime, be covered first by grasses and other succulent plants, then shrubs, and finally young trees of the Deciduous Forest Climax.

Probably the majority of birds met in the field are associated with secondary communities because, in the first place, man-made environments are predominant in the outskirts of cities and towns and, in the second place, secondary communities harbor a rich variety of birds. See Woolfenden and Rohwer (1969) for a study showing the remarkable density of birds in a suburban area and Emlen (1974) for a study in an urban area.

Edge Effect

Some of the highest concentrations of birds anywhere are in the ecotones of seral communities. Here are many species characteristic of the communities involved plus those that are adapted more or less to the ecotonal conditions. Actually the number of species and their respective populations are often much greater in an ecotone than in the communities contributing to it. This phenomenon of increased variety and density of species where communities meet is spoken of as **edge effect.**

The importance of ecotones and edge effect is soon apparent when looking for large numbers of birds. Frequently where such communities adjoin as woods and fields, woods and shores, or riparian woodlands and grasslands, there are shrubby areas that hold a variety of breeding birds that will include, depending on the part of the country, the Gray Catbird *(Dumetella carolinensis),* Northern Mockingbird *(Mimus polyglottos),* Yellow Warbler *(Dendroica petechia),* Chestnut-sided Warbler *(Dendroica pensylvanica),* Prairie Warbler *(Dendroica discolor),* Yellow-breasted Chat *(Icteria virens),* Northern Cardinal, Blue Grosbeak *(Guiraca caerulea),* Lazuli Bunting *(Passerina amoena),* Indigo Bunting, Painted Bunting *(Passerina ciris),* Clay-colored Sparrow *(Spizella pallida),* Field Sparrow *(Spizella pusilla),* and Song Sparrow. Here, in addition to shrub birds, are open-country birds that seek trees and shrubs for nests and perches—e.g., Western Kingbird *(Tyrannus verticalis),* Eastern Kingbird *(Tyrannus tyrannus),* and Scissor-tailed Flycatcher—and woodland birds that often prefer trees or shrubs near openings for nests—Western Flycatcher *(Empidonax difficilis),* and Northern Oriole *(Icterus galbula).*

In the Piedmont of Georgia D.W. Johnston and Odum (1956) found that nearly 40 to 50 percent of the common breeding birds in the region were forest edge birds in their habitat requirements. In Illinois V.R. Johnston (1947) concluded that the forest edge is actually a distinct community with characteristic species not found either in the forest or open-country communities.

As a rule, birds of seral communities have more extensive geographical ranges than those limited to climax communities because their communities are more widely distributed. One may find marsh birds—e.g., American Bittern, Northern Harrier *(Circus cyaneus),* Virginia Rail, Marsh Wren, and Red-winged Blackbird—in several Major Biotic Communities for the reason that marsh seral communities occur in them. The same situation is true with regard to many "edge" birds and birds of artificial communities. In a trip by car from the Atlantic Coast west to the Great Plains or Rocky Mountains, the species one will see along the roadsides and agricultural lands—e.g., Mourning Dove *(Zenaida macroura),* Eastern Bluebird *(Sialia sialis),* American Robin, Chipping Sparrow *(Spizella passerina),* and Common Grackle *(Quiscalus quiscula)* will be monotonously the same despite the fact that the car will pass through Deciduous Forest and Grassland where restricted climax species exist.

A comparatively small number of birds, usually those that nest on cliffs, in holes, cavities, or crevices, or in man-made "substitutes" such as buildings or birdboxes, are not restricted to any one biotic community and range widely in accordance with conditions of climate and other controlling factors. Among these birds are the Turkey Vulture *(Cathartes aura),* Osprey, Peregrine Falcon *(Falco peregrinus),* Common Barn-Owl *(Tyto alba),* Chimney Swift *(Chaetura pelagica),* Belted Kingfisher *(Ceryle alcyon),* Common Raven *(Corvus corax),* Rock Wren *(Salpinctes obsoletus),* and Canyon Wren *(Catherpes mexicanus),* together with eagles, several hawks, phoebes, and swallows.

The guides by Pettingill (1977, 1981), which include descriptions of the more important communities in each state and name some of their characteristic bird species are very helpful in becoming acquainted with biotic communities. The climax communities are usually treated at greater length and include lists of species to be expected. For more extensive information on biotic communities, consult Shelford (1963).

Niches

Within biotic communities, organisms are further distributed according to the niches they occupy. The **niche** of a bird species is its position or role in the community resulting from its structural, physiological, and behavioral adaptations. Do not confuse it with habitat, which is the environment or place. Think of the habitat of a species as its "address" and the niche as its "occupation."

To determine the niche of a species, one must not only discover where the species carries on its activities (e.g., feeding, singing, nesting, and roosting) but also find out the kinds of food it requires, the relationships between the species and other organisms of the community, and the part played by the species in the general functioning of the community as a whole. See Root (1967) for an instructive study.

The problem of determining the niche of a species is invariably complicated by the natural variation of the community in different geographical areas. Thus the niche relationships in the northern part of a species' range may be very different from those in the southern, owing to the differences in vegetation, population densities of associated organisms, and other factors in the community. While a number of bird species in any one community may show similarities in their niche relationships, they seldom, if ever, share identical niches.

Altitudinal Distribution

In mountains biotic communities succeed one another altitudinally with corresponding climatic changes just as they do latitudinally elsewhere on the continent. But due to the steepness of the slopes, mountain communities have a different configuration, which in turn sometimes creates puzzling ecological problems.

About 250 feet (76.2 m) of elevation is equivalent to one degree of latitude—roughly 69 miles (111.02 km) in length northward. In a general way, then, one may expect the same rate of change in climate when ascending a mountain for 250 feet (76.2 m) that one will find when going north on the continent 69 miles. In the high Rocky Mountains, where many peaks reach well above 12,000 feet (111.02 km), a person will experience, during a climb, the same changes as when taking a trip north from the base of the Rockies to the Arctic Tundra. Furthermore, just as one will experience changes in climate, one will also observe essentially the same changes in communities. The great difference between the climb and the trip, besides distance, is the abruptness of change on the mountain compared with the gradualness of change during the trip overland.

Instead of being spread over wide areas, the biotic communities on mountains are usually distinct belts corresponding to the same climatic features. Their width is a matter of footage rather than of mileage. Mountain ecotones, instead of being wide transitions between communities, are exceedingly narrow, the two communities standing in close proximity. Because of this distinct belting of mountain community units, their identification on the basis of plant formations is simpler and more readily seen.

Figure 42, originally suggested by Woodbury (1947), illustrates the altitudinal succession of the Major Biotic Communities on a hypothetical peak of the south-central Rocky Mountains, approximately in eastern Utah. Note that on the warmer southern slope all the belts reach a greater altitude than on the northern.

As indicated in the preceding description of North American Major Biotic Communities, a number of bird species are characteristic of both the latitudinal and mountain communities—species characteristic of both the Transcontinental and Western Montane Coniferous Forests. At the same time, populations of at least the higher communities on a mountain range are often sufficiently isolated from the corresponding communities on another range as to become racially distinct. For instance, there are three subspecies of Rosy Finches *(Leucosticte arctoa)*, each one limited to the Alpine Tundra of a separate group of western mountains. In effect, mountain biotic communities are like islands in that they hold certain bird populations in isolation where each population may develop peculiarities of its own.

For studies on the relationships of birds and other organisms, animal and plant, to mountains in North America, see Brooks (1965), Kendeigh and Fawver (1981), Tanner (1955), and Winternitz (1976).

Life Zones

Some authorities, in dealing with the distribution of birds in North America, follow the concept of **life zones** instead of communities. In place of the Major Biotic Communities they recognize six life zones.

Life zones are temperature zones mapped out in accordance with certain lines—*isotherms*—across the continent that have the same temperature at the same period of the year. Life zones consequently appear on the map as broad transcontinental bands, except in the mountain regions, where they are extremely irregular owing to the effect of elevation on temperature. For each life zone there are certain characteristic species of birds and other organisms, which authorities have designated as zonal "indicators."

The six life zones are divided into a boreal group containing, from north to south, the **Arctic, Hudsonian,** and **Canadian;** and an austral group, comprising from north to south, the **Transition, Upper Austral,** and **Lower Austral.** The austral group is further divided longitudinally, on the basis of prevailing moisture, into three groups: eastern humid (from the Atlantic Coast west to the 100th meridian,

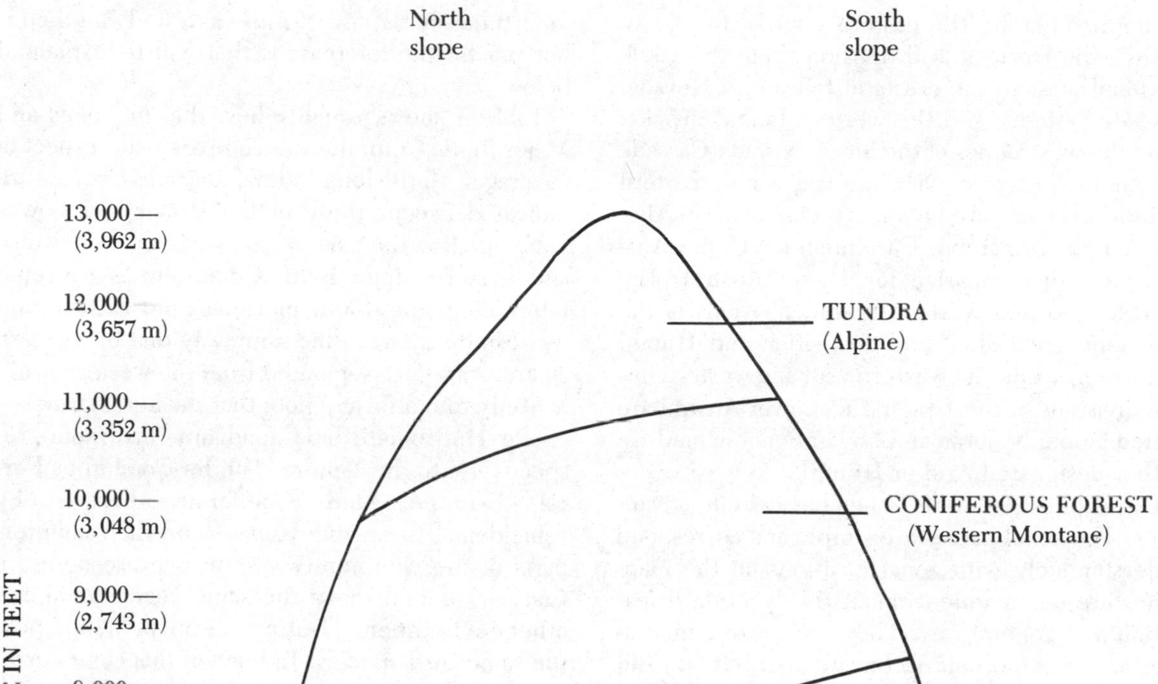

Figure 42 **Altitudinal Succession of Major Biotic Communities**
(Modified from Woodbury, 1947; redrawn from Odum, *Fundamentals of Ecology,* W. B. Saunders Company, 1953.)

or, approximately, the eastern edge of the Great Plains), the western arid division (from the 100th meridian west to the crests of the Sierra Nevada-Cascade system), and the western humid division (from the west slopes of the Sierra Nevada-Cascade system to the coast). The austral zones, eastern humid division, are given special names: Alleghanian for Transition, Carolinian for Upper Austral, and Austroriparian for Lower Austral. The western arid and western humid divisions of the Transition are called Arid Transition and Humid Transition, while the western arid and western humid divisions of the Upper and Lower Austral are named Upper Sonoran and Lower Sonoran, and are further designated Arid or Humid.

The application of life zones has definite advantages. The zonal names are simple and correspond understandably with zonal positions on the map. Zones are a convenient means of denoting distribution in a general way. They recognize temperature as an important factor in distribution. But unfortunately the use of zones as a study method is not practicable for reasons that will be explained below.

Table 5 shows roughly how the life zones and Major Biotic Communities compare with respect to coverage. All the longitudinal divisions of zones are indicated except those of the Upper and Lower Sonoran. For the sake of comparison, the eastern and western Major Biotic Communities are separated along the 100th meridian into eastern and western divisions in the same way that the eastern austral zones are separated from the western arid.

Study the table and note that the boreal zones—Arctic, Hudsonian, and Canadian—correspond, respectively, to the Tundra, Tundra-Coniferous Forest Ecotone, and Coniferous Forest. By coincidence these zones agree with their counterparts of the community system because both are transcontinental, have the same vegetational and other environmental features throughout, and hold the same bird species. In view of this close corre-

Table 5 **Comparison of Life Zones and Major Biotic Communities**

<table>
<tr><th>MAJOR
BIOTIC COMMUNITIES
AND ECOTONES
Eastern North America</th><th colspan="2">LIFE ZONES</th><th>MAJOR
BIOTIC COMMUNITIES
AND ECOTONES
Western North America</th></tr>
<tr><td>Tundra →</td><td colspan="2">Arctic</td><td>← Tundra</td></tr>
<tr><td>Tundra-Coniferous Forest Ecotone →</td><td colspan="2">Hudsonian</td><td>← Tundra-Coniferous Forest Ecotone</td></tr>
<tr><td>Coniferous Forest →</td><td colspan="2">Canadian</td><td>← Coniferous Forest</td></tr>
<tr><td rowspan="2">Grassland (part) →
Deciduous-Coniferous Forests Ecotone →</td><td colspan="2">Transition</td><td rowspan="2">← Grassland (part)
← Moist Coniferous Forest
← Western Oak Woodland (part)
← Sagebrush (part)</td></tr>
<tr><td>Alleghanian</td><td>Transition</td></tr>
<tr><td rowspan="2">Grassland (part) →
Deciduous Forest-Southwestern Pine Woodland Ecotone →</td><td colspan="2">Upper Austral</td><td rowspan="2">← Western Oak Woodland (part)
← Sagebrush (part)
← Grassland (part)
← Pinyon-Juniper Woodland
← Chaparral</td></tr>
<tr><td>Carolinian</td><td>Upper Sonoran</td></tr>
<tr><td rowspan="2">Southeastern Pine Woodland Subclimax →</td><td colspan="2">Lower Austral</td><td rowspan="2">← Grassland (part)
← Mesquite Subclimax
← Scrub Desert</td></tr>
<tr><td>Austroriparian</td><td>Lower Sonoran</td></tr>
</table>

lation one may apply either life zones or communities to the distribution of birds in northern regions and obtain essentially the same picture. Now compare the austral zones with their counterparts and observe that no one austral zone corresponds with any one community or ecotone across the country, mainly because no community or ecotone is transcontinental in the area covered. Actually each austral zone in its cross-country sweep takes in, or cuts through, no fewer than four Major Biotic Communities. There is no species of bird or any other organism remotely characteristic of an austral zone, or any one of its longitudinal divisions. The austral zones, then, expose the main weakness of the life-zone system. Life zones, by following only temperature lines across the country, fail to take cognizance of east-to-west changes in plant formations and other factors affecting distribution. As a consequence, they are misleading in that they infer a continent-wide uniformity of distribution that really exists only across northern regions.

The concept of life zones was developed just before the turn of the century by C. Hart Merriam and has been followed for a long period in this country; the concept of communities, though of earlier origin, has since gained a wide following. (For a history of these concepts, see Kendeigh, 1954.) Most ornithologists now recognize the inadequacy of life zones in mapping distribution on a continental basis, but many are reluctant to abandon the concept entirely in studies of a local nature. The reason is understandable and often justified. In a mountainous area, for example, belts or zones of life very definitely succeed one another altitudinally as the temperature changes; they are usually obvious and readily described. The practice of these ornithologists is to apply the original life-zone terminology while at the same time stressing the importance of plant formations and other factors in holding bird species to each zone. An outstanding advantage of this procedure is that, by using zonal names, one may avoid commitment on the seral status of communities.

For an illustration of how life zones with a modern interpretation may be used successfully in a state, refer to the work by Gabrielson and Jewett (1940, pages 28–41). Also refer to Miller (1951), who related plant formations to life zones in his study of the distribution of California birds, and to V.R. Johnston (1974) and Small (1974), who applied life zones in describing the distribution of birds in the Sierra Nevada.

References

Aldrich, J.W.
1943 Biological survey of the bogs and swamps of northeastern Ohio. *Amer. Midland Nat.* 30:346–402.

Allen, A.A.
1914 The Red-winged Blackbird: A study in the ecology of a cat-tail marsh. *Proc. Linnaean Soc.* New York, Nos. 24–25: 43–128. (A pioneer study of lasting significance.)

Amadon, D.
1966 *Birds around the world: A geographical look at evolution and birds.* Garden City, New York: Natural History Press. (Brief, highly readable coverage of both geographical and ecological distribution.)

American Ornithologists' Union
1983 *Check-list of North American birds.* 6th ed. Published by the Union: available from Allen Press, P.O. Box 368, Lawrence, Kansas 66044.

Anderson, S.H.
1970 The avifaunal composition of Oregon White Oak stands. *Condor* 72:417–423. (Low elevation in the Willamette Valley.)

Austin, G.T.
1970 Breeding birds of Desert Riparian habitat in southern Nevada. *Condor* 72:431–436.

Beckwith, S.L.
1954 Ecological succession on abandoned farm lands and its relation to wildlife management. *Ecol. Monogr.* 24:349–375.

Beecher, W.J.
1942 *Nesting birds and the vegetation substrate.* Chicago Ornithological Society. (Includes good illustrations of the effect of edge.)

Behle, W.H.
1976 Mohave Desert avifauna in the Virgin River Valley of Utah, Nevada, and Arizona. *Condor* 78:40–48.

Berger, A.J.
1981 *Hawaiian birdlife.* 2nd ed. Honolulu: Univ. Press of Hawaii.

Bond, R.R.
1957 Ecological distribution of breeding birds in the upland forests of southern Wisconsin. *Ecol. Monogr.* 27:351–384.

Breckenridge, W.J.
1974 A virgin prairie in Minnesota. In *The bird*

watcher's America, ed. O.S. Pettingill, Jr. Paperback ed. New York: Thomas Y. Crowell. (The birds one may expect in a prairie whose sod has never been broken.)

Brewer, R.
1967 Bird populations of bogs. *Wilson Bull.* 79:371–396.

Brooks, M.
1952 The Allegheny Mountains as a barrier to bird movement. *Auk* 69:192–198. (The author shows that this mountain range exerts a profound influence on bird movement.)
1965 The Appalachians. Boston: Houghton Mifflin.

Carlquist, S.
1965 *Island life: A natural history of the islands of the world.* Garden City, New York: Natural History Press.

Carpenter, J.R.
1940 The Grassland Biome. *Ecol. Monogr.* 10:617–684.

Cogswell, H.L.
1974 The California Chaparral. In *The bird watcher's America,* ed. O.S. Pettingill, Jr. Paperback ed. New York: Thomas Y. Crowell.

Cooke, F.; Ross, R.K.; Schmidt, R.K.; and Pakulak, A.J.
1975 Birds of the Tundra Biome at Cape Churchill and La Pérouse Bay. *Canadian Field-Nat.* 89:413–422.

Cottrille, B.D.
1974 Northern spruce bogs. In *The bird watcher's America,* ed. O.S. Pettingill, Jr. Paperback ed. New York: Thomas Y. Crowell. (A typical spruce bog and some of the problems of studying birds in such a habitat.)

Crosby, G.T.
1972 Spread of the Cattle Egret in the Western Hemisphere. *Bird-Banding* 43:205–212.

Darlington, P.J., Jr.
1957 *Zoogeography: The geographical distribution of animals.* New York: John Wiley & Sons.

Dixon, K.L.
1959 Ecological and distributional relations of desert scrub birds of western Texas. *Condor* 61:397–409.

Emlen, J.T.
1974 An urban bird community in Tucson, Arizona: Derivation, structure, regulation. *Condor* 76:184–197.

Errington, P.L.
1974 An Iowa marsh. In *The bird watcher's America,* ed. O.S. Pettingill, Jr. Paperback ed. New York: Thomas Y. Crowell. (The seasonal succession of birds and other life in a large glacial marsh.)

Evans, E.W.
1978 Nesting responses for Field Sparrows *(Spizella pusilla)* to plant succession on a Michigan old field. *Condor* 80:34–40.

Farb, P.
1963 *Face of North America: The natural history of a continent.* New York: Harper and Row.

Fisher, J.
1954 A history of birds. Boston: Houghton Mifflin. (Chapter 5, "Geographical Distribution.")

Fautin, R.W.
1946 Biotic communities of the northern Desert Shrub Biome in western Utah. *Ecol. Monogr.* 16:251–310.

Franzreb, K.E.
1978 Tree species used by birds in logged and unlogged mixed-coniferous forests. *Wilson Bull.* 90:221–238. (Study made in White Mountains, Arizona.)

French, N.R., and French, J.B.
1974 Rosy Finches of the high Rockies. In *The bird watcher's America,* ed. O.S. Pettingill, Jr. Paperback ed. New York: Thomas Y. Crowell. (A vivid description of habitat conditions in the Alpine Tundra and how different birds cope with it.)

Gabrielson, I.N., and Jewett, S.G.
1940 *Birds of Oregon.* Corvallis: Oregon State College.

Gammell, A.M.
1974 North Dakota prairie. In *The bird watcher's America,* ed. O.S. Pettingill, Jr. Paperback ed. New York: Thomas Y. Crowell. (The large assortment of birds in prairie country and the habitats they occupy.)

Hamilton, T.H.
1962 The habitats of the avifauna of the Mesquite Plains of Texas. *Amer. Midland Nat.* 67:85–105.

Hardy, R.
1945 Breeding birds of the Pigmy Conifers in the Book Cliff region of eastern Utah. *Auk* 62:523–542.

Hayward, C.L.
1948 Biotic communities of the Wasatch Chaparral, Utah. *Ecol. Monogr.* 18:473–506.
1952 Alpine biotic communities of the Uinta Mountains, Utah. *Ecol. Monogr.* 22:93–120.

Hensley, M.
1954 Ecological relations of the breeding bird population of the Desert Biome of Arizona. *Ecol. Monogr.* 24:185–207.

Hochbaum, H.A.
1974 The Delta Marshes of Manitoba. In *The bird watcher's America,* ed. O.S. Pettingill, Jr. Pa-

perback ed. New York: Thomas Y. Crowell. (One of North America's finest marshes for waterfowl.)

Hurley, R.J., and Franks, E.C.

1976 Changes in the breeding ranges of two grassland birds. *Auk* 93:108–115. (Horned Lark and Dickcissel.)

Hylander, C.J.

1966 *Wildlife communities: From the tundra to the tropics in North America.* Boston: Houghton Mifflin. (A nontechnical approach to the subject, aimed at the general reader; well illustrated.)

Ingles, L.G.

1950 Nesting birds of the Willow-Cottonwood Community in California. *Auk* 67:325–332. (A good example of a riparian woodland community.)

Irving, L.

1960 *Birds of Anaktuvuk Pass, Kobuk, and Old Crow: A study in arctic adaptation.* U.S. Natl. Mus. Bull. 217.

Jehl, J.R., and Smith, B.A.

1970 *Birds of the Churchill region, Manitoba.* Manitoba Mus. of Man and Nature, Spec. Publ. 1.

Jaeger, E.C.

1957 *The North American deserts.* Stanford, California: Stanford Univ. Press.

Johnston, D.W., and Odum, E.P.

1956 Breeding bird populations in relation to plant succession on the Piedmont of Georgia. *Ecology* 37:50–62.

Johnston, V.R.

1947 Breeding birds of the forest edge in Illinois. *Condor* 49:45–53.

1974 The Sierra Nevada. In *The bird watcher's America,* ed. O.S. Pettingill, Jr. Paperback ed. New York: Thomas Y. Crowell. (A colorful description of zonal distribution of birds in one of North America's highest mountain ranges.)

Kendeigh, S.C.

1932 A study of Merriam's temperature laws. *Wilson Bull.* 44:129–143.

1941 Birds of a Prairie Community. *Condor* 43:165–174.

1946 Breeding birds of the Beech-Maple-Hemlock Community. *Ecology* 27:226–244.

1947 Bird population studies in the Coniferous Forest Biome during a spruce budworm outbreak. *Ontario Dept. Lands and Forests, Biol. Bull.* 1:1–100.

1948 Bird populations and biotic communities in northern Lower Michigan. *Ecology* 29:101–114.

1954 History and evaluation of various concepts of plant and animal communities in North America. *Ecology* 35:152–171. (An excellent review, with an important list of references.)

1961 *Animal ecology.* Englewood Cliffs, New Jersey: Prentice-Hall.

Kendeigh, S.C., and Fawver, B.J.

1981 Breeding bird populations in the Great Smoky Mountains, Tennessee and North Carolina. *Wilson Bull.* 93:218–242.

Lack, D.

1976 *Island biology, illustrated by the land birds of Jamaica.* Berkeley and Los Angeles: Univ. of Calif. Press. (Recommended reading for anyone interested in insular distribution.)

Laudenslayer, W.F., Jr., and Balda, R.P.

1976 Breeding bird use of a Pinyon-Juniper-Ponderosa Pine Ecotone. *Auk* 93:571–586.

Lowry, W.P.

1967 *Weather and life: An introduction to biometeorology.* New York: Academic Press. (A compendium of information on weather problems related to birds and other life.)

Macnab, J.A.

1958 Biotic aspection in the Coast Range Mountains of northwestern Oregon. *Ecol. Monogr.* 28:21–54.

Marshall, J.T., Jr.

1957 *Birds of the Pine-Oak Woodland in southern Arizona and adjacent Mexico.* Berkeley, California: Pacific Coast Avifauna No. 32. Cooper Ornithological Society.

Martin, N.D.

1960 An analysis of bird populations in relation to forest succession in Algonquin Provincial Park, Ontario. *Ecology* 41:126–140.

Martin, T.E.

1981 Limitation in small habitat islands: Chance or competition? *Auk* 98:715–734. ("Shelter belts" in eastern South Dakota.)

Mayr, E.

1946 History of the North American bird fauna. *Wilson Bull.* 58:3–41.

Miller, A.H.

1951 An analysis of the distribution of the birds of California. *Univ. Calif. Publ. in Zool.* 50:531–644. (See S.C. Kendeigh's review of this paper in *Auk* 69:471–473, 1952.)

1963 Desert adaptations in birds. *Proc. XIIIth Internatl. Ornith. Congr.* pp. 666–674.

Miller, A.H., and Stebbins, R.C.

1964 *The lives of desert animals in Joshua Tree National Monument.* Berkeley: Univ. of Calif. Press.

Mitchell, M.J.

1961 Breeding bird populations in relation to

grassland succession on the Anoka Sand Plain. *Flicker* 33:102–108.

Monson, G.
1974 The Arizona Desert. In *The bird watcher's America*, ed. O.S. Pettingill, Jr. Paperback ed. New York: Thomas Y. Crowell. (The succession of birds and other life in desert country.)

Morse, D.H.
1971 Effects of the arrival of a new species upon habitat utilization by two forest thrushes in Maine. *Wilson Bull.* 83:57–65. (The Wood Thrush "potentially lowers the population sizes" of the Hermit Thrush and Veery, but seldom, if ever, nests in their habitats.)
1977 The occupation of small islands by passerine birds. *Condor* 79:399–412. (Twelve small islands in and around Muscongus Bay, Maine.)

Murray, J.J.
1974 The Great Dismal Swamp. In *The bird watcher's America*, ed. O.S. Pettingill, Jr. Paperback ed. New York: Thomas Y. Crowell. (Bird life in one of the most famous southern wetlands.)

Odum, E.P.
1945 The concept of the biome as applied to the distribution of North American birds. *Wilson Bull.* 57:191–201. (An excellent presentation of the subject plus a noteworthy comparison with the life zone concept.)
1950 Bird populations of the Highlands (North Carolina) Plateau in relation to plant succession and avian invasion. *Ecology* 31:587–605.
1971 Fundamentals of ecology. 3rd ed. Philadelphia: W.B. Saunders. (Highly recommended for general reading.)

Odum, E.P., and Johnston, D.W.
1951 The House Wren breeding in Georgia: An analysis of a range extension. *Auk* 68:357–366.

Pettingill, O.S., Jr.
1977 *A guide to bird finding east of the Mississippi.* 2nd ed. New York: Oxford Univ. Press.
1981 *A guide to bird finding west of the Mississippi.* 2nd ed. New York: Oxford Univ. Press.

Pitelka, F.A.
1941 Distribution of birds in relation to Major Biotic Communities. *Amer. Midland Nat.* 25:113–137.

Power, D.M.
1976 Avifauna richness on the California Channel Islands. *Condor* 78:394–398.

Provost, M.W.
1947 Nesting of birds in the marshes of northwest Iowa. *Amer. Midland Nat.* 38:485–503.

Rand, A.L.
1955 The origin of the land birds of Tristan da Cunha. *Fieldiana: Zoology* 37:139–166.

Rasmussen, D.I.
1941 Biotic communities of Kaibab Plateau, Arizona. *Ecol. Monogr.* 11:229–275.

Reilly, E.M., Jr.
1968 *The Audubon illustrated handbook of American birds.* New York: McGraw-Hill.

Riatt, R.J.
1976 Dynamics of bird communities in the Chihuahuan Desert, New Mexico. *Condor* 78:427–442.

Richards, T.
1974 In northern New Hampshire. In *The bird watcher's America*, ed. O.S. Pettingill, Jr. Paperback ed. New York: Thomas Y. Crowell. (Birds at the southern edge of the Coniferous Forest Major Biotic Community.)

Robertson, W.B., Jr.
1974 The Everglades. In *The bird watcher's America*, ed. O.S. Pettingill, Jr. Paperback ed. New York: Thomas Y. Crowell Company. (Bird life in one of North America's most remarkable marsh-swamps.)

Root, R.B.
1967 The niche exploitation pattern of the Blue-gray Gnatcatcher. *Ecol. Monogr.* 37:317–350. (Shows how the exploitative behavior of one species is organized to achieve "optimal adaptation to the conflicting demands of a changing environment.")

Schmidt-Nielson, K.
1964 *Desert animals: Physiological problems of heat and water.* New York: Oxford Univ. Press.

Serventy, D.L.
1960 Geographical distribution of living birds. In *Biology and comparative physiology of birds*, vol. 1, ed. A.J. Marshall. New York: Academic Press.

Shelford, V.E.
1926 *Naturalists' guide to the Americas.* Baltimore: Williams & Wilkins. (Contains very useful information on the location and distribution of natural areas.)
1963 *The ecology of North America.* Urbana: Univ. of Illinois Press. (Largely an updating of the previous work.)

Shelford, V.E., and Olson, S.
1935 Sere, climax and influent animals with special reference to the Transcontinental Coniferous Forest of North America. *Ecology* 16:375–402.

Shelford, V.E., and Twomey, A.C.
1941 Tundra animal communities in the vicinity of Churchill. *Ecology* 22:47–69.

Shugart, H.H., Jr., and James, D.
1973 Ecological succession of breeding bird popula-

tions in northwestern Arkansas. *Auk* 90: 62–77.

Small, A.
1974 From Monterey to Yosemite. In *The bird watcher's America*, ed. O.S. Pettingill, Jr. Paperback ed. New York: Thomas Y. Crowell. (Description of zonal distribution of birds from sea to Alpine Tundra.)

Smith, R.L.
1966 *Ecology and field biology*. New York: Harper & Row. (A very readable and thorough text; includes a chapter-by-chapter treatment of communities.)

Snyder, D.P
1950 Bird communities in the Coniferous Forest Biome. *Condor* 52: 17–27.

Stewart, R.E., et al.
1952 Seasonal distribution of bird populations at the Patuxent Research Refuge. *Amer. Midland Nat.* 47:257–363.

Tanner, J.T.
1955 The altitudinal distribution of birds in a part of the Great Smoky Mountains. *Migrant* 26:37–40.

Tester, J.R., and Marshall, W.H.
1961 *A study of certain plant and animal interrelations on a native prairie in northwestern Minnesota*. Minnesota Mus. Nat. Hist., Occas. Papers No. 8.

Thompson, L.S.
1978 Species abundance and habitat relations of an insular Montana avifauna. *Condor* 80:1–14. (Sweetgrass Hills east of the Rocky Mountains.)

Tomoff, C.S.
1974 Avian species diversity in Desert Scrub. *Ecology* 55:396–403.

Udvardy, M.D.F.
1958 Ecological and distributional analysis of North American birds. *Condor* 60:50–66. (The author analyzes and compares the North American avifauna with the European. A significant paper requiring careful study by those interested in the origin of North America avifauna.)
1963 Bird faunas of North America. *Proc. XIIIth Internatl. Ornith, Congr.* pp. 1147–1167, (Demonstrates with many maps how passerine species may be grouped as faunas on the basis of their geographical ranges and ecological preferences.)

Van Tyne, J.
1951 The distribution of the Kirtland Warbler *(Dendroica kirtlandii). Proc. Xth Internatl. Ornith. Congr.* pp. 537–544.

Verbeek, N.A.M.
1967 Breeding biology and ecology of the Horned Lark in Alpine Tundra. *Wilson Bull.* 79:208–218.

Vuilleumier, F.
1975 Zoogeography, In *Avian biology*. Vol. 5, eds. D.S. Farner and J.R. King. New York: Academic Press.

Walkinshaw, L.H., and Zimmerman, D.A.
1961 Range expansion of the Brewer Blackbird in eastern North America. *Condor* 63:162–177.

Wallace, A.R.
1876 *The geographical distribution of animals*. New York: Harper & Brothers. (The classic work on the subject, containing the concept of the six zoogeographical regions, adapted from the earlier work of P.L. Sclater.)

Warner, A.C.
1966 Breeding-range expansion of the Scissor-tailed Flycatcher into Missouri and other states. *Wilson Bull.* 78:289–300.

Weller, M.W., and Spatcher, C.S.
1965 *Role of habitat in the distribution and abundance of marsh birds*. Agric. and Home Econ. Exp. Sta. Iowa State Univ., Spec. Rept. No. 43.

Wiens, J.A.
1969 *An approach to the study of ecological relationships among grassland birds*. Amer. Ornith. Union, Ornith. Monogr. No. 8.
1974 Climatic instability and the "ecological saturation" of bird communities in North American grasslands. *Condor* 76:385–400.

Williams, A.B.
1936 The composition and dynamics of a Beech-Maple Climax Community. *Ecol. Monogr.* 6:318–408.

Winternitz, B.L.
1976 Temporal change and habitat preference of some montane breeding birds. *Condor* 78:383–393. (Study on lower slopes of Pikes Peak near Colorado Springs.)

Woodbury, A.M.
1947 Distribution of Pigmy Conifers in Utah and northeastern Arizona. *Ecology* 28:113–126.

Woodin, H.E., and Lindsey, A.A.
1954 Juniper-Pinyon east of the Continental Divide, as analyzed by the line-strip method. *Ecology* 35:473–489.

Woolfenden, G.E., and Rohwer, S.A.
1969 Breeding birds in a Florida suburb. *Bull. Florida State Mus.* 13(1):1–83.

Field Identification

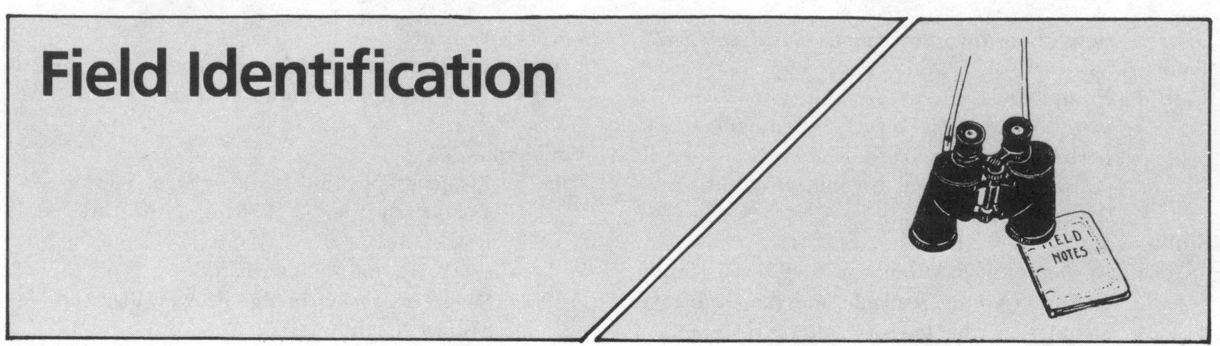

Ability to recognize birds in the field is one of the primary objectives of a beginning course in ornithology. At the first opportunity, start to identify birds and recognize them by appearance and sound.

Equipment

Ownership or access to the following is advised:

Binocular A 7x, 35 standard binocular, with coated optics and center focusing, is recommended for general use. The figures "7x, 35" signify that the binocular has a magnifying power of 7 times (making a bird at 70 feet [21.34 m] seem only 10 feet [3.05 m] away) and has objectives 35 mm in diameter. They also indicate that the binocular has a relative brightness value of 25, determined by dividing the diameter of the objective (35) by the magnifying power (7) and squaring the quotient (5). The term "standard" before binocular denotes a regular model as opposed to a wide-field or any other specially built model. The width of the field covered by the standard 7x, 35 model is 381 feet (116.13 m) at 1,000 yards (304.80 m). A binocular with "coated optics" has both prisms and lenses treated to reduce light reflection. Thus more light passes through the binocular, giving greater brightness when the instrument is used in poor light and cutting down glare when it is pointed in the direction of, but not into, the sun. A coating on the optics increases the relative brightness value by 50 percent; therefore, the brightness value of a 7x, 35 model with coated optics is actually 37 instead of 25. A binocular with "center focusing" is focused by turning a wheel on the central hinge-post after the right eyepiece has been adjusted to compensate for optical differences between the eyes.

Other models of binoculars may be useful for special purposes. All, however, should have coated optics. A 6x, 30 standard binocular, with its low power and consequent wider field, is advantageous when watching birds close at hand, as in woods or thickets. Standard binoculars with higher magnification, such as the 8x, 30 and 9x, 35, are particularly helpful when "reaching" for birds across water and canyons or in mountains. They must be used in good light and clear atmosphere because their brightness value is lower, and they must be held steady, since their field of view is not increased in proportion to their magnification. For one doing bird work in dim light, as at dawn or dusk, or in shadowy places, the 7x, 50 special model is highly desirable. Though necessarily big and bulky to accommodate the large optics, it has the advantage of very great light-transmission qualities. If magnifying power above 9x is required, a telescope is preferable since a binocular above 9x cannot be held steady enough for accurate vision. Wide-field 7x, 35 and 8x, 35 models, which cover 525 and 445 feet (160.02 m; 135.64 m), respectively, at 1,000 yards (914 m) are available. Their principal advantage is in enabling one to follow flying birds more easily. Nearly all standard binoculars are available with individual focusing—adjusting both eyepieces to one's eyes for a particular distance. Although models with individual focusing are inconvenient for looking at birds at varying distances because refocusing is slow, they are more sturdily built and more moisture-proof, hence less likely to be damaged in the field.

Telescope A telescope, supported by a tripod, is indispensable for identification and observation of birds at great distances. Usually one telescope per class is sufficient for ordinary purposes. A recommended telescope is the so-called spotting scope, 60 mm model, either with interchangeable eyepieces giving magnifications of 15x, 25x, 40x, and 60x, or with a lens that zooms from 15 to 60x.

Tape Recorder A portable, fully transistorized recorder, one or more per class, for recording and playing back sounds on tape in cassettes, can be used effectively to bring singing birds into closer view. The technique is to prerecord (e.g., from a phonograph record) those songs of species most likely to inhabit the area under investigation, then play back the songs in suitable habitats. Most any male breeding bird, on hearing the song of his species, will mistake it for the song of a rival male, approach the source, and start singing, thereby revealing himself. This technique is especially effective in getting good views of birds in forests or dense shrubbery. *Caution:* Playbacks of songs should not be repeated more than a few times as they may be disruptive, causing the male to desert the area or distracting attention from their nests. Males of some species play a significant role in the incubation of the eggs and care of the young.

Materials

Certain materials are necessary or desirable for identification work.

Field Guide Among the available guides to the identification of bird species in North America north of Mexico, the following are convenient in size to carry in the field and are especially useful for the reasons indicated.

Birds of North America: A Guide to Field Identification, by C.S. Robbins, B. Bruun, and H.S. Zim; illustrated by A. Singer. (New York: Golden Press. 1983). Each species drawn in full color with pertinent text and range map on opposite page; some similar species also compared in double-page, full-color spreads.

Field Guide to the Birds of North America. Washington, D.C.: National Geographic Society, 1983. Chief Consultants, J.L. Dunn and E.A.T. Blom; General Consultant, G.E. Watson; Consultant on Songbirds, J.P. O'Neill; S.L. Scott, Editor. Illustrated by various artists. Each species drawn in full color with pertinent text and range map on opposite page. Seven double-page, full-color spreads of birds in flight: two of ducks, two of shorebirds, one of gulls, and two of female hawks.

The Audubon Society Master Guide to Birding. Three volumes: 1, Loons to Sandpipers; 2, Gulls to Dippers; 3, Old World Warblers to Sparrows. Edited by J. Farrand, Jr. (New York: Alfred K. Knopf. 1983). Most species photographed alive in full color; others drawn in full color by various artists. Text by expert field ornithologists. Range maps accompany the illustrations on opposite pages.

For east of the Rocky Mountains: *A Field Guide to the Birds,* by R.T. Peterson (Boston: Houghton Mifflin. 1980). Each species drawn in full color with principal clues to identification on opposite page; range maps grouped together in a section before index.

For west of the Great Plains: *A Field Guide to Western Birds,* by R.T. Peterson (Boston: Houghton Mifflin, 1961). Many plates drawn in full color, comparing similar species; a few species illustrated in black and white only; concise text pointing out essential field characters, giving range, habitat, and description of nest and eggs of each species. A section on the birds of the Hawaiian Islands, is included.

Daily Field Checklist This is a list of the birds regularly found in the region where the course is undertaken. Use new copy each day in the field. The format is small so that it folds and fits into a pocket, field guide, or notebook and contains spaces for checking each species seen or heard and for indicating the observer or observers, locality or localities, hours in the field, and weather conditions. Each copy, properly checked and filled out at the conclusion of the day in the field, serves as a permanent record. If a checklist is not already available for the region where the course is undertaken, a suitable one should be prepared.

Pocket Notebook This notebook should have a durable cover (aluminum optional), be of a size to fit into a pocket, and have loose leaves so that they can be interchanged, and contain as many leaves as there are species likely to be observed.

Procedure

For each field trip, take a binocular, field guide, daily field checklist, pocket notebook, and pencil. As each species is identified, check its name on the checklist and write the name at the top of a blank page in the pocket notebook. Reserve this page thereafter for all information obtained on the species.

At the outset of the trip, bear in mind that identification of birds in the field is much more than matching their color patterns with pictures and descriptions in the field guide. The majority of birds encountered will be under such circumstances as to preclude the possibility of noting their colors. Many birds in the distance will be silhouettes; many will be forms dashing through foliage; and many will be heard only. Successful identification involves knowing *when* to expect, *what* to notice, and *how* to observe.

Much of the initial difficulty in identification is lack of observational acuity in detecting subtle differences in shape or motion of small forms and in noting distinctions in quality and pitch of sounds in the high octaves. Thus, in the beginning, many birds will look alike and sound alike. When the instructor readily recognizes one species after the other and they all seem alike, do not blame personal vision and hearing and conclude that the task of identification is beyond mastering. Any such conclusion, though understandable, is erroneous. Observational acuity, like any skill, is acquired through conscientious practice and repeated experience. Be diligent—and very soon a recognition of the more common birds will become second nature.

Identification Clues

Rapid progress in recognizing different species is effected by paying special attention to the many kinds of identification clues. Some clues depend on a thorough knowledge of the distribution of birds by season and by habitat or biotic community. For instance, the winter season is a clue to certain species, a marsh community to others. An important group of clues is provided by the birds themselves even when they are not close enough to details of form and color pattern. The general shape and posture of different birds, the way they fly or feed, and their flocking habits are among the many examples.

The drawings and comments on the following pages present a wide variety of clues. Before taking the first trip, study the drawings carefully, comparing and contrasting the birds depicted, then watch for these clues and know what to look for when seeing the birds.

Shape and Posture The four birds in Figure 43 illustrate a situation where different species appear together in lighting conditions so poor as to prevent seeing either their colors or their markings. At first glance the birds look much alike; closer inspection reveals several distinctions. They vary in size, the smallest being half the size of the largest. Their bills, though all acute, differ in length and thick-

Figure 43

ness. The tail of one bird is exceedingly long; that of another quite stubby. These are clues obtained by comparing outlines of birds while they are together and useful in combination with other clues (e.g., call notes, peculiarities of behavior), noted at the same time.

Locomotion A source of many clues. When on the ground, many birds simply walk, many others—primarily arboreal species—hop by moving both legs together. In water some birds swim while pumping their heads and necks back and forth. The mode of flight (Fig. 44) is an especially important clue to identification. The flight of a meadowlark is an alternation between sailing with the wings spread and directed slightly downward and flying with rapid wing-beats. A flicker, like other woodpeckers, has an undulating flight, rising upward

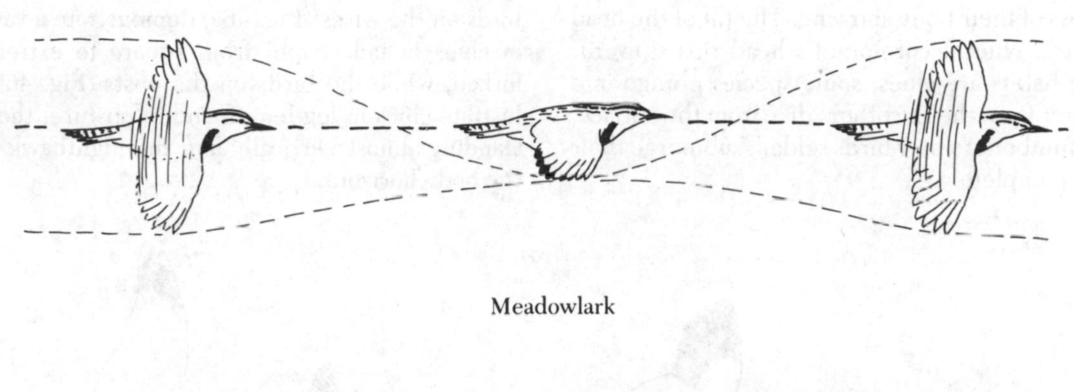

Meadowlark

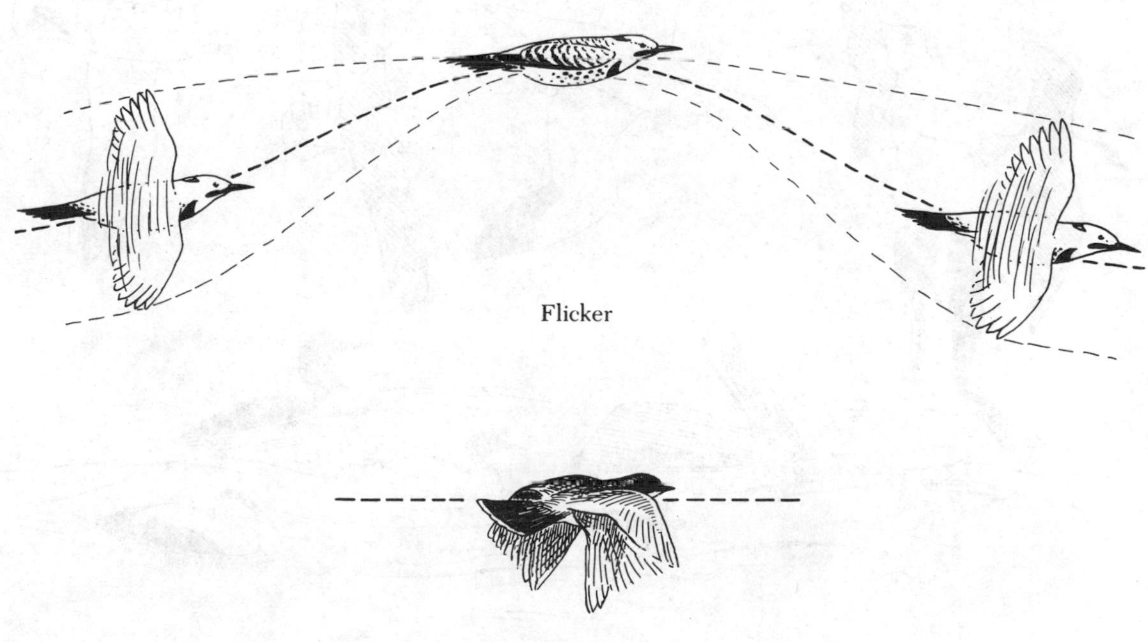

Flicker

Eastern Kingbird

Figure 44 **Modes of Flight**

after several quick wing-strokes and then dipping downward. The Eastern Kingbird flies in a straight line, with a continuously quivering wing action.

Birds of a Sea Coast The general color pattern of wings (Fig. 45) reveals clues: dark wing tips or white patches and stripes are excellent identification marks. The way a bird rests on the water gives other clues: pelicans and most ducks sit high on the water; loons and cormorants farther down in the water with less of their body showing. The tilt of the head may be a clue: a cormorant's head tips upward. Diving habits are clues: some species plunge into the water from the air; others dive from the surface; and a number of water birds seldom submerge their bodies completely.

Wire and Fencepost Sitters The fact that these birds sit on wires and fenceposts is a clue — they are open-country birds. Forest birds, for instance, would not be seen in any such situation. Even at a distance these birds show a variety of recognition marks — stripes through the eyes, over the head, and across the breast; spots or streaks on the under parts; terminal bands on the tail. Their general color pattern provides clues; some are uniformly dark or light; some are dark above and light below. The birds on the wires (Fig. 46a) demonstrate a variety of clues in tail shape, from square to extremely forked, while the birds on the posts (Fig. 46b) illustrate clues in leg length and in posture, the owl standing almost vertically and the nighthawk with the body horizontal.

Figure 45 **Birds of a Sea Coast**
In the air, left to right: Parastic Jaeger, Black Skimmer, Common Tern, Gannet, Fulmar, Herring Gull. On the water, left to right: Brown Pelican, Cormorant, Common Merganser, Common Loon, Common Eider.

(a)

(b)

Figure 46 **Wire and Fencepost Sitters**
(a) Left to right, top wire: Eastern Bluebird, Cedar Waxwing, American Kestrel, Scissor-tailed Flycatcher, American Robin, Eastern Kingbird, Red-winged Blackbird. Bottom wire: Loggerhead Shrike, Ruby-throated Hummingbird, Purple Martin, Barn Swallow, Tree Swallow, European Starling. (b) Left to right: Burrowing Owl, Common Nighthawk, Meadowlark, Horned Lark, Upland Sandpiper.

Birds of a Shore The shore is a good habitat clue to all the birds shown in Figure 47. Birds of such a habitat are waders, characterized by long legs and often by long necks. A few have peculiar teetering and bobbing actions. Variations in bill length and shape are especially helpful clues to the identification of shorebirds.

Birds of a Deciduous Forest Knowing what birds to expect in any given area the year round and during the different seasons is helpful. Furthermore, as birds spend much of their time feeding and have different adaptations for securing their food, their feeding habits are sometimes good clues to recognition. In a forest (Fig. 48) woodpeckers, nuthatches, and creepers constantly search for food on tree trunks; tanagers and vireos in the forest canopy; ovenbirds on the forest floor; and other birds in a variety of places.

Birds of a Marsh Becoming familiar with different biotic communities and learning what birds to expect in them is always an aid. Certain birds in a marsh community (Fig. 49) are particularly difficult to find because they are usually obscured by the vegetation. If one were not aware of the different kinds of marsh birds, some of them might be overlooked. Two marsh dwellers, the Red-winged Blackbird and Northern Harrier, are readily identifiable by the banner-marks on wings and tail, respectively.

Figure 47 **Birds of a Shore**

Left to right, front row: Dowitcher, Greater Yellowlegs, Semipalmated Sandpiper, Ruddy Turnstone. Middle row: Common Snipe, Killdeer, Semipalmated Plover, Spotted Sandpiper, Black-necked Stilt, American Avocet, Whimbrel, Marbled Godwit, American Oystercatcher. In the background: Great Blue Heron, Black-crowned Night-Heron, Glossy Ibis.

Figure 48 **Birds of a Deciduous Forest**
From the bottom, left side of trunk: Red-bellied Woodpecker, Ovenbird, Scarlet Tanager, Brown Creeper, Screech-Owl, White-breasted Nuthatch. Right side of trunk: American Redstart, Least Flycatcher, American Robin, Black-capped Chickadee, Red-eyed Vireo, Blue Jay, Ruffed Grouse.

Figure 49 **Birds of a Marsh**
From the left, perching or standing: Red-winged Blackbird, Marsh Wren, American Bittern, American Coot, Common Moorhen, Virginia Rail. In the water: Pied-billed Grebe. In the air: Northern Harrier, Black Tern.

Outlines of Flying Birds Frequently all that can be seen of birds in flight are their outlines, which are, except for wing motions, the principal clues to identification. Note particularly some of the clues indicated in Figure 50: (1) the shape of the wings, whether pointed or rounded, narrow or broad, slotted or unslotted; (2) the length of the neck and tail in proportion to body length; (3) the shape of the tail; (4) the position of the feet, whether or not they are extended posteriorly beyond the body and tail. In Figure 50(b) observe how the outline of a bat compares with the outlines of the birds.

Songs and Call Notes Learning to recognize birds by their songs and call notes is most difficult for the beginner for the simple reason that one's ear is not yet trained to distinguish the extraordinarily fine tones given by so many different birds. Nevertheless, learning to recognize songs and call notes is essential because, on field trips, more birds are heard than seen. It is safe to say that experienced ornithologists depend more on identifying birds by sound than by sight and that, as a rule, over two thirds of the birds ornithologists record in the field are only heard.

Conscientious and repeated listening to bird sounds will eventually lead to their ready identification. The following methods may aid in recording and remembering sounds:

Figure 50 **Outlines of Flying Birds**
(a) 1. Tundra Swan. 2. Turkey Vulture. 3. Brown Pelican. 4. Bald Eagle. 5. Canada Goose. 6. Osprey. 7. Peregrine Falcon. 8. American Crow. 9. Sandhill Crane. 10. Common Raven. 11. Common Loon. 12. Cooper's Hawk. 13. Great Blue Heron. 14. Red-tailed Hawk. 15. Double-crested Cormorant. 16. Herring Gull.

(b)

Figure 50 **Outlines of Flying Birds (Continued)**
(b) 1. Redhead. 2. Common Merganser. 3. Chimney Swift. 4. Pied-billed Grebe. 5. Forster's Tern. 6. Big Brown Bat. 7. Black-bellied Plover. 8. American Coot. 9. Virginia Rail. 10. Screech-Owl. 11. Black-billed Cuckoo. 12. Mourning Dove. 13. Common Nighthawk. 14. Belted Kingfisher. 15. Purple Finch. 16. Common Crackle.

1. Record the sounds phonetically—i.e., by the use of phrases that seem to resemble the sounds. Such phrases are easily remembered, particularly if they are coherent. Examples follow:

SWEET SWEET SWEET I SWITCH YOU (Song of the Chestnut-sided Warbler)
CHUCK BURR (Call of the Scarlet Tanager)
O GURGLEEEE (Song of the Red-winged Blackbird)

2. Record and remember the songs by using diagrammatic symbols. Use any symbols that seem logical. The following are from *The Book of Bird Life* by A.A. Allen (New York: D. Van Nostrand. 1961).

Black-capped Chickadee..............:

Eastern Wood-Pewee..................:

Eastern Meadowlark....................:

White-throated Sparrow...............:

Dark-eyed Junco ....:

Field Sparrow...........................:

Ruby-crowned Kinglet ....:

Veery......................................:

The phonetical rendering of these same sounds would be as follows:

Black-capped Chickadee
PHE-BE-BE
Eastern Wood-Pewee
PEE-A-WEE
Eastern Meadowlark
SPRING-IS-HERE
White-throated Sparrow
POOR-SAM-PEABODY, PEABODY, PEABODY
Dark-eyed Junco
SWEET, SWEET, SWEET, SWEET, SWEET, SWEET, SWEET
Field Sparrow
HERE, HERE, HERE, HERE, SWEET, SWEET, SWEET
Ruby-crowned Kinglet
SEE-SEE-SEE-, JUST LOOK AT ME, JUST LOOK AT ME, JUST LOOK AT ME, SEE-SEE-SEE
Veery
TUREE, AREE, AREE, AREE, AREE

According to Allen's method, a line indicates a whistle-like sound, the thickness of the line varying with the intensity of the sound. Thus, a thin line means a whispering whistle, a broad line is a clear whistle, and a series of small circles a trill, A small dash represents a clear note, a wavy line a tremulous note, etc. A line running upward points out a rise in pitch; a line running downward, a drop in pitch. In a continuous song the lines are connected; in a song interrupted by silent intervals the lines are correspondingly broken.

3. Use the method of recording bird sounds by A.A. Saunders in his *A Guide to Bird Songs* (Garden City, New York; Doubleday, 1951). He diagrammed a song as a series of lines, chiefly horizontal. In his words, "Each line represents one note of a song. Its horizontal length represents the period of time the note occupies. Its vertical height,

in relation to other notes in the song, represents the note's pitch. Its heaviness represents its loudness or intensity." Saunders then explains in detail how he applied his method to various types of songs. The main text of his book gives a concise description, illustrated by diagrams when possible, of the songs of most of the land birds and a number of water birds in the United States east of the Great Plains.

4. Remember the songs by using descriptive words. Among words commonly used are the following:

abrupt
alarming
ascending

buzzing
broken
bell-like
babbling

continuous
canary-like
carol-like
chipping
chirping
chattering
cheeping
chucking
churring
choppy
cooing

chinking
cat-like
clear
coarse

discordant
double-toned

effervescent
energetic
explosive

flowing
faint
flute-like
forced
fife-like

grating
gurgling
guttural
grasshopper-like
gabbling
gibbering
gushing

hurried
harsh
hollow
husky
hooting
hiccupping

instrumental
insect-like
intoned

jabbering

long
low
loud
liquid
lively
lispy
laughing

mewing
moderate
melodious
metallic
modulated
musical
mellow

nasal
noisy

owl-like

piping
pealing
piercing
puttering
plaintive
penetrating
peeping

quivering
quacking
quawking

rapid
rich
ringing
rattling
rasping
resonant
reedy
raucous
reverberant
repetitive
rollicking

short
slow
shrill
soft

screaming
strident
sipping
sighing
screeching
sibilant
sweet
sputtering
smacking
squawking
squealing
staccato
sonorous
stentorian
squeaking
strained
slurring

thin
trilling
twittering
tinkling
tremulous
throaty
ticking
tinking
throbbing

ululating
unmusical

varied
vibratory
ventriloquial
voluble

whining
weak
warbling
whistling
watery
whirring
whooping

yelping

5. Remember bird songs and become familiar with them by recording them in the field with a portable tape recorder and then reviewing them, or reviewing them from phonograph records. Heard repeatedly, the song will gradually become impressed upon the memory.

Activity

After each field trip go over the notes in the pocket notebook and rearrange the pages so that the species will be in order—preferably taxonomic order, as in the field guide. Then add any observations or impressions that seem important or desirable for future reference. Do not recopy field notes unless absolutely necessary; mistakes may result. And never destroy the original record. Notes are more valuable in their original form, even though they are not always neatly written in the field. As pages are filled, they may be withdrawn and filed, to save increasing the bulk of the notebook. Add new pages in their place. The importance of note-taking cannot be too greatly emphasized. In the first place, putting down observations and impressions in writing is the most dependable means of remembering what one has seen or heard. In the second place, notes are the most reliable record of field work and will almost invariably be useful at some future time.

Many phonograph records, as well as cassettes, of actual bird songs are available. For a list, write the Cornell Laboratory of Ornithology, Sapsucker Woods, Ithaca, New York 14850, or see *Fifty Years of Bird Sound Publication in North America: 1931–1981* by J. Boswall and D. Couzens (1982, *American Birds* 36:924–943).

Supplementary Guides

Besides the regular guide to identification of species, there are two other kinds of guides that may be useful.

For the Identification of Birds' Nest: *A Guide to Birds' Nests of 285 Species Found Breeding in the United States East of the Mississippi River* and *A Field Guide to Western Birds' Nests of 520 Species Found Breeding in the United States West of the Mississippi River* by H.H. Harrison (Boston: Houghton Mifflin. 1975 and 1979, respectively). *A Field Guide to the Nests, Eggs, and Nestlings of North American Birds* by C. Harrison. (Toronto: Collins.1978). All three books provide a vast array of useful information in addition to matters of identification.

For Locating Birds: *A Guide to Bird Finding East of the Mississippi* and *A Guide to Bird Finding West of the Mississippi*, by O.S. Pettingill, Jr. (New York: Oxford Univ. Press. 1979 and 1981, respectively). The volumes include detailed instructions for reaching the outstanding places for birds in the contiguous United States and tell what the places are particularly noted for and when they should be visited.

Behavior

by Jack P. Hailman

What a bird does and how it responds to its surroundings constitute its **behavior,** which is as adaptive as its anatomy and physiology. The study of behavior is **ethology,** although some behaviorists consider this term to be merely a subset of behavioral studies concentrating on field observations.

The Levels and Determinants of Behavior

Levels of Behavior

Any consistent way of behaving is a **behavorial pattern,** and behavioral patterns occur at different levels of behavior. The lowest level consists of simple motor acts such as flapping flight and soaring, which are called **action patterns,** whereas consistencies such as fall migration are behavioral patterns at a higher level.

Determinants of Behavior

The ethologist first describes behavior and then attempts to analyze the factors that determine it. These factors fall into four major categories:

Behavioral Control "Causation"—refers to the immediate external stimuli and internal motivational states that bring about behavior, as in weather patterns that trigger migratory flights and the preparatory state for migration that itself depends upon prior external factors such as shortening day-length.

Behavioral Ontogeny "Epigenesis"—refers to the genetic and experiential factors that guide individual development, such as in the development of passerine song, where genetically based dispositions are developed through hearing other birds sing and practicing song.

Behavioral Perpetuation Refers to the maintenance of behavioral patterns over successive generations in a population through genetic and traditional transmission with possible subsequent selection. Many behavioral patterns are adaptive, as shown by the fact that unrelated species show similar solutions to similar environmental problems; for example, cliff-nesting sea birds building nests from which eggs are not easily displaced. Natural selection may help to perpetuate such traits—birds building deviant nests may lose eggs over the cliff face and hence leave fewer offspring.

Behavioral Phylogeny Usually refers to the evolutionary history of a species, although where learned traditions are involved their history is also important, as in the origin and spread of new song types in a population.

Types of Behavior

More or less natural groupings constitute eight major categories of behavior: spatial, maintenance, nutritional, agonistic, sexual, nesting, parental, and interspecific. Within each category there are both action patterns and higher levels of behavior, the former being emphasized here. A distinction is sometimes drawn between **individual** behavior (such as preening) and **social** behavior (such as fighting) but this division is not fundamental. Preening another bird (allopreening) is a social behavior, for example. Behavior is also sometimes divided into reproductive and nonreproductive activities, but this division is also somewhat arbitrary. Holding a winter territory might have little to do with reproduction, but conversely it might secure a nesting area for the following spring.

Ethogram

Descriptive compilations of behavioral patterns of a given species are called **ethograms,** but it is doubtful that a complete ethogram has ever been compiled. The Mallard *(Anas platyrhynchos)* has more than 100 patterns, and almost 200 have been identified in the Scrub Jay *(Aphelocoma coerulescens);* the Black-capped Chickadee *(Parus atricapillus)* gives more than a dozen different calls, not to mention its other behavioral patterns. The first ethogram so called was by Makkink (1936) on the European Avocet *(Recurvirostra avosetta).* Two well known ethograms are of the Song Sparrow *(Melospiza melodia)* by Nice (1937, 1943) and the Herring Gull *(Larus argentatus)* by Tinbergen (1953).

Spatial Behavior

Movements in space range from action patterns such as terrestrial gaits and flight to higher levels of behavior such as home ranges and migration. Static postures such as perching and standing are included because these result from specific movements.

Locomotion Birds move across a substrate such as the ground by hopping (legs move together) or walking and running (legs move alternately). In running both feet are off the ground at some part of the locomotory cycle. Many species use several gaits depending upon the speed of movement, slope and evenness of the substrate, and material under foot. Other gaits include skipping (legs not moved exactly together, the interval of left-right being different from that of right-left) and sidestepping (one foot brought alongside the other, and then the other extended). Birds show *aboreal* (in branches of trees), *aquatic* (floating, swimming on the surface, diving, swimming under water, etc.) and *aerial* movements. Only insects, bats, and birds are capable of powered flight, but many kinds of animals are capable of gliding. Birds such as frigatebirds, shearwaters, and some vultures are experts in sustained gliding, or soaring, traveling great distances without a wing-beat. A bird about to fly often gives incipient takeoff movements of crouching, tail-flicking, and wing-flicking: so-called **flight intention movements.**

Higher Levels of Social Behavior Sandpipers and other social species fly in tight, well-coordinated flocks, sometimes further developed as specializations for foraging and antipredator reactions. Many passerines have exaggerated flight-intention movements that warn social companions, as in the flashing of white outer tail feathers by the Dark-eyed Junco *(Junco hyemalis).* Many other kinds of **social signals** are also evolved from simple locomotory action patterns.

Individuals often travel habitual paths and have favorite perches as well as specific loci for other activities. Most temperate species are migratory, withdrawing from polar parts of their ranges, some going to the tropics or even the opposite hemisphere to avoid the rigors of winter. Dark-eyed Juncos nesting in the Great Smoky Mountains migrate to lower elevations in the winter, and many sea birds have complicated patterns of movement over of the globe.

Maintenance Behavior

A bird, like an airplane, requires frequent preventive maintenance for its proper functioning on the ground and in the air. Some maintenance acts are directed toward the bird's exterior (preening), others relate to the internal state (stretching), still others are concerned with applying substances to the bird's surface (bathing), and some are direct responses to environmental factors (panting). Maintenance activities, like locomotory acts, appear to be a rich source for the evolution of display elements.

Manipulation of Body Surfaces Birds scratch their heads in two ways. Most species other than passerines simply extend the leg under or alongside the body so that the foot (usually the middle toe) reaches the lowered head: **underwing** (or "direct") **scratching.** Most passerines droop the wing and extend the leg above it toward the head so that the tarsal joint ("heel") is lateral to the wing: **overwing** (or "indirect") **scratching.** However, some nonpasserines (e.g., plovers) scratch over the wing and a few passerines scratch under the wing. Head-rubbing on branches or other substrates occurs in some species. Sometimes scratching accompanies yawning, and hence is presumably elicited by irritation of the Eustachian tubes; other times it seems to replace preening where the bill cannot reach.

Bill-scratching is a movement similar to underwing head-scratching. A few parrots that scratch the head over the wing nevertheless scratch the bill under the wing. Passerines and some other birds also exhibit **bill-wiping,** in which the side of the bill is stropped quickly from base to tip against a branch or other substrate, and then the movement is repeated on the opposite side. Gulls sometimes bore their bills into the sand, and there may be still other methods of bill care.

Care of the foot is usually accomplished by picking or chewing with the bill, many parrots raising the foot to the bill as in bill-scratching, other species lowering the head to pick at the foot on the perch or substrate. Gulls, terns, and some other birds periodically bend down and seem to inspect the feet.

A higher level activity composed of various action patterns is **preening,** including nibbling with the bill-tips, which may remove foreign matter and ectoparasites. Drawing the feather from base to tip through the mandibles appears to align the barbs and fasten together the barbules of adjacent barbs, thus maintaining the integrity of the feather surface. Birds noticeably fluff feathers during preening and often assume characteristic postures.

In oiling the feathers, a fatty substance is extracted by the mandibles from the uropygial (oil) gland at the base of the tail, and then spread over feathers with the mandibles. The bird also rubs its head against oiled areas and may further spread oil by rubbing the now-oiled head over other, unoiled areas. Long-necked birds often rub the head directly on the oil gland. Some passerines transfer oil to the foot with the bill, then comb the head feathers with the foot in a movement that resembles head-scratching. The precise significance of oiling is not fully understood: oiling probably helps to waterproof feathers but unequivocal experimental evidence is lacking.

Comfort Movements Comfort movements seem to be elicited by internal factors such as muscle stiffness or subtle surface factors such as feathers being out of place. "Comfort movement" may not be a good term, as the observer cannot judge what "comfort" the bird may derive, but that is the traditional name.

Stretching Most birds have two kinds of wing-stretching: the *wings-up stretch,* in which the carpal joints are raised up above the level of the back, and the *wing-leg stretch,* in which the wing is extended on one side along with the leg on the same side. *Bill-stretching* resembles mammalian yawning, but there is no evidence that the bird's lungs are actively filled with air as in true yawning. Perched passerines also show *toe-stretching,* usually involving the middle of the three forward-pointing toes. *Wing-flapping* is common among ducks and other waterfowl while floating on the surface and consists of rising almost vertically from the water while vigorously beating the wings several times. All these stretching activities may maintain muscle tone for movement, much in the way a human runner stretches out methodically before a race.

Shaking Shaking of the feathers may restore the integrity of the plumage surface, as opposed to preening acts that restore individual feather integrity. Shaking usually begins with erection of the feathers, which are relowered after the shake. *Head shakes* usually involve raising the head so that the axis of the bill is nearly in line with the backbone, and then vigorously rotating the head back and forth about this axis. *Body shakes* involve a similar rotary motion around the vertebral axis, often accompanied by *wing shuffling,* which seems to seat the folded wings alongside the body; the entire sequence is sometimes called **feather setting.**

Sleeping Sleeping could also be considered a maintenance activity, although its physiological consequences in birds have not been well studied. Sleeping postures vary, a common one involving retraction of one leg up into the belly feathers, with the head rotated and the bill inserted into the scapular feathers. Some birds—e.g., pigeons—sleep with the head retracted into the shoulders and the bill resting on the upper breast. Mallards sleep either way. Sleeping birds noticeably fluff the feathers, which further helps to conserve body heat. Details of sleeping vary greatly. Some water birds sleep while afloat but others never do so; large species either stand or squat on land; passerines typically perch crosswise on a small branch; goatsuckers align themselves along the axis of a large branch; woodpeckers cling vertically to a surface; and some parrots hang upside down. Certain swifts may sleep at high altitudes on the wing, where they have been detected at night by radar.

Anointing Most birds *water-bathe,* a high-level activity composed of complex action patterns, such as dipping the head into the water and raising it, erecting the feathers, beating the wings, and so on. Bathing might remove dirt but probably is more important in discouraging ectoparasites and serving other fuctions. Terrestrial species bathe in shallow puddles or even in the morning dew on leaves. Swifts, swallows, and some other highly aerial birds, as well as hummingbirds, kingfishers, and some owls, plummet into water or skim across its surface. Swifts also bathe in falling rain. Some ducks have elaborate somersaulting behavior in which they turn upside down while diving completely under water.

Dust-bathing Only certain species engage in dust-bathing, among them the House Sparrow *(Passer domesticus)* and wrens, both of which also water-bathe. Goatsuckers and gallinaceous birds such as quail dust-bathe almost exclusively. Dust-bathing includes action patterns—e.g., scratching in loose soil, crouching in a depression with raised feathers, and throwing dust over the body with the wings. Dusting presumably helps discourage ectoparasites and might have other uses.

Anting Perhaps the most bizarre maintenance activity of birds is anting, behavior described by John James Audubon in the Wild Turkey *(Meleagris gallopavo)* and known from a surprising variety of species. The commonest form, which Whitaker (1957) called *passive anting,* is to lie on an anthill with raised feathers and sometimes spread wings, often accompanied by complex tumbling movements. In *active anting* the bird picks up the individual ants in its bill tip, apparently crushes them, and wipes them on the body surface with peculiar vibratory motions. Anting also has been observed with a variety of other objects—burning matches, cigarette smoke, moth balls, hair tonic, and others. Birds seem partial to ants of the genus *Formica;* the formic acid extruded by ants may repel feather mites.

Response to Climate Variables In cold birds **fluff the plumage,** which increases the boundary layer of still air, reducing loss of body heat by convection. Birds also **stand on one leg** to conserve heat, tucking the other leg into the belly feathers to warm it. Gulls can decrease blood flow to the legs, letting the feet fall to just above freezing. Sleeping postures incorporate similar heat-conserving aspects.

A nearly universal response to heat is **panting** with open bill and raised tongue, losing heat through evaporation. Some birds also shunt air through the respiratory system by **gular flapping,** rhythmic pulsations of the throat. Wetting the body, seeking shade, perching in a breeze, and other responses to heat stress also occur. The Wood Stork *(Mycteria americana)* defecates on its well-vascularized tarsi, cooling the circulating blood by evaporation.

In **sunbathing** birds fluff or ruffle feathers and may spread their wings while lying on the ground. A so-called involuntary type consists of immediate immobility when emerging from shade into sunlight. Radiation may provide feather care by stimulating growth, discouraging ectoparasites, and so forth, but the bird risks overheating, so afterward flies immediately to shade or a bathing place.

Similar to sunbathing, **sunning** usually involves only open wings and perpendicular orientation toward the sun, but sunning is not followed by retreat to the shade or bathing in water. Sunning therefore appears to be bodily warming, often performed by vultures, woodpeckers, and other birds in early morning. Hawks frequently perch with their ventral or dorsal sides facing the low sun, perhaps a lesser manifestation of sunning.

In so-called **wing-drying** by anhingas and cormorants the wings and tails are spread in sunlight after fishing, but the behavior may also function to warm the bird. Anhingas *(Anhinga anhinga)* wave their wings and tail when sunlight is not available, behavior consistent with both the feather-drying and body-warming interpretations.

Nutritional Behavior

A flying machine requires fuel as well as maintenance. Particularly the diversity in avian bills reflects the wide range of specific foraging habits. **Carnivores** eat other animals and **herbivores** eat plants, but many further distinctions can be made. **Raptors** (hawks and owls) eat largely other vertebrates; **insectivores** (many passerines) eat insects and other small arthropods; **nectivores** (hummingbirds, also sunbirds, Nectariniidae) feed on nectar; **frugivores** (toucans) feed on fruit; and there are fish-eating, crab-eating, seed-eating, and a host of other specific-foraging birds. Some general groups of foraging acts are location, procurement, transport, preparation, storage, and ingestion of food.

Water-related Behavior Water balance is as crucial as energy balance, most birds requiring intake of free water to supplement that from metabolism and in food. Most birds drink with a *scoop-and-tilt* method, in which water is scooped in the lower mandible and allowed to run down the throat as the head is tilted backward. Pigeons, doves, and a few other birds actively pump water by inserting their bills somewhat as drinking straws. Birds conserve water by excreting concentrated urine, and many aquatic birds, particularly sea birds, have special nasal glands (see page 101) that extrude a concentrated salt solution. A head-shaking action throws the salty water from the upper mandible.

Locating Food Nectivores and frugivores may search wide areas to find particular plants in flower or fruit. Hummingbirds visit flowers systematically so as to avoid repeats at those whose nectar supply is diminished. Among special action patterns are those used in turning fallen leaves to locate seeds (and other items) in the litter beneath. The White-throated Sparrow *(Zonotrichia albicollis)* and related species throw leaves backward with a two-footed scratch, some thrushes extend one leg and circle the foot in the litter, gallinaceous birds scratch with one foot, the Brown Thrasher *(Toxostoma rufum)* picks up one leaf with the bill and throws it, and the House Sparrow inserts its bill in the litter and quickly shakes its head back and forth. Sapsuckers drill holes, and later return to lap up the running sap and to feed on the insects attracted to it. Many rodent-eating raptors, *(Buteo* hawks) and scavengers (vultures, gulls) fly over large areas of suitable habitat to find food. Various species of herons, hawks, flycatchers, and shrikes wait motionless for prey. Small insectivorous passerines hop through branches, move from tree to tree, hover at the ends of branches, and so on. Special probing and extracting acts are often closely linked to the procurement of prey. The American Woodcock *(Scolopax minor)* uses its long bill to pull earthworms from soft soil. Some woodpeckers drill holes directly to the prey, but others tear off large pieces of bark to expose larvae. Many birds flush prey, such as Cattle Egrets *(Bubulcus ibis)*, which walk across grassy fields or follow cows that are similarly flushing insects. Some gulls paddle their feet alternately on mudflats, apparently causing aquatic invertebrates to surface. The Northern Mockingbird *(Mimus polyglottos)* employs jerky movements called wing-flashing, which either reflects light upon or casts a shadow over insects, making them move. The Snowy Egret *(Egretta thula)* extends its bright yellow foot in the water and stirs it in a circle, luring fish.

Procuring Food Predatory birds have some sort of *strike* behavior that follows prey detection. Some herons simply strike with the bill, but others stalk prey stealthily or chase prey by running after it. The Rough-legged Hawk *(Buteo lagopus)*, American Kestrel *(Falco sparverius)*, and other raptors may hover above their prey, then drop upon it with talons outstretched.

Location and procurement of food are often integrated, as in Black Skimmers *(Rynchops niger)*, which insert an elongated lower mandible into the water while flying, thus cutting a long wake while skimming over the surface. When striking prey, the lower mandible rebounds backward, much as a wise baseball player gives a bit with his hand when catching a hard-thrown ball. As the head is thus going down and back, the mandibles are closed on the prey—and only then is the head raised. Skimmers do *not* jerk prey upward like a fisherman scooping with a dipnet, despite such popular claims. The Wood Stork inserts its open bill in murky water and snaps it closed when feeling a fish touch the interior. Many sandpipers probe into mud or sand to find aquatic invertebrates tactually.

Birds feeding on small aquatic plants and animals have sieve mechanisms that trap food. Dabbling or puddle ducks *(Anas)* stand in shallow water or swim while holding the flattened bill parallel with the surface and turning the head back and forth to move water laterally through the bill. Flamingos sift with head down and the distal part of the bent bill upside down (see illustration on p. 127), moving water through the bill with a sucking action performed with throat muscles. Spoonbills have spatulate distal modifications of the bill specialized for sieve feeding.

Tool-use for procurement of food is known in the Woodpecker Finch *(Cactospiza pallida)* of the Galapagos Islands. It breaks off a cactus spine, which it holds in the bill to extract insects from holes and crevices much in the way a woodpecker uses its long tongue.

Actual grasping of large food items is usually performed with the mandibles, but most raptors grasp with the outstretched talons to catch small birds and rodents; the Osprey *(Pandion haliaetus)* similarly catches fish. The Anhinga is unusual in spearing fish with its bill, usually its lower mandible; and pelicans scoop fish into their gular sacs.

Transporting Food Birds have few options for transporting food and many species simply eat at the procurement site. Raptors fly to a perch where the prey is torn into pieces for ingestion, transporting prey with the talons. Some raptors, especially with small prey, and nearly all other birds carry food in the bill.

Preparing Food Most seed-eating birds husk seeds by mandibular movements, positioning the seed with the tongue; chickadees take a large seed to a perch, hold it with the toes and pound it open with the bill. Fruit-eaters swallow the fruit whole, eat the fleshy part while dropping the pit by manipulations with the mandibles and tongue, or peck away fleshy parts—depending upon the species and kind of fruit. Birds that eat nuts such as acorns often hold them with the feet while pecking, much as chickadees open seeds, or wedge the nut into bark furrows or other holding places for opening.

The European Oystercatcher *(Haematopus ostralegus)* has two methods for opening shellfish. One is to insert its laterally flattened bill into the opening between the valves (shells) of bivalves, and then sever the adductor muscle. The other is simply to hammer at the shell wedged into rocks; an equivalent technique used by gulls is to fly up and drop shellfish upon rocks.

Rocks are also used to open eggs. The Black-breasted Buzzard *(Hamirostra melanosternon)* of Australia drops stones from the air onto emu eggs, and the Egyptian Vulture *(Neophron percnopterus)* throws stones onto ostrich eggs from a standing position (see chapter title illustration on p. 212).

Carnivorous birds often kill prey, raptors biting through the vertebral column of mammalian and avian prey. Passerines commonly hold prey such as caterpillars in their bills while wiping or banging them against a branch. Such actions may also wipe noxious secretions or stinging hairs from the prey. Scrub Jays sharply strike small toads against a hard surface.

Storing Food Shrikes impale grasshoppers and other invertebrates on the thorns of thorny bushes; the Loggerhead Shrike *(Lanius ludovicianus)* now commonly uses the barbs of barbed-wire fences. Special *impaling movements* involve holding the prey in the bill while pulling it back onto the thorn.

Following the usage applied to mammals, **larder-storing** (hoarding) is the storing of many nuts in one location, as opposed to **scatter-storing.** The Acorn Woodpecker *(Melanerpes formicivorus)* drills many individual, acorn-sized holes in a dead tree for its larder. Corvids (jays, nutcrackers) bury nuts individually, as well as hiding them in the furrows of tree trunks. Burying by the Scrub Jay consists of excavation by several digging actions with the bill, insertion of the acorn with the bill tips, tapping the acorn in a pecking-like motion, filling the hole, and then covering the site with a leaf or other handy material. Nutcrackers have a serial memory and tend to dig up nuts in the order in which they were buried. There is also a tendency to bury nuts in certain kinds of sites, such as at the base of a grass clump, so that searching the sites later has a better than random probability of success.

Ingesting Food The bird's tongue is helpful in moving food back into the throat. Some live prey is swallowed whole but may involve special action patterns—e.g., throwing the prey into the air and catching it in the open throat. Sometimes the Anhinga throws its speared fish up off of the bill, catches it in the mandibles, and then throws it up again for ingestion.

Agonistic Behavior

Activities such as threatening, fighting, fleeing, and submitting are categorized as **agonistic behavior.**

Physical Interactions Actual fighting is common in early phases of spacing and establishing dominance relationships. Most species use jabs or pecks, but phrases such as "pecking order" may overemphasize this aspect. Raptors may strike one another in the air with their talons, and some species wrestle on the ground by grasping the opponent with the feet. The bill is also used to grasp, pulling on the opponent or pushing it into the ground. Gulls, swans, and other species also slap the opponent with the wings.

Spatial interactions are important in agonistic behavior. One bird may approach another to fight, but if the first flees instead of standing its ground, the second often gives chase. Often the aggressor will *supplant* an opponent, occupying the spot it formerly held without further chasing. (Such behavior is also called *replacing,* but the term *displacement* is avoided because it has other meanings.)

Communicative Interactions Readiness to fight is communicated by threat displays, specific vocalizations, "confident" approach, and other actions. **Threat displays** are action patterns—often taken to include specific vocalizations—evolved for communication. The process by which displays are evolved from noncommunicative action patterns is called **ritualization.** Many threat displays show obvious origins from action patterns of fighting. The "upright" threat posture of gulls, for example, consists of drawing the body high with bill pointed down in readiness to strike, and wings held out a little at the carpal joints (wrists) in readiness to slap. Juncos and related species, on the other hand, hold the body horizontally in threat, with the body feathers fluffed and with the head directed toward the opponent and the bill open in preparation for biting.

A threatened bird fights back (confident approach), gives ground, or shows communicative behavior different from that of the threatening bird. Charles Darwin noticed that such return displays are often the "opposite" of threat, which he called the "principle of antithesis." The objective of such **appeasement displays** appears to signal to the threatening bird acknowledgment of its dominance, and in return the appeasing bird may be allowed to remain without overt challenge. Appeasement displays are antithetical or opposite to threat in appearance so to be clearly understood at a glance. Thus, in juncos appeasement consists of standing very high, with the body feathers sleeked and even the crest feathers raised, which makes the bird appear high and thin—opposite to the low, thick threat posture. Appeasement signals are evolutionarily derived largely from fleeing actions.

Higher Levels of Agonistic Behavior One evident manifestation of spacing is the regular interval between flocking birds such as swallows perched on a utility wire. The minimum distance separating birds is called the **individual distance** and may be no more than the length a bird can reach to peck its neighbor; in other cases the individual distance may be greater. There is a whole continuum of spacing distances up to large areas, called **territories,** that have spatial boundaries (see the chapter on "Territory"). Territories are defended by special agonistic displays called **advertisement,** such as particular postures in part and particular vocalizations (see the chapter on "Song"), which are broadcast generally where any potential intruder may receive them.

Priority of access to some resource (such as food) or place is **dominance.** Dominance-subordinance relations are often established by fighting and reinforced through display. Dominance may depend upon the resource contested or other factors, but in many social species dominance is general or "absolute," specifying a *matrix* of relation or **dominance hierarchy.** The extreme case is found in flocks of domestic hens, which exhibit a **linear** (or near-linear) **hierarchy.** The α-bird (alpha, first letter of the Greek alphabet) dominates all hens, the β-bird (beta, second letter) dominates all but the α-bird, and so on to the Ω-bird (omega, last letter), which is subordinate to all birds. In species with large flocks dominance is rarely linear, instead containing **intransitive triads:** x dominates y, y dominates z, but z dominates x. The situation is therefore like the scissors-paper-rock game of children. Large flocks may have a few birds at the top and bottom forming linear subhierarchies, but intermediate birds are of nearly equal dominance, forming many unstable intransitive triads.

Sexual Behavior

For interactions between mates, see the chapters on "Song" and "Mating." A few notes specifically on behavior suffice here. Advertisement, which may be part of territorial defense by song, also attracts potential mates. **Courtship** or **reproductive displays** are communicative postures, movements, and vocalizations. **Courtship feeding** is widespread and consists of the male feeding his mate or at least sharing his food with her, acts which may help maintain the pair-bond and demonstrate to the female the foraging efficiency of her mate. **Copulation** usually consists of the juxtaposing of cloacal openings for transfer of sperm, but male ducks and a few other species have an intromittent organ for copulation (see Appendix A, p. 376).

Some courtship display serves to bring together potential mates. The male's courtship display may also serve as one basis upon which the females select mates individually, although the quality of the male's territory is important, and the male's displays may help to show off important aspects of his real estate. Stimulation of the mate and synchronization of reproductive development within the pair may also be accomplished through display. Special displays such as inciting by female Mallards and hiccupping by female Scrub Jays are given when a male other than the mate intrudes upon the pair, and some male displays may function similarly in bond maintenance of the pair. Most species have special **precopulatory displays,** including reverse mounting (each member of the pair mounting the other) in sexually monomorphic birds like European Starlings *(Sturnus vulgaris).* Of obscure function are **postcopulatory displays,** which are quite elaborate in grebes and dabbling ducks.

Nesting Behavior

As there are separate chapters on "Nests and Nest-building" and "Eggs, Egg-laying, and Incubation," this section concerns only certain categories of nesting action patterns.

Nest-site displays may be used to agree upon the site, as in choking of gulls and vocalizations by chickadees at potential nesting holes. *Material-gathering acts* may take place, such as carrying materials in the bill, gathering mud on the feet, clutching other materials in the toes and claws, and, in the case of a lovebird *(Agapornis roseicollis),* even bringing bits of plant material tucked under the feathers to the nest site. *Building* the nest involves a host of action patterns (to be discussed later in the aforementioned chapter). Many species have a nest lining that is distinctly different in materials and construction from the rest of the nest. *Incubation* consists of a parent (usually the female, in some cases either parent, and in a few species only the male) sitting on the eggs to warm them with body heat. Mound birds (Megapodiidae) of new Guinea and Australia incubate by heat of fermentation in a nest that is basically a huge leaf pile tended by the male. Conventional incubation involves acts such as turning the eggs with the bill and settling on the eggs with breast feathers erect. If both sexes incubate, there may be exchanges of display at *nest-relief. Egg-retrieval* in ground-nesters consists of rolling back a displaced egg with the underside of the lower mandible.

Parental Behavior

For a detailed description of the action patterns involved in parental behavior, refer to the later chapters on the "Young and Their Development" and "Parental Care." Only a few kinds of action patterns are briefly surveyed here.

Many species dispose of eggshells by eating them, carrying them away, or trampling them into the nest; the white interior of eggshells may attract visually hunting predators. Nestlings and chicks are warmed by body heat in behavior called **brooding** to distinguish it from incubating eggs, and parents may cool young by shading them with their own bodies. The Egyptian Plover *(Pluvianus aegyptius)* regurgitates water on its chicks to cool them through evaporation. In some species a parent (usually the mother) leads young to appropriate foraging spots. Many species feed young by bringing food such as insects (most songbirds), regurgitating food that has been partially digested (gulls), or regurgitating food actually manufactured in the parent's body (the crop milk of pigeons and doves). The desert-nesting sandgrouse *(Pterocles)* fly to the nest with wettened breast feathers from which the young strip water with their bills (see p. 36). Ospreys lead their young from the nest with fish. They later drop the fish into the water instead of providing them directly, and so in essence *teach* fishing techniques to the fledglings. Young birds also *play,* by picking up and manipulating objects.

Interspecific Behavior

Some interspecific interactions involve eggs or young, whereas others are more relevant to the adults themselves. Such behavior is surveyed conveniently under parasites (various kinds), competitors, predators, and commensals.

Parasites There is apparently no special behavior dealing with internal parasites.

Surface Parasites Microparasites (feather mites) are discouraged through various maintenance activities discussed previously. Surface macroparasites such as ticks are often rubbed against perches.

Brood Parasites Birds that lay their eggs in nests of other species. The host's response is often desertion and subsequent renesting elsewhere; a few simply cover over the eggs and build a new nest on top. Oropendolas (several kinds) and caciques *(Cacicus)* parasitized by the Giant Cowbird *(Scaphidura oryzivora)* actively chase away the parasites, although at some places they apparently welcome parasitism because the foster-nestling cowbirds eat botflies that burrow into the host nestlings.

Kleptoparasites Birds that steal food from the host, a well known example being the Bald Eagle *(Haliaeetus leucocephalus),* which chases Ospreys to make them drop fish. Frigatebirds and gulls are also common kleptoparasites. Host responses are mainly evasive in nature.

Competitors Interspecific fighting occurs at food sources among vultures and other scavengers at a carcass and among small birds at a feeding tray. The House Wren *(Troglodytes aedon)* punctures holes in the eggs of other species nesting in the vicinity, perhaps to decrease local competitors. Clapper Rails *(Rallus longirostris)* prey upon eggs of the Laughing Gull *(Larus atricilla),* and gulls may in turn be predators on rail chicks.

Predators Adults usually hide or flee from their own predators but often challenge the different predators at their nests. Predators on adults include *aerial* forms such as raptors and *terrestrial* forms such as mammalian carnivores and large snakes; sea birds on the surface may be taken by sharks, jacks, and other predatory fishes. Many species allow large animals to approach only to a minimum flight distance. Special vocalizations, the so-called **alarm calls,** may differ depending upon whether the potential predator is in the sky or on the ground. Flocking birds often bunch tightly and fly erratically in the presence of raptors, as in many sandpiper species; European Starling flocks may turn back upon the raptor. Bunching seems to deny the predator stragglers that they can single out for a strike. Small birds may take flight upon alarm, even if the predator is aerial; others freeze in place or fly a short distance to hide. A perched hawk or owl may be *mobbed* by numerous species simultaneously, which surround the predator, call repeatedly, fly to stay near it, assume characteristic postures—and even stoop at the predator and strike it. Snakes are also mobbed and even bitten on the tail. American Crows *(Corvus brachyrhynchos)* are notorious for mobbing Great Horned Owls *(Bubo virginianus)* at any time of year. The crows in turn are mobbed by smaller birds, particularly during nesting.

The foregoing responses are also given to *nest predators;* some birds also have special **distraction displays,** especially after the eggs hatch. The familiar type is **injury-feigning,** such as the "broken-wing" display of the Killdeer *(Charadrius voci-*

ferus). More subtle is **rodent-run** distraction display, in which songbirds hold their bodies horizontally, draw the head back toward the shoulders, depress the tail and run rapidly across the ground. As with injury-feigning, the predator is distracted from the nest by the apparent prey.

Commensals The Sandwich Tern *(Sterna sandvicensis)* commonly nests in large colonies of Royal Terns *(S. maxima)* and may benefit from their alarms and other predator responses. Commensal relationships are common in foraging, as when mergansers surround and drive fish toward shore, where Snowy Egrets await their arrival. Strikes by the wading egrets in turn probably help disrupt the fish school and hence allow the mergansers to feed more readily. Cattle Egrets follow large mammals such as ungulates and hippos in their native Africa and domestic cattle in North and South America. The egrets not only eat the flushed insects but may also take ectoparasites off their mammalian commensals.

Principles of Behavior

Major ethological principles and concepts relate to the four classes of behavioral determinants mentioned at the beginning of this chapter: control, ontogeny, perpetuation, and phylogeny.

Control

What a bird does is controlled by both internal and external factors. Internal factors are collectively called the **internal motivational state** (drive, tendency, etc.), which includes hormone balance, proprioceptive neural feedback, and complicated states of the central nervous system. Internal factors must exist because birds react differently to the same stimuli at different times and behavior changes even when the external situation remains constant. The **external stimulus situation** includes specific objects such as another bird's display, general variables such as weather, and complicated environmental contexts.

Single-factor Control Rarely is behavior controlled by an internal or external factor alone. For example, much avian song may be truly spontaneous, and due to high blood titers of male hormones, yet other singing is directly stimulated by seeing another bird or hearing its song while in a motivational state to sing. Early ethologists believed that motivation increased continually in the absence of eliciting stimuli until it exceeded a threshold, when the act was performed "in vacuum" (thus **vacuum activity**). However, there is little evidence for such spontaneous performances of ordinarily elicited acts. Conversely, some acts may be elicited by stimuli regardless of the motivational state, as in *spinal reflexes* (jerking one's hand off a hot surface). Ducking branches in flight and fleeing from a falling limb are complicated examples of reflexive-like avian behavior.

External Stimulus Situation A simple stimulus is a **sign stimulus,** such as red flower color eliciting approach by hummingbirds. (Because most insects are blind to red, they will not be attracted to deplete the nectar reserves evolved to attract hummingbirds.) A conspecific sign stimulus evolved for social communication is a **releaser,** such as the red spot on the Herring Gull's bill, which elicits and directs the begging pecks of its hungry chicks. Laughing Gull chicks also peck at red, the color of the parent's releaser. However, the chicks peck most strongly when the background near the nest is greenish or yellowish—the hues of fresh and dried marsh grass. Such effects of background and other sign stimuli are probably important in controlling behavior. Long-term and general stimuli also have profound effects, as in the gradual shortening of autumn days causing migrating birds to exhibit restlessness. Birds in this stimulus-induced state then launch into migration when stimulated by specific weather variables.

Internal Motivational State The broadest internal states control the bird's annual cycle. A typical north-temperate male passerine on its wintering grounds first changes to become migratory in the spring. Its changing internal state brings about a cluster of territorial and courtship behavior, followed by nesting, incubation of eggs (not all males incubate), care and feeding of nestlings after hatching, care and feeding of fledglings after the young leave the nest, and then back to a migratory state in the fall. Such general states are produced partly by *hormone balance,* which in turn is influenced by external factors of day-length, stimuli from the mate, eggs and young, and others. Testosterone injection often increases fighting in males; progesterone, primed with estrogen, helps bring about incubation in the Ringed Turtle-Dove *(Steptopelia risoria);* and prolactin facilitates feeding of squabs.

The interaction of external stimuli with broad internal states was demonstrated in the Black-headed Gull *(Larus ridibundus).* Settling on the eggs differs between the start of laying (one egg in the nest) and later incubation (full clutch of three). Beer (1961–63) added two eggs during laying and removed two during incubation and found the settling behavior in both cases to be intermediate. This elegant experiment demonstrates that both the broad internal motivational state and the immediate external stimulus situation affect behavior.

Displacement Activities Birds may suddenly switch action patterns without evident changes in external stimuli. Early ethologists thought that motivational "energy" was displaced from one behavioral system to another; hence the term *displacement activity* for sudden, seemingly out-of-context acts. Conflicting lower levels of motivation that could not find proper outlet were also thought to produce harmless but inappropriate displacement activities. The simple displacement principle now has been replaced by more detailed concepts as follows.

Transitional Actions Lind (1959) showed that many acts share some motor element in common, so behavior shifts from one act to another through this common *transitional action*. Birds often turn their heads back when hearing a noise, the turn also occurring when preening the back and wing; so frequently a bird makes a few preening movements before turning back. The initial action may bring about some external stimulus that elicits the following act; or, the internal proprioceptive feedback from the initial act may facilitate the following one.

Redirected Activities Male Herring Gulls at the territorial boundary sometimes stop their fighting and display to pull at grass with their bills, originally called "displacement nest-building." Bastock et al. (1953) showed that the territorial gull is stimulated to fight by the opponent's presence, but also inhibited from crossing into the neighboring territory, so he redirects his attack at the grass—pecking down, grabbing, and pulling at it as he would at an opponent. Grass-pulling probably serves as a display that the gull is ready to attack should the opponent come across the territorial border. Such *redirected activities* occur in humans too, as when they pound the table rather than a person toward whom they feel aggressive.

Disinhibition Stimulation of a given act may simultaneously inhibit other acts, so that the bird does one thing instead of trying to do two or more things simultaneously and hence poorly. If the inhibition is removed, the disinhibited act will occur, but how such disinhibition occurs is not fully understood. Perhaps mutually antagonistic acts (fighting and fleeing) cancel one another when simultaneously stimulated, so their inhibitions on other activities are removed, leading to the performance of seemingly irrelevant acts such as preening. Desultory preening does occur during agonistic interactions in terns.

The disinhibition concept says only that a previously inhibited act *can* occur, not why a specific act such as preening does occur. Perhaps relevant stimuli are always present in the form of ectoparasites, foreign material, ruffled feathers, and so forth. A hierarchy of **behavioral priority** has been proposed, such that activities like preening, which can be done any time, have low priority and are continually inhibited by other activities. Usually preening occurs only after important things like feeding are finished or when stimuli are especially strong, as in rain. Such low-priority behavioral patterns are occasionally disinhibited by conflicts between high-priority, but mutually exclusive, acts.

Ontogeny

Ontogenetic development also depends upon internal and external factors. Developmental change depends upon two sets of internal factors (the phenotype of the bird and its genes) and external factors that dictate experiences.

Behavioral Genetics The genes of an offspring (collectively its **genotype**) are determined by the genes of its parents. Offspring of different genotypes reared in identical environments may develop differences in behavioral or morphological traits (different **phenotypes**), which must be due to the different genotypes. Therefore differences in traits are genetic or genetically based, but traits per se are not inherited. One should not expect simple Mendelian ratios in behavioral genetics; instead, each gene usually affects many traits—**pleiotropism**—and each trait is usually influenced by many different genes—**epistasis.**

Genetic analysis of behavior requires advanced approaches of quantitative genetics. Available genetical analyses come mainly from commercially valuable traits of domestic species (such as egg production in hens and meat of turkeys), but genetically inbred lines of domestic fowl exhibit many different behavioral patterns. Furthermore, interspecific hybrids of waterfowl show many genetic effects in behavior. Hybrid offspring may show some behavior like one parent species, other traits like the other parent, and still other traits that are intermediate.

Learning Learning is an obvious way in which the environment may affect behavioral development.

Habituation Decreasing response to stimuli having no consequences. Habituation refers only to permanent changes rather than to temporary satiation that spontaneously recovers, but there may be a continuum between these extremes, as shown by Hinde's (1954) studies of mobbing in Chaffinches *(Fringilla coelebs)*. One component of the loss of response was specific to the stimulus (did not affect mobbing to other stimuli) and another specific to the response (affected mobbing to any stimulus). Some mobbing soon recovered, indicating processes of stimulus satiation and motor fatigue, but full mobbing never recovered over a long period, indicating true habituation.

Imprinting A dramatic form of learning in the young of precocial species (domestic fowl and waterfowl). Chicks and ducklings approach and follow the first relatively large, moving object they see, usually the mother. As a result, they learn her characteristics and choose her over other objects that initially they would not have distinguished. Some authors restrict the term "imprinting" to such early learning in precocial birds, but others use it to refer to any rapid learning as a result of exposure without any obvious reward.

Conditioning All the types of conditioning studied in psychology experiments have been demonstrated with birds, usually using food as *reward* or positive reinforcement and mild electric shock as *punishment* or negative reinforcement. All **trial-and-error learning** in wild birds is basically operant conditioning, as when nest-building improves with experience. Baby chicks improve pecking accuracy at grain partly through trial-and-error, and food habits are commonly affected by such learning.

More complicated processes of avian learning (including insight, latent learning, and learning to learn) seem to be compounded from basic processes of habituation, imprinting, and classical and operant conditioning.

Imitation Also known in birds, as when Klopfer (1959, 1961) allowed Great Tits *(Parus major)* to watch their companions open colored sunflower seeds that had been doctored with bitter-tasting quinine or aspirin. The tits discarded ill-tasting seeds, and observing companions when later tested refused even to open the colored seeds. Chaffinches, on the other hand, failed to learn by observing their companions.

Development Some species appear to imprint to parental calls and exhibit other kinds of learning in the egg, although a suggestion that the embryo chick's head movements (resulting from heartbeat) facilitate later pecking is probably farfetched. At least behavioral development can be studied from hatching onward to reveal much about interactions between the bird and its environment.

The roles of genotype and experience are shown in Dilger's (1962) study of interspecific hybrids of lovebirds. Fischer's Lovebird *(Agapornis personata fischeri)* carries nesting materials in its mandibles, but the Peach-faced Lovebird *(A. roseicollis)* tucks several pieces under the back feathers, holding them firmly by the feather muscles as it flies to the nest. Hybrids tried in vain to tuck strips under the back feathers, while occasionally carrying one in the bill. Over several nesting cycles the birds learned to carry almost exclusively with the bill.

Begging of gull chicks illustrates several points about behavioral development (Hailman, 1967). Newly hatched Laughing Gull chicks peck at their parents' red bills, accuracy improving rapidly and reaching a maximum in two days. Some improvement may be due to practice and some to subtle experience of standing, which strengthens leg muscles and improves general coordination. Chicks first peck at a variety of things (grassblades and the parents' red legs) but soon restrict pecking to the parental bill. The newly hatched chick pecks equally at models of its parent and a different species, but chooses its own species after experience with its parents. Herring Gull chicks experimentally trained to peck at a Laughing Gull model became conditioned to the "wrong" species. In sum, the chick hatches with sufficient motor skills and perceptual preferences to beg effectively, but perfects begging rapidly through experience.

Just because every member of a species behaves similarly does not prove behavior is "genetic" or "instinctive." Each individual may learn about the same things, as each is reared in about the same environment. Week-old gull chicks beg only to models of their own species, but species-specific preference develops purely by learning. Only experimental analysis can reveal how the bird with its particular genotype interacts with its environment to structure ontogenetic development.

Perpetuation

Many species show behavioral traits that seem especially suited to their environments. For example, cliff-nesting alcids, boobies, gulls, and terns all show behavioral patterns that help reduce the danger of eggs or chicks falling from the nesting ledge. Such convergence of similar traits in unrelated birds show that these are **adaptations** to their environments. Not all useful traits are necessarily passed on from parent to offspring by genetics, however, as **traditions** are also important. Norton-Griffiths (1967, 1969) showed that both members of the pair in European Oystercatchers use the same method opening bivalves. When he switched eggs of "smasher" and "severer" parents the offspring grew up to use the method of their *foster* (not genetic) parents. Passerine song, food habits, and migration routes are other examples of behavior in which traditions play an important role in the perpetuation of traits from generation to generation.

Phylogeny

Behavior does not fossilize like bones, so tracing the evolutionary history of a behavioral pattern relies largely upon the *comparative method* of noting similarities among living species. Displays are believed to have been evolved from noncommunicative behavioral acts, which may still exist in the same species, so that comparison of similarities within as well as among species also helps in tracing the phylogeny of behavior. Behavioral patterns in different species that are similar because of their common ancestry are called *homologous* (e.g., long-call displays of gull species), whereas those that are similar due to convergent evolution are termed *analogous* (e.g., leaf-turning methods of sparrows and domestic fowl).

Selected Studies

Although professional research projects on avian behavior may require months or years for completion, many observational studies can be accomplished in days, hours, or even a few minutes. However, even the behavior of very common species such as the American Robin (*Turdus migratorius*) and House Sparrow have been only partially described.

Four groups of related studies follow: observations at a zoo, observations in an urban or suburban yard or park, observations at special places where groups of birds may be found, and simple experiments that can be performed without highly technical equipment. In each case the most important equipment is a prepared mind, keen eye aided by a binocular if possible, and a field book for recording copious notes on behavior. No description of a behavioral act is complete without explicit notes on all its components: movements of the head relative to the body, whether the bill is open or closed, extension or withdrawal of the neck, posture and movements of the wings, posture and movement of the tail, whether the legs are extended or flexed, which parts of the plumage are fluffed and which are sleeked, and so on. Also note the bird's position relative to relevant parts of its surroundings, including other individuals. Sketch frequently, make counts whenever possible, and record time durations with a stopwatch or sweep second hand on a wristwatch. Each observation is a datum. Watch the same behavioral pattern repeatedly to be certain of its details and to assess the way in which it varies from performance to performance.

1. Zoo Studies

A convenient place to begin observing behavior is the nearest zoo because the captive birds are caged or wing-clipped and hence can be kept under constant observation.

Procedure First try sketching several species to become aware of anatomical features. Then begin describing different postures and action patterns by writing down details in a notebook. Study simple patterns that birds repeat, such as the movements of taking flight, and simple postures, such as sleeping positions, paying attention to how the tail is held or moved, and which feathered parts are fluffed and which are compressed. Sketch postures and movements frequently and catalogue as many different patterns as possible. Give descriptive names to the patterns, such as "head-forward posture," rather than names implying an interpretation, such as "threat display." Avoid anthropomorphic descriptions and names, which impart human motives and characteristics to birds.

2. Backyard Studies

Once having practiced describing behavior of captive birds, look nearby in urban and suburban areas for wild species. Many behavioral patterns can be studied from a window, using the building as a blind.

Procedure Establish a **feeding station,** providing a mixture of seeds, fruits, peanut butter, and suet to attract a variety of species. Even in summer when natural food is often plentiful, several species will visit feeding trays. Remember to scatter seeds on the ground for species like juncos that prefer to forage there. Note the choice of food: some species such as the Northern Mockingbird will take fruit, others such as woodpeckers will prefer suet, and many species will take seeds. If it is difficult to see clearly which kinds of seeds are chosen, separate types of seed and place them in separate piles. Sparrows generally select small seeds such as millet, goldfinches and siskins are partial to thistle seeds, and Evening Grosbeaks *(Coccothraustes vespertinus)* prefer sunflower seeds. Many species eat a variety of seeds, so make a sample count of how many they take of each kind.

Observe how birds prepare and transport food. Heavy-billed species, such as the Northern Cardinal *(Cardinalis cardinalis),* crush sunflower seeds in their bills, positioning the seed with the tongue; others such as chickadees carry one seed off to open it as described previously. Notice that Blue Jays *(Cyanocitta cristata)* may repeatedly pick up one sunflower seed at a time, maneuver it into the throat pouch, and collect many seeds before flying off with a bulging throat.

Birds feeding on the ground will turn leaf litter in various ways to find the seeds beneath, as described previously. White-throated Sparrows, House Sparrows, and Brown Thrashers, for example, each use different methods. After a bird has eaten, watch for acts such as bill-wiping. Follow the bird with a binocular to see where it goes and what it does after eating. Resting and maintenance activities often follow feeding.

A feeding tray is also a good place to study agonistic behavior. Keep track of which species supplant which from the feeder to see if consistent dominance relations exist between given pairs of species. Take detailed notes and sketch agonistic displays, noting which displays are followed by attack and which by fleeing so as to interpret threat and appeasement. Determine whether the acts directed toward individuals of the same and other species are similar or different.

Maintenance behavior can be observed by setting up a **bird bath.** (Do not use a metal lid as it heats up in sunlight.) An old milk carton full of water with a small hole punched in the bottom will drip into the bath, drawing birds' attention to the water. Drinking may be the commonest behavior at the bath. Notice that Mourning Doves *(Zenaida macroura)* and their relatives drink differently from most other birds.

Bathing is complicated behavior, consisting of many different action patterns. Observe ducking of the head, beating of the wings, and similar acts, noting which feathers are raised during which phases of bathing. Continue to watch the bird after it leaves the bath, and note other maintenance activities—such as preening and oiling, head-scratching, feather erection and depression with associated shaking, and stretching movements.

The backyard is also an excellent place to study territoriality and reproductive activities, Many specific behavioral patterns can be observed in conjunction with studies described in the chapters on "Territory," "Nests and Nest-building," "Eggs, Egg-laying, and Incubation," "Young and Their Development,"and "Parental Care." When studying behavior it is desirable to mark individuals uniquely, as with color bands, but many birds can be recognized individually by abnormally unpigmented feathers, missing flight feathers, plumage stains, or more subtle characteristics. Pay particular attention to behavioral spatial details, as in the exact song perches and flight paths used when defending a territory. Note exact postures in agonistic and courtship displays and record details of movements involved in behavior such as building the nest. Also watch for special behavioral acts like courtship-feeding and removal of the nestlings' fecal sacs by the parents. Many important ethological studies have been conducted close to home without the need to mount expeditions to exotic places.

3. Other Field Studies

Field notes on behavioral patterns may be taken nearly anywhere birds are present, including city streets full of Common Pigeons *(Columba livia)*. However, birds will be more plentiful and tame in specially protected areas—e.g., local sanctuaries and National Wildlife Refuges—where many studies of avian behavior have been done. Any patch of woodland or field provides opportunities to study behavior similar to those just described, but the following are specific opportunities for special studies.

Procedure (a) The Red-winged Blackbird *(Agelaius phoeniceus)* nests in wet areas throughout the conterminous United States and much of Canada. Locate an appropriate marsh and watch the males take up territories in the spring. In the northern states and Canada, where the species does not winter, males arrive first in the spring and females a little later once the territories have been established. Since the species is polygynous (see p. 278), try to locate all the nests on territories of several males to see if males with larger or better-quality territories secure more mates than other males. Describe the site and context of the male's "rough-out/squeak" display in which the red epaulets are erected and the feathers of the body are "roughed out" while the bird vocalizes.

(b) Dabbling, or puddle, ducks tend to nest in northern states and winter in the southern ones, so nearly everywhere a lake or pond can be found on which ducks reside at some time of year. Watch feeding patterns, of which most species have a variety, such as dabbling on the water's surface, ducking the head below the surface, and tipping up in shallow water to reach vegetation rooted on the bottom. Bay ducks and sea ducks dive more frequently than dabbling ducks. Time the interval between diving and resurfacing and also the interval upon the water until the next dive.

Courtship and other reproductive displays in ducks occur mainly in the winter months and are among the most dramatic and well-studied displays of birds. Watch for *"courtship parties,"* as in the Mallard, when several males surround a female and give nearly simultaneous displays. Describe and sketch the displays over and over again, as most are rapid and complicated action patterns. Also watch pairs, which may be temporary liaisons in some species. Although the birds are not fertile, copulation in winter is frequent in Mallards for reasons not entirely understood. Note precopulatory display and especially some of the dramatic postcopulatory displays of dabbling ducks.

(c) Some kind of colonial species nests in virtually every state. Colonies of Great Blue Herons *(Ardea herodias)* can be found inland, for example, and in coastal regions colonies of gulls and terns abound. Prepare thoroughly by reading a relevant reference such as Tinbergen's (1953) *The Herring Gull's*

World. Set up a blind in a colony where several nesting sites can be observed from one place (see Appendix A, p. 363), but be sure to have a companion enter the blind too and then depart so that the birds will settle down more readily. Try to distinguish individuals by their markings and other variations. Study details of behavior appropriate to the time of the visit: territorial defense, courtship, nest-building, incubation, and care of the young. Remain sufficiently long to note the same behavioral patterns repeatedly, but remember that after several hours one becomes physically cramped and mentally tired, so it is better to break observations up into a number of different periods.

(d) Many species, especially in winter, form huge communal roosts overnight. An urban example is the European Starling; American Crows form large roosts in surburban areas; and in mid-southern states blackbirds of various species congregate in roosts of millions of birds in woodlands and swamps. Note the time of leaving the roosts in the morning and returning in the evening; measure the ambient light by pointing a photometer directly upward. (For general purposes a photograpic light meter will do, although many are not properly balanced for the spectral distribution of light at low intensities.) When birds return to the roost in the evening, note which spots are taken first and any agonistic exchanges over perching places.

4. Some Simple Field Experiments

Elaborate and well-controlled experiments in the field are difficult and time-consuming, but much can be learned from some simple preliminary experiments. The following are just a few examples to suggest a wider range of similar manipulations that could be carried out.

Procedure (a) Watch the reaction of birds to the playback of their species' song. A simple cassette machine will do, there being no need for a professional-grade tape recorder. Nor is it necessary to record the birds in the field; either purchase a commercial cassette of bird songs, now widely available in conjunction with field identification guides, or record songs from a commercial record of bird vocalizations. Locate a singing male in the field and play the recording of its species' song. Time how long it takes the male to approach the sound source, whether he sings in response, and what other behavior he shows. If females are attracted to the recording, describe their behavior.

(b) Many species will take nesting material provided within their territories by the observer. For passerine birds one can provide an array of colored yarn or coarse thread cut to lengths of 10–20 cm. Flag the supply offered with a strip of aluminum foil to attract the attention of the nesting birds. Color choice of nesting materials in gulls has been studied by dyeing straw with food coloring. Either offer one of each color to see the order in which they are chosen in repeated tests, or provide piles of each color to see which pile disappears most rapidly. One can also vary other aspects, such as the length of pieces and the diameter of the material.

(c) The reactions of small songbirds to potential predators can be studied by placing a model out in the open. A tethered owl or hawk is the best stimulus, but a stuffed mount is often very effective until the birds habituate to it. Even papier-mâché models often elicit some mobbing. Watch which species are attracted to the site and how they behave. Count the frequency of call notes and the dives at the stimulus. Pay special attention to differences in reaction among the various species attracted and to differences between the sexes in species where sex can be told in the field. Mobbing can also be elicited in many species by using a pet snake, tethered on a tight-fitting Velcro collar just back of the head and allowed to move on a length of monofilament fishing line.

(d) Studies of agonistic behavior can be made by placing a large mirror in the territory of a bird. Note the displays and any calls used to challenge the image. As the phantom opponent does not retreat, fighting reactions often ensue. See if the bird tries to go behind the mirror to find the opponent. Watch over a long period, or make repeated trials with the mirror to see if the bird will eventually habituate to the image.

Presentation of Results Regardless of which study is undertaken, write a paper on all observations. (Consult Appendix B for directions on preparing the manuscript.) Always be certain to include the ecological setting and locality, dates, times, and other contextual information. Each behavioral pattern should be described in detail, stating explicitly how each is distinguished from any others described that are similar. An important part of behavioral description is the variation from performance to performance, separating those aspects of behavior that are variable from those that are relatively fixed. Measurements, counts, and other quantitative aspects should be tabulated where possible, and sketches should be provided as valuable parts of behavioral description. When writing a report, remember never to rely on memory; always consult details in the field notes, which constitute the recorded data of the study.

References

Andrew, R.J.
1956 Intention movements of flight in certain passerines, and their use in systematics. *Behaviour* 10:179–204.
1961 The displays given by passerines in courtship and reproductive fighting: A review. *Ibis* 103a:315–348, 549–579.

Bastock, M.; Morris, D.; and Moynihan, M.
1953 Some comments on conflict and thwarting in animals. *Behaviour* 6:66–84. (First paper to challenge the concept of "displacement activities.")

Batista, L.F.
1973 Leaf bathing in three species of Emberizines. *Wilson Bull*. 85:346–348.

Bayer, R.D.
1982 How important are bird colonies as information centers? *Auk* 99:31–40. (Challenges the evidence that colonies promote communication about the location of food among birds.)

Beer, C.G.
1961–1963 Incubation and nest building behaviour of Black-headed Gulls. *Behaviour* 18:62–106, 19:283–304, 21:13–77, and 21:155–176.

Bowman, R.I.
1961 Morphological differentiation and adaptation in the Galapagos finches. *Univ. of Calif. Publ. in Zool.* 58:1–302.

Brockman, H.J., and Barnard, C.J.
1979 Kleptoparasitism in birds. *Animal Behaviour* 27:487–514.

Burtt, E.H., Jr., and Hailman, J.P.
1978 Head-scratching among North American wood-warblers (Parulidae). *Ibis* 120:153–170.

Cade, T.J.
1973 Sun-bathing as a thermoregulatory aid in birds. *Condor* 75:106–133. (Concerns the behavior called "sunning" in the text.)

Chisolm, A.H.
1954 The use by birds of "tools" or "instruments." *Ibis* 96:380–383.

Cullen, E.
1957 Adaptations in the Kittiwake to cliff-nesting. *Ibis* 99:275–302.

Curio, E.
1978 The adaptive significance of avian mobbing. I. Teleonomic hypotheses and predictions. *Z. Tierpsychol*. 48:175–183.

Daanje, A.
1950 On locomotory movements in birds and the intention movements derived from them. *Behaviour* 3:48–98. (The original description of intention movements.)

Darling, F.F.
1938 *Bird flocks and the breeding cycle*. London: Cambridge Univ. Press.

Dilger, W.C.
1960 The comparative ethology of the African parrot genus *Agapornis*. *Z. Tierpsychol*. 17:649–685.
1962 The behavior of lovebirds. *Sci. Amer.* 206:88–98.

Ficken, M.S.
1977 Avian play. *Auk* 94:573–582.

Ficken, M.S., and Ficken, R.W.
1962 The comparative ethology of the wood warblers: A review. *Living Bird* 1:103–122.

Ficken, R.W., and Ficken, M.S.
1966 A review of some aspects of avian field ethology. *Auk* 83:637–661.

Ficken, R.W.; Ficken, M.S.; and Hailman, J.P.
1978 Differential aggression in genetically different morphs of the White-throated Sparrow *(Zonotrichia albicollis)*. *Z. Tierpsychol*. 46:43–57. (The white-striped morph is more aggressive than the tan-striped morphs, differing genetically by replacement of an entire chromosome.)

Friedmann, H.
1929 *The cowbirds, a study in the biology of social parasitism*. Springfield, Illinois: Charles C. Thomas. (Concerns the phenomenon called "brood parasitism" in the text.)

Hailman, J.P.
1967 *The ontogeny of an instinct: The pecking responses in chicks of the Laughing Gull (Larus*

atricilla L.) and related species. Behaviour Supplement 15. Leiden: E.J. Brill. (Challenges the older concept of instinct.)

1977 *Optical signals: Animal communication and light*. Bloomington: Indiana Univ. Press. (Many examples concerning displays and display-coloration of birds.)

Hardy, J.W.

1974 Behavior and its evolution in Neotropical jays *(Cissilopha). Bird-Banding* 45:253–268.

Harrison, C.J.O.

1965 Allopreening as agnostic behaviour. *Behaviour* 24:116–208.

Hauser, D.C.

1957 Some observations on sun-bathing in birds. *Wilson Bull*. 69:78–90.

Heinroth, O.

1910 Beiträge zur Biologie, namentlich Ethologie und Physiologie der Anatiden. *Verh. Vth Internatl. Ornith. Kongr.* pp. 589–702. (The original descriptions of imprinting and displays in ducks and the first paper to use the term "ethology" in its modern sense.)

Hinde, R.A.

1954 Factors governing the changes in strength of a partially inborn response, as shown by the mobbing behaviour of the Chaffinch *(Fringilla coelebs). Proc. Royal Soc. (London), B*, 142:306–331; 142:331–358; and (1960) 153:398–420 (three part paper).

1966 *Animal behaviour: A synthesis of ethology and comparative psychology*. New York: McGraw-Hill. (For the advanced student of ethology.)

Hochbaum, H.A.

1955 *Travels and traditions of waterfowl*. Minneapolis: Univ. of Minnesota Press.

Huxley, J.S.

1914 The courtship habits of the Great Crested Grebe *(Podiceps cristatus);* with an addition to the theory of sexual selection. *Proc. Zool. Soc. (London)*, pp. 491–562. (One of the classic papers of avian ethology.)

Johnsgard, P.A.

1965 *Handbook of waterfowl behavior*. Ithaca, New York: Cornell Univ. Press. (The first book ever to summarize comprehensively the behavior patterns in a major avian family.)

Kahl, M.P.

1972 Comparative ethology of the Cicioniidae, the wood storks (genera *Mycteria* and *Ibis*). *Ibis* 114:15–29.

Klopfer, P.H.

1959 Social interactions and discrimination learning with special reference to feeding behaviour in birds. *Behaviour* 14:282–299.

1961 Observational learning in birds. *Behaviour* 97:71–80.

Klopfer, P.H., and Hailman, J.P.

1965 Habitat selection in birds. In *Advances in the study of behavior,* vol. 1, eds. D.S. Lehrman, R.A. Hinde, and E. Shaw. New York: Academic Press.

Leger, D.W., and Nelson, J.L.

1982 Effects of contextual information on behavior of *Calidris* sandpipers following alarm calls. *Wilson Bull*. 94:322–328. (Playbacks elicited stronger responses from sandpipers near cover where predators could be hiding than from birds in the open, even when the latter were closer to the speaker.)

Lehrman, D.S.

1959 Hormonal responses to external stimuli in birds. *Ibis* 101:478–496.

1964 Control of behavior cycles in reproduction. In *Social behavior and organization among vertebrates,* ed. W. Etkin. Chicago: Univ. of Chicago Press.

Lind, H.

1959 The activation of an instinct caused by a "transitional action." *Behaviour* 14:123–135. (Proposes an explanation of how birds shift suddenly from one behavioral pattern to an unrelated one.)

Lorenz, K.

1937 The companion in the bird's world. *Auk* 54:245–273. (Translation of the author's classic 1935 paper.)

Makkink, G.F.

1936 An attempt at an ethogram of the European Avocet with ethological and psychological remarks. *Ardea* 25:1–60.

Marler, P., and Hamilton, W.J., III

1966 *Mechanisms of animal behavior.* New York: John Wiley & Sons. (Concerned to a large extent with the physiological processes of behavior; recommended to the advanced student.)

Maron, J.L.

1982 Shell-dropping behavior of Western Gulls *(Larus occidentalis). Auk* 99:565–569.

McKinney, F.

1965 The comfort movements of Anatidae. *Behaviour* 25:120–220.

Morse, D.H.

1980 *Behavioral mechanisms in ecology.* Cambridge: Harvard Univ. Press. (A modern text with much material on behavioral ecology of birds.)

Moynihan, M.
1955a Remarks on the original sources of displays. *Auk* 72:240–246.
1955b Types of hostile display. *Auk* 72:247–259.

Nice, M.M.
1937 Studies in the life history of the Song Sparrow, I. *Trans Linnaean Soc. New York* 4:i–vi; 1–247.
1943 Studies in the life history of the Song Sparrow, II. *Trans. Linnaean Soc. New York* 6:i–viii; 1–328.
1962 Development of behavior in precocial birds. *Trans. Linnaean Soc. New York* 8:i–xii; 1–211. (Follows the appearance of stereotyped motor movements from hatching onward in a variety of hand-reared species.)

Noble, G.K.
1936 Courtship and sexual selection of the Flicker *(Colaptes auratus luteus). Auk 53:269–282.*

Norton-Griffiths, M.N.
1967 Some ecological aspects of the feeding behaviour of the Oystercatcher *Haematopus ostralegus* on the edible mussel *Mytilus edulis. Behaviour* 109:412–424.
1969 The organization, control and development of parental feeding in the Oystercatcher *(Haematopus ostralegus). Ibis* 34:55–114.

Orians, G.H., and Christman, G.M.
1968 A comparative study of the behavior of the Red-winged, Tricolored, and Yellow-headed Blackbirds. *Univ. Calif. Publ. in Zool.* 84:i–vi; 1–85.

Patterson, I.J.
1965 Timing and spacing of broods in the Black-headed Gull *Larus ridibundus. Ibis* 107:433–459.

Pavlov, I.P.
1927 *Conditioned reflexes.* Translated by G.V. Anrep. London: Oxford Univ. Press.

Payne, R.B.
1977 The ecology of brood parasitism in birds. *Ann. Rev. Ecol. System.* 8:1–28.

Pennycwick, C.J.
1972 *Animal Flight.* London: Edward Arnold.

Pettingill, O.S., Jr.
1964 Penguins ashore at the Falkland Islands. *Living Bird* 3:45–64.

Potter, E.F.
1970 Anting in wild birds, its frequency and probable purpose. *Auk* 87:692–713.

Potter, E.F., and Hauser, D.C.
1974 Relationship of anting and sunbathing to molting in wild birds. *Auk* 91:537–563.

Sebeok, T.A.
1977 *How animals communicate.* Bloomington: Indiana Univ. Press. (A compendium with chapters on visual, acoustic, and other signals and review chapters on birds and other animals.)

Shedd, D.H.
1982 Seasonal variation and function of mobbing and related antipredator behaviors of the American Robin *(Turdus migratorius). Auk* 99:342–346.

Simmons, K.E.L.
1957a A review of anting behaviour of passerine birds. *Brit. Birds* 50:401–424.
1957b The taxonomic significance of the head-scratching methods of birds. *Ibis* 99:178–181.
1964 Feather maintenance. In *A new dictionary of birds,* ed. A.L. Thomson. New York: McGraw-Hill. (Concerns preening, oiling, and related maintenance activities.)

Skinner, B.F.
1938 *The behavior of organisms.* New York: Appleton-Century. (The classic book on operant conditioning.)

Smith, N.G.
1968 The advantage of being parasitized. *Nature* 219:690–694. (Reports how oropendola and cacique colonies protected from maggots by predatory wasps reject eggs from Giant Cowbird brood parasites. Colonies not so protected allow the brood parasites, whose young pick botfly maggots off the nestling oropendolas and caciques.)

Spalding, D.
1872 Instinct with original observations on young animals. (Reprinted in *Brit. Jour. Anim. Behaviour,* 2:2–11.) (The first report of imprinting, as discovered in hand-reared domestic chicks.)

Stillwell, T., and Hailman, J.P.
1978 Spatial, semantic, and evolutionary analysis of an animal signal: Inciting by female Mallards. *Semiotica* 23:193–228.

Thomson, A.L., ed.
1964 *A new dictionary of birds.* New York: McGraw-Hill. (Contains separate articles, each by an authority, on learning, imprinting, sign stimulus, releaser, fixed action pattern, redirection, displacement activity, ritualization, development of behavior, counting, play, and tameness.)

Thorpe, W.H.
1961 *Bird-song: The biology of vocal communication and expression in birds.* Cambridge: Univ. Press.
1963 *Learning and instinct in animals.* 2nd ed. London: Methuen. (Contains a thorough review of the literature on the subject.)

Tinbergen, N.

1948 Social releasers and the experimental method required for their study. *Wilson Bull.* 60:6–51.

1951 *The study of instinct.* London: Oxford Univ. Press.

1952 Derived activities: Their causation, biological significance, origin and emancipation during evolution. *Quart. Rev. Biol.* 27:1–32. (The classic paper on the ritualization of displays.)

1953 *The Herring Gull's world: A study of the social behaviour of birds.* London: Collins.

1959 Comparative studies of the behaviour of gulls: A progress report. *Behaviour* 15:1–70.

Tinbergen, N.; Broekhuysen, G.J.; Feeks, V.; Houghton, J.G.W.; Kruuk, H.; and Szule, E.

1962 Egg shell removal of the Black-headed Gull, *Larus ridibundus* L., a behaviour component of camouflage. *Behaviour* 19:74–117.

Tinbergen, N., and Perdeck, A.C.

1950 On the stimulus situation releasing the begging response in the newly-hatched Herring Gull chick *(Larus a. argentatus* Pontopp.). *Behaviour* 3:1–38.

van Iersel, J.J.A., and Bol, A.

1958 Preening of two tern species: A study of displacement activities. *Behaviour* 13:1–88.

van Lawick-Goodall, J., and van Lawick-Goodall, H.

1966 Use of tools by the Egyptian Vulture *Neophron percnopterus. Nature,* 212:1468–1469.

van Tets, G.F.

1965 A comparative study of some social communication patterns in the Pelecaniformes. *Amer. Ornith. Union, Ornith. Monogr. No. 2.*

Whitaker, L.M.

1957 A résumé of anting, with particular reference to a captive Orchard Oriole. *Wilson Bull.* 69:195–262.

Whitman, C.O.

1919 The behavior of pigeons. *Carnegie Inst. of Washington Publ.* 257:1–161. (The classic work published posthumously by Whitman's student Wallace Craig.)

Migration

By Sidney A. Gauthreaux, Jr.

In the broadest sense any movement between two areas may be called **migration,** but spatial movements can be classified more usefully into four functional types: maintenance, dispersal, migratory, and irruptive.

Maintenance Movements There are three different types: (1) appetitive responses to hunger and thirst in the absence of food and water at a location, (2) escape responses to avoid predation and find shelter, and (3) roosting flights between feeding areas and sleeping sites.

Dispersal Movements There are two types: natal and breeding (Greenwood, 1980; Greenwood and Harvey, 1982).

Natal Dispersal A young bird moves away from its birthplace in search of a location that has a higher mean favorability for survival, for reproduction, or for both. Once a suitable location is found, site fixation occurs.

Breeding Disperal In adult birds dispersal movements may be initiated between successive breeding sites, particularly if a previous breeding attempt failed. Such movements may be shown by an individual, a pair, or a group of individuals.

Dispersal usually involves movements of individuals between local populations (gene flow), but it may result in the colonization of unoccupied habitats and the establishment of a portion of the gene pool of the original population at a new location (range extension). Dispersal is usually a one-way movement, without a return movement to the original location (breeding place or birthplace). When regular return movements are involved, migration is the more appropriate term for the pattern of movement.

Migratory Movements These movements enable individuals to exploit favorable geographical areas and habitats. Such movements enable individuals to reproduce at locations where and during periods when food is maximal and to escape from locations where and when food is minimal or absent.

Seasonal Migrations Responses to changes in food availability brought about by the annual climatic cycle.

Zoogeographical Migrations Changes in the range of a species that occur when environmental changes or climatic cycles have considerably greater time constants—e.g., glacial and interglacial periods, ecological succession.

Irruptive or Invasion Movements Movements that occur irregularly at intervals of a few years; triggered by high population densities in conjunction with a failure in the food supply. Although some variability exists, the movements tend to occur every other year, are synchronous among several species of seed-eating birds, and are in response to a circum-boreally synchronized pattern of seed-crop fluctuations in certain high-latitude tree species (Bock and Lepthien, 1976). Demographically, young and females usually comprise the bulk of the individuals that move the greatest distances—a condition similar to that found in natal dispersal and partial migration (see Gauthreaux, 1982). Like regular migratory movements, irruptive movements are frequently oriented toward the west and southwest, and the birds may show abnormal restlessness and even deposit fat stores prior to departure.

Seasonal Migration

Since the beginning of recorded history the seasonal migrations of birds have attracted widespread attention and intrigued the imagination. They have likewise been of considerable interest to biologists, and myriad studies have sought answers to the *hows* and *whys* of bird migration. Many of these studies have been summarized in books devoted entirely to the subject (e.g., Brewster, 1886; Clarke, 1912; Coward, 1912; Cooke, 1915; Thompson, 1926; Wetmore, 1926; Tinbergen, 1949; Lincoln, 1950; Rudebeck, 1950; Hochbaum, 1955; Dorst, 1962; Griffin, 1964; Schuz, 1971; Bykhovskii, 1974; Kumari, 1975). The amount of data that has been collected and the quantity of published findings on all aspects of bird migration are staggering, and even the most comprehensive reviews have been able to provide little more than a sketchy overview of the subject.

The Study of Migration

Although Aristotle is said to have recorded the arrival and departure of migrating birds and speculated on the causes of migration, nearly all information on the seasonal movements of birds has been gathered since the middle of the nineteenth century and most of it within the last few decades. Today there is an enormous quantity of literature on bird migration resulting from extensive laboratory studies of physiological and environmental control; the analysis of migrant birds mist-netted or otherwise captured, birds killed during migration by television towers and other man-made hazards, and returns from banded birds; mathematical calculations based on the kinds and numbers of birds observed through a telescope as they fly across the face of the moon; correlation of meteorological data with known migratory activity; deductions derived from direct field observations, radar surveillance, and tracking by radiotelemetry; and experiments on homing and direction-finding (see Able, 1981).

Laboratory Methods

Two techniques have been developed to study the "migratory" behavior of caged migrants in the laboratory: one uses what is frequently called an *activity cage,* and the other technique requires a special type of activity cage called an *orientation cage*. The activity cage involves the electronic or mechanical recording of the intense fluttering and perch hopping shown by caged migratory birds at night during migratory periods (Eyster, 1954). Cage designs vary, but most have some type of microswitch that is closed when a bird hops on a perch or on the floor of the cage. The switch closure sends an impulse to an event recorder or an electronic counter. Other designs have used the mechanical displacement associated with hopping on a perch to scratch smoked paper on a slowly revolving drum or disk. Whatever the design, the technique has made possible the measurement of the intensity, duration, and timing of migratory restlessness (Zugunruhe) in many species of birds, and the technique has enabled researchers to gain insight into how diverse factors—e.g., temperature, photoperiod, hormones, and nutrition—influence the migratory behavior of birds (see Farner, 1955; Berthold, 1975; Gwinner 1975, 1977).

The second technique emphasizes the fact that when migratory birds are placed in circular cages and given appropriate sensory information, they may show intense locomotor activity that is often oriented in a seasonally appropriate direction. Many different cage designs have been used (Kramer, 1950, 1952; Emlen, 1975). One type consists of a circular wall (clear or opaque) with a plexiglass bottom and some kind of netting or screen for the top. A concealed observer below the cage records the direction of the bird's activity (see Kramer, 1950, 1952). Other circular orientation cages have several radial perches or have the floor divided into several treadles. Because each radial perch or treadle activates a separate counter or channel of an event recorder, the circular distribution of migratory restlessness can be determined automatically.

Another cage, developed by Emlen and Emlen (1966), is inexpensive, easily constructed, and records the bird's activity without costly counters and recorders. The cage consists of a shallow cone of blotter paper placed in an aluminum custard pan. Hardware cloth covers the top of the cone and an ink pad is at the base of the cone. As a bird hops on the side of the cone it leaves footprints. The circular distribution of the footprints can be analyzed once a scale of footprint densities has been quantified by the investigator. By manipulating the cues available to the birds in orientation cages, it is possible to investigate a number of questions related to the sensory basis of migratory orientation.

Field Methods

Most studies of bird migration involve field work with no more "equipment" than a pair of binoculars to aid in identification and a notebook or tape recorder to record observations. Two straightforward methods of gathering field data on migration are censusing of grounded migrants and direct observations of birds actively migrating. Daily field censuses can be used to gather information on the timing, routes, and habitat preferences of different species, sexes, and age classes during migration. In some species, sex, or age, or both can be determined easily in the field (e.g., the White-crowned Sparrow, *Zonotrichia leucophrys,* for age differences), and valuable information on the differential migration (routes, timing, and geographical distances) of sexes and age classes can be gathered during a single season.

In addition to field censuses one can watch birds actively migrating and record several aspects of inflight behavior. An observer with binoculars (or a telescope) and a compass can record the rate of passage, estimate the altitude of flight, and accurately determine the direction of flight for a number of species that migrate during daylight hours (see Lowery and Newman, 1963). Additional information on flocking and vocalizations during flight can also be obtained easily.

Most bird migration occurs under the cover of darkness, and consequently the techniques for studying nocturnal migration are a bit more complex than those used for daytime migration. Investigators have developed several diverse methods to study nocturnal migration: flight-call-counting, moon-watching, ceilometer-watching, and various radar techniques.

Flight Calls Historically, flight calls from migrants flying overhead at night have provided some evidence of types of birds migrating and their abundance. Although the "whits," "seeps," "chips," and other call notes given by migrants aloft at night are difficult to identify because of the difficulty of associating a particular bird with a particular sound, some species groups and many species have characteristic nocturnal call notes. The number of call notes heard per unit of time has been suggested as a means of quantifying nocturnal migration, but the rate of calling is strongly biased by four factors: (1) the type of species migrating; (2) the altitude of flight; (3) the weather, particularly the amount of cloud cover; and (4) the time of night. Near dawn the rate of calling usually increases, but it also does so whenever birds are forced to land earlier in the evening. Because of the biases associated with flight calling (see Graber, 1968), one should use this technique cautiously when studying migration.

Moon-watching First done in the late 1800s (Scott, 1881), this technique was not used extensively until the work of Lowery (1951). During spring and fall on mostly clear nights around the time of the full moon it is possible to use a 20x or 30x spotting scope to observe the silhouettes of migrating birds crossing in front of the disc of the moon (see Fig. 51). The disc of the moon can be treated as a clock face so that entry and exit points can be given in clock-face coordinates (e.g., bird entered at nine o'clock and exited at three o'clock). More details of the technique can be found in Lowery (1951), and Nisbet (1959, 1961) has reduced the elaborate mathematical processing required to analyze moon-watching data to tabular form, so the data can be analyzed with no great effort. Moonwatching produces estimates of the number of birds crossing a mile of front (1.6 km) per hour, the *migration traffic rate,* as well as the directions of flight. The major shortcoming of the moon-watching technique is that it cannot be used on cloudy nights when the moon is obscured or during the period of the new moon. The ceilometer technique does not suffer these limitations.

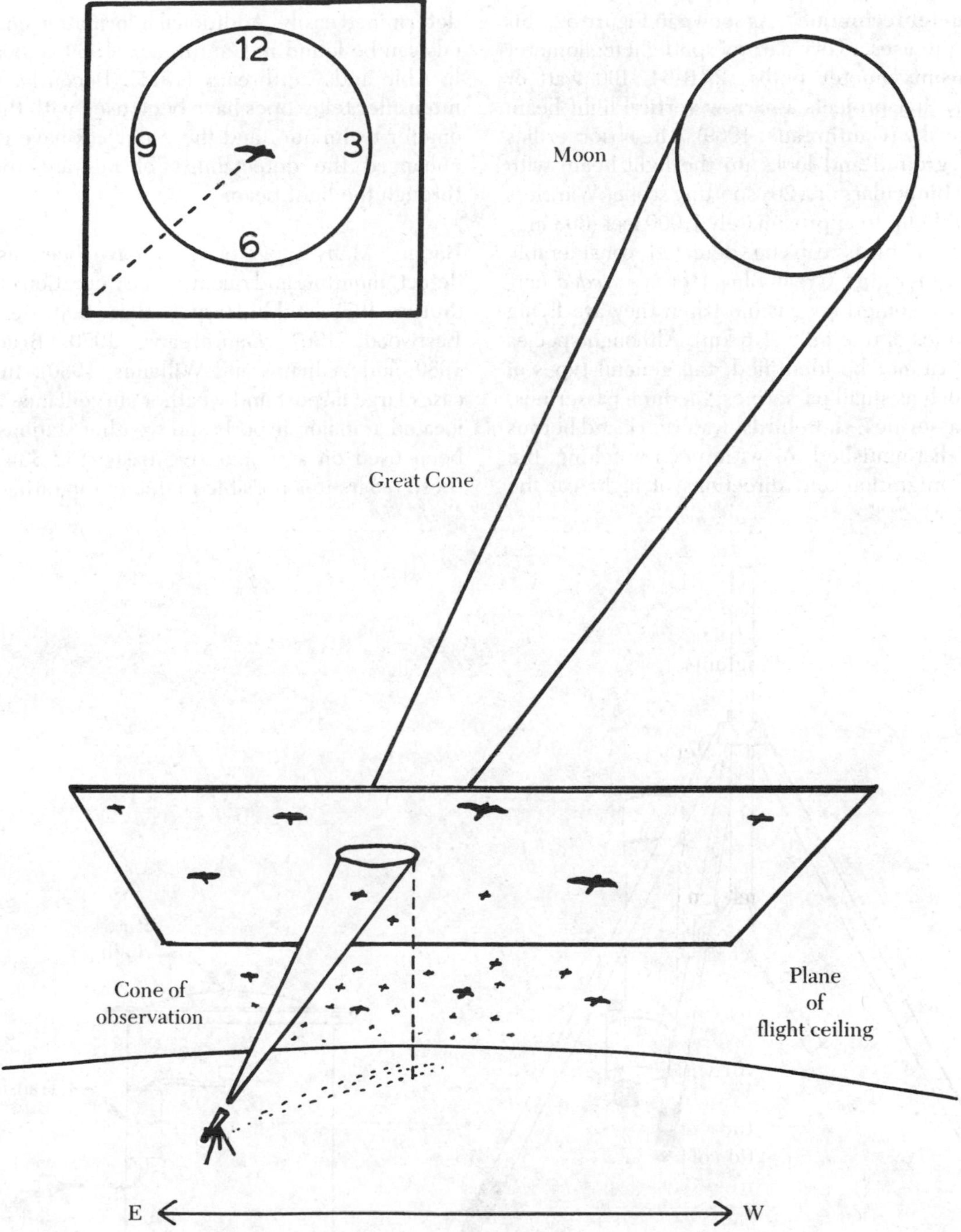

Figure 51 **Moon-watching Technique**
An observer looking through a telescope directed toward the moon will see silhouettes of migrating birds as they pass before the disc of the moon. Directions are recorded in clock-face coordinates (e.g., in at 7:30 and out 1:30). The quantification of migration depends on the elevation and azimuth of the moon relative to the directions of flight. (Redrawn from Lowery 1951:374.)

Ceilometer technique As shown in Figure 52, this technique uses a very narrow spotlight (ceilometer or transmissometer bulbs, PAR-64, 100 watt or greater) that projects a narrow vertical light beam into the sky (Gauthreaux, 1969). The observer lies on the ground and looks up the light beam with 10x, 50 binoculars or a 20x spotting scope. Warblers are visible up to approximately 1,000 feet (303 m.), and larger birds can be detected considerably higher (migrating Great Blue Herons, *Ardea herodias,* are sometimes visible when they are flying at altitudes of one mile (1.6 km). Although species usually cannot be identified, the general types of birds such as small passerines, medium passerines, large passerines, shorebirds, waterfowl, and herons can be distinguished. As with moon-watching, the rate of migration and directions of flight can be determined easily. Additional information on methods can be found in Gauthreaux (1969, 1980a) and in Able and Gauthreaux (1975). Recently, image intensifier telescopes have been used with the ceilometer technique, and these devices have greatly enhanced the detectability of migrants passing through the light beam.

Radar Many types of radar have been used to detect, monitor, and quantify the migration of birds (Sutter, 1957a and b; Harper, 1958; Bellrose, 1967; Eastwood, 1967; Gauthreaux, 1970; Bruderer, 1980; and Williams and Williams, 1980). In most cases large airport and weather surveillance radars located at major airports and weather stations have been used on a cooperative basis (Fig 53). With these radars it is possible to detect migrating birds

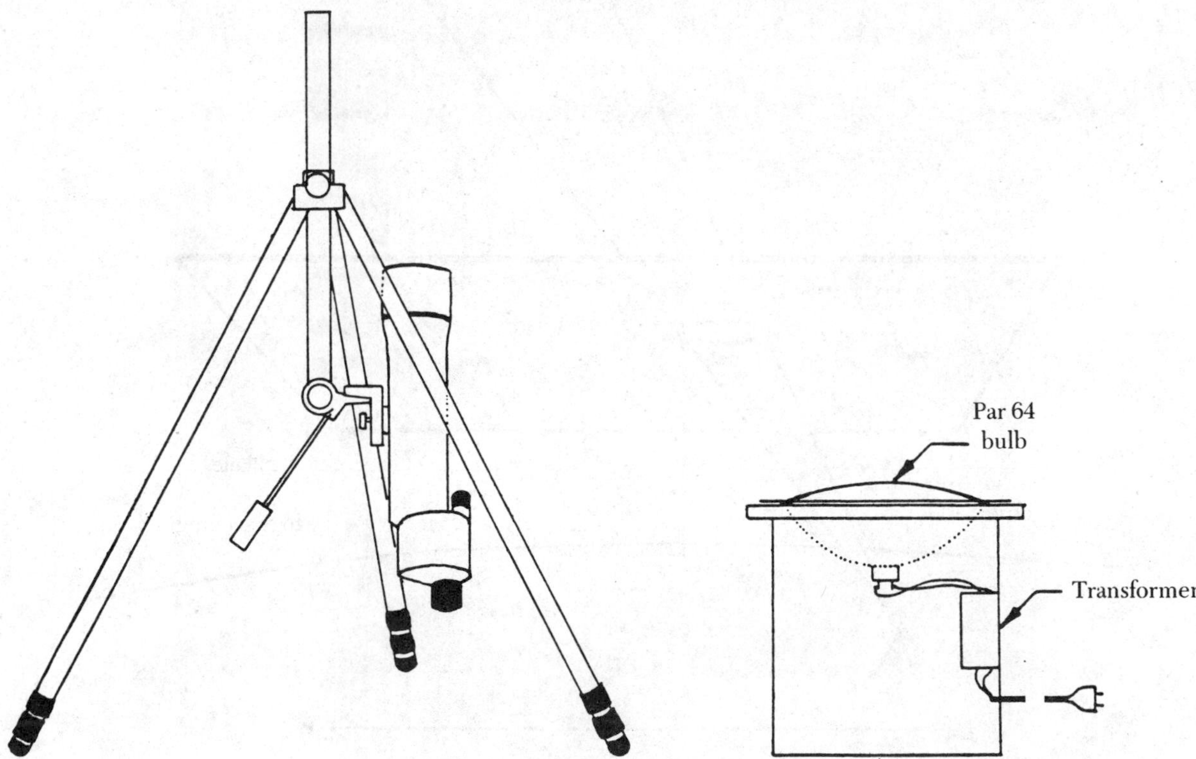

Figure 52 **Ceilometer Technique**
A telescope (or binocular) is mounted vertically on a tripod and aligned in the vertical lighted cone. When the telescope is positioned 10 meters from the ceilometer, a low-level interference zone of brilliantly illuminated insects and dust particles is avoided. The spread of the ceilometer beam is about 7 degrees; the field width of a 20 × 60 telescope is 2.5 degree.

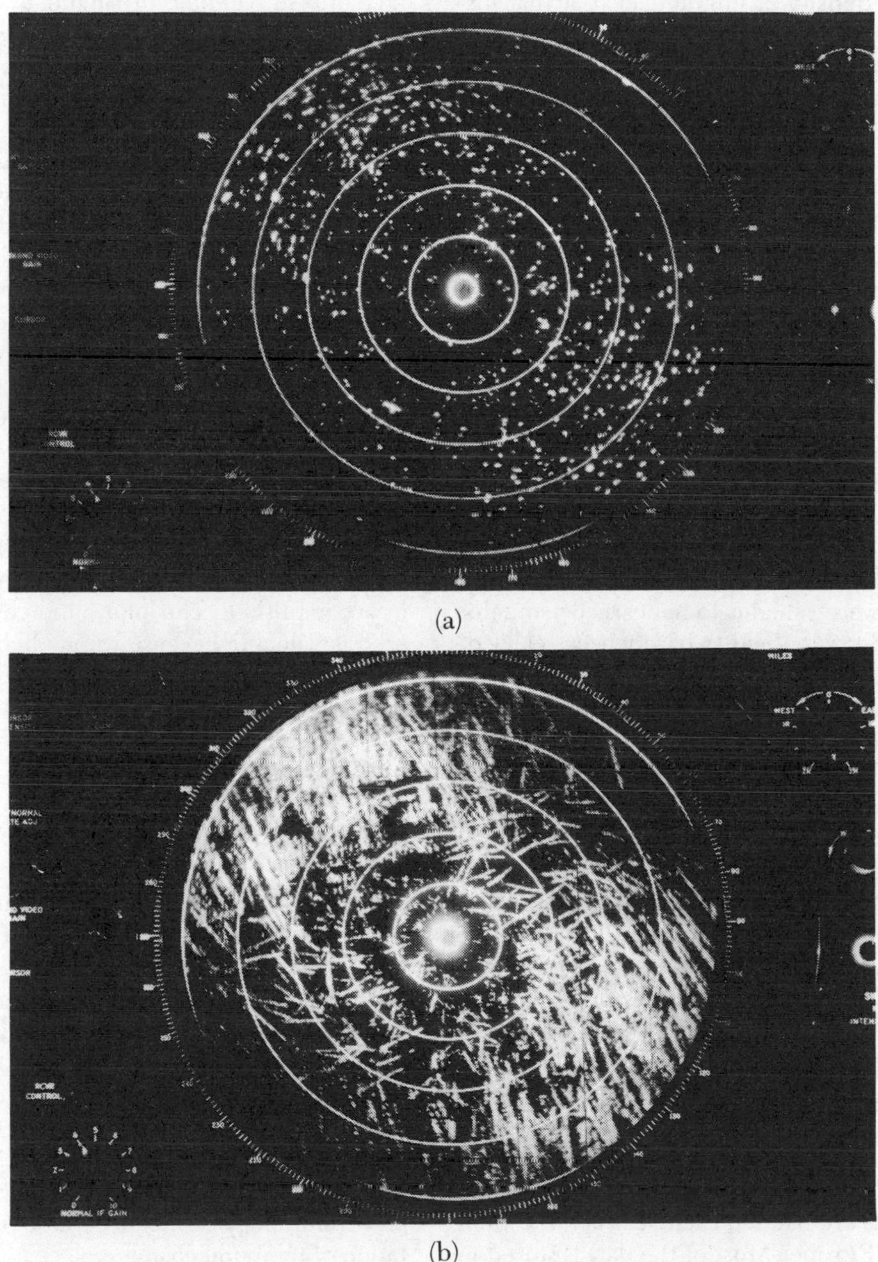

Figure 53 **Radar**

Photographs of the radar screen of the ASR-4 airport surveillance radar at Greenville, South Carolina, showing two distinct altitudinal strata of birds in different wind fields. (a) Taken 20 November 1971 at 19:46 EST on 10 nautical mile range with the moving target indicator (MTI) on. (b) A time exposure (19:47 – 19:49 EST) with identical radar settings.

out to distances of 100 nautical miles (185 km), provided the density of migrants aloft at that distance is sufficient to reflect a detectable return signal. Atmospheric conditions must also be just right to bend the radar beam so that it can sample lower altitudes at such distances. Echoes from birds appear on the radar screen as dots of varying size. A dot may represent a single larger bird or several small birds. The radar displays can be filmed, or photographed, or both, and the data analyzed in the laboratory at a later time. Several studies have used smaller marine radars to monitor bird migration within a few kilometers (e.g., Williams and Williams, 1980), while other investigators have used sophisticated tracking radars to study the flight behavior of single migrants over a distance of a few kilometers (Able, 1982a and b; Bruderer, 1971; Emlen, 1974; Bruderer and Steidinger, 1972; Emlen and Demong, 1978). Even though considerable information on bird migration has been gathered with radar, much more remains to be learned using this technique. The potential has barely been realized.

Radiotelemetry The expense and logistics of using this method to study bird migration are great but not prohibitive. The value of this technique is that the same individual can be followed for extended periods over several days during its migratory journey (see Cochran et al., 1967). Telemetry devices can be made that weigh only a couple of grams, so even small passerines can be followed in migration. Eventually the costs associated with this technique will decrease, the electronics will become more sophisticated, and even physiological studies of actively migrating birds will be possible.

Bird Banding Bird banding is second only to birding as an endeavor of serious amateurs and field ornithologists. Consequently several hundred thousand birds are banded each year in the United States and in Europe. Most of the data is stored in computer files, and in many instances the data are in need of further analysis. Banding data are valuable because they can be used to determine the dynamics of seasonal distribution of different age classes and sexes of a species (e.g., Moore, 1976). Because the birds are in hand, additional information on fat classes (Helms and Drury, 1960), weight, molt, and other characteristics related to migratory condition can be measured. The major disadvantage of bird banding, or ringing as they refer to it in Europe, is the low rates of recovery of banded birds. Although the rates of recovery are higher in Europe because of the number of observatories and banding stations, they are nonetheless often too low for a statistical treatment of the data.

Although not really a technique for studying bird migration in the usual sense, birds migrating at night frequently collide with television towers, tall buildings, and lighthouses, and the victims constitute a valuable source of information for students of migration. Major migration disasters occur infrequently, but several studies have been based on regular searches for collision victims at man-made obstructions (e.g., Tordoff and Mengel, 1956; Brewer and Ellis, 1958; Caldwell et al., 1963; Avery et al., 1977; Weir, 1977; and Crawford, 1981). This technique like many of the others that have been discussed in this section is straightforward and easily accomplished. The information that can be gathered is valuable because it may help answer some of the many puzzling questions about bird migration that remain, and most likely it will raise even more.

Causes of Migration

Movement from one location to another is a fundamental biological response to adversity, and all populations are spatially fluid in some measure. Individual animals move so that they can survive and reproduce in the most appropriate place at the most appropriate time. When a location elsewhere has a higher mean favorability than the present location, natural selection will favor those traits that serve to maximize the expectancy of arriving at the new location. Spatial movements by animals, then, are evolutionary strategies that express the expectation of temporal change.

Avian migration systems, like the migration patterns of animals in general, are adaptations to temporal and spatial changes in the physical and biotic environment. Because habitats change in suitability and resource availability through time, birds that have specific habitat requirements must change location and seek new habitats elsewhere if they are to survive and reproduce. The spatial movements of birds, whether maintenance, dispersal, or migratory are evolutionary strategies that express the expectation of temporal change in conditions at the present locality. If the resources in a suitable habitat can support a certain number of individuals during a period of high resource availability and only a fraction of that number during a period of low resource availability, then during the latter period some of the individuals in the population must move and seek other suitable habitats or perish.

Climatic and Meteorological Causes

The climatic and meteorological factors responsible for changing the suitability and resource availability of a habitat through time may be, in large part, the same factors that shape the spatial and temporal characteristics of movements from or to that habitat. Climatic changes can influence the spatial distribution of suitable habitats as well as the absolute resource availability within those habitats. By doing so, climatic changes either directly or indirectly influence the direction and distance of the movement patterns. Climatic changes can also dictate the heterogeneity of habitats in time (length of favorable and unfavorable periods, and length of time a location remains suitable) and influence the **phenology** (timing and rate) of the movements. Thus by tracking climatic and meteorological changes through movement, birds can maximize the expectancy of finding a new suitable habitat or returning to a suitable habitat after a period of unfavorability. Depending on the rate and nature of environmental changes and the life span of the individuals, several types of "migration" patterns can be realized (e.g., zoogeographic migrations mediated through dispersal, partial migrations where only some individuals in the population migrate, irruptive movements, and short- and long-distance seasonal migrations). Thus the evolution of different avian migration strategies is dependent in large part on the diversity of environmental changes that occur over different temporal and spatial scales (see Gauthreaux, 1980c). Because of this diversity:

1. Some species migrate in directions other than north–south.

2. Some species migrate irrespective of day-length, as in the tropics.

3. Some species migrate when the temperature is mild and the food supply ample, and others when the opposite conditions are true.

4. Some species migrate as a result of seasonal alternation in rainfall and drought.

5. In certain species some populations migrate while others do not.

6. In some populations some individuals migrate while others do not.

7. Some individuals migrate in some years but not in others.

Theoretical Origins

A number of authorities have sought the ultimate causes of migration—*why* questions about the evolution and function of migration. The theories of the "origin" of bird migration can be classified into two major categories: those that invoked some major event "forcing" birds to migrate (e.g., factors such as Pleistocene glaciation and continental drift) and those that detailed the step-by-step processes whereby sedentary species became migratory. The earlier theories suggested that migration originated through the inheritance of habits that evolved under conditions no longer present, and these theories have largely been replaced by more modern evolutionary theories. Major events such as glacial changes certainly have influenced the geographical ranges of migratory birds, but bird migration was not "caused" by these events. Likewise, the lengthy discussions on the ancestral homes of migrants have provided little insight into the processes that have operated and are operating in the evolution of bird migration.

Modern theories consider the ecological circumstances that favor the evolution of migration, and nearly all of these theories stem from the work of Lack (1954), which emphasized the importance of food as an ultimate factor in the evolution of migratory movements. Lack stressed the importance of food availability in relation to actual or potential competition, and not the absolute amount of food. In addition, he advanced a cost-benefit approach in his treatment of migration strategies. If the costs of remaining at a given location outweigh the benefits in terms of survival and reproduction, then natural selection will favor movement to another location. Likewise, if the costs of moving outweigh the benefits of staying, then natural selection will favor the animal's remaining at its original location.

Modern theories stress intraspecific and interspecific competition for food, shelter, nest sites, and other resources as important ultimate factors in the evolution of migration (Cohen, 1967; Cox, 1968; von Haartman, 1968; Baker, 1978; Gauthreaux, 1978a; Fretwell, 1980). Current approaches also emphasize the costs and benefits of migration in terms of relative fitness (survival and reproduction) and the allocation of time to reproductive- versus strictly survival-related activities (Greenberg, 1980).

Having discussed the ultimate factors responsible for the evolution of bird migration, the subject of proximate factors—*how* questions about the mechanisms and development of migratory behavior—must be addressed.

Control and Physiology of Migration

Spring and fall migratory movements are closely tied to the favorability of temperate zone habitats during the annual climatic cycle. As a result of natural selection, relatively precise proximal control mechanisms have evolved to initiate and terminate migratory behavior. In addition, the migratory state also involves marked changes in metabolic physiology, needed to supply the energy required for migratory flight and to provide energy reserves for occasions when foraging is difficult or impossible.

Although trapping and netting of migrants have produced some information on migratory physiology, the studies of caged migratory birds in the laboratory have yielded more detailed information on the physiological mechanisms that control the migratory condition. Caged migrants usually show intense fluttering, hopping, jumping, and whirling at the time of day and during the periods of the year when they would normally migrate. As mentioned earlier, this activity is called **migratory restlessness (Zugunruhe)**, and is measurable quantitatively in cages equipped with microswitches on perches and other electro-mechanical devices. It has served as a useful indicator of migratory condition in a large number of studies.

Just prior to migration, individuals of many species deposit lipid reserves (Blem, 1980). They accomplish this by eating amounts of food in excess of their daily needs—**hyperphagia**—and thereby storing energy in the form of subcutaneous fat. Fat deposition and migratory restlessness are usually coupled and alternate with molt and gonadal development during the annual cycle. This pattern persists in constant photoperiod (12 hr light : 12 hr dark) in laboratory studies for several years, demonstrating an endogenous **circannual rhythm** (Gwinner, 1975, 1977). In those species that respond to photoperiod, fat deposition and migratory restlessness are directly induced by increasing day-length in spring. In autumn the initiation of endogenous events that eventually promote autumnal migratory behavior is also dependent on increasing day-length the previous spring. The circannual rhythm of migratory behavior is thus entrained by increasing photoperiod during the spring.

Seasonal changes in photosensitivity and photorefractoriness are thought to be the consequences of changing temporal relations between two **circadian systems.** One system involves the rhythm of prolactin release from the pituitary gland and the other involves the daily rhythm of plasma corticosteroid hormone. The changing circadian relations of these two hormones account for the seasonal regulation of migration and seem to be a part of the circannual mechanism as well (see Meier and Fivizzani, 1980, for more detailed information).

Once the endogenous physiological mechanisms for migration have been activated, a bird will migrate unless inhibited from doing so. Only when conditions physically hamper bird flight (very hard rain and excessively strong head winds) will migration be interrupted. When lipid stores run low, a bird will remain at a suitable location for a few days until its lipid reserves have been replenished (Cherry, 1982).

For a summary of information on the nature and mechanisms of periodic preparation and stimulus for migration, see Farner (1955, 1960), Berthold (1975), and Meier and Fivizzani (1980).

Geographical Distribution of Migrants

A wealth of information on the distribution of North American migrant birds can be gleaned from the pages of *American Birds* and the range maps in several field guides. Additional information on the geographical pattern of the breeding density of certain migrant species can be obtained from the "Breeding Bird Survey" coordinated by the Nongame Section, Migratory Bird and Habitat Research Laboratory, the United States Fish and Wildlife Service, Laurel, Maryland 20811.

The breeding distribution of North American passerines wintering primarily in the neotropics has been examined by many, including MacArthur (1959). He found that the eastern deciduous forests contained far more neotropical migrants than northern coniferous forests and grasslands, and he correlated the differences with the contrasts between winter and summer food supplies in a given habitat. In a partial re-analysis of MacArthur's (1959) findings Willson (1976) showed that:

1. North American neotropical migrants are less prevalent in grasslands than in forests, but the proportion of neotropical migrants in deciduous forests does not differ significantly from that in coniferous forests.

2. Coniferous forests have relatively fewer year-round resident individuals than grasslands or deciduous forests, and grasslands and coniferous forests have slightly fewer resident species than deciduous forests.

3. Most neotropical migrant birds breed primarily in deciduous forests, and most of those that breed in coniferous forests are parulines (American wood warblers).

In the northeastern deciduous forests, on the average, 62 percent of the breeding species and 75 percent of the individuals are migrants. In the northern coniferous forests 80 percent of the breeding species and 94 percent of the individuals are migrants. In the grasslands 76 percent of the breeding species and 73 percent of the individuals are migrants.

A similar analysis to MacArthur's (1959) has been made for European migrant birds by Herrera (1978), and it shows a conspicuous latitudinal and climatic gradient in the proportion of passerine tropical migrants, with more migrants breeding in northernmost areas and proportionately more residents breeding in the south. Similar analyses for waterfowl and shorebirds are not available.

Migrant individuals appear to be more prevalent wherever large variations in food supply, caused by the prevailing climate, occur during the breeding season (see O'Connor, 1981). Breeding areas where climatic variance is great and where resident birds have relatively low winter survival are ideal areas for migrants to breed. Breeding areas where climatic fluctuations are less marked and where residents have higher winter survival are poorer areas for migrants to breed. Thus most migrant individuals tend to breed in areas of maximum unused resources.

In general there is considerably more bird migration in the eastern two-thirds of the United States than in the West (Lowery, 1951; Lowery and Newman, 1955, 1966). One reason for this is that more migrants (species and individuals) breed in the East, but another reason is that even though a number of land-bird migrants breed considerably farther west and north of the eastern forests of the United States, they migrate through the eastern states (e.g., Philadelphia Vireo—*Vireo philadelphicus*). Approximately 33 species of land-bird migrants conform to this pattern.

Because of the distribution of ornithologists, the bird migration systems of the North Temperate Zone have been studied more than those for any other region. Approximately 400 species of birds are tropical migrants with their breeding ranges restricted to the North Temperate Zone and their nonbreeding ranges in the tropics (Karr, 1980). Of this number, 147 winter in the Neotropics, 118 winter in Africa, and 142 winter in the Oriental region. Thus the relatively wet and nonseasonal regions of the tropics of Southeast Asia and the New World have about 25 percent more nonbreeding species than the drier or highly seasonal areas of the Indian subcontinent and Africa.

In Africa migrants occupy savanna and dry woodlands, whereas in the Neotropics, migrants occupy the highlands, pine woodlands, rain forests, and extensive areas of second growth. The majority of

Neotropical migrants winter in greater numbers in the northern portions of Central America and the West Indies, with the number of wintering species and individuals decreasing progressively from Mexico toward the equator (Terborgh, 1980) and from Florida to the Lesser Antilles (Terborgh and Faaborg, 1980). Even long-distance migrants travel only as far as necessary to locate suitable survival areas during the nonbreeding period.

When the Palearctic-Ethiopian and the Nearctic-Neotropical migration systems are compared (Keast, 1980), one finds that migrants moving toward Africa must contend with the Alps, the Mediterranean, the Sahara, and for those moving from Siberia to Africa, the Himalayan Mountains. In contrast, many of the migrants moving between North and South America have only the Gulf of Mexico and the Caribbean to cross. Possibly because of the barriers to movement, the pace of migration from the Palearctic appears to be slower than that of the movements from the Nearctic.

Seasonal Timing of Migration

Much of what we know about the seasonal timing of bird migration in North America comes from the work of field observers and bird banders, and their findings are regularly summarized in the spring and fall issues of *American Birds*. Virtually every state has a check-list or bird book containing information on the seasonal occurrences of migrant birds.

Saunders (1959) compared the variation in the timing of spring arrivals among 50 different species with the mean 40-year arrival dates and found that in late, cold springs migrants arrived later than in early, warm springs. Gauthreaux and LeGrand (1975) associated the advancement or retardation of the seasonal timing of migration with year-to-year changes in continental wind patterns. Robbins et al. (1983) have summarized considerable data on the seasonal timing of bird migration for most North American species. This information is presented on range maps for each species as isochronal lines that show the average first-arrival date, where birds migrating to the north may be seen about the first of March, April, May, and June. Preston (1966) analyzed mathematically the timing of spring and fall migration and found that in general those species that move north early in the season (e.g., waterfowl and sparrows) return south late in the fall. Preston discussed evidence that shows breeding birds occupy their summer habitat as soon as it is habitable and depart as soon as they have finished breeding. The standard deviation of the timing of a species' migration is less in spring than in fall, hence the birds are better synchronized in spring. Within a species, individuals of different age and sex classes may migrate at different times. In the spring, males of most species arrive before the females, and adults arrive on the breeding grounds before the immatures (see Gauthreaux, 1982), but in the fall the pattern varies greatly depending on the species (see Murray, 1966; Gauthreaux, 1982).

Alerstam and Hogstedt (1982) have proposed that the condition and extent of survival (nonbreeding) habitats in comparison with breeding habitats could be important factors determining when fall migration occurs in different bird species. If this theory is true, those species having a surplus of survival habitats should leave the breeding grounds as soon as possible after breeding is completed and molt on the nonbreeding grounds, and those species having a surplus of breeding habitats should exploit their breeding habitats as long as possible and molt before departure for the nonbreeding grounds. Alerstam and Hogstedt present evidence that these expectations generally do hold true.

A number of factors must be considered in discussing the seasonal timing of migration. The most important of these are vegetational development in the spring, food availability, and climatic factors in spring and fall. Weydemeyer (1973), in a 48-year study of spring arrivals of migrants in Montana, found that ranges in dates of arrival were greatest during late March and April and least in late May and June. Slagsvold (1976), working in Norway, found that for the country as a whole there was a six-day delay in bird arrival for each ten-day delay in vegetation development. The arrival of migrants at higher latitudes and altitudes was faster than the development of vegetation. Slagsvold also found that earlier arriving species varied considerably in arrival date at a particular locality from year to year, but late arriving species had much less variation in arrival time. Pinkowski and Bajorek (1976) examined the spring arrival dates of 29 common or conspicuous migrants and summer resident species in southern Michigan over a seven-year period. They concluded that granivorous, omnivorous, and aquatic species tend to arrive earlier than strictly insectivorous species and, supporting Slagsvold's findings, that earlier arriving species have a greater variance in arrival time than the later arriving species. Buskirk (1980) has analyzed the timing of spring and fall trans-Gulf migration and found that the greatest frequencies of high magnitude flights across the Gulf of Mexico coincide with improved flight conditions for Gulf crossings. The climatic events on the breeding grounds as well as those en route are important selective pressures in determining the phenology of migratory flights.

Daily Timing of Migration

The majority of small birds, including most passerines, migrate at night, and most waterfowl and shorebirds migrate both at night and during the day. Raptors, several woodpeckers, swallows, several corvids, bluebirds, and blackbirds migrate during daylight hours. Nearly all birds that soar during migration migrate during the daylight hours when thermals are the best developed over land. The determination of whether individuals of a species migrate typically at night, during the day, or at both times is sometimes difficult. Laboratory studies of migratory restlessness in caged birds is one way to determine the daily timing of migration. Another way is to examine the species composition of migrants killed by colliding with television and other towers (Crawford, 1981), buildings, powerlines, lighthouses, and the like (see Weir, 1977 for a review). Except for very high altitude flights, migrants that move during the daylight hours can be easily observed. However, according to data gathered by surveillance radars at several localities in the United States and Canada, considerably more birds migrate at night than during the day.

At least two explanations have been advanced for the development of nocturnal migrations: (1) movement by night affords birds, which normally live in thick vegetational cover and rarely take long flights away from it, the protection of darkness against their diurnal predators; and (2) movement by night affords birds the opportunity of using all the daylight hours for feeding, thereby enabling them to build up sufficient energy reserves for sustained long-distance flights.

A number of studies have shown the temporal pattern of nocturnal migration (e.g., Lowery, 1951; Sutter, 1957a; Harper, 1958; Gauthreaux, 1971). The initiation of nocturnal migration generally occurs about 30 to 45 minutes after sunset; the number of migrants aloft increases rapidly, peaking between 10 and 11 P. M. Thereafter, the number of migrants aloft decreases steadily until dawn, indicating that migrants are landing at night. Daytime passerine migration is initiated near dawn (sometimes earlier), peaks around 10 A. M. and declines to minimal density shortly after noon (Sutter, 1957b; Gehring, 1963; Gauthreaux, 1978b).

Routes of Bird Migration

Many species breeding in North America north of Mexico have their winter ranges far south and southeast, in southern Mexico, Central America, and South America. To reach their winter ranges, most of these species proceed at night from their breeding ranges to the southern United States in broad fronts without notable regard to topographical features. In North America radar studies of bird migration have been conducted at many locations (see Gauthreaux, 1978c, for a summary). Although there are many geographical gaps in the coverage and some studies have concentrated on a particular group (e.g., waterfowl), a continental pattern of bird migration in North America is beginning to emerge. Radar and direct visual observations confirm that in central, southern Canada, spring movements are towards the northwest and fall movements are toward the southeast; but in the eastern two-thirds of the United States the movements are toward the north and northeast in spring and toward the south and southwest in fall (Gauthreaux, 1980b). Bellrose (1964, 1980) has shown that most waterfowl in the Mississippi Valley move more north–south, with eastward and westward deviations depending on topographic factors (lakes, marshlands, and river systems).

Effects of Wind Direction on Migration

Wind direction exerts a strong influence on the direction and timing of migration (Gauthreaux and Able, 1970; Able, 1974; Alerstam, 1976). The routes birds fly appear to be determined, at least in part, by the prevailing wind patterns in North America during spring and fall (Gauthreaux, 1980b; Alerstam, 1981). For example, in northwestern South

Carolina in spring the prevailing winds blow to the northeast, and the average distribution of the directions of nocturnal migration on calm nights—i.e., when wind directions are not an influencing factor—is toward the northeast (29.5 degrees). Thus, in spring the preferred direction of migrants closely matches the prevailing wind direction. In fall the winds in the same area usually blow toward the southwest, and the average distribution of the directions of nocturnal migration on calm nights is toward the southwest (231.5 degrees). These data were gathered using direct visual means—moon-watching (Lowery, 1951) and ceilometer-watching (Gauthreaux, 1969)—but data gathered from radar confirm the pattern (see Gauthreaux, 1978b, for details).

Elliptical Courses of Migration

Many species migrate in an elliptical pattern, taking more easterly courses during their fall migration and more westerly courses during spring. In the fall the Blackpoll Warbler *(Dendroica striata)* travels from its vast transcontinental breeding range to New England, where most individuals migrate directly across the western Atlantic Ocean to winter in northern South America (Nisbet, 1970). In the spring it migrates northward through the Caribbean and moves through Florida and across the eastern half of the Gulf of Mexico on the way to its breeding grounds (Stevenson, 1957). The Lesser Golden-Plover *(Pluvialis dominica)* and the Greater Shearwater *(Puffinus gravis)* are two additional species of many that show elliptical courses (see Bellrose and Graber, 1963; Graber, 1968; Alerstam, 1981).

Some birds move eastward or westward before going south. For migrating birds moving eastward and westward at mid and high latitudes, the great circle route is considerably shorter than the rhumb line (fixed compass direction) route (Alerstam, 1981). Ruffs *(Philomachus pugnax)* flying from breeding areas in East Siberia follow a great circle route to their wintering areas in western part of the Sahel (a distance of 6250 mi or 10,060 km) and pass through northwest Europe. This route is 18 percent shorter than the rhumb line route. In spring the birds move back a bit south of their autumn route, passing across Italy and Central Europe. Additional details on avian migration routes can be found in Baker (1978) and Alerstam (1981).

Because of the increase in the number of bird watchers and banders and the widening interest in birds found far from their normal migratory routes, "accidentals" and "stragglers," there is now new and exciting information on the geographical distribution of migrating birds during the spring and fall. This information reveals that virtually every fall large numbers of birds from the West and Southwest show up in the eastern United States well to the east and even north of their "normal" migratory goals. Among a few species that breed in the western United States—e.g., the Western Kingbird *(Tyrannus verticalis)* and the Scissor-tailed Flycatcher *(Muscivora forficata)*—some individuals in the fall move eastward across the Gulf states to winter in Florida (see McAtee et al., 1944), whereas most of the population moves directly south to Mexico. In the spring fewer western vagrants are recorded in the East, but records indicate that some occur every spring. What is perhaps even stranger than western migrants in the East are the scores of eastern migrants that have been recorded in the West, particularly in coastal California, every spring and fall over the last three decades (DeBenedictis, 1971; Roberson, 1980). Some eastern species of migrants have become regular in their occurrence in the West—e.g., the Black-and-white Warbler *(Mniotilta varia)*, Tennessee Warbler *(Vermivora peregrina)*, and Rose-breasted Grosbeak *(Pheucticus ludovicianus)*. Why these "mistakes" happen is a topic of considerable debate. Some explanations emphasize that wind drift during storms may be responsible, while other hypotheses empahsize genetically based navigational errors during migration (e.g., mirror image misorientation—see Diamond, 1982).

Although much has been said of flyways and special routes during migration, extensive banding and recapture data suggest that considerable mixing of migrants takes place between breeding areas and wintering areas. It is not unusual for birds banded at a single location to be recaptured at widely separated locations in subsequent seasons. Likewise, it is not unusual to capture at a single location birds that were originally banded at several widely separated locations (e.g., Kushman et al., 1982). Thus the individuals of a species as well as different species vary greatly in their course of migration and their use of routes. A few species move south over one route and return by another; a few species move eastward or westward before going south; and a few species have spectacularly long routes that take them as far south as southern South America.

For example, the Bobolink *(Dolichonyx oryzivorus)* is the only New World songbird that breeds exclusively in the Northern Hemisphere and winters exclusively south of the equator in two distinct populations—one in southeastern South America and the other in southern Peru and northern Chile (Pettingill, 1983). In the fall vast numbers migrate through the southern Atlantic States in late August and early September and reach peninsular Florida by mid-September. Pettingill points out, however, that not all Bobolinks use the Florida route. Many leave North America from Virginia to Nova Scotia and cross the western Atlantic Ocean and the eastern Caribbean to arrive in northern South America where they join birds that migrated via the Florida Peninsula and across the Caribbean. The Bobolinks then proceed to southwestern Brazil and Paraguay by the beginning of November and reach northeastern Argentina and northwestern Uruguay by early January. The Bobolinks that winter in southern Peru and northern Chile probably migrate from North America west of the Continental Divide, head south through western Mexico, then cross the eastern Pacific over the Galapagos (where a few may land) to the coast of Ecuador and northern Peru. Once on the mainland they fly southward along the west slope of the Andes until they reach their destination.

Altitude of Migration

Radar studies of bird migration show that most songbirds fly at night below 0.25–0.43 mile (400–700 m) above ground level and 90 percent are below 0.62–1.24 miles (1,000–2,000 m) (Nisbet, 1963; Eastwood and Rider, 1965; Able, 1970; Bellrose, 1971; Blokpoel, 1971a, 1971b; Bruderer and Steidinger, 1972; Gauthreaux, 1972). The maximun altitudes reported in these studies are between 1.86–3.9 miles (3,000–6,300 m). The mean altitude of migration changes throughout the night. After take-off, the maximum is reached in about two hours, and thereafter the mean altitude declines slowly as birds begin to terminate their flight for the night (Able, 1970). Birds fly higher by night than by day. Lack (1960) first noted this tendency, by radar, among migrants crossing the southern North Sea between England and the Continent. Later, Eastwood and Rider (1965) at the Bushy Hill station in England proved that the tendency is significant. From their considerable data they were able to show that 80 percent of the birds fly below 0.95 mile (1,524 m) at night and 80 percent below 0.66 mile (1,066 m) during the day. They further demonstrated that migrating birds, in a 24-hour day, have a tendency to fly at the lowest altitudes in the afternoon and the highest just before midnight. In general, the larger the bird species and the faster its air speed, the higher it flies during migration for minimum cost of transport (minimum expenditure of energy; Tucker, 1975).

Most radars cannot detect birds very close to the ground (but some shipboard navigation radars can), and consequently the *minimum* altitude of nocturnal migration displayed on radar usually cannot be measured accurately. Studies using direct visual means to detect migrating birds as they pass through a narrow vertical beam of light (Gauthreaux, 1969) suggest that a considerable number of birds fly within 328 feet (100 m) of the ground at night. This is particularly so within an hour after the initiation of nocturnal migration and at the time birds are landing during the night. On some misty, cloudy nights tremendous numbers of call notes from migrants aloft can be heard, and on many of these occasions the distance of the call notes overhead indicates the birds are flying within a few feet of the ground.

Daytime migration usually occurs at altitudes below 984 feet (300 m), and quite often flocks of daytime migrants can be seen moving just above tree level. This, however, is not always the case. In spring when migrants are arriving on the northern coast of the Gulf of Mexico during daylight hours after trans-Gulf flight, approximately 75 percent are at altitudes between 0.62 and 1.86 miles (1,000 and 3,000 m) (Gauthreaux, 1972). Other radar studies have shown that migrants often fly quite high when crossing large water gaps.

Shorebirds migrating over Nova Scotia in autumn are at a median altitude of 1.1 mile (1,700 m) and 10 percent are between 2.4 and 4.1 miles (3,900 m and 6,650 m) (Richardson, 1979). Most of the migrants that cross the west Atlantic on their "southward" migrations do so between 0.62 and 0.93 mile (1,000 and 1,500 m) above the ocean (Richardson, 1976; Williams et al., 1977). The median altitude of migration may reach 2.5–3.1 miles (4,000–5,000 m) over Puerto Rico (passerines) and over Antigua and Barbados (shorebirds), and occasionally some birds have been detected flying 4.2 miles (6,800 m) over Puerto Rico. An excellent discussion of the factors that influence the altitude of migration can be found in Alerstam (1981, pp. 42–46).

When migrating birds encounter headwinds and cannot land, they fly at considerably lower altitudes, sometimes just a few meters above the ground or near the surface of the water. On such occasions tremendous numbers of birds can perish when they strike radio and television towers, telephone and transmission lines, and even buildings. Nearly all such disasters occur at night, but sometimes migrants are killed during the day when they are flying low because of poor visibility (fog, haze, dense cloud cover, and drizzle).

Weather Influences on Migration

The amount of bird migration on a given day is strongly influenced by weather conditions; consequently, seasonal trends in weather patterns may substantially advance or delay the arrival of migrants on their breeding grounds (see "Seasonal Timing of Migration," p. 242). The departure of migrants from their breeding areas is also strongly dependent on the passage of frontal systems (synoptic weather), and the timing and intensity of cold fronts. It is generally accepted that in spring more migration occurs on the west side of a high pressure system, behind a warm front, and before a cold front. (See Table 6 and Figure 54, which show the relationships between synoptic weather patterns—low and high pressure systems and cold and warm

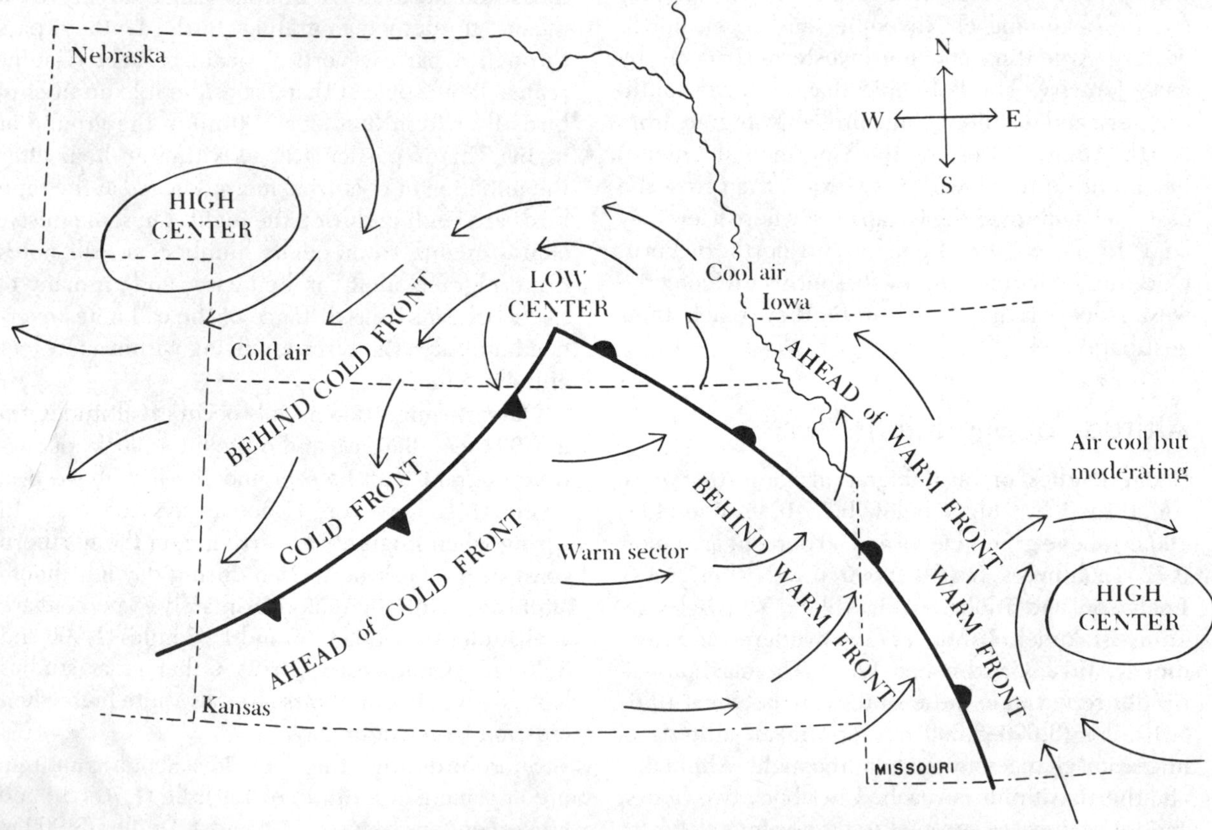

Figure 54 **Pattern of Fronts and Wind Directions**

Table 6 **Relation of Weather Elements to Migration**

⟶ Direction of air masses ⟶

⟵ Sequence of weather elements in a given area ⟵

	HIGH			**LOW**		
	BEHIND COLD FRONT	**COLD FRONT**	**AHEAD OF COLD FRONT**	**BEHIND WARM FRONT**	**WARM FRONT**	**AHEAD OF WARM FRONT**
Barometric Pressure	Rapidly rising ←	Low ←	Dropping steadily ←	Rising slightly ←	Low ←	Dropping rapidly
Wind Direction	From north-west and west; strong and steady ←	From west; strong and gusty ←	Easterly; slight ←	From southwest ←	From south-east and south ←	Easterly; increasing
Tempera-ture	Dropping ←	Unchanged ←	Rising ←	Rising ←	Rising slowly ←	Rising slightly
Weather Aspect	Clearing ←	Low clouds and heavy rain ←	Cloudy; occasional fog or rain ←	Showers, but gener-ally clearing ←	Cloudy; drizzle or showers ←	Cloudy; drizzle or showers

⟵ Sequence of migration in the same area ⟵

Spring Migration	Migration stopped ←	Migration stopped ←	Migration slowed but may be heavy ←	Migration in full force ←	Migration begins ←	No migration
Fall Migration	Migration begins ←	Migration stopped ←	Migration slowed ←	Migration slowed ←	Migration in full force ←	Migration in full force

fronts.) In fall very large migrations occur just after a cold front on the east side of a high pressure system. The relationship between synoptic weather patterns and migration intensity has been known for well over a century (see Lack, 1960), but the specific weather factors or combination of factors that influence the number of birds migrating on a given date have only recently received attention.

In the last several years a number of studies have concentrated on the influence of specific weather factors on migration (see Richardson, 1978, for a detailed review of this subject). Because specific weather factors interact in complex ways, multivariate statistical analyses have been used, and the results of studies using such analyses show unusual agreement. Of all the weather factors examined, wind direction and speed and temperature are significantly correlated with the quantity of spring and fall migration. These weather factors are consistently important and they are, of course, significantly intercorrelated. Other weather factors—e.g., cloud cover, cloud height, relative humidity, precipitation, and barometric pressure—also covary with each other and with wind and temperature, but these additional factors explain only a small amount of the day-to-day variance in the amount of migration.

Another point regarding the influence of weather on the quantity of bird migration should be mentioned. The amount of day-to-day variation in the quantity of migration explained by weather on the average is 50–60 percent. The remaining variation undoubtedly is due to the internal conditions of the migrants (physiological readiness) and the actual number of grounded migrants in an area (Gauthreaux, 1978c; Alerstam, 1981).

In general, the largest migrations in spring occur with winds from the south and southwest, which bring increasing temperatures. Relative humidity increases, and barometric pressure falls. In fall the largest migrations occur with winds from the northwest and north, which usually bring colder temperatures, lower humidity, and higher barometric pressure. Hard precipitation usually temporarily halts migration during both seasons. The close association between heavy migration and following winds relates to the flying time and energy costs for the migratory journey. By flying with basically following winds, birds can realize high ground speeds with a minimum energy expenditure, and by avoiding precipitation, migrants can avoid the stresses of increased wing loading.

Poor visibility, low cloud ceiling, and drizzle are conditions not conducive to very large migratory movements, but such weather conditions are frequently associated with numbers of migrants colliding with man-made objects (television tower guy lines, electrical transmission and distribution lines, and buildings). It is puzzling why tremendous numbers of migrants can collide with such objects when weather conditions are poor for flying and migration should have been inhibited or interrupted. Sometimes birds initiate a migration with favorable weather and move into areas where the weather has deteriorated—e.g., a stalled cold front in fall. Migrants may also be forced to fly in bad weather when they are crossing unfavorable areas (see "Altitude of Migration," p. 245). Occasionally disasters occur under ideal weather conditions for migration, but these are exceptional events (Avery et al., 1977).

Exceptional weather conditions with unusually high winds often deflect migrants from their usual routes and at the same time carry many nonmigrating birds from their regular ranges. The hurricanes or heavy northeast storms that occasionally move north along the eastern Atlantic seaboard, by the counterclockwise direction of their winds, force many south-bound migrants and sea birds far inland. As a result, ducks, geese, and gulls are reported in great abundance in places where they do not ordinarily occur, and even such sea birds as petrels show up far inland. Two very severe northeast storms in December, 1927, and January, 1966, bore spectacular numbers of Northern Lapwings (*Vanellus vanellus*) over the north Atlantic from their migration route in western Europe to the vicinity of Newfoundland. The paper by Bagg (1967) shows in detail, with a series of weather maps, how the great storms brought the lapwings to North America.

Fuel Economy and Flight Speed

Fuel Economy

Most nonsoaring birds store fat as a source of energy for their migratory journey. For its mass, fat produces the highest oxidizing energy of any fuel source. Because of weight, migrants generally deposit only the amount of fat they will need for a particular leg of their migratory journey. Up to 50 percent of the take-off mass of a bird about to embark on a long-distance flight may be fat. In undertaking long-distance migrations, many small land birds tend to begin with short flights and complete them with longer flights. Indirect evidence of this procedure was reported by Caldwell et al. (1963) after comparing the fat reserves of six species of tropical-wintering North American passerines killed during fall migration by television towers, one near the Florida Gulf Coast and the other in central Michigan. All the birds killed by the Florida tower showed significantly greater amounts of fat than those killed in Michigan, strongly suggesting that these migrants began with low to moderate fat reserves that allowed only short flights and then increased their reserves until they had acquired a maximum amount for the long, nonstop flights, such as across the Gulf of Mexico. European birds migrating south across Africa build up fat reserves of 30 to 40 percent of their body weight by the time they set out across the Sahara (Ward, 1963).

For small passerines such as thrushes and warblers, the relative weight loss due to the consumption of fat during flight is about 0.7 percent body mass per flying hour. In larger birds such as ducks, geese, and swans, the weight loss is 1 to 1.5 percent body mass per flying hour. Thus given the amount of fat, it is possible to calculate the time a bird can fly until all fuel is used. The time is dependent on whether the bird is flying at the **maximum range speed** or the **minimum power speed** (Alerstam, 1981). Migrants seem to minimize energy consumption per unit of distance covered (fly maximum range) rather than to minimize the absolute rate of energy consumption (fly minimum power). For Blackpoll Warblers flying across the western Atlantic Ocean in fall, the time of flight until fuel is exhausted varies from 70 to 100 hours if the birds fly maximum range. If they fly at minimum power, the time increases to about 130 hours.

Flight Speed

Most passerine birds fly at mean **air speeds** of 29.5 to 42.7 feet (9 to 13 m) per second. Stronger fliers, such as ducks, fly at air speeds of 55.8 to 68.9 feet (17 to 21 m) per second. Depending on the wind speed and direction, birds can adjust their air speeds to fly at more constant **ground speeds**—speeds with relation to the earth. During migration, according to radar surveillance by Bellrose (1967), Bruderer (1971), Tucker and Schmidt-Koenig (1971), Emlen (1974), and Schnell and Hellack (1979), birds appear to reduce their flight speed somewhat proportionately to the increase in favorable wind speed. Apparently they adjust their flight efforts in relation to the degree of wind assistance or resistance. Thus the ground speeds of migrants tend to remain fairly constant despite variations in wind speed, whereas the ground speeds of birds in the daily activity flights, as shown by Schnell (1965), are definitely influenced by winds.

There is much additional evidence, obtained by means other than radar, of the bird's ability for sustained speed during long distances of migration. Cochran et al. (1967), using radiotelemetry, tracked migrating *Hylocichla (Catharus)* thrushes nearly all night from Illinois northward into Michigan, Wisconsin, and Minnesota. Although they found considerable variation, most flights were at air speeds between 36.1 and 51.2 feet (11 to 15.6 m) per second. These were usually less than ground speeds and thus suggested that the birds were aided by favorable winds. One of the most remarkable records of a long, sustained flight is that of a banded Ruddy Turnstone *(Arenaria interpres)* released by Max C. Thompson at St. George Island, one of the Pribilofs in the Bering Sea, on August 27, 1965, and shot four days later, on August 31, at French Frigate Shoals in a steady bee-line flight. Its average speed was 575 miles (925 km) a day.

In the late fall long-distance migrating ducks commonly pass from breeding area to winter quarters in a short series of mass movements, each of which carries them many hundreds of miles in one continuous flight. They start each flight immediately after the passage of a cold front when temperature has dropped and the sky is clear, but they may overtake bad weather as they proceed. Sometimes, owing to the triggering effect of extremely low temperatures resulting from a strong flow of polar air, the mass movements are spectacular both in numbers of birds involved and distances covered. Bellrose (1957) documented one such migration in 1955 that moved with unusual rapidity from the Great Plains of Canada to the marshes of southern Louisiana. The exodus began from Canada on October 31; early on November 1 the flight was in full force through the Dakotas; and on November 2 the vanguard had reached northern Tennessee and Arkansas shortly after sunrise and Louisiana later in the day. Many thousands of ducks made the flight from Canada to southern Louisiana, a distance of 1,200 to 2,000 miles (1,931 to 3,218 km), in two days, or roughly 35 to 50 hours, at an average speed of 40 miles (64.3 km) per hour. No doubt some of the birds covered the distance without stopping, accomplishing their migration in one flight.

By making longer flights as they near their destinations, birds gradually speed up their migrations. Cooke (1904) provided evidence for this acceleration when he analyzed migration dates of

North American species, mostly passerines, approaching their northern nesting areas. "Sixteen species," he wrote, "maintain a daily average of 40 miles from southern Minnesota to southern Manitoba, and from this point about 12 species travel to Lake Athabasca at an average of 72 miles a day, 5 others to Great Slave Lake at 116 miles a day, and 5 more to Alaska at 150 miles a day."

From all these studies and reports several generalizations on the rate of migratory travel emerge. Strong winds affect ground speed of birds in their daily activity flights more than in migration. Birds make long, sustained flights, usually at increasingly higher speeds at higher altitudes. Spring migration proceeds at a greater rate with less time loss than fall migration. Larger birds, such as ducks, accomplish their migrations in a short series of a few mass movements, occasionally in one nonstop mass movement. Small land birds, however, tend to begin their migrations in many short flights, gradually building fat reserves for long, nonstop flights, thereby accelerating their migrations. Additional details on the theoretical aspects of bird flight during migration can be found in Pennycuick (1978), and Alerstam (1981).

Homing, Navigation, and Orientation

One of the most fascinating aspects of bird study is how birds find their way during homing and migration. Considerable attention has been directed to laboratory studies of direction finding in birds, as well as to field studies of migratory orientation and homing using visual means, surveillance and tracking radars, and radiotelemetry. Excellent reviews of avian homing, navigation, and orientation, in Keeton (1974, 1980, 1981), Emlen (1975), and Able (1980), and three symposium volumes in the last decade (Galler et al., 1972; Schmidt-Koenig and Keeton, 1978; Papi and Wallraff, 1982) attest to the amount of research being done in these areas.

Definitions

Before homing, navigation, and orientation can be discussed, the definitions of these and other related terms must be addressed as a general background. The terms are diverse and sometimes confusing. **Homing** has been used as a general term for the return of a bird to its "home" after some form of **displacement.** Home is usually the bird's birthplace or breeding site, but even returning to a familiar point along a bird's migration route following displacement has been considered a form of homing by some authors. Displacement is usually caused by a storm, wind drift, or the like; but in experiments, displacement is traditionally transportation by automobile or airplane.

Homing ability has been subdivided into three types by Griffin (1952): **piloting, compass orientation,** and **true navigation.** *Piloting* is the ability to orient to a goal by referring to familiar landmarks, but Griffin also includes random and systematic search for familiar landmarks in this category. In this form of homing the bird learns landmarks through experience and associates them with the location of home.

Compass orientation is the ability to orient in a given direction without reference to landmarks. If, for instance, a bird flew south of its nesting area, then it would return to it by flying north, even though it had flown well beyond its familiar area or into an area devoid of conspicuous guideposts. Compass orientation permits a bird to venture into areas that are unfamiliar without taking the time to learn the landmarks in the area.

True navigation is the ability to determine from a new and unfamiliar location the direction and distance of a goal. In other words, a bird must find out where it is in relation to where it should be or wants to be. This is regarded as the most complex of homing capabilities and has been conclusively demonstrated only in a small number of species. Once direction and distance of the goal are known, a bird must select the correct compass bearing, and once underway, maintain it. The process of finding and maintaining a preferred compass direction is of course compass orientation, and the compass can be any nonchanging reference system that provides accurate directional information.

What birds have what type of homing ability? This question has been hotly debated, and two theoretical papers that relate to the matter have reached different conclusions regarding the homing ability that is most "primitive" and hence most widespread among birds. Possibly because of Griffin's (1952) scheme of classifying homing abilities, Bellrose (1972) suggested that use of landmarks (piloting) was the most primitive means of orientation. He considered compass orientation to be a derived condition and bicoordinate, or true, navigation to be a very complex derived condition.

In their theoretical model of the phylogenetic and ontogenic development of bird orientation, Wiltschko and Wiltschko (1978) suggest that navigation may be phylogenetically older than migratory orientation. They postulate that compass orientation enables a sedentary bird, homing pigeon, or migratory bird in a nonmigratory state to become familiar with the distribution of landmarks and to integrate them directionally, thus forming a "mosaic map" for the orientation of movements within a home range. The bird applies the same mechanisms to factors such as gradients. This results in a bird's learning a grid of coordinates and through integration forming a "navigational map" that enables it to home from an unfamiliar area by extrapolation.

Wiltschko and Wiltschko suggest that a migratory bird uses genetically fixed information about the direction and distance of its migration (**vector navigation**—see Schmidt-Koenig, 1970). Vector navigation differs from the processes normally termed true navigation, because the goal is not a specific place but a more or less extensive area. The Wiltschkos expect "that the migratory direction is genetically fixed in relation to the compass system which was already developed within the home range before migration evolved, provided that it was not only for local use, but was sufficient to allow the determination of the migratory direction also in distant parts of the world." Although this may be true, a sound argument can be made that true navigation ability is more likely to occur naturally in migratory birds than in sedentary species.

Migratory birds have many occasions when they can be displaced naturally to unfamiliar areas during their travels. Sedentary or nonmigratory birds on the other hand have little occasion to travel far from their home (except possibly during one way dispersal) and consequently would have few or no occasions when they would be displaced naturally. One sedentary species, the Common Pigeon or Rock Dove *(Columba livia)*, has undergone considerable artificial selection by man to produce homing pigeons that probably use true navigation. It is so far the only nonmigratory bird species thoroughly investigated that can use compass orientation and show true navigation. In contrast, a number of migratory species have been shown to have the ability to use compass orientation (see reviews by Emlen, 1975; Able, 1980), and many of these species have also homed to familiar areas from distant, unfamiliar localities following artificial displacement (e.g., Laysan Albatross—*Diomedea immutabilis;* Manx Shearwater—*Puffinus puffinus;* Leach's Storm-Petrel—*Oceanodroma leucorhoa;* Northern Gannet—*Sula bassanus;* Sooty and Brown Noddy Terns—*Sterna fuscata* and *Anous stolidus;* Bank Swallow and Purple Martin—*Riparia riparia* and *Progne subis;* Bobolink—*Dolichonyx oryzivorus;* and Golden-crowned and White-crowned Sparrows —*Zonotrichia atricapilla* and *Z. leucophrys).* Some of these cases of homing were probably the result of random search until familiar landmarks were encountered, but some were likely cases of true navigation.

Unfortunately, all experimental investigations of homing suffer from an uncontrolled bias: they require that animals be displaced artificially. Although an individual animal can accumulate information during the artifical "outward journey" that may be used in homing, we know virtually nothing about natural "outward journeys" or displacement (i.e., how it occurs, when it occurs, or what birds do during natural displacement). Able (1980) correctly cautions that until the **home range** (familiar area) of an individual is known and the investigator has eliminated the possibility that the individual is in some way in direct contact with "home" through some sensory modality, true navigation should not be invoked in cases of homing, even rapid homing from great distances.

Types of Compass Navigation

Although the matter of true navigation is hotly debated, there is general agreement that migratory birds have the ability to orient using a compass. The more familiar orientational cues used by birds as simple compasses include the sun (Kramer, 1950; Kramer 1952; Moore, 1978); the stars (Sauer, 1957; Emlen, 1967a and b); and the earth's magnetic field (Merkel and Wiltschko, 1965; Emlen et al., 1976).

Sun The position of the sun in relation to the bird's biological clock (circadian rhythm) provides compass information, and the position of the setting sun can also be used by nocturnal migrants as a compass at about the time they begin their flights. Presumably this information is transferred to the stars, topographic features, or even to the wind direction for the maintenance of flight direction during the night. When nocturnal migrants are forced to fly during the daylight hours, and sometimes into a second night while crossing the Gulf of Mexico in spring, they have occasion to use both sun and star compasses. Two studies have demonstrated that nocturnal migrants can use time compensated, sun-compass orientation (Saint-Paul, 1953; Able and Dillon, 1977).

Stars The positions of the stars coupled with an accurate chronometer can tell people their latitude and longitude (the sun, of course, can be used as well), but no evidence exists to support this ability in birds. The evidence is overwhelming that the stars function only as a simple compass for birds. Time compensation is not required, because the pattern of stars in the night sky can provide directional information (in the Northern Hemisphere, for example, Polaris, the North Star, is always north). The use of the sun and star compasses depends on their visibility, and during periods of solid cloud cover, these compasses may have little utility.

Magnetic Field There is, however, increasing evidence that birds have the ability to determine compass directions using the earth's magnetic field. The polarity of the field is not important. In the Northern Hemisphere, north is the direction in which the magnetic and gravity vectors form the most acute angle, just as the dip of an unmarked magnetic needle in a compass indicates "north" north of the equator. On the equator no dip is present, and presumably the magnetic compass would give no information to birds. In the Southern Hemisphere the dip would be toward the south.

Magnetic storms as a function of sunspot and solar flare activity have affected the spread of flight directions in migratory birds at night, the homing performance of pigeons, and the orientation of Ring-billed Gull *(Larus delawarensis)* chicks under overcast. Deposits of magnetite (a ferromagnetic mineral) in the head and neck of homing pigeons and in the neck of White-crowned Sparrows suggest that the receptor for magnetic information might be in the region of the head, but no information on a specific receptor or how that receptor might work is known. Once again, speculation abounds.

The sun, stars, and magnetic compasses are not totally independent mechanisms. Evidence that the star compass is calibrated by the magnetic compass every few days has been provided by Wiltschko and Wiltschko (1975), but Emlen (1970) has shown that birds learn the star compass by viewing the rotation of the stellar pattern. A detailed discussion of these and related issues can be found in Able (1980).

Polarization Other familiar sources of compass information have been demonstrated or suggested for migratory birds, based on their discovery in laboratory studies of the homing pigeon. Pigeons can detect the polarization of light, and Able (1982b) has shown that White-throated Sparrows *(Zonotrichia albicollis)* respond to changes in the axis of skylight polarization during the period between sunset and darkness by changing the orientation of their subsequent migratory restlessness at night in orientation cages. This ability means that migrants can possibly continue using a derivative of the sun compass on partially overcast days, when the sun's disk is hidden from view but some blue sky remains, or after sunset and before sunrise when only a horizon glow from the sun is present. The position of the sun can be derived from the polarization pattern.

Laboratory studies have also shown that pigeons can detect barometric pressure, infrasound (frequencies below 1 Hertz down to 0.05 Hertz), ultraviolet light, and odors. In the latter case a substantial literature exists supporting an olfactory navigation hypothesis for pigeon homing, particularly studies of pigeon homing in Italy (see Papi, 1982). Recent studies have also shown that orientation abilities differ with respect to age and in some cases even sex. Future studies will have to take these factors into consideration in an effort to reduce some of the variance so inherent in orientation studies.

Field studies of migratory orientation have used direct visual means (e.g., moon-watching—Lowery, 1951; ceilometer observations—Gauthreaux, 1969; and vertical telescope during the day—Lowery and Newman, 1963) and radar (e.g., surveillance radar—Eastwood, 1967; Gauthreaux, 1970; Williams and Williams, 1980; and tracking radar studies of individual birds—Emlen, 1974; Emlen and Demong, 1978; Able, 1982a). These studies have generated an enormous literature on the orientation of birds during migration. Gauthreaux (1980b) has summarized much of the directional data for most of North America, and Bruderer (1980) has done the same for Europe. The directions of movements are strongly influenced by wind patterns and topography. In the Western Hemisphere during the spring migrants move toward the northwest, south of 25 degrees latitude under the influence of prevailing easterly winds. Between 25 and 30 degrees north latitude most spring migrants move northward, and above 30 degrees north latitude most migrants fly toward the northeast. On the eastern edge of the prairie, movements are oriented more northerly, and in southern Manitoba and Saskatchewan, most birds once again orient toward the northwest. In fall the direction of migration is toward the southeast from south-central Canada and most of the central United States. The movements in the eastern United States tend to be more southerly and even southwesterly. A distinct movement to the southeast directed over the western Atlantic Ocean is also present in the Northeast and in the Canadian Maritimes. Below 30 degrees north latitude, fall flights are biased toward the southwest by the prevailing easterly winds. Because of the prevailing westerly winds, migration along the Atlantic Coast can be spectacular in the fall, and occasionally large numbers of migrants can be seen coming ashore from over the ocean after presumably being drifted by the northwest winds behind a cold front the previous night. Similar flights may occur inland, suggesting that birds may correct for drift sustained during nocturnal migration by flying generally into the wind the next morning. These flights could be a form of homing to get back on course after a wind-induced deviation (see Pettingill, 1964; Gauthreaux, 1978b).

Radar studies have provided considerable descriptive information on the direction-finding behavior of migrant birds. Migrating birds can maintain straight tracks in opaque clouds, they can fly straight courses in the seasonally appropriate direction without seeing stars or the sun (but this ability may fail after several days of overcast), they frequently show *reversed* flights in response to sudden changes in weather and a sudden reversal in wind direction, they may respond to conspicuous landmarks by changing their flight directions, and in some cases they show a greater spread in flight directions during magnetic storms. Considerably more information of this type is needed if meaningful hypotheses about the mechanisms of direction-finding are to be formulated and tested in the field and in the laboratory in the future.

References

Able, K.P.

1970 A radar study of the altitude of nocturnal passerine migration. *Bird-Banding* 41:282–290.

1974 Environmental influences in the orientation of free-flying nocturnal bird migrants. *Animal Behaviour* 22:224–238.

1980 Mechanisms of orientation, navigation, and homing. In *Animal migration, orientation, and navigation*, ed. S.A. Gauthreaux, Jr. New York: Academic Press.

1981 Field studies of bird migration — a brief overview and some unanswered questions. *Continental Birdlife* 2:101–110.

1982a Field studies of avian nocturnal migratory orientation I. Interaction of sun, wind and stars as directional cues. *Animal Behaviour* 30:761–767.

1982b Skylight polarization patterns at dusk influence migratory orientation in birds. *Nature* 299:550–551.

Able, K.P., and Dillon, P.M.

1977 Sun compass orientation in a nocturnal migrant, the White-throated Sparrow. *Condor* 79:393–395.

Able, K.P., and Gauthreaux, S.A., Jr.

1975 Quantification of nocturnal passerine migration with a portable ceilometer. Condor 77:92–96.

Alerstam, T.

1976 *Bird migration in relation to wind and topography.* Thesis. Sweden: Univ. of Lund.

1981 The course and timing of bird migration. In *Animal migration*, ed. D.J. Aidley. Cambridge: Cambridge Univ. Press.

Alerstam, T., and Hogstedt, G.

1982 Bird migration and reproduction in relation to

habitats for survival and breeding. *Ornis Scandinavica* 13:25–37.

Avery, M.; Springer, P.F.; and Cassell, J.F.
1977 Weather influences on nocturnal bird mortality at a North Dakota tower. *Wilson Bull*. 89:291–299.

Bagg, A.M.
1967 Factors affecting the occurrence of the Eurasian Lapwing in eastern North America. *Living Bird* 6:87–121.

Baker, R.R.
1978 *The evolutionary ecology of animal migration*. New York: Holmes & Meier.

Bellrose, F.C.
1957 *A spectacular waterfowl migration through central North America*. Illinois Dept. Registration and Education, Nat. Hist. Surv. Div., Biol. Notes No. 36.
1964 Radar studies of waterfowl migration. *Trans. 29th N. Amer. Wildl. Conf.*, pp. 128–143.
1967 Radar in orientation research. *Proc. XIVth Internatl. Ornith. Congr.*, 1966, pp. 281–309.
1971 The distribution of nocturnal migrants in the air space. *Auk* 88:397–424.
1972 Possible steps in the evolutionary development of bird navigation. *Jour. Theor. Biol*. 6:76–117.
1980 *Ducks, geese and swans of North America*. 3rd ed. Harrisburg, Pennsylvania: Stackpole Books.

Bellrose, F.C., and Graber, R.R.
1963 A radar study of the flight directions of nocturnal migrants. *Proc. XIIIth Internatl. Ornith. Congr.*, 1962, pp. 362–389.

Berthold, P.
1975 Migration: Control and metabolic physiology. In *Avian biology,* vol. 5, eds. D.S. Farner and J.R. King. New York: Academic Press.

Blem, C.R.
1980 The energetics of migration. In *Animal migration, orientation, and navigation,* ed. S.A. Gauthreaux, Jr. New York: Academic Press.

Blokpoel, H.
1971a The M33C Track Radar (3–cm) as a tool to study height and density of bird migration. In *Studies of bird hazards to aircraft*. *Canadian Wildl. Serv. Rept. Ser. No. 14,* pp. 77–94.
1971b A preliminary study on height and density of nocturnal fall migration. In *Studies of bird hazards to Aircraft*. *Canadian Wildl. Serv. Rept. No. 14,* pp. 95–104.

Bock, C.E., and Lepthien, L.W.
1976 Synchronous eruptions of boreal seed-eating birds. *Amer. Nat*. 110:559–571.

Brewer, R., and Ellis, J.A.
1958 An analysis of migrating birds killed at a television tower in east-central Illinois, September 1955–May 1957. *Auk* 75:400–414.

Brewster, W.
1886 *Bird migration*. Mem. Nuttall Ornith. Club, No. 1.

Bruderer, B.
1971 Radarbeobachtungen über den Fruhlingszug im Schweizerischen Mittelland. *Ornith. Beob.* 68:89–158.
1980 Radar data on the orientation of migratory birds in Europe. *Proc. XVIIth Internatl. Ornith. Congr., 1978,* pp. 547–552.

Bruderer, B., and Steidinger, P.
1972 Methods of quantitative and qualitative analysis of bird migration with tracking radar. In *Animal orientation and navigation,* eds. R.E. Galler, K. Schmidt-Koenig, G.J. Jacobs, and R.E. Belleville. NASA SP-268. Washington, D.C.: U.S. Government Printing Office.

Buskirk, W.H.
1980 Influence of meterological patterns and Trans-Gulf migration on the calendars of latitudinal migrants. In *Migrant birds in the neotropics,* eds. A. Keast and E. Morton. Washington, D.C.: Smithsonian Institution Press.

Bykhovskii, B.E., ed.
1974 *Bird migration: Ecological and physiological factors*. New York: John Wiley & Sons.

Caldwell, L.D.; Odum, E.P.; and Marshall, S.G.
1963 Comparison of fat levels in migrating birds killed at central Michigan and a Florida Gulf Coast television tower. *Wilson Bull*. 75:428–434.

Cherry, J.D.
1982 Fat deposition and length of stopover of migrant White-crowned Sparrows. *Auk* 99:725–732.

Clarke, W.E.
1912 *Studies in bird migration*. 2 vols. London: Gurney & Jackson.

Cochran, W.W.; Montgomery, G.C.; and Graber, R.R.
1967 Migratory flights of *Hylocichla* thrushes in spring: A radiotelemetry study. Living Bird 6:213–225.

Cohen, D.
1967 Optimization of seasonal migratory behavior. *Amer. Nat*. 101:5–17.

Cook, W.W.
1904 Some new facts about the migration of birds. In *Yearbook of the United States Department of Agriculture—1903.*
1915 *Bird migration*. U.S. Dept. Agric. Bull. No. 185.

Coward, T.A.
1912 *The migration of birds*. London and New York: Cambridge Univ. Press.

Cox, G.W.
1968 The role of competition in the evolution of migration. *Evolution* 22:180–192.

Crawford, R.L.
1981 Bird casualties at a Leon County, Florida TV tower: A 25-year migration study. *Bull. Tall Timbers Res. Sta. No. 22.*

DeBenedictis, P.
1971 Wood warblers and vireos in California: The nature of the accidental. *Calif. Birds 2(4):111–128.*

Diamond, J.M.
1982 Mirror–image navigational errors in migrating birds. *Nature* 295:277–278.

Dorst, J.
1962 *The migrations of birds*. Boston: Houghton Mifflin.

Eastwood, E.
1967 *Radar ornithology.* London: Methuen.

Eastwood, E., and Rider, G.C.
1965 Some radar measurements of the altitude of bird flight. *Brit. Birds* 58:393–426.

Emlen, S.T.
1967a Migratory orientation in the Indigo Bunting, *Passerina cyanea,* part I: The evidence for use of celestial cues. *Auk* 84:309–342.
1967b Migratory orientation in the Indigo Bunting, *Passerina cyanea*, part II: Mechanism of celestial orientation. *Auk* 84:463–489.
1970 Celestial rotation: Its importance in the development of migratory orientation. *Science* 170:1198–1201.
1974 Problems in identifying bird species by radar signature analysis: Intra-specific variability. In *Proc. Conf. on the Biological Aspects of the Bird/Aircraft Collision Problem,* ed. S.A. Gauthreaux, Jr. Clemson, South Carolina: Clemson Univ.
1975 Migration: Orientation and navigation. In *Avian biology,* vol. 5, eds. D.S. Farner and J.R. King. New York: Academic Press.

Emlen, S.T., and Demong, N.J.
1978 Orientation strategies used by free-flying bird migrants: A radar tracking study. In *Animal migration, navigation, and homing,* Proc. Symp. Tubingen, West Germany, eds. K. Schmidt-Koenig and W.T. Keeton. New York: Springer-Verlag.

Emlen, S.T., and Emlen, J.T.
1966 A technique for recording migratory orientation of captive birds. *Auk* 83:361–367.

Emlen, S.T.; Wiltschko, W.; Demong, N.J.; Wiltschko, R.; and Bergman, S.
1976 Magnetic direction finding: Evidence for its use in migratory Indigo Buntings. *Science* 193:505–508.

Eyster, M.B.
1954 Quantitative measurement of the influence of photoperiod, temperature, and season on the activity of captive songbirds. *Ecological Monogr.* 24:1–28.

Farner, D.S.
1955 The annual stimulus for migration: Experimental and physiologic aspects. In *Recent studies in avian biology,* ed. A. Wolfson, Urbana: Univ. of Illinois Press.
1960 Metabolic adaptations in migration. *Pro. XIIth Internatl. Ornith. Congr.,* 1958, pp. 197–208.

Fretwell, S.D.
1980 Evolution of migration in relation to factors regulating bird numbers. In *Migrant birds of the neotropics*, eds. A. Keast and E. Morton. Washington, D.C.: Smithsonian Institution Press.

Galler, R.E.; Schmidt-Koenig, K.; Jacobs, G.J.; and Belleville, R.E., eds.
1972 *Animal orientation and navigation*. NASA SP-268. Washington, D.C.: U.S. Government Printing Office.

Gauthreaux, S.A., Jr.
1969 A portable ceilometer technique for studying low-level migration. *Bird-Banding* 40:309–320.
1970 Weather radar quantification of bird migration. *BioScience* 20:17–20.
1971 A radar and direct visual study of passerine spring migration in southern Louisiana. *Auk* 88:343–365.
1972 Behavioral responses of migrating birds to daylight and darkness: A radar and direct visual study. *Wilson Bull*. 84:136–148.
1978a The ecological significance of behavioral dominance. In *Perspectives in ethology,* vol. 3, eds. P.P.G. Bateson and P.H. Klopfer. New York: Plenum.
1978b The importance of daytime flights of nocturnal migrants: Redetermined migration following displacement. In *Animal migration, navigation, and homing*, Proc. Symp. Tubingen, West Germany, eds. K. Schmidt-Koenig and W.T. Keeton. New York: Springer-Verlag.
1978c Migratory behavior and flight patterns. In *Impacts of transmission lines on birds in flight*. Oak Ridge Associated Universities Publication No. 142, Oak Ridge, Tennessee.
1980a *Direct visual and radar methods for the detection, quantification, and prediction of bird migration*. Special Publication No. 2. Clemson Univ.: Dept. of Zool.

1980b The influence of global climatological factors on the evolution of bird migratory pathways. *Proc. XVIIth Internatl. Ornith. Congr., 1978,* pp. 517–525.

1980c The influences of long-term and short-term climatic changes on the dispersal and migration of organisms. In *Animal migration, orientation, and navigation,* ed. S.A. Gauthreaux, Jr. New York: Academic Press.

1982 The ecology and evolution of avian migration systems. In *Avian biology,* vol. 6, eds. D.S. Farner, J.R. King, and K.C. Parkes. New York: Academic Press.

Gauthreaux, S.A., Jr., and Able, K.P.
1970 Wind and the direction of nocturnal songbird migration. *Nature* 228:476–477.

Gauthreaux, S.A., Jr., and LeGrand, H.E., Jr.
1975 The changing seasons: Spring migration 1975. *Amer. Birds* 29:820–826.

Gehring, W.
1963 Radar- und Feldbeobachtungen über den Verlauf des Vogelzuges im Schweizerischen Mittelland: Der Tagzug im Herbst (1957–1961). *Ornith. Beob.* 60:35–68.

Graber, R.R.
1968 Nocturnal migration in Illinois: Different points of view. *Wilson Bull.* 80:36–71.

Greenberg, R.
1980 Demographic aspects of long-distance migration. In *Migrant birds of the neotropics,* eds. A. Keast and E. Morton. Washington, D.C.: Smithsonian Institution Press.

Greenwood, P.J.
1980 Mating systems, philopatry and dispersal in birds and mammals. *Animal Behaviour* 28:1140–1162.

Greenwood, P.J., and Harvey, P.H.
1982 The natal and breeding dispersal of birds. *Annu. Rev. Ecol. Syst.* 13:1–21.

Griffin, D.R.
1952 Bird navigation. *Biol. Rev. Cambridge Phil. Soc.* 27:359–400.

1964 *Bird migration.* Garden City, New York: Doubleday. (Reprinted in 1974 by Dover Publications, New York.)

Gwinner, E.
1975 Circadian and circannual rhythms in birds. In *Avian biology,* vol. 5, eds. D.S. Farner and J.R. King. New York: Academic Press.

1977 Circadian and circannual rhythms in birds. *Annu. Rev. Ecol. Syst.* 8:381–405.

Harper, W.G.
1958 Detection of birds by centrimetric radar: A cause of radar "angels." *Proc. Royal Soc. London B* 149:484–502.

Helms, C.W., and Drury, W.H., Jr.
1960 Winter and migratory weight and fat field studies on some North American buntings. *Bird-Banding* 31:1–40.

Herrera, C.M.
1978 On the breeding distribution pattern of European migrant birds: MacArthur's theme reexamined. *Auk* 95:496–509.

Hochbaum, H.A.
1955 *Travels and traditions of waterfowl.* Minneapolis: Univ. Minnesota Press.

Karr, J.R.
1980 Patterns in the migration systems between the north temperate zone and the tropics. In *Migrant birds in the neotropics,* eds. A. Keast and E. Morton. Washington, D.C.: Smithsonian Institution Press.

Keast, A.
1980 Synthesis: Ecological basis and evolution of the Nearctic-Neotropical bird migration system. In *Migrant birds in the neotropics,* eds. A. Keast and E. Morton. Washington, D.C.: Smithsonian Institution Press.

Keeton, W.T.
1974 The orientational and navigational basis of homing in birds. *Advances in the Study of Behavior* 5:47–132.

1980 Avian orientation and navigation: New developments in an old mystery. *Proc. XVIIth Internatl. Ornith. Congr., 1978,* pp. 137–157.

1981 The orientation and navigation in birds. In *Animal migration,* ed. D.J. Aidley. Cambridge: Cambridge Univ. Press.

Kramer, G.
1950 Weitere Analyse der Faktoren, welche die Zugaktivität des gekäfigten Vogels Orientieren. *Naturwissenschaften* 37:377–378.

1952 Experiments on bird orientation. *Ibis* 94:265–285.

Kumari, E.
1975 *Lindude ränne.* Tallin USSR: Kirjastus Valgus. (English summary.)

Kushman, J.A.; Bass, O.L., Jr.; and McEwan, L.C.
1982 Wintering waterfowl in the Everglades estuaries. *Amer. Birds* 36:815–819.

Lack, D.
1954 *The natural regulation of animal numbers.* London and New York: Oxford Univ. Press (Clarendon).

1960 Migration across the North Sea studied by radar: Part 2, the spring departure 1956–1959. *Ibis* 102:26–57.

Lincoln, F.C.
1950 *Migration of birds.* U.S. Dept. Interior, Fish

and Wildlife Service Circ. 16. (Commercial edition published in 1952 by Doubleday, Garden City, New York.)

Lowery, G.H., Jr.

1951 A quantitative study of the nocturnal migration of birds. *Univ. Kansas Publ. Mus. Nat. Hist.* 3:361–472.

Lowery, G.H., Jr., and Newman, R.J.

1955 Direct visual studies of nocturnal bird migration. In *Recent studies in avian biology,* ed. A. Wolfson, Urbana: Univ. of Illinois Press.

1963 *Studying bird migration with a telescope.* Spec. Publ. Mus. Zool. Baton Rouge: Louisiana State Univ.

1966 A continentwide view of bird migration on four nights in October. *Auk* 83:547–586.

MacArthur, R.H.

1959 On the breeding distribution pattern of North American migrant birds. *Auk* 76:318–325.

McAtee, W.L.; Burleigh, T.D.; Lowrey, G.H., Jr.; and Stoddard, H.L.

1944 Eastward migration through the Gulf States. *Wilson Bull.* 56:152–160.

Meier, A.H., and Fivizzani, A.J.

1980 Physiology of migration. In *Animal migration, orientation, and navigation,* ed. S.A. Gauthreaux, Jr. New York: Academic Press.

Merkel, F.W., and Wiltschko, W.

1965 Magnetismus und Richtungsfinden zugunruhiger Rotkehlchen *(Erithacus rubecula). Vogelwarte* 23:71–77.

Moore, F.R.

1976 The dynamics of seasonal distribution of Great Lakes Herring Gulls. *Bird-Banding* 47:141–159.

1978 Sunset and the orientation of a nocturnal migrant bird. *Nature* 274:154–156.

Murray, B.G., Jr.

1966 Migration of age sex classes of passerines on the Atlantic coast in autumn. *Auk* 83:352–360.

Nisbet, I.C.T.

1959 Calculations of flight directions of birds observed crossing the face of the moon. *Wilson Bull.* 71:237–243.

1961 Studying migration by moon-watching. *Bird-Migration* 2:38–42.

1963 Measurements with radar of the height of nocturnal migration over Cape Cod, Massachusetts. *Bird-Banding* 34:57–67.

1970 Autumn migration of the Blackpoll Warbler: Evidence for long flight provided by regional survey. *Bird-Banding* 41:207–240.

O'Connor, R.J.

1981 Comparisons between migrant and non-migrant birds in Britain. In *Animal migration,* ed. D.J. Aidley. Cambridge: Cambridge Univ. Press.

Papi, F.

1982 Olfaction and homing in pigeons: Ten years of experiments. In *Avian navigation.* International Symposium on Avian Navigation at Tirrenia (Pisa), September 11–14, 1981, eds. F. Papi and H.G. Wallraff. New York: Springer-Verlag.

Papi F., and Wallraff, H.G., eds.

1982 *Avian navigation.* International Symposium on Avian Navigation at Tirrenia (Pisa), September 11–14, 1981. New York: Springer-Verlag.

Pennycuick, C.J.

1978 Fifteen testable predictions about bird flight. *Oikos* 30:165–176.

Pettingill, O.S., Jr.

1964 Spring migration at Point Pelee. *Audubon Mag.* 66:78–80.

1983 Winter of the Bobolink. *Audubon* 85(1):102–109.

Pinkowski, B.C., and Bajorek, R.A.

1976 Vernal migration patterns of certain avian species in southern Michigan. *Jack-Pine Warbler* 54:62–68.

Preston, F.W.

1966 The mathematical representation of migration. *Ecology* 47:375–392.

Richardson, W.J.

1976 Autumn migration over Puerto Rico and the western Atlantic: A radar study. *Ibis* 118:309–332.

1978 Timing and amount of bird migration in relation to weather: A review. *Oikos* 30:224–272.

1979 Southeastward shorebird migration over Nova Scotia and New Brunswick in autumn: A radar study. *Canadian Jour. Zool.* 57:107–124.

Robbins, C.S.; Bruun, B.; and Zim, H.S.

1983 *Birds of North America: A guide to field identification.* New York: Golden Press.

Roberson, D.

1980 *Rare birds of the West Coast. Patterns and Theories on Vagrancy,* pp. xxvi–xxxii. Pacific Grove, California: Woodcock Publications.

Rudebeck, G.

1950 Studies on bird migration. *Vår Fågelvärd Suppl.* 1:1–148.

Saint-Paul, U. von

1953 Nachweis der Sonnenorientierung bei nächtlich ziehenden Vögeln. *Behaviour* 6:1–7.

Sauer, F.

1957 Die Sternorientierung nächtlich ziehender Grasmücken *(Sylvia atricapilla, borin* und *curruca). Z. Tierpsychol.* 14:29–70.

Saunders, A.A.
1959 Forty years of spring migration in southern Connecticut. *Wilson Bull*. 71:208–219.

Schmidt-Koenig, K.
1970 Ein Versuch, theoretisch mögliche Navigationsverfahren von Vögeln zu klassifizieren und relevante sinnesphysiologische Probleme zu umreissen. *Verh. Dtsch. Zool. Ges.* 64:243–245.

Schmidt-Koenig, K., and Keeton, W.T. eds.
1978 *Animal migration, navigation, and homing*. Proc. Symp. Tubingen, West Germany. New York: Springer-Verlag.

Schnell, G.D.
1965 Recording the flight-speed of birds by Doppler radar. *Living Bird* 4:79–87.

Schnell, G.D., and Hellack, J.J.
1979 Bird flight speeds in nature: Optimized or a compromise? *Amer. Nat*. 113:53–66.

Schüz, E.
1971 *Grundriss der Vogelzugskunde*. Berlin: Paul Parey.

Scott, W.E.D.
1881 Migration of birds at night. *Bull. Nuttall Ornith. Club,* 6:188.

Slagsvold, T.
1976 Arrival of birds from spring migration in relation to vegetational development. *Norwegian Jour. Zool*. 24:161–173.

Stevenson, H.M.
1957 The relative magnitude of the trans-Gulf and circum-Gulf spring migrations. *Wilson Bull*. 69:39–77.

Sutter, E.
1957a Radar als Hilfsmittel der Vogelzugforschung. *Ornith. Beob*. 54:70–96
1957b Radar-Beobachtungen über den Verlauf des nächtlichen Vogelzuges. *Rev. Suisse Zool*. 64:294 303.

Terborgh, J.W.
1980 The conservation status of neotropical migrants: Present and future. In *Migrant birds in the neotropics,* eds. A. Keast and E. Morton. Washington, D.C.: Smithsonian Institution Press.

Terborgh, J.W., and Faaborg, J.R.
1980 Factors affecting the distribution and abundance of North American migrants in the eastern Caribbean region. In *Migrant birds in the neotropics,* eds. A. Keast and E. Morton. Washington, D.C.: Smithsonian Institution Press.

Thompson, A.L.
1926 *Problems of bird migration*. London: Witherby.

Tinbergen, L.
1949 *Vogels onder Weg*. Amsterdam: Scheltema and Holkema N.V.

Tordoff, H.B., and Mengel, R.M.
1956 Studies of birds killed in nocturnal migration. *Univ. Kansas Publ. Mus. Nat. Hist*. 10:1–44.

Tucker, V.A.
1975 Flight energetics. In *Avian physiology,* ed. M. Peaker. London: Academic Press.

Tucker, V.A., and Schmidt-Koenig, K.
1971 Flight speeds of birds in relation to energetics and wind directions. *Auk* 88:97–107.

von Haartman, L.
1968 The evolution of resident versus migratory habit in birds. Some considerations. *Ornis Fennica* 45:1–7.

Ward, P.
1963 Lipid levels in birds preparing to cross the Sahara. *Ibis* 105:109–111.

Weir, R.D.
1977 *Annotated bibliography of bird kills at manmade obstacles: A review of the state of the art and solutions*. Dept. Fish Env., Env. Manag. Serv., Canadian Wildl. Serv.

Wetmore, A.
1926 *The migrations of birds*. Cambridge, Massachusetts: Harvard Univ. Press.

Weydemeyer, W.
1973 The spring migration pattern at Fortine, Montana. *Condor* 75:400–413.

Williams, T.C., and Williams, J.M.
1980 A Peterson's guide to radar ornithology? *Amer. Birds* 34:738–741.

Williams, T.C.; Williams, J.M.; Ireland, L.C.; and Teal, J.M.
1977 Autumnal bird migration over the western north Atlantic Ocean. *Amer. Birds* 31:251–267.

Willson, M.F.
1976 The breeding distribution of North American migrant birds: A critique of MacArthur (1959). *Wilson Bull*. 88:582–587.

Wiltschko, W., and Wiltschko, R.
1975 The interaction of stars and magnetic field in the orientation system of night migrating birds. I. Autumn experiments with European warblers (gen. *Sylvia*). *Z. Tierpsychol*. 37:337–355.
1978 A theoretical model for migratory orientation and homing in birds. *Oikos* 30:177–187.

Territory

Any area defended by a bird against individuals of its own species is **territory.** Most bird species show at least some type of territorialism.

Classification of Territory

Territory may be roughly classified into two main categories—**breeding territory** and **nonbreeding territory.** Each category includes several types.

I. **Breeding Territory**

A. **Mating, nesting, and feeding area for adults and young.** The commonest type of territory and characteristic of most passerine birds. Once a pair of birds establishes a territory, they customarily remain on it until the young are independent. References: Best (1977); Darley et al. (1977); Erickson (1938); Gullion (1953); Hickey (1940); Kendeigh (1941); MacQueen (1950); Michener and Michener (1935); Nolan (1978); Potter (1972); Southern (1958); Zimmerman (1971).

B. **Mating and nesting (but not feeding) area.** Characteristic of a few species. Closely related to and sometimes not easily distinguished from Type A. References: Drum (1939); MacQueen (1950); Nero (1956); Venables and Lack (1934).

C. **Mating area only.** A "court" or "singing ground" apart from the nest, maintained by the male in certain species. "Arenas," "tournament grounds," or "leks," where two or more males of a particular species gather for mating purposes, are considered breeding territories of this type even though the males do not defend the areas in the strict sense. The nesting area, maintained by the female only, is not considered breeding territory. References: Alison (1975); Hochbaum (1981); Pettingill (1936); Pitelka (1942); D.W. Snow (1968).

D. **Restricted mating and nesting area.** Found in colonial species or in a few solitary nesting species. Different from Type B in that the territory is in the immediate vicinity of the nest. References: C.R. Brown (1979); Emlen (1954); Kuerzi (1941); Tuck (1960).

II. **Nonbreeding Territory**

A. **Feeding territories.** Feeding areas outside the breeding territory but nevertheless defended. References: Lyon (1976); Michael (1935); Pitelka (1942).

B. **Winter territories.** Areas defended through the winter months, particularly by permanent-resident birds that may or may not use the same territories during the breeding season. References: Erickson (1938); Kilham (1958); Michener and Michener (1935); Schwartz (1964).

C. **Roosting territories.** Specific areas used for night roosting. Evidence suggests that certain species defend such areas. Reference: Rankin and Rankin (1940).

Not all species show the types of territorialism outlined above. Some show combinations of two or more types; a few show no evidence of any type. Furthermore, a species may show a **strong** territorialism, or a **weak** territorialism, or a species may vary as to its territorialism under different ecological conditions.

For a discussion of the different types of territory, consult the papers by Mayr (1935) and Nice (1941) from which these classifications have been adapted. The following information concerns breeding territory only.

Establishment and Maintenance of Territory

The male of the species usually establishes a breeding territory, advertises it, and defends it against other males of the same species. The female may or may not participate in territorial defense; if she does participate, her role is generally less active and is usually directed toward other females of the species. The primary purpose of the defense is the territory itself, not the sex-partner, nor the nest and young. Competition for territory is theoretically intraspecific, not interspecific, although interspecific competition sometimes occurs between closely related species. See Gorton (1977) for a good example.

The methods by which males establish their territories involve agonistic behaviors—threat displays, physical encounters, appeasement displays, pursuit-flying, and singing or other vocalizations. All such agonistic activities continue intensively while the females are building nests and laying eggs. Generally older males show more aggression than younger and consequently succeed in establishing their territories in areas with more optimal habitat.

How the territories are maintained after the eggs are laid depends on the particular species concerned, the type of territory, and population density. As a rule, in passerine species with Breeding Territory Type A, any two neighboring males, once their mates are incubating full clutches of eggs, tend to moderate their aggressive behavior. Encounters between them become briefer and less frequent. More often when they meet at the common boundary, they jockey into position for encounters without going through with them. After the eggs in the nest have hatched, the males increasingly orient their attention toward the defense of their mates and young with the result that territory as a defended area begins to break down.

It is not unusual for males maintaining Breeding Territory Type A, even as nesting is under way, to leave their territories and disappear out of sight and hearing. In the Prairie Warbler *(Dendroica discolor)* Nolan (1978) called these departures "explorations"; they "covered considerable distances, lasted for minutes or hours rather than seconds, frequently appeared stealthy, and rarely led to observed encounters."

Some sedentary (nonmigratory) species maintain all-purpose, permanent territories, living on them and defending them the year-round. Permanent and monogamous pairs of Florida Scrub Jays *(Aphelocoma c. coerulescens)* occupy the same territories year after year for the life of the males (Woolfenden, 1973). The Acorn Woodpecker *(Melanerpes formicivorus)* in California lives from one year to the next in groups on a territory defended from conspecifics of the other groups. (MacRoberts and MacRoberts, 1976).

Pairs of migratory birds may often return to the same territory. Gray Catbirds *(Dumetella carolinensis)* in Ontario reused their territories if they were successful in the preceding season (Darley et al., 1977). Age was not a factor in their return although males showed stronger fidelity. Colonially nesting birds, primarily those maintaining Breeding Territory Type D, are notably tenaceous to colony sites, though not necessarily to the exact position of territories. Of the 436 Ring-billed Gulls *(Larus delawarensis)* captured and banded as breeding adults by Southern (1977) in a Great Lakes colony, 90 percent returned from previous nesting.

Size of Territory

A combination of factors dictates the size of a breeding bird's territory: (1) The size of the species—the larger it is, the larger the territory it requires. (2) Density of the species' population—the greater the population is, the smaller each individual territory. (3) Behavior of the species—whether it is strongly or mildly aggressive; whether it nests in colonies or solitarily. (4) Available habitat—whether it is extensive or restricted. (5) Food supply—whether it is ample or sparse.

Extremes in the size of territory are demonstrated on the one hand by the Bald Eagle *(Haliaeetus leucocephalus)* and on the other by the murres (*Uria* spp.). The eagle maintains a territory of about one square mile (2.59 sq. km) (Broley, 1947); the murre, nesting in a colony on a rocky ledge overlooking the sea, has an average territory of about one square foot (0.09 m^2), so small that occupant birds sometimes touch one another (Tuck, 1960). Territories of many other colonial sea birds, such as Gentoo Penguins *(Pygoscelis papua)*, are just large enough to prevent the incubating birds from reaching one another (Pettingill, 1975).

Occasionally, species of similar size in similar habitats may have vastly different space requirements. For example, 13 pairs of Least Flycatchers *(Empidonax minimus)* in a Michigan woods had territories averaging 0.18 acre (0.07 ha) (MacQueen, 1950), yet 15 pairs of Black-capped Chickadees *(Parus atricapillus)* in a New York woods had territories averaging 13.2 acres (5.34 ha) (Odum, 1941). Territory in a forest may involve only a particular vertical section. Thus the territory of a Red-eyed Vireo *(Vireo olivaceus)*, even though it averages 1.7 acres (0.69 ha) of forest, actually includes just the upper canopy down to the edge of the lower (Southern, 1958).

Functions of Territory

Breeding territory may assure essential cover, nesting materials, and food for developing young; protect the nest, sex-partner, and young against the despotism of other males; serve as a means of instigating the sexual bond by attracting the female through the singing of the male; and facilitate the remating of sex-partners from the previous breeding season.

It should be realized that all the functions, attributed to territory, are based largely on speculation. According to the consensus of most authorities, different bird species in their evolution become adapted to particular habitats and within their respective habitats, each species through its aggressive behavior spaces itself out (Tinbergen, 1957). This control of population density by avoiding crowding allows for survival factors such as a reserved food supply. On a 40-acre (16.19 ha) tract of coniferous forest in northern Maine, Stewart and Aldrich (1951) killed 81 percent of the territorial passerine species, mostly paruline warblers, and during the succeeding days from June 15 to July 8 kept the numbers reduced to about that level. When they were through, they had eliminated twice as many males as were present when the experiment began. The same procedure was repeated the next season in the same area (Hensley and Cope, 1951) with similar results. Both experiments show that the potential for crowding exists in an area and suggest that only the agonistic activities of territorial males keep an area from being overpopulated. Lack (1954, 1965) dissented from this view, believing that males space themselves out by *avoiding* occupied areas rather than by aggressive behavior. He also believed that food supply is not a factor in maintaining territory, since in his opinion, few territorial birds feed and gather food for their young solely on their own territories.

Tinbergen (1957) stressed the point that once a bird establishes its territory it becomes conditioned or closely attached to it with many resulting advantages besides reserving an adequate food supply. Close attachment, for example, enables the bird to become intimately acquainted with suitable cover for ready escape from predators, with sources of nest material, with places for bathing or dusting—all certain to enhance its survival.

For more information on the subject of territory in a variety of species, consult the July number of *The Ibis* for 1956 (Number 3, Volume 98) in which there are 18 papers on territory in different species of Old and New World birds, mainly by British and American authors.

Home Range

Home range is the total area that a bird habitually occupies. It may be the same as territory if the bird defends the whole area. It may be territory and the area outside that the bird frequents for food, water, bathing, dusting, etc. Or it may contain no territory—i.e., in cases where territory has broken down after the breeding season.

The home range of species with Breeding Territory Types B and D comprises vastly more area outside the territories than within, whereas the home range of species with Breeding Territory Type A is little larger than the territory itself. Any home range has indefinite boundaries that can be determined only by observing the extent and frequency of the individual bird's movements.

For some studies on home range, read Balda and Bateman (1972), Dixon (1937), and MacDonald (1968).

Selected Studies

Find several adjacent nests of one species in a given area and determine the associated territories. If possible, find the nests early in the breeding cycle—preferably at the time of nest-building or during the egg-laying period—when the birds defend their territories vigorously, making them more evident. Prepare a rough map of the area; indicate the topographical features, distribution of vegetational types, conspicuous landmarks (roads, fences, houses, large tree stubs, boulders, etc.), and the location of each nest.

Procedure Follow the activities of each nesting male, marking on the map the points where singing, threat displays, pursuit-flying, and fighting take place. In determining the extent of the male's territory, play tape recordings of another singing male and also affix in full view either a stuffed specimen or a realistic model of a male in full breeding plumage. Presumably the points where the occupant male fails to show aggression toward the recordings or visual representation of another male will mark the boundary of his territory. (For other methods of determining territory, see the paper by Odum and Kuenzler, 1955.) Follow the activities of the female, looking for any evidence of intraspecific competition. If possible, capture both sexes and mark them for individual identification. (See Appendix A for methods of capturing and marking.)

Make a series of observations throughout the nesting cycle, taking full notes on all territorial activities, and recording the date, length, and time of day of each activity. After gaining experience in recognizing territories already established, attempt to find territories being established.

Presentation of Results The marked areas on the map will show, at least roughly, the extent of each territory. Now prepare a finished map similar in content to the first one, but show only the boundary of each territory by drawing a line through the extreme points where the males appeared. Prepare a paper as directed in Appendix B based on a careful analysis of notes taken. Give attention to some of the following important problems:

1. Size of territory estimated in square feet or acres.

2. Physical and biological characteristics of territory. Illustrate with at least one photograph.

3. Establishment and defense of territory (with emphasis on the methods employed and the role of the sexes).

4. Location of nest (or nests, if the species is polygamous) in the territory and noteworthy relationships between nest location and territorial activity.

5. Length of time during the breeding season that the bird maintains territory.

6. Changes in extent of territory and in behavior of the adults during second and third nestings of a season.

7. Maintenance of territory at times other than the breeding season.

When working on territiories of the same species in the same area in subsequent years, attempt to find out whether or not the same individuals reoccupy the territories.

References

Alison, R.M.
1975 Breeding biology and behavior of the Oldsquaw *(Clangula hyemalis L.). Amer. Ornith. Union, Ornith. Monogr.* No. 18 (Breeding Territory Type C.)

Balda, R.P., and Bateman, G.C.
1972 The breeding biology of the Pinon Jay. *Living Bird* 11:5–42. [In Arizona a permanent flock of 250 Pinyon Jays maintained a home range of 8 square miles (2.59 km^2) and within it their tra-

ditional breeding area of 230 acres (93.1 hectares).]

Best, L.B.
1977 Territory quality and mating success in the Field Sparrow *(Spizella pusilla). Condor* 79:192–204. (Breeding Territory Type A.)

Broley, C.L.
1947 Migration and nesting of Florida Bald Eagles. *Wilson Bull*. 59:3–20.

Brown, C.R.
1979 Territoriality in the Purple Martin. *Wilson Bull*. 91:583–591. (Breeding Territory Type D.)

Brown, J.L.
1969 Territorial behavior and population regulation in birds: A review and re-evaluation. *Wilson Bull*. 81:293–329. (The hypothesis that territorial behavior has limiting effects on population densities and prevents over population is rejected.)

Burger, J.
1980 Territory size differences in relation to reproductive stage and type of intruder in Herring Gulls *(Larus argentatus). Auk* 97:733–741. (Three defended areas were discernible for each pair: a primary territory, a secondary territory, and a "unique" territory.)

Darley, J.A.; Scott, D.M.; and Taylor, N.K.
1977 Effects of age, sex, and breeding success on site fidelity of Gray Catbirds. *Bird-Banding* 48:145–151. (Breeding Territory Type A.)

Dixon, J.B.
1937 The Golden Eagle in San Diego County, California. *Condor* 39:49–56. (The average area—i.e., home ranges—to support a pair of Golden Eagles comprises close to 36 square miles (93.24 km^2).

Drum, M.
1939 Territorial studies on the Eastern Goldfinch. *Wilson Bull*. 51:69–77. (Breeding Territory Type B.)

Emlen, J.T., Jr.
1954 Territory, nest building, and pair formation in the Cliff Swallow. *Auk* 71:16–35. (Breeding Territory Type D.)

Erickson, M.A.
1938 Territory, annual cycle, and numbers in a population of Wren-tits *(Chamaea fasciata). Univ. Calif. Publ. in Zool*. 42:247–334. (Breeding Territory Type A and Nonbreeding Territory Type B.)

Gorton, R.E., Jr.
1977 Territorial interactions in sympatric Song Sparrow and Bewick's Wren populations. *Auk* 94:701–708.

Gullion, G.W.
1953 Territorial behavior of the American Coot. *Condor* 55:169–186. (Breeding Territory Type A.)

Hensley, M.M., and Cope, J.B.
1951 Further data on removal and repopulation of the breeding birds in a spruce-fir forest community. *Auk* 68:483–493.

Hickey, J.J.
1940 Territorial aspects of the American Redstart. *Auk* 57:255–256. (Breeding Territory Type A.)

Hinde, R.A.
1956 The biological significance of the territories of birds. *Ibis* 98:340–369.

Hochbaum, H.A.
1981 *The Canvasback on a prairie marsh*. 3rd ed. Lincoln: Univ. of Nebraska Press. (Breeding Territory Type C. Chapter 5, pages 56–87, contains an excellent account of territory and territorial problems in ducks.)

Howard, H.E.
1948 *Territory in bird life*. London: Collins. (A republication of Howard's 1920 classic work, with an introduction by Julian Huxley and James Fisher.)

Kendeigh, S.C.
1941 Territorial and mating behavior of the House Wren. *Illinois Biol. Monogr.*, 18:1–20. (Breeding Territory Type A.)

Kilham, L.
1958 Territorial behavior of wintering Red-headed Woodpeckers. *Wilson Bull*. 70:347–358.

Kuerzi, R.G.
1941 Life history studies of the Tree Swallow. *Proc. Linnaean Soc. New York* Nos. 52–53: 1–52. (Breeding Territory Type D.)

Lack, D.
1954 *The natural regulation of animal numbers*. London: Oxford Univ. Press.
1965 *The life of the Robin*. 4th ed. London: H.F. and G. Witherby. (Concerns *Erithacus rubecula;* Chapter 11 deals with the significance of territory.)

Lyon, D.L.
1976 A montane hummingbird territorial system in Oaxaca, Mexico. *Wilson Bull*. 88:280–299. (Feeding territories involving six species of hummingbirds.)

MacDonald, S.D.
1968 The courtship and territorial behavior of Franklin's race of the Spruce Grouse. *Living Bird*

7:5–25. (The male's home range "is too large to have distinct boundaries" and "firm territorial lines are established only in areas of interaction with an adjacent male.")

MacQueen, P.M.
1950 Territory and song in the Least Flycatcher. *Wilson Bull.* 62:194–205. (Breeding Territory Types A and B.)

MacRoberts, M.H., and MacRoberts, B.R.
1976 Social organization and behavior of the Acorn Woodpecker in central coastal California. *Amer. Ornith. Union, Ornith. Monogr.* No. 21.

Mayr, E.
1935 Bernard Altum and the territory theory. *Proc. innaean Soc. New York,* Nos. 45–46: 24–38.

Michael, C.W.
1935 Feeding habits of the Black-bellied Plover in winter. *Condor* 37:169. (Nonbreeding Territory Type A.)

Michener, H., and Michener, J.R.
1935 Mockingbirds, their territories and individualities. *Condor* 37:97–140. (Breeding Territory Type A and Nonbreeding Territory Type B.)

Nero, R.W.
1956 A behavior study of the Red-winged Blackbird. II. Territoriality. *Wilson Bull.* 68:129–150. (Breeding Territory Type B.)

Nice, M.M.
1941 The role of territory in bird life. *Amer. Midland Nat.* 26:441–487.

Nolan, V., Jr.
1978 The ecology and behavior of the Prairie Warbler *Dendroica discolor. Amer. Ornith. Union, Ornith. Monogr.* No. 26. (Breeding Territory Type A. Recommended reading for any student of territoriality.)

Odum, E.P.
1941 Annual cycle of the Black-capped Chickadee-1. *Auk* 58:314–333.

Odum, E.P., and Kuenzler, E.J.
1955 Measurement of territory and home range size in birds. *Auk* 72:128–137. (The authors make a distinction between *maximum* territory (defended area) or home range (in case the area is not defended) and *utilized* territory; and they suggest two new methods of measuring and expressing maximum territory or home range.)

Orians, G.H. and Willson, M.F.
1964 Interspecific territories of birds. *Ecology* 45:736–745. (An important paper on the subject.)

Penny, R.L.
1968 Territorial and social behavior in the Adélie Penguin. *American Geophysical Union, Antarctic Res. Ser.* 12:83–131. (An excellent, detailed study of territorial behavior in a colonial nesting species.)

Pettingill, O.S., Jr.
1936 The American Woodcock *Philohela minor* (Gmelin). *Mem. Boston Soc. Nat. Hist.* 9:167–391. (Breeding Territory Type C.)
1975 *Another penguin summer.* New York: Charles Scribner's Sons.

Pitelka, F.A.
1942 Territoriality and related problems in North American hummingbirds. *Condor* 44:189–204. (Breeding Territory Type C, but modified to include feeding area. The author suggests the possibility that Nonbreeding Territory Type A is also maintained.)

Potter, P.E.
1972 Territorial behavior in Savannah Sparrows in southeastern Michigan. *Wilson Bull.* 84:48–59. (Breeding Territory Type A.)

Rankin, M.N., and Rankin, D.H.
1940 Additional notes on the roosting habits of the Tree-creeper. *Brit. Birds* 34:46–60. (Nonbreeding Territory Type C.)

Schwartz, P.
1964 The Northern Waterthrush in Venezuela. *Living Bird* 3:169–184. (Nonbreeding Territory Type B.)

Snow, B.K.
1974 The Plumbeous Heron of the Galapagos. *Living Bird* 13:51–72. (A sedentary species forming permanent pair-bonds. Each pair member defends a separate section of the foreshore as a feeding territory, which may also include the nest site.)

Snow, D.W.
1968 The singing assemblies of Little Hermits. *Living Bird* 7:47–55. (The tropical hummingbird, *Phaethornis longuemareus,* has singing grounds—Breeding Territory Type C.)

Southern, W.E.
1958 Nesting of the Red-eyed Vireo in the Douglas Lake Region, Michigan. *Jack-Pine Warbler* 36:105–130, 185–207. (Breeding Territory Type A.)
1977 Colony selection and colony site tenacity in Ring-billed Gulls at a stable colony. *Auk* 94:469–478.

Stewart, R.E., and Aldrich, J.W.
1951 Removal and repopulation of breeding birds in a spruce fir forest community. *Auk* 68: 471–482.

Stokes, A.A., ed.
1974 *Territory.* New York: Dowden, Hutchinson & Ross. (Nineteen of 26 selections concern birds.)

Tinbergen, N.
1953 *The Herring Gull's world: A study of the social behaviour of birds.* London: Collins. (Chapters 9 and 10 are particularly pertinent to territorial behavior.)
1957 The functions of territory. *Bird Study* 4:14–27.

Tuck, L.M.
1960 *The murres: Their distribution, populations and biology: A study of the genus Uria.* Canadian Wildlife Series, 1. Ottawa: Canadian Wildlife Service. (Breeding Territory Type D.)

Venables, L.S.V., and Lack, D.L.
1934 Territory and the Great Crested Grebe. *Brit. Birds* 28:191–198. (Breeding Territory Type B.)

Woolfenden, G.E.
1973 Nesting and survival in a population of Florida Scrub Jays. *Living Bird* 12:25–49.

Zimmerman, J.L.
1971 The territory and its density dependent effect in *Spiza americana*. *Auk* 88:591–612. (Breeding Territory Type A. Two factors are important in the Dickcissel's selction of habitat: presence of song perches and sufficiently dense cover.)

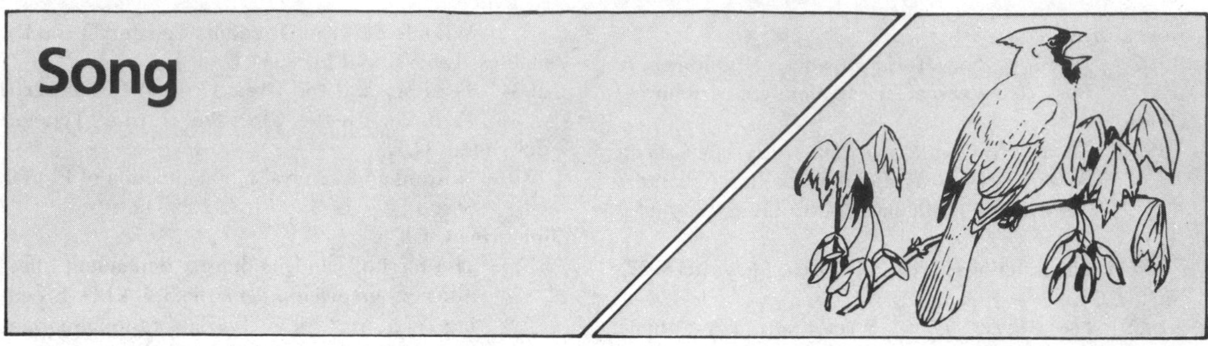

Song

Song is a vocal display in which one or more sounds are consistently repeated in a specific pattern. It is given mainly by males, usually during the breeding season. Other bird vocalizations are collectively termed **call notes** or simply **calls.** The distinction between songs and calls is most evident in songbirds (oscines) belonging to some forty "higher" families of the order Passeriformes; it is not clear in all bird species.

Functions of Song

One of the primary functions of song is to advertise territory. A tape recording of a male's song, played over a loudspeaker within a territory, elicits immediate aggression from the territory holder. If males are removed from their territories and replaced with a tape recorder and an array of loudspeakers, the advances of conspecific males seeking territories are deterred (Krebs et al., 1978).

Another primary function of song is to attract and sexually stimulate females of the same species. Females of captive Brown-headed Cowbirds *(Molothrus ater)* and Swamp Sparrows *(Melospiza georgiana)* will perform precopulatory displays in response to the appropriate conspecific songs (King and West, 1977; Searcy and Marler, 1981). Males often correlate the amount of their singing with activities of females. Brown Thrashers *(Toxostoma rufum)* nearly cease singing when they mate; the polygynous Marsh Wrens *(Cistothorus palustris)* increase their singing dramatically when they begin displaying for additional mates (Verner, 1965).

Substitutes for Song

Like all vocalizations, songs originate in a part of the respiratory tract called the syrinx (see "Anatomy and Physiology," pages 75 and 93).

Many birds, particularly among some of the lower orders, produce sounds by the wings, bill, tail, or other parts of the body. These sounds enhance the effects of songs or function as songs. For example, the Common Nighthawk *(Chordeiles minor),* while in the air, utters a rasping *peent* at wide intervals and then, as it checks a quick dive earthward, lets the air rush through its wing feathers, producing a loud, hollow *whoooom*. The Ruffed Grouse *(Bonasa umbellus)* on a log makes a solely mechanical sound by "drumming"—beating the air with its wings, the tempo increasing until the pulsating sounds come together in a roar. Woodpeckers drum by hammering with their bills, preferably on a hollow stump or some other object that provides resonance. Very few species of birds fail to produce sounds that are either songs in the strict sense or at least substitutes for songs.

Male Songs

Songs among different species vary, widely ranging from a repetition of one syllable to a highly complex series of sounds. The "best" songsters—i.e., those giving loud, extended songs with a strong musical quality—are among songbirds in which the males are usually dull in color and closely resemble the females. Such males, it is believed, depend more on sounds than they do on appearance for the identification of their species and stimulation of mates.

Duration of Singing

Most species come into full song with the establishment of territory. While singing may decrease in intensity or cease altogether during the short mating period, most birds resume singing with considerable vigor during nest-building, egg-laying, and incubation. When the period of caring for the young is reached, singing almost invariably wanes. Unless the pair rears a second brood, singing usually stops altogether between the period of caring for the young and the end of the postnuptial (prebasic) molt (Saunders, 1948a).

Some species resume singing in the fall after the postnuptial molt, though the songs are far fewer in number per day and tend to be incomplete (Saunders, 1948b). With the advent of winter, singing usually ceases altogether except in a few permanent-resident species, such as the Carolina Wren *(Thryothorus ludovicianus),* European Robin *(Erithacus rubecula),* and Northern Cardinal *(Cardinalis cardinalis),* which sing more or less the year round. Most migratory species sing irregularly during migration; those which do not usually arrive silently on their breeding grounds and are not likely to start singing until at least a few days have passed (Saunders, 1954).

Singing in Relation to Habitat

While they are singing, males of many species make themselves conspicuous by standing or perching on prominent objects. Very often an individual will choose one or more favorite perches for much of his singing.

Males of a number of species, especially those in open country, give **flight songs,** achieving conspicuousness by singing on the wing. Their flight songs are almost always accompanied by aerial maneuvers that increase conspicuousness. In some species flight songs include mechanical sounds from the action of flight feathers against the air. For instance, the American Woodcock *(Scolopax minor)* produces whistling sounds with its wings as it spirals skyward, then chippers vocally during its abrupt, zigzag descent. In addition to its song the Flappet Lark *(Mirafra rufocinnamomea)* of Africa often flies over grassy clearings or bare areas in open woodland and makes flapping sounds with its wings; local males make similar flapping sounds, while males in neighboring populations have different patterns of wing-beats (Payne, 1978). The only sound made by the Common Snipe *(Gallinago gallinago),* as it circles high over a marsh, is a "winnowing" from the air filtering through its tail feathers.

Singing in Relation to Nests

As a rule, species do not sing near their nests. When males approach their nests for one purpose or another, they are inclined to be silent or to sing much more softly. Males in a few species will, however, sing while on their nests. Male Rose-breasted and Black-headed Grosbeaks *(Pheucticus ludovicianus* and *P. melanocephalus)* occasionally sing while incubating and brooding.

Amount of Singing

Males of certain species give an impressive number of songs during the daylight hours. In one day a Song Sparrow *(Melospiza melodia)* gave as many as 2,305 songs (Nice, 1943), and a Kirtland's Warbler *(Dendroica kirtlandii)* gave 2,212 songs (Mayfield, 1960). But probably a Red-eyed Vireo *(Vireo olivaceus)* holds the record with 22,197 songs in nearly 10 hours (de Kiriline, 1954).

Relation of Singing to Light

A few species, such as the Red-eyed Vireo, sing more or less continuously all day, but a large proportion of diurnal species sing more energetically —i.e., give more songs per hour—in the early morning and in the evening when there is less light. In the morning and sometimes in the evening a small number of species produce **twilight songs,** which differ in minor ways from their regular daytime songs. The Eastern Kingbird *(Tyrannus tyrannus,* Eastern Wood-Pewee *(Contopus virens),* Least Flycatcher *(Empidonax minimus),* and other tyrannid flycatchers sing at a much quickened tempo as day is breaking. A few diurnal species such as the Nightingale *(Luscinia megarhynchos)* and Northern Mockingbird *(Mimus polyglottos)* commonly sing at night—the Nightingale as much at night as in the day, the mockingbird more often on moonlit nights. The American Woodcock performs its spiral flight song in the twilight of the late evening and early morning, and also at night if there is moonlight of sufficient intensity to match the brightness at twilight. When its singing field happens to be near an airport or some other installation that is greatly illuminated, the woodcock may perform through the night.

The amount of light rather than the exact time of day determines the beginning of singing in the morning and the end of singing in the evening. Cloudiness will delay singing in the morning and hasten its cessation in the evening. Species vary as to the amounts of light and darkness that stimulate and inhibit singing. Consequently, in any given area, different species start and stop singing at different times, forming a more or less orderly sequence. By careful study in an area one can predict the order in which the various species will start singing in the morning and stop in the evening. With some species, at least, singing begins earlier in the morning as the breeding season progresses (Nice, 1943).

Nocturnal birds may sing throughout the night. Like diurnal birds, particular amounts of light stimulate and inhibit their singing. They begin singing earlier in the evening and stop later in the morning when evenings and mornings are cloudy. Crepuscular birds sing energetically only under twilight conditions: morning and evening and during the night when there is a moon to provide twilight conditions.

The total effect of amounts of light on singing is dramatically illustrated at the time of a total solar eclipse (Kellogg and Hutchinson, 1964). As totality nears, diurnal species gradually cease singing and crepuscular species begin. In the dim light during totality, nocturnal species join the crepuscular species and the few diurnal species that would normally continue singing into the late twilight.

Effects of Weather on Singing

Weather may influence the amount of singing from day to day. Excessive coolness approaching frost, as in the early morning, or intense heat, as at midday, may inhibit singing, while mild temperatures may encourage it. Wind is unquestionably a disturbing factor and, if strong enough, will stop singing. Although periods of high humidity, as before and after a rain, may induce vigorous singing, heavy precipitation often markedly reduces it.

Song Development

Songbirds of the order Passeriformes produce the most complex of all bird songs. Young songbirds, according to overwhelming evidence from careful studies, learn the songs from hearing adults. Birds thus rival humans—and some cetaceans—in their extent of vocal learning. Consequently there are many parallels in the development of bird songs and human speech (Marler, 1970a).

Songbirds do most of their learning while in a "sensitive period" during the first year of their lives. A male White-crowned Sparrow *(Zonotrichia leucophrys),* for example, learns his song during his first two months (Marler, 1970b). In this sensitive period he commits the song to memory. Then, early in the next spring he begins practicing the song he has memorized. At first it is a **subsong,** quieter, more rambling, and far less structured than the adult song. But ever so gradually he improves the song with practice until it matches the one memorized nearly a year earlier (Marler and Peters, 1981).

Hearing is important during song development. First, the juvenile male must hear the sounds of other males; second, he must be able to hear himself practicing in order to match the "template" he has stored in his brain. Deafened birds, like deafened humans, cannot learn to vocalize correctly.

A number of areas that control singing have been identified in the songbird's forebrain. The most influential song-control nuclei are on the left side; and it is the left hypoglossal nerve from these nuclei that innervates the muscles on the left side of the syrinx, which produces 75 to 90 percent of the sounds in a typical songbird. Just as in humans, the control of vocal communication is neurally lateralized (Nottebohm, 1980).

Although vocal learning attains its peak among the songbirds of Passeriformes, at least one hummingbird (Little Hermit, *Phaethornis longuemareus;* Wiley, 1971) and probably all parrots

(Psittacidae) learn vocalizations. Among the "lower" families (suboscines) of Passeriformes—e.g., the Tyrannidae—there is no evidence of song learning. At least three species, the Eastern Phoebe *(Sayornis phoebe)*, Alder Flycatcher *(Empidonax alnorum)*, and Willow Flycatcher *(Empidonax traillii)*, develop perfectly normal songs even if deprived of hearing them after seven to eight days of life (Kroodsma, 1982). Songbirds, if similarly deprived, might produce songs of considerable complexity but they would be quite unlike anything ever heard in the wild.

Very few studies of song development have been completed in species of other orders. Deafened domestic fowl and doves develop relatively normal vocalizations, and cross-fostered dove species can develop normal songs even in the absence of songs of their own species. Furthermore, hybrid doves produce intermediate songs, suggesting that songs of birds in these orders are more genetically controlled and less plastic than in songbirds (reviewed in Kroodsma, 1982).

Geographical Variation in Dialects

One of the consequences of vocal learning in birds, just as in humans, is the formation of local dialects. Neighboring male songbirds of the same species possess very similar songs, but males singing only a few kilometers distant may have songs quite distinct from those at the first locality.

Microgeographical Variations

Even though this local variation has been documented in nearly every songbird carefully studied, the functions of these dialects remain poorly understood. One of the most exciting, yet most controversial, hypotheses is that local dialects may inhibit dispersal of juveniles and promote adaptation of populations to local environmental conditions (Nottebohm, 1969). If juveniles learn—i.e., memorize—their songs before dispersing from their home locality, then dialects will be preserved only if juveniles sing and breed near their home locality. However, if juveniles are capable of changing their songs after dispersing from their home locality, then dialect boundaries may not necessarily inhibit dispersal. In several species of birds it has actually been shown that young males can disperse from the parental territory and learn songs quite unlike their father's (e.g., see Payne, 1981). This raises the question, as yet unanswered: How often does such dispersal from one dialect to another take place, and does less dispersal occur as a result of the dialects? Consult Baker and Mewaldt (1978, 1981) and Petrinovich et al. (1981).

Macrogeographical Variations

These variations involve distances greater than a juvenile would typically disperse and occur in the sounds of almost all species studied. Among the songbirds, such variation consists to some extent of an accumulation of differences in learned songs. With increasing distances, however, genetic differences between individuals tend to increase as well. California Marsh Wrens, for instance, develop larger repertoires and songs very different in quality from those of New York. Some of these differences persisted in young birds reared in captivity, indicating that their different singing styles are to some extent inherited and not merely learned from adults (see Kroodsma, 1983).

Geographical variation in songs is far less pronounced among birds that do not learn to sing. In tropical *Myiarchus* flycatchers the overall form of the song may remain virtually unchanged over thousands of kilometers, but the frequency ("pitch") of the song may vary with body size. Larger birds at higher altitudes or latitudes produce lower pitched songs than smaller birds elsewhere (Lanyon, 1978). Songs of most nonpasserines probably vary geographically like songs of these flycatchers, yet few nonpasserines have been studied extensively.

In many species the nature of the occupied habitats changes geographically, and the songs in some of these species seem to change with the habitat. Vegetation absorbs sound energy, and different plant forms can determine the type of song that can be broadcast most efficiently in a particular region (Gish and Morton, 1981). Another reason why songs might change geographically could be related to the other species that are present in the community; if songs are selected for maximum species-distinctiveness, adjustments in song form might be made geographically to accommodate different "sound environments." While this remains an attractive hypothesis, no convincing data yet exist that confirm this possibility (Miller, 1982).

Song Repertoires

The songs of different species differ in *quality* and are therefore recognizable by other members of the same species. There is another very striking difference in the singing style among species: the number of different songs that individual males can sing. In many species, including the Common Yellowthroat *(Geothlypis trichas),* Ovenbird *(Seiurus aurocapillus),* Indigo Bunting *(Passerina cyanea),* and White-throated Sparrow *(Zonotrichia albicollis),* each male has a single song form that it repeats with little variation throughout its life. In fact, in each of these four species the songs of neighboring males are so distinct from one another that each song is a good "fingerprint" of the individual. Spectrographic analysis can easily demonstrate this distinction (see "Selected Studies" later in this chapter).

Furthermore, experiments demonstrate that the birds use these individual differences to recognize each other. If the song of a neighbor or a stranger is played over a loudspeaker at the boundary between the resident and neighbor, the resident responds more strongly to the stranger's song than the neighbor's. Such recognition of neighbors allows a territory male to respond selectively to the greater threat of a strange bird, rather than to waste energy responding to a neighbor with which a territory boundary has already been firmly established (see Falls, 1982).

In other species the song repertoire is larger. A male Song Sparrow averages about ten different songs. First he will sing one of his song types for several minutes, then sing another type, and finally after an hour or so return to the first type (Mulligan, 1966). Even the novice student of birds can clearly tell when the Song Sparrow switches song types, but experience is necessary to recognize each of the different song types in the repertoire. A Song Sparrow sings with what is termed **eventual variety.** This can be recorded with different letters representing the different song types in the repertoire. Thus, AAA . . . BBB . . . and so on.

Lengthy tape recordings and detailed analyses of the singing of other species have indicated that individual song repertoires of some species may be even larger than the Song Sparrows. Male Marsh Wrens in Washington State have about 115 different types (Verner, 1975). Male Northern Mockingbirds may sing 100 to 200 different song types (Wildenthal, 1965), and another mimid, the Brown Thrasher, may have as many as 2,000 different songs in each male's repertoire (Boughey and Thompson, 1981).

Functions of Song Repertoires

In most of the species whose males have repertoires of song types all songs appear to serve the same functions of territory advertisement and mate attraction. However, there are exceptions. The best studied are the New World warblers, the Parulines (see Lein, 1978), whose males usually have two song forms. One is predictably used on the territory boundary when males are clashing; the other tends to be used more in the center of the territory. Whether this second form is used primarily in the absence of other males or in the presence of the female is still debated, but there are clearly different contexts in which the two song forms are used.

Interestingly, the songs used on territorial boundaries tend to vary geographically like typical songbird dialects, yet the song form used in the center of the territory is very stereotyped over large distances. All this suggests that, at least in the Blue-winged Warbler *(Vermivora pinus)* and Chestnut-sided Warbler *(Dendroica pensylvanica),* intense interactions among males promote the formation of local song dialects (Kroodsma, 1981).

But why does the size of song repertoires vary among species? Certainly there is a cost in learning a large repertoire, since more brain space must be devoted to the task (Nottebolm, 1980). Several hints to the evolution of larger song repertoires are available. Hartshorne (1973) noted that males of species that sing faster are more likely to sing with **immediate variety,** when successive songs are different—i.e., A B C D E. . . . He thus formulated his **monotony threshold hypothesis,** stating in essence that rapid or continuous songsters will prevent habituation of listeners if successive songs are different from one another. McGregor et al. (1981) also believe that a larger repertoire probably enables a male to obtain and defend a territory of higher quality.

This relationship between repertoires, singing styles, and stimulation has been demonstrated. During playback experiments with Great Tits *(Parus major)*, the response of a male gradually wanes during a repeated presentation of a given song type; nevertheless, the male responds with renewed vigor when another song type is played (Krebs, 1976). Furthermore, males with larger repertoires within a given population tend to have greater reproductive success. Therefore, these large repertoires could be more attractive and more stimulating to females. For instance, female Common Canaries *(Serinus canaria)*, on hearing large song repertoires, are stimulated into breeding condition faster than are females hearing simpler repertoires (Kroodsma, 1976).

Female Songs

The singing ability of females varies widely among species. Apparently females of many species do not produce sounds that can be called songs. In the Song Sparrow (Nice, 1943) the female occasionally sings early in the breeding season prior to nest-building; the song, always given from an elevated perch, is "short, simple, and unmusical." Undoubtedly, quite a number of species are like the Song Sparrow in this respect.

In a few species—e.g., the Northern Cardinal, Rose-breasted Grosbeak, and Black-headed Grosbeak—the female produces songs that are about as elaborate as the male's, and only in a very few species—e.g., the phalaropes in which the female plays the more active role in mating—is the female's song more elaborate than the male's. There are, however, many tropical species in which the female sings simultaneously and on a par with her mate. Called **duetting,** this behavior plays a prominent role in courtship display as well as in territorial defense. Duetting may aid in maintaining the pair-bond, maintaining contact between mates in dense vegetation, and/or serving jointly in courtship display and territorial defense among long-lived species in which a male and female may pair for life (see Farabaugh, 1982, for a review).

Mimicry

Certain parrots (Psittacidae) and mynas (*Acridotheres* spp.) in captivity imitate human speech and other sounds. In nature, however, parrots and mynas restrict their imitative abilities to learning vocalizations of conspecifics.

While a large number of species in the wild sometimes include in their vocal repertoires the call notes and parts of songs of other species, only a very few in the New World are accomplished mimics. The most notorious is without any doubt the Northern Mockingbird. Two males, singing on their nesting territories in northern Lower Michigan, included in their respective repertoires the imitations of 14 and 15 different species, several of which do not reside that far north (Adkisson, 1966). Perhaps both birds had come from southern areas where they had heard the species and learned to imitate them.

Mimicry may be a way of acquiring a large repertoire in populations of low density where conspecific males might not be as available for song learning. Also, Northern Mockingbirds tend to be aggressive toward competing species, and it is thus possible that a male mockingbird, in learning songs of other species, is directly addressing competitors to keep off his territory. This hypothesis needs testing (Baylis, 1982).

Selected Studies

Until recent years the study of bird song depended entirely on one's hearing and the ability to record what one heard by the musical scale, phonetics, and diagrams. Now thanks to modern instruments, it is possible to record bird sounds on magnetic tape (see "Recording Bird Vocalizations" in Appendix A); then, by means of the sound spectrograph, it becomes possible to reproduce the sounds as actual images, or spectrograms, for visual analysis. This brings to the study of bird song a far more accurate method with wide applications.

Since some field guides and most published studies of bird songs include spectrograms, learn how to interpret them. Figure 55 shows representative

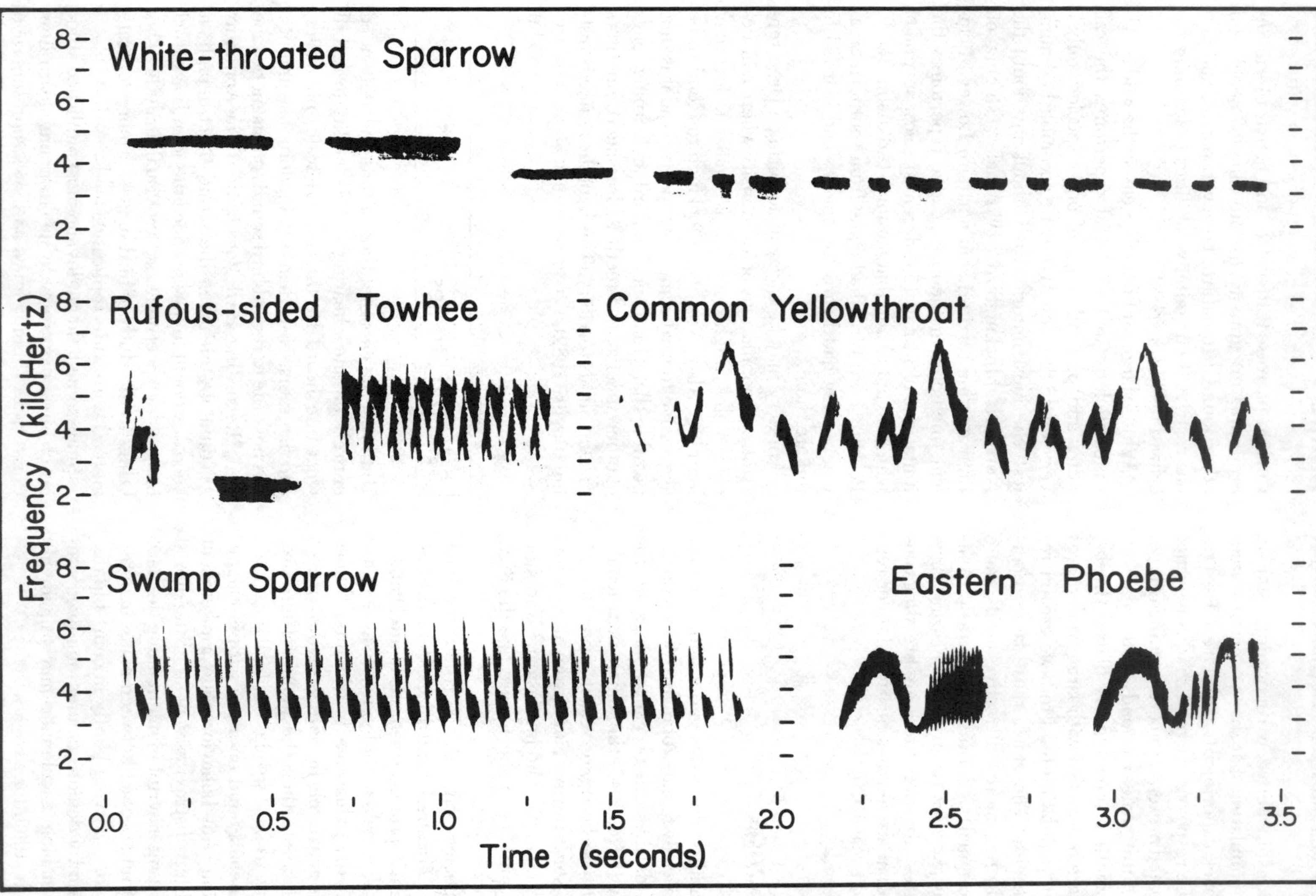

Figure 55 **Song Spectrograms of Five North American Bird Species** (Courtesy of Donald E. Kroodsma.)

songs of five North American passerines. The spectrograms should be read from left to right, with the horizontal axis indicating the time in seconds. The vertical axis is the song **frequency**—i.e., the number of vibrations or "sound waves"—in KiloHertz, or kilocycles per second. The relative loudness of the different parts of the song is indicated by the darkness or density of the notes.

The song of the White-throated Sparrow in Figure 55 consists of pure musical notes or whistles; it begins with two long notes, drops to a lower frequency for another long note, and then concludes with 12 shorter notes, which are arranged in four series of triplets (the "pea-bod-y" portion of the song). All notes are of approximately equal loudness or intensity. The Swamp Sparrow song consists of 21 rapidly repeated identical phrases. The song of the Common Yellowthroat also consists of repeated phrases, but each "witch-i-ty" phrase consists of five different notes and lasts over half a second. The Rufous-sided Towhee *(Pipilo erythrophthalmus)* begins with two introductory notes ("drink-your-") and ends with a series of repeated phrases ("teeeee"). Each male Eastern Phoebe has two song forms, the "fee-bee" and the "fee-b-b-be"; listen closely to easily hear the difference. An interpretation of spectrograms for many other North American birds is given in Bondesen (1977).

Two suggested studies of song follow.

1. Thoroughly Study the Song of One Species.

Procedure Select a species that is fairly common. Begin while territories are being established and continue through the nesting season. If possible, make tape recordings on the spot and spectrograms later for analysis. Otherwise, adopt one of the several methods suggested for learning songs given in the chapter on "Field Identification" in this book.

Presentation of Results Prepare a paper in which all information obtained is carefully analyzed and summarized. (See Appendix B for directions on preparing the manuscript.) Some of the topics that may be covered follow.

1. Types and variations of male song. Describe the typical songs in detail and point out variations in songs both among individuals and within the repertoire of an individual. Note any correlations between types of song and the pairing status of the male, the context in which the song is given (e.g., with male or female), the time of the season, and the time of day.

2. Amount of singing and its relation to different stages of the nesting cycle. Note particularly how males may use different songs at different stages (especially if warblers, Parulinae) or how different song types are used in a singing performance (e.g., Eastern Phoebe). Determine the factors that account for the number of times a male will repeat one of his several song types before switching to another.

3. Song in relation to habitat type, location of the territory, and proximity to nest.

4. Song in relation to light and weather conditions.

5. Song of the female.

2. Determine the Order in Which the Species Start Singing in the Morning.

Procedure Select an area where there is a wide variety of species. Begin early in the nesting season when most species are establishing their territories. With a watch, a photometer such as one commonly used in outdoor photography, and notebook, go to the area on at least six (preferably successive) mornings before there is any evidence of dawn. When the first species starts singing, note the exact time and (with the meter directed toward the east, the source of greatest light) the amount of light in foot candles. Probably the light will be so low that the meter will record only a fraction of one candle. Follow the same procedure when the second species starts to sing, and so on.

Presentation of Results Draw up a report based on observations made. (See Appendix B for directions on preparing a manuscript.) Briefly describe the area in which the birds were heard singing and include any information available on the relative abundance of the various species in the area. For each morning, give weather data—air temperature, wind velocity, humidity, and sky conditions. Then present the results of observations for each morning in tabular form as follows:

Species	Date	Time Civil Twilight Begins	Time of Sunrise	Time of First Song	Minutes Before Sunrise

Record all times in Standard Time. Morning civil twilight begins when the sun is 6 degrees below the horizon and ends at sunrise. In order to determine when civil twilight begins and sunrise occurs in the latitude and longitude where the study is being made, consult the most recent supplement to the *American Ephemeris and Nautical Almanac*.

As a result of several mornings' observations, discuss any correlation noted between the beginning of singing and weather conditions. Show the order in which different species began morning singing.

References

Adkisson, C.S.

1966 The nesting and behavior of Mockingbirds in northern Lower Michigan. *Jack-Pine Warbler* 44:102–116.

1981 Geographic variation in vocalizations and evolution of North American grosbeaks. *Condor* 83:277–288. (One of the few studies of call learning and geographic variation in birds.)

Armstrong, E.A.

1963 *A study of bird song*. London: Oxford Univ. Press. (A thorough and indispensable review of the "older" literature.)

Baker, M.C., and Mewaldt, L.R.

1978 Song dialects as barriers to dispersal in White-crowned Sparrows *(Zonotrichia leucophrys nuttalli)*. *Evolution* 32:712–722.

1981 Response to "Song dialects as barriers to dispersal: A re-evaluation." *Evolution* 35:189–190.

Baylis, J.R.

1982 Mimicry. In *Acoustic communication in birds*, vol. 2, eds. D.E. Kroodsma and E.H. Miller. New York: Academic Press.

Bondesen, P.

1977 *North American bird songs. A world of music*. Klampenborg, Denmark: Scandinavian Science Press Ltd.

Boughey, M.J., and Thompson, N.S.

1981 Song variety in the Brown Thrasher *(Toxostoma rufum)*. *Z. Tierpsychol*. 56:47–58.

Craig, W.

1943 *The song of the Wood Pewee* Myiochanes virens *Linnaeus: A study of bird music*. New York State Mus. Bull. No. 334.

de Kiriline, L.

1954 The voluble singer of the treetops. *Audubon Mag*. 56:109–111.

Falls, J.B.

1982 Individual recognition by sound in birds. In *Acoustic communication in birds,* vol. 2, eds. D.E. Kroodsma and E.H. Miller. New York: Academic Press.

Farabaugh, S.M.

1982 The ecological and social significance of duetting. In *Acoustic communication in birds,* vol. 2, eds. D.E. Kroodsma and E.H. Miller. New York: Academic Press.

Gish, S.L., and Morton, E.S.

1981 Structural adaptations to local habitat acoustics in Carolina Wren songs. *Z. Tierpsychol*. 56:74–84.

Hartshorne, C.

1973 *Born to sing*. Bloomington: Indiana Univ. Press. (Interestingly written, from the viewpoint of a philosopher and biologist.)

Hinde, R.A., ed.

1969 *Bird vocalizations: Their relations to current problems in biology and psychology.* Cambridge: Cambridge Univ. Press. (Contributed chapters cover duetting, individual recognition, tonal quality, and many other topics.)

Jellis, R.

1977 *Bird sounds and their meaning*. London: British Broadcasting. (A good basic book for the beginning student or lay person.).

Kellogg, P.P., and Hutchinson, C.M.
1964 The solar eclipse and bird song. *Living Bird* 3:185–192.

King, A.P., and West, M.J.
1977 Species identification in the North American Cowbird: Appropriate responses to abnormal song. *Science* 195:1002–1004.

Konishi, M.
1963 The role of auditory feedback in the vocal behavior of the Domestic Fowl. *Z. Tierpsychol.* 20:349–367.

Krebs, J.R.
1976 Habituation and song repertoires in the Great Tit. *Behav. Ecol. Sociobiol.* 1:215–227.

Krebs, J.R.; Ashcroft, R.; and Webber, M.I.
1978 Song repertoires and territory defence. *Nature (London)* 271:539–542.

Kroodsma, D.E.
1976 Reproductive development in a female songbird: Differential stimulation by quality of male song. *Science* 192:574–575.
1981 Geographical variation and functions of song types in warblers (Parulidae). *Auk* 98:743–751.
1982 Learning and the ontogeny of sound signals in birds. In *Acoustic communication in birds,* vol. 2, eds. D.E. Kroodsma and E.H. Miller. New York: Academic Press.
1983 Marsh wrenditions. *Natural History* 92:42–47.

Kroodsma, D.E., and Miller, E.H., eds.
1982 *Acoustic communication in birds.* 2 vols. New York: Academic Press. (Chapters in these two volumes cover nearly all aspects of bird vocalizations, including production and perception of song, population and species recognition, and recording techniques.)

Lade, B.I., and Thorpe, W.H.
1964 Dove songs as innately coded patterns of specific behaviour. *Nature (London)* 202:366–368.

Lanyon, W.E.
1978 Revision of the *Myiarchus* Flycatchers of South America. *Bull. Amer. Mus. Nat. Hist.* 161:429–627.

Lanyon, W.E., and Tavolga, W.N., eds.
1960 *Animal sounds and communication.* Washington, D.C.: American Institute of Biological Sciences.

Lein, M.R.
1978 Song variation in a population of Chestnut-sided Warblers *(Dendroica pensylvanica).* Its nature and suggested significance. *Canadian Jour. Zool.* 56:1266–1283.

Marler, P.
1970a Birdsong and speech development: Could there be parallels? *Amer. Scientist* 6:669–673.
1970b A Comparative approach to vocal learning: Song development in White-crowned Sparrows. *Jour. Comp. Physiol. Psychol.* 71 (2): 1–25.

Marler, P., and Peters, S.
1981 Sparrows learn adult song and more from memory. *Science* 213:780–782.

Mayfield, H.
1960 *The Kirtland's Warbler.* Cranbrook Inst. Sci. Bull. No. 40, Bloomfield Hills, Michigan.

McGregor, P.K.; Krebs, J.R.; and Perrins, C.M.
1981 Song repertoires and lifetime reproductive success in the Great Tit *(Parus major). Amer. Nat.* 118:149–159.

Miller, E.H.
1982 Character and variance shift in acoustic signals of birds. In *Acoustic communication in birds,* vol. 1, eds. D.E. Kroodsma and E.H. Miller. New York: Academic Press.

Mulligan, J.A.
1966 Singing behavior and its development in the Song Sparrow. *Univ. Calif. Publ. in Zool.* 81:1–76.

Nice, M.M.
1943 Studies in the life history of the Song Sparrow. II. *Trans. Linnaean Soc. New York* 6:i–viii; 1–328 (Chapter 9–11).

Nolan, V., Jr.
1978 The ecology and behavior of the Prairie Warbler *Dendroica discolor. Amer. Ornith. Union, Ornith. Monogr.* No. 26. (Important information on vocalization in several chapters.)

Nottebohm, F.
1969 The song of the Chingolo, *Zonotrichia capensis,* in Argentina: Description and evaluation of a system of dialects. *Condor* 71:299–315.
1980 Brain pathways for vocal learning in birds: A review of the first ten years. *Progress in Psychobiology and Physiological Psychology* 9:85–124.

Payne, R.B.
1978 Local dialects in the wingflaps of Flappet Larks *Mirafra rufocinnamomea. Ibis* 120:204–207. (It appears that wing-beats are learned by the males and that local dialects result.)
1981 Population structure and social behavior: Models for testing the ecological significance of song dialects in birds. In *Natural selection and social behavior: Recent research and new theory,* eds. R.D. Alexander and D.W. Tinkle. New York: Chiron Press.

Petrinovich, L.; Patterson, T.; and Baptista L.
1981 Song dialects as barriers to dispersal: A re-evaluation. *Evolution* 35:180–188.

Saunders, A.A.
1948a The seasons of bird song—The cessation of song after the nesting season. *Auk* 65:19–30.
1948b The seasons of bird song. Revival of song after the postnuptial molt. *Auk* 65:373–383.
1954 *The lives of wild birds*. Garden City, New York: Doubleday.

Searcy, W.A., and Marler, P.
1981 A test for responsiveness to song structure and programming in female sparrows. *Science* 213:926–928.

Smith, W.J.
1977 *The behavior of communicating: An ethological approach*. Cambridge, Massachusetts: Harvard Univ. Press.

Thielcke, G.A.
1976 *Bird sounds*. Ann Arbor: Univ. of Michigan Press. (An English Translation of the 1970 "Vogelstimmen.")

Thorpe, W.E.
1961 *Bird-song: The biology of vocal communication and expression in birds*. Cambridge: Cambridge Univ. Press.

Vaurie, C.
1946 Early morning song during middle and late summer. *Auk* 63:163–171. (A quantitative study covering 47 days based on 21 bird species.)

Verner, J.
1965 Time budget of the male Long-billed Marsh Wren during the breeding season. *Condor* 67:124–139.
1975 Complex song repertoire of male Long-billed Marsh Wrens in eastern Washington. *Living Bird* 14:263–300.

Wildenthal, J.L.
1965 Structure in primary song of the Mockingbird (*Mimus polyglottos*). *Auk* 82:161–189.

Wiley, R.H.
1971 Song groups in a singing assembly of Little Hermits. *Condor* 73:28–35.

Mating

The pairing of birds is commonly spoken of as **mating.** This term is distinguished from the sex relation, which is **copulation** or **coition,** and from **fertilization,** which is the union of the sperm and ovum.

Preliminary Stages

In migratory passerine and many other species, the spring migration to the breeding grounds usually precedes mating. The sequence of events between migration and mating is outlined below:

Arrival of the Male Males in the majority of species are the first individuals to appear, but, in a number of species, males and females appear at the same time. Occasionally males wander for a brief period after their arrival; as a rule, they immediately establish territory. If they arrive in flocks, they usually remain in flocks for the wandering period. At this time singing is infrequent and subdued.

Establishment of Territory Though the females may still be absent, the males establish territory. When the males are in flocks, they gradually become vociferous and antagonistic toward one another, and the flock breaks up before territory establishment. While they establish territory, they sing frequently and intensely, and they may compete vigorously with other males for specific areas.

Arrival of the Female Like the males, females occasionally wander for a period after arrival. Ordinarily, they go directly to male territories. They may or may not arrive in flocks. Sometimes they join the still unbroken flocks of males and wander with them until solitary behavior finally predominates and they enter male territories.

Mating

Mating usually occurs in association with established territory. When the female arrives on the territory, almost immediately the male receives her by displays and pursues her. Though the male occasionally attempts copulation at this time, he is usually unsuccessful, the female being unreceptive. If the female remains in the male's territory after the first displays and sexual pursuit, mating is accomplished. The two birds then stay more or less within the confines of the territory and in close association. Sometimes the male sings less intensely or stops singing altogether.

After a period of days, the female becomes receptive and copulation occurs. Copulation, once initiated, may be repeated several times a day through subsequent periods of nest-building and egg-laying. The singing of the male is intense; if it decreased in intensity or ceased altogether after mating, it is now resumed.

Mating may not occur in association with territory. In some migratory swans and geese pairing may take place on the wintering grounds, and the birds arrive on their breeding grounds already mated. Pair-bonds may sometimes remain permanent thereafter. In a few sedentary or nonmigratory species pairs stay together throughout the year on their territories although they may not begin copulating until the breeding season approaches.

Kinds of Mating Relations

Three kinds of mating relations occur among birds:

Monogamy The majority of species are monogamous. Faithfulness of the pair to each other is usually constant because of the attachment of the sex-partners to each other, to their territory, and later to their young. However, remating may take place when a partner is lost by death, desertion, eviction from the territory by a rival, or other circumstances.

Polygamy A number of species are normally polygamous, an individual of one sex (usually a male) mating at the same time with two or more individuals of the opposite sex. At times a monogamous species may show polygamy if the sex ratio of a population is unbalanced or, as suggested by Verner (1964), if a male succeeds in establishing a large territory in a habitat particularly attractive to more than one female. There are two conditions of polygamy:

Polygyny A male mates with two or more females. This is exhibited in Ostriches *(Struthio camelus)*, Ring-necked Pheasants *(Phasianus colchicus)*, and several icterines, including the Red-winged Blackbird *(Agelaius phoeniceus)*. Verner and Willson (1966) make the point that polygyny is to be expected more often in species nesting in open country—e.g., prairies and marshes.

Polyandry A female mates with two or more males. The condition is known in the Northern Jacana *(Jacana spinosa)* and Spotted Sandpiper *(Actitis macularia)* and may occur in local populations of species with a preponderance of males.

Promiscuity Though classified as a mating relation, promiscuity is actually copulation without relation. This condition is characteristic of several grouse (e.g., Sage Grouse, *Centrocercus urophasianus;* Greater Prairie-Chicken, *Tympanuchus cupido)*, the Ruff *(Philomachus pugnax)*, many tropical hummingbirds, many manakins (Pipridae), probably bowerbirds (Ptilonorhynchidae), many birds-of-paradise (Paradisaeidae), and the Brown-headed Cowbird *(Molothrus ater)* in some parts of its range.

Duration of Mating

Duration of mating in birds shows the following variations:

1. A pair may not remain mated after the last eggs of the season are laid (as in polygamous birds with breeding territories of Type C) or after the young are reared.

2. Birds that raise two or more broods yearly may remain mated between broods. Two factors account for this variation: close attachment of the birds to each other and to their territory, and overlapping of two broods—i.e., when the female starts a new nest while the male is feeding young of the previous nest, the result is no lapse in breeding activities.

3. Birds that raise two or more broods yearly may change their mates between broods, a variation occurring only occasionally.

4. Birds mated for one season may remate the next. This variation may sometimes be due to the birds' having an attachment to the same territory; at other times it may be due to the birds' having a close attachment to each other.

5. Birds may remain mated, though less closely, through the winter between one breeding season and another. In a few species birds may remain mated for life.

Reproductive Displays

Reproductive displays are behavioral adaptations or signals that facilitate the successful completion of the reproductive cycle. In a general sense they include displays that directly or indirectly promote fertilization, as well as displays that follow fertilization to the conclusion of the reproductive cycle. Reproductive displays may incorporate all the signals—visual, auditory, or tactile, single or in combination—that are available to the species for intraspecific communication.

Many reproductive displays constitute signals, frequently derived from maintenance activities (see

"Behavior," p. 213) that have evolved to the point of being ritualized. These displays may be described according to their predictable characteristics. All such displays are species-typical. Although they are not believed to have evolved as isolating mechanisms, they nevertheless normally succeed in preventing interbreeding with other species.

Each species has a number of displays concerned with reproduction that can be classified according to the order of their occurrence and probable function in the reproductive cycle.

Prefertilization Displays

These include, besides the aggressive behaviors involved in the establishment of territories (see "Territory," p. 260), the activities that serve to bring the sexes together, to promote mate selection, and—in many species—to perform and maintain the pair-bond.

Advertising Displays Sometimes called "courtship" or "epigamic" displays, these activities attract the attention of the opposite sex. In many birds species-specific adornments of plumage (e.g., crests, ruffs, and modified scapulars, tertiaries, and tail coverts), brilliantly colored areas of plumage, lurid gapes and exposed areas of skin, and structural modifications (sometimes highly colored) of bill and face enhance the displays by males. Often eccentric body attitudes and movements of feet, wings, and tail, special sounds both vocal and mechanical, and peculiar aerial and aquatic maneuvers accompany the displays as the performing males direct their most startling adornments and actions toward the nearest and most attentive females. Sometimes they even pursue females. The females' responses, while seemingly indifferent, may consist of feeding movements, preening, wiping the bill, etc.—usually sexually motivated activities.

Solo Displays Performed by solitary males in either monogamous or polygamous species that may or may not be sexually dimorphic—i.e., one sex differing in appearance from the other. The females respond in a much less showy manner.

Mutual Displays Performed by both sexes, almost invariably by monogamous species that are not sexually dimorphic. The displays of each sex are practically identical.

Collective Displays Performed by two or more males of monogamous species simultaneously competing for the attention of a single female or by males of polygamous or promiscuous species, usually sexually dimorphic, performing in "arenas," "tournament grounds," or "leks."

Pair-bonding and Pair-maintaining Displays Activities among monogamous species that assist in forming and strengthening the pair-bond and synchronizing mates for readiness in reproduction. The birds may repeat them indefinitely in association with advertising displays and copulation and later with nesting.

Sexual Pursuit In most species the male pursues the female in the air, but in some species, he pursues on water or land. Sexual pursuit may or may not accompany advertising displays. Presumably the action is either an attempt on the part of the male to copulate or a signal indicating his readiness to copulate.

"Symbolic" Nest-building The male is usually the chief actor even though he does not normally take part in actual nest-building. Symbolic building may consist of picking up, manipulating, and carrying material to the nest without attempting to build; passing material to the sex-partner; and bringing material when coming to the nest to relieve the sex-partner during incubation.

Courtship Feeding Again, the male is usually the chief actor. He gathers food and gives it to the female. He may feed the female prior to copulation or while nest-building, incubating eggs, or brooding young.

Fertilization Displays

There are three types of displays that are directly associated with copulation.

Precopulatory Displays Essentially invitatory or solicitation performances. While often similar in both sexes, they are not necessarily given at the same time. When performing precopulatory displays on the water, a pair of Canada Geese *(Branta canadensis)* face each other and alternately curve their neck backward like a shepherd's crook and dip head and neck well below the surface. The action increases in tempo until the male swims alongside the female, ceases displaying, and proceeds to mount her (Klopman, 1962). In the Herring Gull *(Larus argentatus)* both partners give begging calls and head-tossing movements (Tinbergen, 1953). Among many passerines, the males fluff their body feathers, partially spread the wings, lower and spread the tail, and bow; the females, taking a more submissive role, sleek their body feathers, shiver the wings, elevate the tail, and crouch. Both partners may gape and give special calls heard at no other time.

Copulatory Displays The movements of both sexes from the time the male mounts the female until coition is effected and the male dismounts.

Postcopulatory Displays The actions that take place immediately after the male dismounts. They differ markedly from precopulatory displays and may or may not be mutual. In the Canada Goose (Klopman, 1962) the male raises his breast out of water, throws the neck backward, holds the head somewhat vertically, and keeps the wings partially opened and arched. The female gives the same display and both birds utter "wheezy groans." This mutual display is soon followed, first by both birds stretching their wings and flapping them and then by vigorous preening and bathing. The postcopulatory display of the Herring Gull consists mainly of preening (Tinbergen, 1953). Among passerine birds the display is just as simple, merely fluffing and shaping the feathers and preening.

Postfertilization Displays

These displays are the interactions of the sex-partners during egg-laying, incubation, and care of the young. They include symbolic nest-building and courtship feeding and, in a broader sense, all the interactions between the sex-partners and their young.

Nest-relief Displays Manifest among all species in which both partners incubate the eggs, these displays are the only ones that warrant consideration here. Among sea birds—e.g., penguins, albatrosses, and boobies—where one member of the pair is attentive at the nest for long periods, the arrival of the partner to take over incubation triggers elaborate displays by both birds that include, depending on the species, loud vocalizations, bizarre head and neck motions, and occasionally mutual nibbling—activities that may continue for many minutes. In the Herring Gull (Tinbergen, 1953) the arriving sex-partner often precedes the takeover with mewing calls and a "choking" action, particularly if the occupant bird is slow or reluctant to leave. In passerine species the nest-relief ceremony may be little more than mutual gaping before the incubating bird departs.

Selected Studies

Select one or more species for a study of mating.

Procedure If the species is migratory, begin the study immediately after the conclusion of spring migration; if the species is nonmigratory, begin at "the first sign" of spring. Follow the individuals as closely and as often as possible. (The chief method is to watch them with a binocular from far enough away to avoid disturbance.) Later, when mating has occurred and territories are established, erect blinds in or near territories for close observation of behavior. Try to obtain photographs or make sketches of significant activities.

Presentation of Results For each species prepare a separate report based on the observations made. Record, describe, and discuss in detail as many of the following topics as the acquired information permits. Include photographs or sketches to illustrate text material. Follow the directions for preparing a manuscript in Appendix B.

Preliminary Stages *Arrival of Males:* Time; singly or in flocks; behavior; time when singing and other displays begin. *Establishment of Territory:* Time between arrival and establishment; change in behavior between arrival and establishment. *Arrival of Females:* Time; with or after males; singly, in flocks with males, or mated with males; behavior. (If species is nonmigratory, omit sections relative to arrival; otherwise follow the outline.)

Mating *Meeting of Male and Female:* Time; place; resulting behavior. *Unreceptive Period of Female:* Length of time involved; behavior during;

length of time of song cessation. *Receptive Period of Female:* Length of time involved; number of copulations per day; method of copulation and associated behavior.

Mating Relation Kind; variations.

Duration of Mating Length of time birds remain mated; changes in "strength" of mating bond during the nesting cycle.

Reproductive Displays Types; behavior of both sexes before, during, and after; probable origin and function of the displays.

References

Adams, D.A.
1960 Communal courtship in the Ruffed Grouse, *Bonasa umbellus L. Auk* 77:86–87.

Barash, D.P.
1972 Lek behavior in the Broad-tailed Hummingbird. *Wilson Bull.* 84:202–203.

Bruning, D.F.
1974 Social structure and reproductive behavior in the Greater Rhea. *Living Bird* 13:251–294. (Promiscuity.)

Crawford, R.D.
1977 Polygynous breeding of Short-billed Marsh Wrens. *Auk* 94:359:362. (Evidence is at least occasional.)

Emlen, J.T., Jr.
1954 Territory, nest building, and pair formation in the Cliff Swallow. *Auk* 71:16–35.

Faaborg, J., and Petterson, C.B.
1981 The characteristics and occurrence of cooperative polyandry. *Ibis* 123:447–484.

Ficken, M.S.
1963 Courtship of the American Redstart. *Auk* 80:307–317. (Solo displays.)

Ficken, M.S., and Ficken, R.W.
1962 The comparative ethology of the wood warblers: A review. *Living Bird* 1:103–122. (Solo displays.)

Hayes, H.
1972 Polyandry in the Spotted Sandpiper. *Living Bird* 11:43–57.

Hochbaum, H.A.
1981 The Canvasback on a prairie marsh. 3rd ed. Lincoln: Univ. of Nebraska Press. (Chapter 3 has an excellent account of mating and related activities in the Canvasback and other ducks.)

Jenni, D.A., and Collier, G.
1972 Polyandry in the American Jacana *(Jacana spinosa). Auk* 89:743–765.

Johnsgard, P.A.
1965 Handbook of waterfowl behavior. Ithaca, New York: Cornell Univ. Press.

Kilham, L.
1979 Courtship and the pair-bond of Pileated Woodpeckers. *Auk* 96:587–594. (Among woodpeckers the pair bond is "among the strongest on a year round basis.")

Klopman, R.B.
1962 Sexual behavior in the Canada Goose. *Living Bird* 1:123–129.

Lack, D.
1940 Courtship feeding in birds. *Auk* 57:169–178.

MacDonald, S.D.
1968 The courtship and territorial behavior of Franklin's race of the Spruce Grouse. *Living Bird* 7:5–25. (Solo displays.)
1970 The breeding behavior of the Rock Ptarmigan. *Living Bird* 9:195–238. (Polygamy.)

Marshall, A.J.
1954 *Bower birds: Their displays and breeding cycles.* London: Oxford Univ. Press. (Collective displays.)

Martin, S.G.
1974 Adaptations for polygynous breeding in the Bobolink, *Dolichonyx oryzivorus*. Amer. Zool. 14:109–119.

McAllister, N.M.
1958 Courtship, hostile behavior, nest-establishment, and egg laying in the Eared Grebe *(Podiceps caspicus). Auk* 75:290–311. (Mutual displays.)

McKinney, F.
1970 Displays of four species of Blue-winged Ducks. *Living Bird* 9:29–64. (Northern and Cape Shovelers, Blue-winged and Cinnamon Teal.)

Meanley, B.
1955 A nesting study of the Little Blue Heron in eastern Arkansas. *Wilson Bull.* 67:84–99. (Promiscuity.)

Meyerrieks, A.J.
1960 Comparative breeding biology of four species of North American herons. *Publ. Nuttall Ornith. Club,* No. 2 (Green-backed Heron, Great Blue and Great White Herons, Reddish Egret, Snowy Egret; displays.)

Mock, D.W.
1978 Pair-formation displays of the Great Egret. *Condor* 80:159–172.

Nero, R.W.
1956 A behavior study of the Red-winged Blackbird. I. Mating and nesting activities. *Wilson Bull*. 68:5–37. (Solo displays.)

Nero, R.W., and Emlen, J.T., Jr.
1951 An experimental study of territorial behavior in breeding Red-winged Blackbirds. *Condor* 53:105–116. (Polygamy.)

Nice, M.M.
1930 Do birds usually change mates for the second brood? *Bird-Banding* 1:70–72.

Niebuhr, V.
1981 An investigation of courtship feeding in Herring Gulls, *Larus argentatus*. *Ibis* 123:218–223. (Explanations for the behavior.)

Oring, L.W., and Knudson, M.L.
1972 Monogamy and polyandry in the Spotted Sandpiper. *Living Bird* 11:59–73.

Petrinovich, L., and Patterson, T.L.
1978 Polygyny in the White-crowned Sparrow *(Zonotrichia leucophrys)*. *Condor* 80:99–100.

Rice, D.W., and Kenyon, K.W.
1962 Breeding cycles and behavior of Laysan and Black-footed Albatrosses. *Auk* 79:517–567. (Mutual displays; nest-relief displays.)

Richdale, L.E.
1951 *Sexual behavior in penguins*. Lawrence: Univ. of Kansas Press. (Mutual displays; nest-relief displays.)

Sauer, E.G.F., and Sauer, E.M.
1966 The behavior and ecology of the South African Ostrich. *Living Bird* 5:45–75. (Example of polygyny; solo displays.)

Schamel, D., and Tracy, D.
1977 Polyandry, replacement clutches, and site tenacity in the Red Phalarope *(Phalaropus fulicarius)* at Barrow, Alaska. *Bird-Banding* 48:314–324.

Scott, J.W.
1942 Mating behavior in the Sage Grouse. *Auk* 59:477–498. (Collective displays.)

Sick, H
1967 Courtship behavior in the manakins (Pipridae): A review. *Living Bird* 6:5–22. (Collective displays.)

Simmons, K.E.L.
1967 Ecological adaptations in the life history of the Brown Booby at Ascension Island. *Living Bird* 6:187–212. (Mutual displays; nest-relief displays.)

Sjolander, S., and Agren, G.
1976 Reproductive behavior in the Yellow-billed Loon, *Gavia adamsii*. *Condor* 78:454–463.

Smith, S.M.
1980 Demand behavior: A new interpretation of courtship feeding. *Condor* 82:291–295. (The behavior is "strongly correlated with female dominance over the male.")

Snow, D.W.
1968 The singing assemblies of Little Hermits. *Living Bird* 7:47–55. (Possible promiscuity in the hummingbird, *Phaethornis longuemareus;* collective displays.)
1971 Social organization of the Blue-headed Manakin. *Wilson Bull*. 83:35–38.

Stewart, R.M.; Henderson, R.P.; and Darling, K.
1977 Breeding ecology of the Wilson's Warbler in the High Sierra Nevada, California. *Living Bird* 16:83–102. (Breeding in High Sierra is polygynous; on central coast, monogamous.)

Stiles, F.G., and Wolf, L.L.
1979 Ecology and evolution of lek mating behavior in the Long-tailed Hermit Hummingbird. *Amer. Ornith. Union, Ornith. Monogr.* No. 27.

Tinbergen, N.
1953 *The Herring Gull's world: A study of the social behaviour of birds*. London: Collins. (Part 3, on pair formation and pairing, is especially instructive.)

Verner, J.
1964 Evolution of polygamy in the Long-billed Marsh Wren. *Evolution* 18:252–261. (A suggestion that the size of a male's territory and total amount of emergent vegetation in it are correlated with his success in acquiring mates. Females may rear more young by pairing with a male on a superior territory than with a bachelor in an inferior one. The imbalance in sex ratio is not necessarily the only reason for polygamy.)

Verner, J., and Willson, M.F.
1966 The influence of habitats on mating systems of North American passerine birds. *Ecology* 47:143–147.
1969 Mating systems, sexual dimorphism, and the role of male North American passerine birds in the nesting cycle. *Amer. Ornith. Union, Ornith. Monogr.* No. 9.

Woolfenden, G.E.
1956 Comparative breeding behavior of *Ammospiza caudacuta* and *A. maritima*. *Univ. Kansas Publ. Mus. Nat. Hist.* 10:45–75. (Promiscuity in sharp-tailed sparrows.)

Zimmerman, J.L.
1966 Polygyny in the Dickcissel. *Auk* 83:534–547.

Nests and Nest-building

All receptacles for eggs laid by birds are called **nests**. In each bird species nests are remarkably similar in form and location; among different species they show wide diversity.

The Development of Nests

Early in their acquisition of **homeothermy** ("warm-bloodedness") birds could no longer abandon their eggs, as did their reptilian forebears, leaving them to hatch in the heat of the environment. The embryos within required the steady warmth of the parental body. This imposed on birds the necessity of incubating their eggs—sitting immobile on them for long periods of time. To offset what might well have been lethal exposure to predation and other adversities of the environment, birds simultaneously developed protective measures. One was to select sites for eggs that provided adequate cover and freedom from adversities; another was to build nests that would accommodate and shelter their eggs and themselves. Nest-building has evolved through three phases or types of construction—ground and cavity nests, platform nests, and cupped nests.

Ground and Cavity Nests

From the meager information available on the ancestry and descent of birds, and from what is known about nest-building habits today, one may assume that the earliest nests of birds were on the ground in depressions that the birds scraped out with the bill and feet and molded to the shape of their bodies by repeated turning. Some of the nests may also have been on the floors of natural cavities or cavities (including burrows, holes in trees) that the birds excavated themselves. All such nests were without structure, although they may have been lined with materials—e.g., plant stems, leaves, bits of shell, etc.—gathered from the immediate vicinity and sometimes with feathers from the incubating bird's body.

Among modern birds one finds both ground and cavity nests still in abundant use. Ground nests, for instance, are characteristic of loons, pelicans, gannets, swans, geese, most ducks, grouse, quail, pheasants, shorebirds, gulls, terns, murres, and goatsuckers. Probably the floating nests of grebes and the marsh nests of cranes, rails, gallinules, moorhens, and coots are derived from ground nests. Most birds with ground nests either have cryptic coloration to conceal them while they are on their nests or place their nests on islands or in other areas generally inaccessible to predators. Cavity nests are typical of storm-petrels; a few ducks, such as goldeneyes and mergansers; most vultures; puffins; auklets; parrots; trogons; kingfishers; and woodpeckers.

Platform Nests

The earliest nests to be elevated in vegetation—shrubs, trees, and even marsh plants—were essentially platforms of loosely assembled plant materials, without evident structure, but with a shallow depression for eggs. Platform nests perhaps represent the second stage in nest-building, differing principally from the first stage, ground and cavity nests, in being independent of a uniformly firm surface and comprised entirely of accumulated materials for holding the eggs as well as for lining the nest.

Modern birds that build platform nests include the anhingas, herons, storks, ibises, pigeons, and cuckoos. Platform nests are also built by cormorants, hawks, eagles, and the Osprey *(Pandion haliaetus),* but not always in trees. Some cormorants place their nests consistently on cliffs. A few species of hawks, both the Bald and Golden Eagles *(Haliaeetus leucocephalus* and *Aquila chrysaetos),* and the Osprey may place their huge nests either in trees (including stubs), on cliffs, or even (in the case of the Osprey) on the ground. The Northern Harrier *(Circus cyaneus)* nests regularly on the ground without building a platform.

Cupped Nests

The third and final phase of nest-building was the development of cupped nests, with true structure, consisting of materials arranged and compacted for the bottom and sides and softer materials inside for lining.

In building cupped nests birds succeeded in directly adapting structures to shrubs and trees from the crotches to the tips of branches. Presumably the first cupped nests were statant, supported mainly from below, with the rims standing firmly upright. The majority of hummingbirds and passerine species today build **statant nests.** A few species, such as magpies, have modified them by extending the sides upward and arching over the top in a dome.

Eventually, a number of species suspended their cupped nests from branches by rims and sides, without supporting them from below. These nests took two forms: **pensile nests,** as built by modern vireos, suspended from stiffly woven rims and sides; and **pendulous nests,** as built by modern orioles, suspended from rims and flexibly woven sides, with the deeply cupped lower parts swinging freely.

In the rapid multiplication of passerine species and the consequent increase in competition for living space, many new species moved into heretofore unoccupied niches of the environment and readapted cupped nests to suit the particular situations other than the crotches and branches of trees. Such birds as phoebes and several swallows, using adhesive substances, built their nests **adherent** to cliff walls and similarly vertical surfaces. Larks, pipits, several thrushes, and numerous paruline warblers, blackbirds, and emberizines constructed cupped nests on the ground in forests, in open country, or in marsh vegetation. Titmice, nuthatches, creepers, bluebirds, several swallows and wrens, and the Prothonotary Warbler *(Protonotaria citrea)* put their cupped nests in preformed cavities, or in rare cases—e.g., the Bank Swallow *(Riparia riparia)*—in cavities that they excavated themselves.

The nests of several groups of modern birds defy categorizing as to stages of development. For example, owls rarely, if ever, build nests of any sort. Unless they nest on the ground as do Snowy and Short-eared Owls *(Nyctea scandiaca* and *Asio flammeus),* they appropriate tree nests of other birds or use cavities. Chimney Swifts *(Chaetura pelagica)* construct "half-cupped" nests on vertical surfaces. To make the nesting materials stick together and the nests adhere to vertical surfaces, the birds apply their own saliva. Other species of swifts show wide diversity in nests and the use of saliva in forming them—e.g., the cave swiftlets (*Collocalia* spp.) of Asia form their nests almost entirely of coagulated saliva (see Lack, 1956, for a review of nest-building by world swifts). In eastern Indonesia, Polynesia, New Guinea, and Australia gallinaceous birds called megapodes (Megapodiidae) dig pits in the soil or build mounds of rotting vegetable matter in which they lay their eggs and then cover them, leaving them to be incubated by the surrounding heat. The mounds of one species *(Megapodius freycinet)* may measure up to 60 feet long, 15 feet wide, and 10 feet high (18.28 m by 4.57 m by 3.05 m) (Frith, 1962)—probably the largest "nests" of any bird.

Classification of Nests

It is possible to set up a classification of nest indicating their probable course of development and illustrating the diversity of nest types. Such a classification for North American birds (suggested in part by Herrick, 1911) is presented below with the groups or particular names of species showing each type. Bear in mind that the classification is purely artificial and does not necessarily portray the phylogenetic relationships of the nest-builders.

I. Ground and Cavity Nests. Nests without structure.

A. ***Ground Nests.*** Simple depressions with or without lining. Birds with ground nests either have cryptic coloration or select nest sites inaccessible to predators. Loons, grebes, most waterfowl, vultures, gallinaceous birds, shorebirds, gulls, terns, Black Skimmer *(Rynchops niger),* Short-eared Owl *(Asio flammeus),* goatsuckers.

B. ***Cavity Nests.*** Nests in caves, crevices, burrows, holes in trees, or birdboxes, with or without lining.

1. In preformed or natural cativites. Several ducks and owls, American Kestrel *(Falco sparverius).*
2. In cavities excavated by occupant birds. Leach's Storm-Petrel *(Oceanodroma leucorhoa),* Burrowing Owl *(Athene cunicularia),* kingfishers, woodpeckers.

II. Platform Nests. Nests usually elevated, without structure, consisting of loosely assembled materials with a depression for the eggs. Pelicans, cormorants, Anhinga *(Anhinga anhinga),* bitterns, herons, egrets, ibises, Wood Stork *(Mycteria americana),* most falconiform birds, rails, gallinules, coots, moorhens, pigeons, doves, cuckoos, some owls.

III. Cupped Nests. Nests adapted to crotches and branches of trees, with definite structure, the materials arranged and compacted to form a cup.

A. Used in crotches and branches of trees to which they are adapted.

1. *Statant Cupped Nests.* Nests with rims standing firmly upright, supported mainly from below. The sides may or may not be extended upward and arched over the top in a dome. Hummingbirds and many passerine species.
2. *Suspended Cupped Nests.* Nests not supported from below but from the rims, or sides, or both.
   a. *Pensile.* Nests suspended from the rims and sides; rather stiff. Kinglets and vireos.
   b. *Pendulous.* Nests suspended from the rims and sides; rather flexible and extremely deep-curved with lower part swinging freely. Orioles.

B. Readapted to other situations.

1. *Adherent Nests.* Cupped nests whose sides are attached by adhesive substance to a vertical surface. Swifts; phoebes; Cliff, Cave, and Barn Swallows *(Hirundo pyrrhonata, H. fulva,* and *H. rustica).*
2. *Ground Nests.* Cupped nests on the ground. The sides may or may not be extended upward and arched over the top, making a domed structure. Many passerine species.
3. *Cavity Nests.* Cupped nests in crevices, holes, birdboxes, etc.
   a. In preformed or natural cavities. Great Crested Flycatcher *(Myiarchus crinitus),* Purple Martin *(Progne subis),* Tree and Violet-green Swallows *(Tachycineta bicolor* and *T. thalassina),* titmice, nuthatches, Brown Creeper *(Certhia americana),* most wrens, European Starling *(Sturnus vulgaris),* Prothonotary Warbler *(Protonotaria citrea).*
   b. In cavities excavated by occupant birds. Northern Rough-winged Swallow *(Stelgidopteryx serripennis),* Bank Swallow *(Riparia riparia).*

Identification of Nests

Because nests of many closely allied species are very similar in location, materials, and structure, care must be taken in identifying any nest, with complete certainty, without knowing the bird that constructed it. But nests of certain species, or groups of closely allied species, are sufficiently distinctive to allow identification with reasonable certainty. In attempting to identify nests in the conterminous United States, refer to the guides by H.H. Harrison (1975, 1979).

Nest-building

The various phases of nest-building have been investigated the least of any subject in the breeding cycle of birds. The author's own studies and the works of many others have contributed to the outline of nest-building procedures that follow.

Selection of the Nesting Site

The need of suitable support and concealment, as well as the need of protection from the forces of the environment (e.g., sunlight, wind, cool night temperatures), governs the selection of the nesting site.

Almost invariably a period of searching during which birds move from one potential site to another—usually within the confines of their established territory—precedes the final selection of the nesting site. Chickadees investigate many potential holes before selecting one. In species that build cupped nests in trees the birds try fitting their bodies into crotches between branches; in species building the same type of nests on the ground, the birds scratch small depressions in the surface and attempt to mold them to suit the body contours. Often they accumulate a few nesting materials, and occasionally they partly construct nests at several sites before making the final selection.

The searching for a nest site is first strongly manifest after mating when the males are singing frequently and intensively and the females are receptive to males. They defend the territory strongly during this period. Nest-searching activities frequently appear following each sexual relation. The number of copulations during the days of the searching period is unknown.

Neither has the duration of the searching period been satisfactorily determined. In some cases it lasts from three to five days; in a few cases it is more prolonged. Searching, by no means continuous during the period, occurs at intervals, especially during the early hours of the morning. These intervals of searching increase in length daily as the nest-building drive matures.

The role of each sex in searching for and selecting the nesting sites varies with the species. In birds maintaining territories of Types A, B, and D one or both sexes participate. When both sexes participate, generally the female takes the more aggressive role. In birds with territories of Type C the female alone searches for and selects the nesting site.

The searching period represents the phase in the breeding cycle during which the nest-building drive develops. Even though birds find suitable nesting sites during this period, the selection is not final until the mated pairs arrive together on the sites chosen and demonstrate with mutual displays.

Beginning of Nest-building

In many birds nest-building begins with the initiation of rapid growth in the ovum (Riddle, 1911) and continues more or less intensively while the ovum matures. The first egg is usually laid one to several days after the completion of the nest, but in a few species nest-building continues to some extent after laying has begun. Climatic conditions may influence the beginning of nest-building. High temperatures at the start of the nesting season stimulate nest-building; low temperatures tend to inhibit it (Nice, 1937). Delayed development of vegetation may also delay the building.

Process of Nest-building

The process of nest-building may be roughly divided into three steps: preparing the site or support (e.g., scratching a depression, "cleaning out" a preformed cavity, excavating a new burrow or hole); constructing the floor and sides (i.e., the "outside"); and lining the nest. One or two of the steps do not occur, or occur only in part, in certain species.

Usually, the birds gather all "outside" materials for nests of Types II and III in the vicinity. For nests of Type I, the materials are frequently those within reach of the bird as it stands on the nest. But materials for lining the nests of Types II and III are often sought at great distances, presumably because the special materials are not readily available close at hand. Tree Swallows *(Tachycineta bicolor)* may sometimes fly several miles to a chicken farm to obtain the much-preferred white feathers.

Most birds carry materials in their bill, although diurnal birds of prey—eagles, hawks, etc.—commonly use their feet. The Peach-faced Lovebird *(Agapornis roseicollis)*, an African species, carries nesting materials amidst its feathers (Dilger, 1962).

Each bird builds its nest with a number of stereotyped movements. If the nest is to have structure, the movements required are necessarily more numerous and complex. In constructing a typical cupped statant nest in a tree the passerine bird sits in the center of the cup that is taking shape and performs the following characteristic movements: Pulling a long slender piece of material (e.g., a plant fiber) with its bill inward over the rim and tucking it into the wall under its breast or drawing it alongside against the wall; looping a similar piece of material around a supporting branch by starting it around one side and then reaching around the opposite side and drawing it back so that it encircles the branch—and sometimes repeating the performance until the material encircles the branch several times; inserting short pieces of material into the rim or wall with jabs of the bill; shaping the cup by squatting in it while alternately fluffing and compressing the body feathers; pressing its head, tail, and partly opened wings down against the rim of the cup, while alternately pushing with one foot and then the other against the floor and sides of the cup; turning frequently between and sometimes during any of the above movements.

Learning plays a role in nest-building to the extent that a bird must determine, for instance, by "trial and error" which materials are the most suitable for the structure and the lining. The principal movements in nest-building are complex but generally stereotyped motor patterns that may have evolved from maintenance activities, acts associated with foraging, and other behavioral patterns. For example, C.J.O. Harrison (1967) suggested that the behavior called "sideways-building," in which the bird pulls materials into its nest and draws it alongside, is derived from "sideways-throwing," an irrelevant behavior common in a number of shorebirds and gulls when the bird picks up a leaf, stick, shell, or any other small object and flicks it backward on one side with a quick motion of the head.

Participation and Behavior of the Sexes

In a great many species the female builds the nest alone. In many others the female builds the nest with the assistance of the male. The female gathers the materials, works them into place, and molds the depression. The male's assistance usually amounts to gathering materials with or without the female and, on returning with them, passing them to the female for use in the nest.

In a number of species the role of the sexes in nest-building does not conform to the above. The male may build the entire nest, although this is rare; build certain parts of it, such as the floor and sides, or excavate the burrow or hole; share all nest-building activities with the female. Within species the role of the sexes is subject to variation, but the variation is seldom as extreme as that among species.

There is apparently no dependable correlation among species between coloration of the male and his nest-building proclivities nor is there a correlation between his nest-building proclivities and his part in subsequent incubation. There is, on the other hand, a rather positive correlation between a male's nest-building proclivities and his participation in subsequent parental care—i.e., if a male assists in building, he will also assist in feeding the young.

During nest-building the male customarily sings as vigorously as at any other time during the breeding cycle. Copulation occurs throughout nest-building, though the frequency is not known.

Length of Time Involved

The total length of time involved in nest-building is difficult to determine because seldom is the beginning observed. Nests, because they are constructed in protected sites, are usually discovered as a result of building activities already under way. Furthermore, the cessation is sometimes indeterminable. Nest-building may stop suddenly a day or more before egg-laying, or it may continue for some time after egg-laying has begun.

The *number of days* is used as a measure of the time involved in the construction of nests. Due to the irregularities of nest-building in most species, no more accurate measure can be applied unless the observer has sufficient persistence and good fortune to follow daily activities from start to finish, in which case to count the total *number of hours* involved.

Remarkably few records are available to show even the number of days involved in nest-building. Most passerine birds may require about six (Allen, 1961), three for constructing the outside of the nest and three more for finishing the interior and lining it. But judging by the few precise records at hand, the length of time is subject to wide interspecific and intraspecific variation.

Interspecific Variation Interspecific variation is accounted for in part by two factors: type of nest and climate.

Type of Nest Obviously, certain nests that are more elaborate than others require more time for construction. To build their long, pendulous nests, the Altamira Oriole *(Icterus gularis)* in Mexico took "at least 18 and perhaps as many as 26 days" (Sutton and Pettingill, 1943) and the Chestnut–headed Oropendola *(Psarocolius wagleri)* in the Canal Zone, Panama, "about one month" (Chapman, 1928).

Climate Species in the tropics take more time than closely allied species in the temperate regions. Great Kiskadees *(Pitangus sulphuratus)*, large tyrannids in Mexico, took 24 days (Pettingill, 1942) whereas most tyrannids farther north take only three to rarely more than 13 days for finishing their nests.

Intraspecific Variation Two factors account in part for the intraspecific variation in time required for nest-building.

Weather Conditions Cool or inclement weather may retard nest-building. On the other hand, rainfall may stimulate it, at least in the case of the European Wren *(Troglodytes troglodytes)*, because it makes the nesting material more flexible and easier to manipulate (Armstrong, 1955).

Renesting An individual, or pair, building a second or third nest in the same season takes less time.

Though the length of time taken in nest-building is measured in number of days, nest-building does not proceed steadily during those days. Usually only the early parts of the day are involved, at which time there are periods of building (i.e., attentive periods) and periods of no building (i.e., inattentive periods). As nest-building progresses, the attentive periods lengthen and inattentive periods shorten. At the height of nest-building the average length of attentive and inattentive periods often quite closely corresponds to the average length of such periods during incubation.

A few birds modify or repair their nests after they have laid eggs and begun incubation, or even later. Bald Eagles frequently add materials to their large aeries throughout the nesting season. The marsh-dwelling rails, gallinules, moorhens, and coots are noted for their ability to build up their nests whenever the water rises to keep the flood from their eggs. The King Rail *(Rallus elegans)* may elevate its eggs in this way by as much as a foot, 30.48 cm, (Meanley, 1953).

Nest-building in Young Birds

The nest-building abilities of young birds need investigation. From meager evidence it appears that younger birds may build their nests as quickly and expertly as older birds (Nice, 1943). This is perhaps not true among species that build elaborate nests.

In rare instances young from a nest built early in a season assist adults in building a second nest later in the season. Independent attempts at nest-building have been seen among young birds-of-the-year.

False Nests

Certain species of birds build structures called "nests" that are not true nests because they are not constructed for the purpose of containing eggs.

Multiple Nests

These are nests constructed by males, notably wrens, in the vicinity of the regular nest. They resemble the regular nest, but they are not completely constructed and are without lining. Possibly the incomplete nests tend to direct attention of predators away from the regular nest or to accustom predators to so many nests as to discourage further search. In the House Wren *(Troglodytes aedon)*, which often nests in boxes, the male may fill all those available with his nests as if to prevent their occupancy by competitors.

Refuge "Nests"

A few birds, especially burrow- and hole-nesters, often create "nests," similar to their regular nests, for roosting and for shelter during unfavorable weather. They may be constructed at any time of year but particularly in the fall. Some birds merely use cavities already available.

Re-use of Nests

A few species of birds, notably some of the large hawks, the Bald and Golden Eagles, and the Osprey, use the same nest year after year. When the same nest is used, it is generally either "repaired" and enlarged each year, or a completely new nest is built on top of the old.

Certain species appropriate the nests used by other species in the preceding season. Sometimes the new tenants alter the nests to suit their own requirements. A Mourning Dove *(Zenaida macroura)*, for instance, may build its platform nest on top of the cupped nest of an American Robin *(Turdus migratorius)*. The new tenants, however, may accept the nest as it is. This is true of a Great Horned Owl *(Bubo virginianus)* taking over a hawk or eagle nest and of the Solitary Sandpiper *(Tringa solitaria)* using the cupped tree nest of an American Robin, Rusty Blackbird *(Euphagus carolinus)*, or Eastern Kingbird *(Tyrannus tyrannus)*.

Among passerine birds, as well as among many other groups, it is exceptional for pairs to renest in the same nest or at the same nesting site.

Protection of Nests

Nests are often concealed through the choice of nesting sites. If they are not concealed, then they may be inaccessible as when nests are placed on small islands, on the shelves of cliffs, and near the ends of slender branches. The theory that snake skins, sometimes found in nests, serve as a means of frightening enemies has been discredited (see Rand, 1953).

Nests may gain protection against predators when they are placed close together in a colony, or singly in a colony of other species, because there are many more occupant birds on the alert to attack and take deterrent action. Nests in tropical regions also gain protection when they are placed—deliberately by some species—in the immediate vicinity of aggressive social ants and wasps, or near the nests of larger, more aggressive bird species. In Mexico Pettingill (1942) found one nest each of the Social Flycatcher *(Myiozetetes similis)* and the larger Great Kiskadee in a bull's horn acacia, a shrub about 12 feet (3.66 m) high, that was tenanted by countless thousands of small ants ready to bite and sting. Although the ants did not annoy the nest occupants, they viciously swarmed over anyone touching the shrub. Both nests obviously benefited from the protection afforded by the bellicose ants, while the nest of the Social Flycatcher benefited further by being close to the nest of the larger and more belligerent Great Kiskadee.

Nest Fauna

Nests of birds, particularly those in burrows, cavities, or birdboxes or under the eaves of buildings, are snug havens for many small invertebrates, especially arthropods—mites, spiders, insects, etc. Most of them are visitors stopping only for shelter. (For an idea of the variety of insects that may be found in birds' nests, see the checklist by Hicks, 1959.) But along with the visitors are a number of flies and fleas that pass their life cycles specifically in birds' nests and at one stage are parasitic on the birds. They are to be distinguished from the obligate ectoparasites on the feathers and/or other parts of birds.

Chief among the nest parasites, from the viewpoint of their serious effect on the occupant birds, are blowflies *(Protocalliphora* spp.*)*, whose larvae live in the nest cup and, beginning at twilight, attach themselves on the nestlings for a blood meal during the night. If the nest infestation by larvae is heavy, the nestlings can be greatly weakened, even killed, from loss of blood.

For extensive information on the subject of nest fauna, consult Boyd (1951), George and Mitchell (1948), Herman (1937), Hill and Work (1947), Kenaga (1961), Mason (1944), Nolan (1955), and Rothschild and Clay (1957).

The North American Nest Record Program

The North American Nest Record Program, headquartered at the Cornell Laboratory of Ornithology, collects, processes, and stores information on the nesting of birds throughout the United States and Canada. When a nest is found, the species, location, habitat, and reproductive history are reported by the observer on a card specially prepared by the Program at no extra charge. More than 250,000 cards have been received since the Program was initiated in 1965. Data obtained are available to any qualified amateur or professional ornithologist interested in avian reproduction. Anyone wishing to participate in the Program by supplying data on cards, or to use the available data for their research, should write to the Laboratory (Sapsucker Woods, Ithaca, New York 14850) for a folder describing the Program and giving instructions for application.

Selected Studies

Select a species for the study of nests and nest-building and concentrate on the activities of one pair. Do not disturb the nests. (In Appendix A, pages 372–375, read the federal regulations that prohibit taking or otherwise destroying nests of protected species.)

Procedure When possible, begin the study soon after the territory is established, thus observing the selection of the nesting site and the start of nest-building. Follow all activities of the pair through the searching and nest-building periods. Practice extreme caution in watching the birds during the nest-building, since they are not strongly attached to the nesting site and will abandon it on the slightest provocation. Therefore, use a binocular or telescope and remain as far away as possible.

Take full notes on the activities of both sexes; time their respective attentive and inattentive periods, and the periods when the male is singing. When the birds are not present at the nest, take photographs of stages in construction, and make careful notes on the appearance of the nest at each stage.

Obtain full data on the completed nest, particularly the following:

Measurements *If ground nest without structure:* inside diameter (i.e., diameter of the depression); inside depth (i.e., depth of the depression). *If cavity nest without structure:* length, or depth, of cavity from lower edge of entrance to floor; diameter (or diameters) of the entrance and diameter (or diameters) of the part of cavity containing the eggs. *If elevated nest or readapted ground nest:* outside diameter, outside depth; inside diameter, inside depth. (If nests are domed, consider outside depth to be from top of arch, and inside depth from the under surface of arch.) *If readapted cavity nest:* use the measurements indicated above for directly adaptive cavity nests plus the measurements of the inside depth and inside diameter of the nest structure built in the cavity.

Location Take notes on the location and support of the nest, naming all dominant vegetation in the vicinity and observing the condition and protective value. If the nest is in a tree, measure its height from the ground. Use an altimeter or clinometer (see Appendix A, p. 367) in case the height is too great for accurate determination by measuring stick or tape. Also measure or estimate the distance of the nest from the main trunk of the tree. If the nest is in a cavity, measure the distance from the lower edge of the entrance to the ground and note the direction (using a compass) that the entrance faces. Photograph the environment, with the nest occupying a central position in the background.

Description of the Nest Take two or three photographs that show the support of the nest and the appearance of the nest close up. Disturb the surrounding cover as little as possible. When taking the close-up pictures of an open nest, place the camera partly above and partly to the side so that the far inside wall and near outside wall will show.

After the young have left, the nest should be taken to the laboratory and studied in detail. Its structure should be analyzed, all materials identified, and the relative quantity of each material determined.

If, after a nest is completed, it is for some reason destroyed or deserted, attempt to follow the same pair and gather evidence of renesting. The following data are particularly desirable: (1) Time involved between the destruction or desertion of the first nest and the completion of the second. (2) Variation among individuals in the selection of the second nesting site and manner of building. (3)

Proximity of the first and second nests. Compare the data with the findings of other investigators.

Presentation of Results Draw up a report on the nests and nest-building of the pair observed. Prepare the manuscript as directed in Appendix B. An outline of suggested topics is given below.

Territory Type; brief description; date discovered.

Selection of Nesting Site Time and date of searching period; relation to weather conditions; role and activities of sexes, including number and duration of singing periods of male and number of copulatory acts per day; number and description of nest-building attempts; time of day and total length of time (in hours) devoted to searching; searching as related to territory and territorial behavior.

Beginning and Duration of Nest-building Brief statements of date of beginning and the duration. Discussion of weather, vegetation, and other ecological conditions as related to the time of beginning.

Nest-building Detailed chronological account of the stages and mechanics of nest-building. This should include the role of the sexes, number of copulations per day, times when the male sings, manner of gathering materials, sources of materials, and number of material-gathering trips. Discuss the length of time involved separately. Include the hours of the day when nest-building takes place, a table summarizing attentive and inattentive periods, and statements indicating the total amount of time spent in building the nest. (In deriving the table, follow the directions for the table on incubation periods given in the next chapter.) Use photographs to illustrate stages in construction.

Description of Nest *Location:* Describe in full, and give height from ground, if elevated. Use photographs to illustrate the nesting site. Discuss protective factors, etc. *Measurements:* Include the measurements as previously directed. *Structure:* Note any details of structure not determined when watching the nest under construction. Analyze the materials, using a wheel diagram to show relative quantities of different materials. Include photographs of the nest.

Miscellaneous Observations Present any observations made on renesting, building ability of younger birds, presence of false nests, protection afforded the nest, and nest fauna.

References

Allen, A.A.

1961 *The book of bird life*. 2nd ed. New York: D. Van Nostrand. (Chapter 16 for information on finding nests and identifying them. A key to the nests of birds of northeastern North America is given on pages 317–324.)

Armstrong, E.A.

1955 *The Wren*. London: Collins. (Chapter 9: Nest-building.)

Austin, G.T.

1974 Nesting success of the Cactus Wren in relation to nest orientation. *Condor* 76:216–217. (Nests face predominant wind direction during warm part of breeding season.)

Beer, C.G.

1966 Incubation and nest-building behavior of the Black-headed Gull. *Behaviour* 26:189–214. (A good study of nest-building from the ethological viewpoint).

Blanchard, B.D.

1941 The White-crowned Sparrows *(Zonotrichia leucophrys)* of the Pacific seaboard: Environment and annual cycle. *Univ. Calif. Publ. in Zool*. 46:1–78. (Compression of nesting cycle in northern latitudes as compared with southern.)

Boyd, E.M.

1951 The external parasites of birds: A review. *Wilson Bull*. 63:363–369.

Brackbill, H.

1950 Successive nest sites of individual birds of eight species. *Bird-Banding* 21:6–8. (Data on 40 nests indicate fairly fixed nesting heights for some individuals and highly variable ones for others.)

1952 A joint nesting of Cardinals and Song Sparrows. *Auk* 69:302–307.

Chapman, F.M.

1928 The nesting habits of Wagler's Oropendola *(Zarhynchus wagleri)* on Barro Colorado Island. *Bull. Amer. Mus. Nat. Hist*. 58:123–166.

Collias, N.E., and Collias, E.C.

1964 Evolution of nest-building in the weaverbirds

(Ploceidae). *Univ. Calif. Publ. in Zool.* 73:i–viii; 1–239. (Concerned with some two dozen selected and representative species of weaver-birds that show great range in variation in their nests. An illuminating paper on the manner in which modes of nest-building have evolved and the functions they serve.)

Davis, C.M.
1978 A nesting study of the Brown Creeper. *Living Bird* 17:237–263. (Nests built primarily by females, although both sexes may carry nesting materials.)

Dilger, W.C.
1962 Methods and objectives of ethology. *Living Bird* 1:83–92.

Emlen, J.T., Jr.
1954 Territory, nest building, and pair formation in the Cliff Swallow. *Auk* 71:16–35.

Frith, H.J.
1962 *The Mallee-fowl: The bird that builds an incubator.* Sydney: Angus & Robertson.

George, J.L., and Mitchell, R.T.
1948 Notes on two species of Calliphoridae (Diptera) parasitizing nestling birds. *Auk* 65:549–552.

Harrison, C.J.O.
1967 Sideways-throwing and sideways-building in birds. *Ibis* 109:539–551.

Harrison, H.H.
1975 *A field guide to birds' nests of 285 species found breeding in the United States east of the Mississippi River.* Boston: Houghton Mifflin.
1979 *A field guide to western birds' nests of 520 species found breeding in the United States west of the Mississippi River.* Boston: Houghton Mifflin.

Herman, C.M.
1937 Notes on hippoboscid flies. *Bird-Banding* 8:161–166.

Herrick, F.H.
1911 Nests and nest-building in birds. *Jour. Animal Behavior* 1:159–192; 244–277; 336–373.
1934 *The American Eagle*. New York: D. Appleton-Century. (Nest-building and re-use of nests year after year.)

Hicks, E.A.
[1959] *Check-list and bibliography on the occurrence of insects in birds' nests.* Ames: Iowa State College Press. (Contains two checklists: one of insects found in birds' nests and one of birds in whose nests insects have been found. Both lists contain many hundreds of references to a 68-page bibliography.)

Hill, H.M., and Work, T.H.
1947 Protocalliphora larvae infesting nestling birds of prey. *Condor* 49:74–75.

Kenaga, E.E.
1961 Some insect parasites associated with the Eastern Bluebird in Michigan. *Bird-Banding* 32:91–94. (One nest contained 2,300 insects, dependent directly or indirectly on the nestlings.)

Kendeigh, S.C.
1952 Parental care and its evolution in birds. 2nd corrected printing, 1955. *Illinois Biol. Monogr.* 22:i–x; 1–356. (Contains a considerable amount of information on nest-building in many species.)

Kilgore, D.L., Jr., and Knudson K.L.
1977 Analysis of materials in Cliff and Barn Swallow nests: Relationships between mud selection and nest architecture. *Wilson Bull.* 89:562–571. (Significant differences in mud composition.)

Kilham, L.
1977 Nesting behavior of Yellow-bellied Sapsuckers. *Wilson Bull.* 89:310–324. (Role of males in excavating cavities.)

Kuerzi, R.G.
1941 Life history studies of the Tree Swallow. *Proc. Linnaean Soc. New York* Nos. 52–53: 1–52. (Nest-building.)

Lack, D.
1956 *Swifts in a tower.* London: Methuen.

Laskey, A.R.
1950 A courting Carolina Wren building over nestlings. *Bird-Banding* 21:1–6.

Lawrence, L. de K.
1953 Nesting life and behaviour of the Red-eyed Vireo. *Canadian Field-Nat.* 67:47–77. (Nest-building.)
1967 A comparative life-history study of four species of woodpeckers. *Amer. Ornith. Union, Ornith. Monogr.* No. 5. (Excavating nests and related behavior.)

Lea, R.B.
1942 A study of the nesting habits of the Cedar Waxwing. *Wilson Bull.* 54:225–237. (Nest-building.)

Low, S.H.
1934 Nest distribution and survival ratio of Tree Swallows. *Bird-Banding* 5:24–30. (Information concerning re-use of nests.)

Martin, D.J.
1973 Selected aspects of Burrowing Owl ecology and behavior. *Condor* 75:446–456. (Returning males occupy same burrows of previous season,

but females show no strong attachment to any particular burrow.)

Mason, E.A.

1944 Parasitism by Protocalliphora and management of cavity-nesting birds. *Jour. Wildlife Mgmt.* 8:232–247.

McClaren, M.A.

1975 Breeding biology of the Boreal Chickadee. *Wilson Bull.* 87:344–354. (Nest selection and building.)

Meanley, B.

1953 Nesting of the King Rail in the Arkansas rice fields. *Auk* 70:261–269.

Nauman, E.D.

1930 The nesting habits of the Baltimore Oriole. *Wilson Bull.* 42:295–296. (Information on length of time in building.)

Nice, M.M.

1937 Studies in the life history of the Song Sparrow, I. *Trans. Linnaean Soc. New York* 4: i–vi; 1–247. (Chapter 10.)

1943 Studies in the life history of the Song Sparrow, II. *Trans. Linnaean Soc. New York* 6: i–viii; 1–328. (Chapter 17.)

Nickell, W.P.

1943 Secondary uses of birds' nests. *Jack-Pine Warbler* 21:48–54.

Nolan, V., Jr.

1955 Invertebrate nest associates of the Prairie Warbler. *Auk* 72:55–61.

1978 The ecology and behavior of the Prairie Warbler *Dendroica discolor. Amer. Ornith. Union, Ornith. Monogr.* No. 26. (Chapters 12–17 especially important on nest selection and building.)

Odum, E.P.

1941 Annual cycle of the Black-capped Chickadee-2. *Auk* 58:518–535. (Nest-building, role of sexes, etc.)

Pettingill, O.S., Jr.

1942 The birds of a bull's horn acacia. *Wilson Bull.* 54:89–96. (Nest-insect relationships; comparison of length of building periods between temperate and tropical species; small species nesting in vicinity of larger species.)

Rand, A.L.

1953 Use of snake skins in birds' nests. *Chicago Acad. Sci. Nat. Hist. Miscellanea*, No. 125.

Riddle, O.

1911 On the formation, significance and chemistry of the white and yellow yolk of ova. *Jour Morph.* 22:455–490.

Rothschild, M., and Clay, T.

1957 *Fleas, flukes, and cuckoos: A study of bird parasites*. 3rd ed. London: Collins. (Chapter 14: The Fauna of Birds' Nests.)

Schaefer, V.H.

1976 Geographic variation in the placement and structure of oriole nests. *Condor* 78:443–448. (Mainly in response to differences in vegetation but also in response to variation in environmental conditions.)

1980 Geographic variation in the insulative qualities of nests of the Northern Oriole. *Wilson Bull.* 92:466–474. (Significant correlation with local temperature.)

Skowron, C., and Kern, M.

1980 The insulation in nests of selected North American songbirds. *Auk* 97:816–824.

Skutch, A.F.

1940 Social and sleeping habits of Central American wrens. *Auk* 57:293–312. (Nests used as places for sleeping.)

Smith, A.G.

1968 The advantage of being parasitized. *Nature* 219:690–694. (Nesting tropical icterines given protection by hymenopterans nestings in the same trees.)

Stoddard, H.L.

1931 *The Bobwhite Quail: Its habits, preservation and increase*. New York: Charles Scribner's Sons. (Nests, nesting sites, and nest-building in Chapter 2.)

Strecker, J.K.

1926 On the use, by birds, of snakes' sloughs as nesting material. *Auk* 43:501–507.

Summers-Smith, J.D.

1963 *The House Sparrows*. London: Collins. (Chapter 6: Nests and Nest-building.)

Suthard, J.

1927 On the usage of snake exuviae as nesting material. *Auk* 44:264–265.

Sutton, G.M., and Pettingill, O.S., Jr.

1943 The Alta Mira Oriole and its nest. *Condor* 45:125–132. (Nest-building; comparison of nest-building periods in several species.)

Tinbergen, N.

1935 Field observations of East Greenland birds, I. The behaviour of the Red-necked Phalarope *(Phalaropus labatus L.)* in spring. *Ardea* 24:1–42. (Searching; nest-building; role of sexes.)

1953 *The Herring Gull's world: A study of the social behaviour of birds*. London: Collins. (Chapter 15 concerns nest-building.)

Verner, J., and Engelsen, G.H.

1970 Territories, multiple nest-building, and polygyny

in the Long-billed Marsh Wren. *Auk* 87: 557–567.

Walsberg, G.E.

1981 Nest-site selection and the radiative envir onment of the Warbling Vireo. *Condor* 83: 86–88. (Effective selection for thermal advantage.)

Welter, W.A.

1935 The natural history of the Long-billed Marsh Wren. *Wilson Bull*. 47:3–24. (Nest-building, multiple nests.)

Williams, L.

1942 Interrelations in a nesting group of four species of birds. *Wilson Bull*. 54:238–249. [Western Flycatcher, Brown Creeper, Bewick's Wren, Dark-eyed (Oregon) Junco. Use of man-made structures as nest sites, re-use of nests, roosting in nests, etc.]

Woolfendon, G.E.

1973 Nesting and survival in a population of Florida Scrub Jays. *Living Bird* 12:25–49. (Both pair members construct nests.)

Eggs, Egg-laying, and Incubation

All species of birds lay eggs, and all with few exceptions incubate them with the heat of their own bodies.

Size, Shape, and Coloration

Size

The largest eggs are laid by the largest birds and the smallest by the smallest, but this proportion of egg size to body weight does not apply to all birds. The largest egg known, that of the extinct elephant bird *(Aepyornis)*, measures 14.5 by 9.5 inches (36.83 cm by 24.13 cm) and is assumed to have held two gallons (7.57 l) of fluid that may have weighed as much as 27 pounds (12.25 kg)—perhaps less than 3 percent of the adult bird's weight; and the egg of an Ostrich *(Struthio camelus)* measures roughly 7 by 5.5 inches (17.78 cm by 13.97 cm) and weighs nearly 3 pounds (1.36 kg)—about 1.7 percent of the body weight. At the opposite extreme, the egg of one of the smallest hummingbirds measures 13 by 8 millimeters and weighs 0.5 gram—around 10 percent of the parent's weight. From these figures, one may deduce correctly that the larger birds lay proportionately smaller eggs. But there are exceptions. The chicken-size Kiwi *(Apteryx* spp.*)* produces an egg which, measuring nearly 5.5 by 3.5 inches (13.97 cm by 8.89 cm) and weighing approximately a pound (2.54 kg) constitutes as much as 25 percent of the body weight.

Sometimes bird species that are the same size, such as the Common Snipe *(Gallinago gallinago)* and the American Robin *(Turdus migratorius)*, lay eggs differing greatly in size. The egg of the Common Snipe, a shorebird, is larger because it yields a well-developed downy young bird or chick whereas the egg of the robin, a passerine bird, produces only a helpless nestling. Within species, the size of eggs normally varies in accordance with age, the younger birds in their first year of nesting laying smaller eggs.

Shape

Like the eggs of the Domestic Fowl *(Gallus gallus)*, the eggs of most birds are somewhat rounded at one end and bluntly pointed at the other. The shape may vary from this in two directions: toward the two ends matching each other in shape or toward the two ends becoming respectively more pointed and more rounded. Thus the eggs of Ostriches, owls, and kingfishers are spherical, those of grebes are equally pointed at both ends, and those of hummingbirds are elongate and equally blunt at both ends. In the other direction the eggs of shorebirds, quail, and murres *(Uria* spp.*)* are remarkably pointed at one end, large and rounded at the other. Such exaggerated shapes are probably adaptations: in the shorebirds, so that the normal clutch of four large eggs may be incubated with the pointed ends toward the nest's center, thereby occupying minimum space; in quail, so that the normally large clutch of 10 or more eggs will take up less space; and in the murres, so that the single egg of each pair of birds on a bare nesting ledge will turn in a circle when disturbed and not roll off to its destruction.

Coloration

The shells of eggs, even though always penetrated by countless microscopic pores, vary widely in surface texture from being smooth, as in most birds, to being on the one hand quite glossy, as in woodpeckers, and even spectacularly shiny, as in the tinamous (Tinamidae), a New World group of birds, or being on the other hand noticeably rough or chalky, as in cormorants.

Many eggs of widely different species are white, but the majority of eggs show a great range in colors from faint to intense. In some species the eggs have simply a uniform **ground color** from buff to reddish brown or from pale blue to deep blue-green. This color is deeply suffused in the calcareous material forming in the shell. In other species the eggs have in the outermost stratum of the shell an array of pigments, with or without the underlying ground color, that provide the **markings**—blotches, scrawls, streaks, speckles, etc. Quite often the markings tend to be concentrated in a wreath around the large end, since, in the egg's final descent through the oviduct, the large end comes first, picking up the major supply of pigments from the cellular walls.

Because reptiles, the forebears of birds, probably produced white eggs, one may assume that the eggs of the earliest birds were similarly white. Then, in the course of avian evolution, the eggs acquired first the ground color and second the markings for adaptive purposes.

Heavy pigmentation is characteristic of eggs in open nests, particularly in ground nests, where it serves a dual function: to shield the embryos within from intensive solar radiation and to provide cryptic patterns for concealing the eggs from the view of predators. Light pigmentation or none at all is characteristic of eggs in cavity nests. Bluebirds *(Sialia* spp.*)* which nest in cavities have pale blue eggs, while their familial relatives, robins and thrushes, which construct open nests, have vividly blue eggs. Not all cavity-nesting birds lay pale or colorless eggs. Those of chickadees and nuthatches are spotted and, inexplicably, the Marsh Wren *(Cistothorus palustris)* lays chocolate brown eggs and the Sedge Wren *(Cistothorus platensis)* nearly white eggs in the dark interiors of their similarly ball-shaped nests.

Identification of Eggs

An examination of a collection of birds' eggs shows the wide range in size and shape and the almost infinite variety of colors and markings. At the same time it reveals the hopelessness of attempting either a classification or a key based on their differences. Relatively few eggs are as distinct as the species producing them. Within a species there can be remarkable diversity in color. A collection of murre eggs from one colony will show every possible combination of ground colors and markings. Between species that are closely related the intergradation is sometimes so perfect as to exclude distinction. It is advised, therefore, never to attempt positive identification of eggs without direct knowledge of the birds laying them. For descriptive information and aids to identification, refer to C. Harrison (1978), H.H. Harrison (1975, 1978), and Reed (1904).

Number of Eggs in a Clutch

A **clutch,** or **set,** of eggs is the total number of eggs laid by one bird in one nesting. A few groups of species characteristically lay one egg (e.g., albatrosses, shearwaters, storm-petrels, diving petrels, tropicbirds, frigatebirds, most alcids), a few other groups lay two (loons, goatsuckers, most pigeons, hummingbirds), and most shorebirds lay four. With the exception of shorebirds, practically all other species normally laying more than two eggs show marked variation in clutch size.

As a rule, the greater the characteristic size of a clutch in a species, the greater the variation. Nice (1943) showed that clutch size in the Song Sparrow *(Melospiza melodia)* is influenced by the following factors:

1. *Age*. Young females may lay smaller clutches the first season.

2. *Weather conditions*. Cold weather may reduce the size of a clutch.

3. *Time of season*. Smaller clutches may be laid at the end of the season by birds that have laid larger clutches earlier. This was confirmed by von Haartman (1967) in his studies of the Pied Flycatcher *(Ficedula hypoleuca)* nesting in Finland. See also Howard (1967) for seasonal variation in the American Robin.

4. *Individual variation*. Some adults lay typically smaller clutches than others.

The same or closely allied species, nesting in areas widely separated geographically or ecologically, may show significant differences in clutch sizes. Thus birds nesting on sea islands may have smaller clutches than those on continents (e.g., see Crowell and Rothstein, 1981; Pettingill, 1960), and birds nesting on sea coasts or in the tropics generally produce fewer eggs per clutch than those in the interior of continents or in temperate regions. There are at least two explanations for this phenomenon. Clutch size, in the view of Lack (1954, 1966, 1968), is adapted to the amount of available food that the parents can provide for the resulting brood so that all of its members may survive. In other words, the more food for the young because of more daylight for foraging, as in temperate regions, the larger the number of eggs that can be laid. Cody (1966) theorized that the stability of environment determines the clutch size. Predator-free conditions as on sea islands and a more uniform climate as on sea islands, along sea coasts, and in the tropics favor a smaller clutch because the resulting young are subjected to fewer risks. Instability, conversely, means larger clutches. Cody cited as an example the Bay-breasted, Cape May, and Tennessee Warblers *(Dendroica castanea*, D. *tigrina*, and *Vermivora peregrina)*, all of which depend on an unstable food supply, such as the irregular outbreaks of spruce budworm, and consequently lay clutches of 5–6, 5–7, and 5–6 eggs, respectively—more eggs per clutch than most other parulines whose normal clutch is 4, rarely 5.

Among passerine species generally there tends to be a positive correlation between the size of the clutch and the type of nest, the clutches averaging larger in cavity nests than in all other types.

Occasionally nests of certain ducks and a few other birds in one area contain many more eggs than in a normal clutch. This is the result of two or more females laying in the same nest.

A few wild birds continue laying beyond the usual number of their clutch when eggs are removed from the nests (e.g., see Phillips, 1887). Such birds are called **indeterminate egg-layers.** Most birds are **determinate egg-layers**—they lay a definite number of eggs per clutch. If the eggs are removed, the birds desert the nests after they have laid the normal number.

Number of Clutches per Breeding Season

Most species characteristically lay only one clutch a year since the time remaining in the breeding season, after the young are reared, is not long enough for repeating the nesting cycle. For a few species of large birds whose young develop at an exceptionally slow rate, even one year is not long enough. This is the case with the Wandering Albatross *(Diomedea exulans)* and California Condor *(Gymnogyps californianus)*, which nest every other year (Tickell, 1968; Koford, 1953), and the King Penguin *(Aptenodytes patagonica)*, which nests twice in three years (Stonehouse, 1960).

Among many passerine birds and pigeons and in a few other species two or more clutches a year are common, particularly in regions where ecological conditions favor a long or unending season for breeding. Species, for example, nesting in southern United States produce more clutches per year than other species in northern United States and Canada; and, similarly, individuals of some species, whose breeding range extends from southern United States into Canada, produce more clutches per year in the south than in the north. This same situation applies to a few sea birds nesting on islands with an equable year-round climate and a continuously available food supply. On Ascension Island in the tropical Atlantic where there is little seasonal change, the Sooty Tern *(Sterna fuscata)* breeds at intervals of 9.6 months or roughly five times in every four-year period (Chapin, 1954; Ashmole, 1963), but on other islands with more seasonal conditions it breeds annually.

Probably all but a few species will replace the first clutch if it is destroyed, although no thorough investigation has ever been made to determine the "limits" of clutch replacement following successive destructions. Presumably the more clutches per season the species characteristically lays, the more clutches it will replace if they are destroyed. The time required to replace the clutch is worthy of study. Nice (1943) found that, in the case of the Song Sparrow, the first egg of the new clutch is laid five days after destruction of a clutch.

Egg-laying

Egg-laying, like nest-building, is a phase of the breeding cycle in need of careful investigation. Several topics particularly worthy of attention follow. Many of the statements are based on findings in a few passerine species and others and must not be construed as being applicable to all birds.

Start of Egg-laying

Generally egg-laying begins when the nest is completed. Actually egg-laying may start sometimes before and frequently from one to several days after the nest is completed.

Time of Egg-laying

Birds usually lay very early in the morning. Sometimes the female enters the nest the evening before; ordinarily she does so between the break of day and sunrise. She lays one egg each morning—i.e., every twenty-four hours—until the clutch is complete. The female sits on the nest for a brief time, before and after laying, for an average of perhaps fifty minutes. The time on the nest tends to lengthen by a few minutes with each egg laid. Full incubation behavior does not usually begin until she has completed the clutch.

Behavior of the Sexes

During the egg-laying period—i.e., the period of days during which the female completes the clutch—her behavior is in marked contrast to that during nest-building. She spends a considerable amount of time in leisurely preening, especially following the interval spent on the nest for egg-laying; she moves casually over the territory and searches for food; she rests for relatively long intervals. She seemingly ignores the nest and nesting site between layings.

Except in birds with breeding territories of Type C the male is usually in close attendance upon the female when she is not on the nest. He accompanies her to the nest and rejoins her when she leaves. He customarily sings vigorously during the egg-laying period. Copulations are frequent at any time of day except immediately after egg-laying.

In birds with territories of Type C the female may join the male in the mating area after egg-laying and remain for an undetermined period. Elsewhere than in the mating area the sexes are not in contact either during egg-laying or during subsequent phases of the breeding cycle.

Incubation

Incubation—the process by which the bird applies its body heat to the eggs—and the accompanying behavior have been given more attention than many other phases of the breeding cycle, but the subject is far from exhausted; information is still lacking for a wide variety of species.

Development of the Incubation Patch

In most species of birds prior to incubation one or more **incubation patches** develop on the ventral surface of the body. Each patch consists of a feather-free area with thickened skin and a rich supply of blood vessels to facilitate the transfer of heat from the body of the incubating bird to the eggs. In the Song Sparrow, the feathers where the patch is to develop are lost four to six days before the bird lays the first egg (Nice, 1937). The patches persist through the incubation period and into the early part of the brooding period; the feathers grow back again with the acquisition of the winter (basic) plumage. The majority of species, including passerines, have only one patch, situated in the median apterium; but some species have two patches, one in each lateral apterium; and other species have three patches, one in each lateral apterium and one in the median. The paper by Bailey (1952) shows how the incubation patch develops in passerine birds.

Start of Incubation

As a rule, the bird begins sitting on the nest before the last egg is laid. Sometimes, however, it may begin from one to three days after the last egg is laid, or sometimes, as in several lower orders, it may begin after the laying of the first egg. Sitting on its nest, before the clutch is complete, may or may not involve fully warming the eggs, particularly if the incubation patch is not developed; thus its presence on the nest may not always indicate that incubation has actually begun. Incubation is actually under way only when the bird applies *maximum* body heat to the eggs.

Participation and Behavior of the Sexes

Among all bird species the role of the sexes in incubation is subject to every conceivable variation. Skutch (1957), after analyzing this diversity from his own observations and published information, prepared a synopsis or "key" to incubation patterns. He concluded, as Kendeigh (1952) had concluded earlier, that because the prevailing mode of incubation is by both sexes, it is therefore probably the primitive method. From this initial pattern all the other modes have diverged, from greater male to greater female participation to sole participation by one sex or the other.

In North American species incubation by the female only is by far the more common pattern. Although the male and female share incubation in a number of species, incubation by the male only is very rare.

There is some correlation between the coloration of the male and his role in incubation. If he is more colorful than the female, he usually never sits on the eggs; if his coloration is similar to that of the female, he sometimes participates; if he is less colorful than the female, he alone sits on the eggs.

There is also some correlation between the presence or absence of incubation patches and the participation of one or both sexes in incubation. If only one sex has patches, it usually seems to be true that only the sex with the patches incubates; if both sexes have patches, both incubate. In certain passerine species the males are without incubation patches; yet they may appear on the nest, perhaps to provide shade or protection from predators, though it is unlikely that they are ever in sufficient contact with the eggs to give them any heat.

In most passerine species the behavior of the male during incubation follows the same general pattern. He moves leisurely over the territory, feeds, and rests. He sings as vigorously, or even more vigorously, as during the previous periods of territory establishment, nest-building, and egg-laying. He appears to be fully aware of the location of the nest and the incubating female. Sometimes he accompanies the female to the nest. If he does not incubate, he frequently visits the nest carrying food (courtship feeding), or nesting material (symbolic nest-building), or simply approaches the immediate vicinity of the nest to "warn" (i.e., "signal") or to "call off" the female. If he incubates, he comes to the nest with more or less regularity to relieve the female. The arrival of the male at the nest, whether for courtship feeding, symbolic nest-building, or nest-relief, initiates some form of mutual display by both sexes that is peculiar to the species.

Among all birds the behavior of the incubating bird, regardless of sex, is quite similar. While the bird is on the nest there are moments of restlessness, during which it changes sitting position, rises and settles, moves and turns the eggs with the bill, tampers with the nesting material and overhanging cover, and pokes at the bottom of the nest, sometimes with a trembling motion (see Hartshorne, 1962). Generally the incubating bird is vocally quiet, with few exceptions (see "Song"), although it sometimes responds in subdued tones to the calls of the sex-partner. Departure and return are deliberate and secretive. If the bird is a ground nester, it walks away for a short distance and then flies up; if an elevated nester, it flies off quickly in a downward direction and continues near the ground for a short distance, occasionally taking advantage of concealment provided by underbrush and lower branches of trees. The method of return resembles the method of departure.

Only a few species deliberately cover their eggs when they leave the nest. Waterfowl cover their eggs with down from the nest's lining. Plucked from their lower breast earlier, the down serves to insulate as well as conceal the eggs. The Kittlitz's Sandplover *(Charadrius pecuarius)* of Africa, when taking leave of its nest, hastily kicks sand over its eggs until they are completely out of sight (Conway and Bell, 1968).

While off the nest the incubating bird commonly seeks and accompanies the sex-partner if the sex-partner does not replace the incubating bird on the nest. Off-the-nest periods are devoted mainly to feeding, preening, and bathing or dusting in areas seldom near the nest. Copulation during the incubation period is infrequent.

The reactions of incubating birds to intrusions, human or other, vary markedly among different species. Reactions also vary within a species and show changes during the progress of incubation. Some species, particularly those with cryptic coloration or wellconcealed nests, permit the close approach of an intruder; others leave the nest in haste far ahead of the intruder. Some individuals allow themselves to be touched with the hand, while others of the same species flush before the hand touches them. Birds which, early in incubation, flush well in advance of the intruder may, toward hatching, sit more closely, remaining on the nest until almost touched.

A few nonpasserine species are able to move their eggs. The Herring Gull *(Larus argentatus)*, and probably many other gulls as well as terns, can retrieve an egg inadvertently kicked out of the nest by putting its bill over the egg and rolling it back into the nest (Tinbergen, 1953). The Clapper Rail *(Rallus longirostris)* can retrieve an egg similarly displaced by picking it up between its mandibles (Pettingill, 1938) and so can the Virginia Rail *(Rallus limicola)*, as shown in a film taken by O. S. Pettingill, Jr. In instances of this sort the eggs are within reach of birds sitting on their nests. Undoubtedly the birds are incapable of retrieving eggs that are beyond their reach. Common Nighthawks *(Chordeiles minor)*, and probably other goatsuckers, can move their eggs several feet to a new nest site if for some reason the original site becomes unfavorable (Gross, 1940; Sutton and Spencer, 1949). They accomplish this by rolling the eggs in front of their feet (Weller, 1958). A female Pileated Woodpecker *(Dryocopus pileatus)* in Florida, whose cavity nest in a tree became exposed when the tree broke off at the nest hole, made three trips from the nest hole to an unknown destination, each time carrying one of her three eggs in her bill (see Truslow, 1967, for details on the action and photographs).

Birds do not recognize their eggs as their own, even toward the end of the incubation period. If one substitutes eggs of different color but approximately the same size, the incubating bird will readily accept them. Some birds such as gulls and albatrosses will "incubate" electric light bulbs or other odd objects in place of their eggs for an indefinite period. The persisting adage that a bird will desert its nest if its eggs are touched or handled obviously has no foundation in fact. It is the disturbance by an intruder at the nest that causes the owner's desertion.

Why birds do not recognize their eggs as their own is understandable. Eggs are stationary; therefore, there has been no "need" for birds to evolve a means of keeping track of them. As Tinbergen (1953) has suggested, all a bird needs to recognize or know about its eggs is their location, the nest, not their properties.

Length of Time Involved

Incubation, once begun, usually continues until the last egg is hatched. The number of days involved in incubation varies with different species, more time being required for those species producing well-developed young or chicks. Within species incubation time varies slightly. For example, eggs subjected to an excessive amount of cooling while the incubating bird is off the nest for extended periods are slower to hatch. Discussion of several factors affecting the length of incubation is found in the work of Kendeigh (1940).

The **incubation period** of an egg is the time between the start of a regular, uninterrupted incubation and the emergence of the young. Some of the longest incubation periods on record are 75–82 days in the Wandering Albatross *(Diomedea exulans)*, 75–80 in the Brown Kiwi *(Apteryx australis)*, and 54–64 in the King and Emperor Penguins *(Aptenodytes patagonica* and *A. fosteri)*. The eggs of the Mallee Fowl *(Leipoa ocellata)*, an Australian megapode, after being laid and buried in a mound of decaying vegetable matter, need to incubate from 49–90 days, depending on the temperature of the mound (Frith, 1962). In no species of birds are incubation periods normally shorter than 11 days; authenticated periods of fewer than 11 days are very rare (Nice, 1954).

In families of North American birds that include species of greatly different sizes—e.g., in the Ardeidae, Anatidae, Accipitridae, Scolopacidae, Laridae, Strigidae, Picidae, and Corvidae—the incubation periods are correspondingly more wide-ranging because the embryos of the larger species require more time for development than the smaller.

But the incubation periods are not always correlated with the size of the birds. In families of passerine birds, species nesting in cavities have longer incubation periods than species of similar size nesting in open cupped nests. The Eastern Bluebird *(Sialia sialis)*, for example, has an incubation period of about 14 days, whereas the incubation period of *Catharus* thrushes averages 12 days. Between families of birds the incubation periods may vary widely even though the birds may be similar in size. The Leach's Storm-Petrel *(Oceanodroma leucorhoa)*, a burrow-nesting sea bird, has an incubation period of 41–42 days, about twice as long as that of the similarly sized Least Tern *(Sterna antillarum)*. Lack (1968) attributes these differences to the greater protection from predators offered by nest sites and hence the less "need" for a quickened pace in the nesting cycle.

The accepted method for determining the incubation period is to measure the time in days—in hours when possible—from the laying of the last egg to the time when all eggs in the clutch have hatched. But this method, while practical, is not entirely accurate, as Kendeigh (1963) points out, because "the first eggs of a clutch may receive some heat and undergo a certain amount of development before the clutch is completed or full incubation begins. The eggs tend, as a consequence, to hatch in the order laid, but the intervals between hatching of the eggs are shorter than the intervals between their laying so that the time that elapses between laying of an egg and its hatching progressively decreases with each additional egg in the clutch." Kendeigh also describes a more precise method of measurement by recording mechanically the total heat applied to the eggs from the time they are laid until hatching. For details, see his paper.

Studies of some species—for example, the European Wren *(Troglodytes troglodytes)*, see Armstrong, 1955; and the Marsh Wren, see Verner, 1965—have revealed that incubation periods shorten as the season advances, possibly because the air temperature averages warmer, preventing the eggs from losing heat when the incubating bird is off the nest. There may be other reasons, such as an increased and more available food supply, enabling the incubating bird to spend less time in foraging and more time on the nest.

Bear in mind that, in most North American species, the incubation period is "known" in the sense that there are a few (sometimes only one or two) records obtained by the method just defined. Many "records" for some species are unreliable estimates or guesses (Nice, 1954). For figures on the incubation periods in different species, refer to C. Harrison (1978), Reilly (1968), and Terres (1980). The great number of species for which the incubation period is either unknown or simply estimated is impressive.

Incubation Rhythm

At intervals during incubation the incubating bird leaves the nest. The eggs are either left uncovered temporarily or are covered in the meantime by the sex-partner. Thus the incubating bird has periods of **attentiveness** (i.e., periods *on* the nest) alternating with periods of **inattentiveness** (i.e., periods *off* the nest). This alternation of periods is spoken of as the **incubation rhythm.**

The frequency and length of these periods presents a fascinating study because of the variations involved. When both sexes incubate, the male and female may take turns at attentiveness for almost equal periods or one sex may be attentive more often and for longer periods than the other. The length of attentive periods ranges from a few minutes in some species to several days in others. In the Wandering Albatross attentive periods may be as short as two days to as long as 38 days (Tickell, 1968). When one sex incubates, as does the female in many passerine species, the periods of attentiveness show great extremes, ranging from numerous brief periods per day to one period many hours in length. If the female passerine's attentive periods are long, usually the male is feeding her. The sex attentive for the night period (in the case of diurnal birds) is not known in many species. In at least two passerine species (see Erickson, 1938; and Weston, 1947) only the female is known to incubate at night, even though both sexes share in covering the eggs during the day. In woodpeckers the male commonly incubates at night.

Much more time during the day is spent by the incubating bird in attentive than in inattentive periods. From observational data on the incubation rhythm of 137 individuals representing 82 species of tropical birds—mostly passerine, in which the female alone incubates without receiving much if any food from her mate—Skutch (1962) found that 101 kept their eggs covered from 60 to 80 percent of the time. This he regarded as "average or normal constancy." Constancy above 80 percent was shown chiefly by birds that were well fed by their mates or could obtain food in ready abundance when they left their nests. In temperate regions most passerine species, in which the female alone incubates without being fed appreciably by the male, normally cover their eggs within the same range of constancy as in the tropics. For example, in Michigan female Scarlet Tanagers *(Piranga olivacea)* revealed a constancy of 77 percent (Prescott, 1964). Also in Michigan one female American Robin, watched continuously at the nest on two days from daybreak to darkness, was attentive 77 percent of the time one day and 80 percent the other (Pettingill, 1963). In north-central Alaska where, during the nesting season, there is continuous daylight save for a few hours of civil twilight between sunset and sunrise, female American Tree Sparrows *(Spizella arborea)* were attentive 76 to 77 percent of the day (Weeden, 1966). When the male feeds the female frequently while she incubates, she covers the eggs for a greater percent of the time. This is the case with the female Cedar Waxwing *(Bombycilla cedrorum)* whose attentiveness rarely falls below 90 percent (Putnam, 1949).

Attentive periods tend to be longer toward evening and in the early morning when it is cooler and, for insectivorous birds, there is less food readily available. With the increase in daytime air temperature, attentive periods shorten and inattentive periods increase in number, resulting in much less total time spent on the nest. Climatic conditions may influence the length of attentive and inattentive periods. Warm weather usually shortens attentive periods; excessively hot weather prolongs attentive periods if nests are exposed and require the shade of the bird's body; cool weather almost invariably lengthens attentive periods. Attentive periods may be consistently long throughout the day if the incubating bird is fed by its mate. The female Cedar Waxwing has attentive periods often extending more than two hours (Putnam, 1949).

From his studies of incubation rhythm in the House Wren *(Trolodytes aedon)*, Kendeigh (1952) concluded that the length of the attentive periods is much more variable than the length of the inattentive and that the relationship between the two periods is probably a psychological one "since with shorter attentive periods, less rest and less food are required, while with longer periods there are needs to be more rest and more food."

Restlessness on the nest, described earlier, is much reduced during long attentive periods, as toward evening and during the early morning. Individual birds vary greatly in the amount of restlessness, some being much more active than others. Beer (1965), in paying special attention to this behavior in the Black-billed Gulls *(Larus bulleri)* of New Zealand, noted that, when the clutch of eggs contained the "optimum" number, the birds sat on the nest more quietly (with "more uninterrupted sitting") than when the clutch was smaller.

Abnormalities of Incubation

When eggs fail to hatch, birds will continue to incubate for a variable length of time. Most birds, according to Skutch (1962), continue for at least 50 percent longer than the normal period and they may even continue for twice or even three times longer. An exception, he points out, are pigeons, some of which will not continue to incubate "even a day beyond the usual time of hatching."

Brood Parasitism

At least five groups of birds in the world contain brood parasites—bird species adapted to laying their eggs in the nest of another (host) species that incubates the eggs and rears the young (see Payne, 1977, for an overall review). Early works suggested that parasitism evolves when a species' normal breeding biology degenerates to the point where it turns to parasitism for survival. But Hamilton and Orians (1965) believe that such reasoning conflicts with modern evolutionary theory and that hypotheses for the development of parasitism have yet to explain why the initial shift from a normal breeding mode is in itself adaptive.

Ducks

Among ducks, two species in North America, the Redhead *(Aythya americana)* and Ruddy Duck *(Oxyura jamaicensis)*, often lay their eggs in nests of the same species or different species. Redhead females are of three kinds: wholly parasitic, partially parasitic, or never parasitic (Weller, 1959). In South America the Black-headed Duck *(Heteronetta atricapilla)* parasitizes, besides other ducks, such unrelated birds as coots and ibises; it is entirely parasitic (Weller, 1968).

Honey-guides

Among the 14 species of honey-guides (Indicatoridae) of Africa and southeastern Asia, at least five African species are known to lay a white egg in the nest of a host. These hosts are usually a hole-nesting species such as a barbet (Capitonidae) or woodpecker, which also have white eggs (see Friedmann, 1955, 1968a).

Weaverbirds

Among the weaverbirds (Ploceidae), about nine African species of *Vidua* parasitize mostly estrildine finches. Some are limited to a single host species and mimic the egg, nestling gape, juvenal plumage and vocalizations of their respective hosts. Although the young are divergent, the adults of some *Vidua* species are remarkably similar and pose a taxonomic nightmare; in some cases the most reliable species differences are the songs, which include components learned from the host species (Payne 1973). Another ploceid, the Cuckoo Finch *(Anomalospiza imberbis)*, has evolved parasitism independently and victimizes a range of sylviid warblers (Friedmann, 1960).

Cuckoos

Among the cuckoos, some 50 species in the Old World and three in the New World tropics lay their eggs, sometimes quite small in proportion to body size, in the nests of usually smaller species, mainly passerine. The Common Cuckoo *(Cuculus canorus)* of Eurasia is an extraordinarily specialized brood parasite (Chance, 1940; Baker, 1942). Its eggs often match the small size and coloration of its host's eggs. In a given area the population of Common Cuckoos may be comprised of several groups, each producing eggs that resemble those of a different host. It is believed that each group parasitizes only the host species it mimics, that the mimicry evolved because hosts reject nonmimetic eggs, and that female cuckoos acquire recognition of their proper host by imprinting on their foster parents.

But this often repeated classic example of coevolution has many uncertainties (Lack 1963). There is little evidence for imprinting and the apparent egg matching by cuckoos may be due to selective ejection of nonmimetic eggs by hosts, rather than to finely tuned host selection by the cuckoos (Rothstein, 1971). For brood parasitism in other cuckoos, some of which mimic ravens and hawks and specialize on large corvids, see Baker (1942) and Friedmann (1948, 1956, 1964, 1968b).

Cowbirds

Among the New World cowbirds are five parasitic species. Three of these are generalists but two neotropical species are as highly specialized as are any avian parasites. Among the specialists, the crow-sized Giant Cowbird *(Scaphidura oryzivora)* parasitizes other large icterines, such as communal oropendolas (N.G. Smith, 1968), while the Screaming Cowbird *(Molothrus rufoaxillaris)* parasitizes a single species—the Bay-winged Cowbird *(Molothrus badius)*, which is the only nonparasitic cowbird (Friedmann, 1929). Although adult Screaming and Bay-winged Cowbirds are dissimiliar, their eggs and nestlings are virtually identical (Fraga, 1979), and because of this, the parasitism went unrecognized for years after these species were described.

The Brown-headed Cowbird *(Molothrus ater)* is the only parasitic bird widely distributed in parts of North America north of Mexico. This cowbird and its South American counterpart, the Shiny Cowbird *(Molothrus bonariensis)*, are the most generalized of all avian parasites. Brown-headed Cowbirds are known to have parasitized at least 216 species. Although many of these are rare or even accidental hosts such as gulls, ducks, or hawks, 139 species are known to have actually reared cowbirds (Friedmann, 1963; Friedmann et al., 1977). The species most often victimized include the vireos, tanagers, *Catharus* and *Hylocichla* thrushes, several of the smaller tyrannid flycatchers, a few icterines, and many warblers, buntings, sparrows, and finches.

Various workers (Friedmann, 1929; Hann, 1941; Mayfield, 1960; Van Tyne, in Mayfield, 1960) made extensive observations and reviewed the literature on the cowbird's procedure in parasitizing a nest. Although some nests may be found by simply searching through likely vegetation (Norman and Robertson, 1975), the female usually finds nests by first noticing building activities and subsequently makes frequent "trips of inspection" while her prospective host is absent. Usually after her host has laid two or more eggs and before her host begins incubation, she enters the nest, about a half hour before sunrise while the nest is still unoccupied, and deposits her egg in a few seconds. If she removes one or more eggs, as is often the case, she does so later in the day, the day before, or the day after, but not at the time of laying. She carries away the egg by piercing it with her bill and later eats it. Norris (1947) shows excellent pictures of cowbirds visiting host nests.

The female cowbird's interest in the nest may continue after the host begins incubation (Mayfield, 1961), even to the extent of her removing additional eggs, although her approach to the nest is likely to be discouraged by the host's behavior, which is more agonistic than at the time of egg-laying. There are a few instances (Tate, 1967) reported of a cowbird's removing the nestling of the host. Such behavior could account for some of the losses of nestlings from a parasitized nest.

Most parasitized nests contain one cowbird egg, but a third of them have two or three. One Wood Thrush *(Hylocichla mustelina)* nest had 12 eggs (Friedmann, 1963). Some of this multiple parasitism is undoubtedly due to more than one female, but since female cowbirds in some areas are territorial (Dufty, 1982), individual females probably often lay repeatedly in the same nest. Although occasional females may specialize on one host species (e.g., Walkinshaw, 1949), most appear to parasitize several species (Friedmann, 1929; McGeen and McGeen, 1968). The fecundity of cowbirds is of great interest, but its study poses formidable problems. Payne (1976, 1977) estimated that females lay 10–24 eggs a year. More recent work (Scott and Ankney, 1979, 1980) suggests that 40 or even 50 eggs is a likely figure and that females lay on about 70 percent of the days during the breeding season. Furthermore, this remarkable fecundity seems to incur little or no physiological cost (Ankney and Scott, 1980). Even calcium does not become depleted from the female's body despite that fact cowbirds have unusually thick eggshells (Blankespoor et al., 1982), an adaptation shown by other parasitic birds.

Because the cowbird lacks a nest that can serve as a focal point for studies of reproductive activities, much about its breeding system is poorly known. Cowbirds appear to be monogamous in New York (Dufty, 1982) but promiscuous in Kansas (Elliott, 1980). Other studies, including some on captive birds (e.g., Darley, 1978; West et al., 1981) have provided support for almost every possible mating system. Cowbirds occupy larger home ranges or territories than virtually any other breeding passerine, averaging about 50 acres (20.24 ha) in New York and 168 acres (68.01 ha) in California. Nearly all cowbirds show extraordinary mobility during the breeding season, with those in the Sierra Nevada being perhaps the most extreme. Radiotelemetry shows that these mountain cowbirds abandon breeding ranges each morning and "commute" up to 4.5 miles (7.24 km) to feed socially at the few suitable foraging sites in the region (Stephen I. Rothstein, Jared Verner, and Ernest E. Stevens, unpublished).

Experiments with mounted birds showed that some species are especially aggressive toward cowbirds, seemingly recognizing them as a special threat (Robertson and Norman, 1977). It is unclear whether this recognition is learned from prior experience with cowbirds or is in part adapted. Although cowbird eggs usually differ from host eggs, most species accept them. Rothstein (1975a) systematically investigated egg rejection by experimentally simulating parasitism with real and artificial cowbird eggs. The species tested fell into two groups, "accepter" and "rejecter" species. The former accept cowbird eggs at nearly 100 percent of their nests, whereas the latter show the opposite inclination; no clearly intermediate species were found. Many accepters are heavily parasitized and can easily differentiate between their eggs and those of the cowbird, so their lack of rejection is unexpected. Rejecter species, on the other hand, show three means of rejection: ejection of the cowbird egg; desertion of the entire nest; and burial of the cowbird egg with nesting material and re-use of the nest.

Ejection was by far the most common means of rejection in Rothstein's experiments and was shown by Gray Catbirds *(Dumetella carolinensis)*, American Robins, Northern Orioles *(Icterus galbula)*,

Eastern Kingbirds *(Tyrannus tyrannus)*, and other species. Interestingly, ejection is observed less commonly in natural parasitism than are egg burial and nest desertion. This discrepancy may exist because ejection leaves no evidence that a nest was ever parasitized, unlike desertion and egg burial.

The two latter responses usually do not constitute clear evidence of reactions to cowbird eggs because entire clutches are affected. Birds will, for example, desert nests for many reasons, including visits by humans, and eggs may be buried simply because they are laid before the nest is completed. However, controlled experiments and careful observations have shown that nest desertion (Rothstein, 1976a) by Cedar Waxwings and egg burial (Clark and Robertson, 1981) by Yellow Warblers *(Dendroica petechia)* are direct responses to cowbird eggs. In one instance a Yellow Warbler built a six-storied nest that contained 11 buried cowbird eggs (Berger, 1955).

Rejecter species recognize their own eggs and do not simply act against any egg that constitutes a minority of the clutch (Rothstein, 1975b). This recognition is probably learned via the host females imprinting on the first egg she lays (Rothstein, 1978). Host choice by cowbirds does not seem particularly efficient as some rejecter species are parasitized (Rothstein, 1976b). The full extent of such parasitism is hard to determine because rejecters remove most cowbird eggs before observers see them.

Selected Studies

Select, if possible, a nest under construction and follow the succession of events through egg-laying and incubation. If the nest is in a cavity or a tunnel in the ground, consult Demong and Emlen (1975), De Weese et al. (1975), Erskine (1959), and Jackson (1976) for useful devices and methods. For monitoring nesting activities, consult Cooper and Afton (1981). Do not disturb the nest itself and handle the eggs with great care. Bear in mind that it is unlawful to take or otherwise destroy any nest or eggs of a protected species. See Appendix A for federal regulations.

Procedure Much valuable information can be obtained during the egg-laying period. Note the start of egg-laying (exact hour of the day when the first egg is laid) and note its relation to stage in nest-building and to environmental factors. Determine the exact time when the other eggs of the clutch are laid, the time spent on the nest by the female at each laying, and the first evidence of actual incubation. (After each egg is laid, mark it indelibly so that it may be recognized at the time of hatching.) Follow as closely as possible the activities and behavior of the pair from day to day.

Throughout incubation, make detailed direct observations on the participation and behavior of the sexes. Attempt to determine the incubation period by measuring the time in hours from the laying of the last egg in the clutch to the hatching of the last egg. Gather specific data on the incubation rhythm at different times of the day, at different stages of incubation, and under different weather conditions. In order to yield significant results each observation period should be at least *two hours* in length. Use the following methods for organizing data in tabular form. They were first applied by Pitelka (1940) and later modified by him (1941). Adjust this method to suit the particular relationships of the sexes in the species studied.

For each observation period, record:

Date
Stage of incubation (i.e., day of incubation)
Time of day; also total hours and minutes
Air temperature at beginning and ending
Wind velocity
Weather conditions

Attentive period of both sexes:
- Total number
- Average time in minutes
- Extremes in minutes
- Percentage of total time

Inattentive periods of both sexes:
- Total number
- Average time in minutes
- Extremes in minutes
- Percentage of total time

Attentive and inattentive periods of male:
(Repeat procedure)

Attentive and inattentive periods of female:
(Repeat procedure)

In addition to tabulating the data as directed, take notes on the activities and behavior of the sexes during each of the attentive and inattentive periods. Qualitative as well as quantitative data are desirable. If possible, use a portable tape recorder in dictating the notes and transcribe them later. This method allows the observer to keep an eye on activities and not to miss any action while writing notes.

For a complete all-day record of nest attentiveness, the use of an automatic recorder is desirable, provided it can be adapted to the type of nest. For most nests a thermocouple and potentiometer can be used (see Kendeigh, 1952, for details). By means of a very thin, flexible wire threaded through the nest so that it lies just above or on the eggs, the device records the heat generated by the bird when it settles on the eggs to incubate them and instantly records the drop in temperature when it leaves the eggs. For ground nests and hole nests, use the itograph (also see Kendeigh, 1952). This consists of a double set of perch contacts at the nest's edge or entrance attached to a set of dry-cell batteries and an electromagnet with a pen that registers all the trips to the nest. For nests that do not show an appreciable amount of light through the bottoms and sides, use a small (three-quarters of an inch in diameter) photo-resistor positioned in the floor of a nest and connected by an extension cord through the floor to an adjustable resistor and recording element (see Weeden, 1966). This device registers rapidly the decrease of light when the bird sits on the nest and the increase when it gets off. It also registers the bird's restlessness as it rises up and resettles.

Should the nest and eggs be destroyed, strive to follow the pair in order to learn the length of time required to start the laying of a new clutch. All too often study of a pair ceases if disaster comes to the nest. The result is that surprisingly little is known of what the occupant birds do thereafter.

Pay attention to the size of the clutch and search the literature to determine whether or not the size is average. If it is not average, attempt to determine why.

Full description of the eggs is sometimes advisable, particularly if contemplating an extensive study of the species or some special problem that concerns the eggs. (See Nice, 1937, for a discussion of problems connected with the eggs of the Song Sparrow.) Use the following directions for gathering the descriptive information: *Coloration*. Determine the colors of each egg by comparing it directly with a color chart (see Appendix A for available color charts). Note the distribution of color and markings. *Measurements*. Measure the eggs, using dividers, or calipers, and a millimeter ruler. Length and greatest width are the two measurements taken. Extreme accuracy is essential, since variations, especially in small eggs, are slight. *Weights*. Weigh the eggs in grams to the second decimal place. It is important to note the state of incubation, since eggs become gradually lighter as incubation proceeds. (Sometimes eggs will lose as much as one-third of their original weight.) Weighing eggs in the field is generally unsatisfactory because it is difficult to be accurate. Proper balance is affected by air currents and off-level position of the scales. If eggs must be weighed during incubation, it should be done in the laboratory while substitute eggs are placed in the nest to prevent desertion by the adults. Obviously, eggs must be weighed as quickly as possible to avoid extreme cooling. Caution: Handle with great care. Eggs are generally thin-shelled. The slightest cracking of the shell will very likely inhibit the growth of the embryo.

If the species selected for study is known to be victimized by the Brown-headed Cowbird, watch for first evidences of cowbird relationships: how and when the cowbird discovers the nests; time when the cowbird lays its eggs. Should parasitism actually result, observe the reactions of the hosts to the eggs and any evidences of harm that the eggs may have caused.

Photographs showing activities of adults during incubation (e.g., courtship-feeding, turning eggs, incubating positions, reactions to intrusions) are useful as illustrations in the presentation of results.

Presentation of Results Draw up a report on the results obtained from a study of egg-laying and incubation. Prepare the manuscript as directed in Appendix B. An outline of suggested topics is given below.

Egg-laying Date of start. Relation of start to construction of nest and to environmental factors. Time of laying of each egg and interval between layings. Amount of time spent on nest by female at each laying, and accompanying activities. Time of first evidence of incubation. Behavior of the sexes: activities of the female off the nest and reaction to male; number and times of copulations; activities of the male, including number of singing periods and vigor of singing; territorial activities of the pair; reactions of the pair to intrusions.

The Clutch *Size:* Number of eggs finally laid. Reason for size of clutch if abnormal. *Number of clutches per season:* Present whatever data have been obtained. *Description of clutch:* Present descriptive material if significant. Compare with information in the literature.

Incubation *Start of incubation:* Relation of start to size of clutch at the time. Brief statement of duration. If the eggs are of a large species and incubation is already in progress, it is sometimes possible to determine the state of incubation. For a method see Weller (1956). *Participation and behavior of the sexes:* Brief statement of sexes participating, followed by detailed accounts, (1) Activities of male: territorial behavior; singing; relation to nest, incubation, and incubating female (visits, courtship-feeding, symbolic building, calls to female and resulting responses, nest-relief displays, reactions to intrusion). (2) Activities of incubating bird: general behavior and activities; manner of leaving nest and returning; behavior and activities while off the nest. (3) Length of time involved: total length of incubation (in hours) for each egg. Present tables for each observation on incubation rhythm, followed by tables of similar design summarizing all the data from all observations. (See Pettingill, 1963, for a chart showing the actions and attentiveness of an incubating bird during a whole day.) After the tables, discuss the information, noting variations in rhythm at different times of the day, at different stages of incubation, and under different weather conditions. Graphs may be used to illustrate certain points in the discussion. Indicate which sex incubates at night. *Abnormalities of incubation*.

Brood Parasitism Present whatever information has been obtained on cowbird relationships.

References

Ankney, C.D., and Scott, D.M.
1980 Changes in nutrient reserves and diet of breeding Brown-headed Cowbirds. *Auk* 97:684–696.

Armstrong, E.A.
1955 *The Wren*. London: Collins.

Ashmole, N.P.
1963 The biology of the Wideawake or Sooty Tern *Sterna fuscata* on Ascension Island. *Ibis* 103b:297–364.

Bailey, R.E.
1952 The incubation patch of passerine birds. *Condor* 54:121–136.

Baker, E.C.S.
1942 *Cuckoo problems*. London: H.F. & G. Witherby.

Beer, C.G.
1965 Clutch size and incubation behavior in Black-billed Gulls *(Larus bulleri)*. *Auk* 82:118.

Berger, A.J.
1951 The cowbird and certain host species in Michigan. *Wilson Bull*. 63:26–34.

Best, L.B.
1978 Field Sparrow reproductive success and nesting ecology. *Auk* 95:9–22. (Major causes of nest failure: snake predation, desertion.)

Blankespoor, G.W.; Oolman, J.; and Uthe, C.
1982 Eggshell strength and cowbird parasitism of Red-winged Blackbirds. *Auk* 99:363–365.

Brackbill, H.
1952 Three-brooded American Robins. *Bird-Banding* 23:29.

Bruning, D.F.
1974 Social structure and reproductive behavior in the Greater Rhea. *Living Bird* 13:251–294. (Male constructs nests, incubates the eggs, and rears young.)

Chance, E.P.
1940 *The truth about the cuckoo*. London: Country Life.

Chapin, J.P.
1954 The calendar of Wideawake Fair. *Auk* 71:1–15.

Clark, K.L., and Robertson, R.J.
1981 Cowbird parasitism and evaluation of anti-parasite strategies in the Yellow Warbler. *Wilson Bull*. 93:249–258.

Cody, M.L.
1966 A general theory of clutch size. *Evolution* 20:174–184.

Conway, W.G., and Bell, J.
1968 Observations on the behavior of Kittlitz's Sandplovers at the New York Zoological Park. *Living Bird* 7:57–70.

Cooper, J.A., and Afton, A.D.
1981 A multiple sensor system for monitoring avian nesting behavior. *Wilson Bull*. 93:325–333.

Coulter, M.C.
1980 Stones: An important incubation stimulus for gulls and terns. *Auk* 97:898–899. (Stones rolled into nests a stimulus for incubating clutches smaller than normal size.)

Cox, G.W.
1960 A life history of the Mourning Warbler. *Wilson Bull.* 72:5–28. (Incubation period, attentiveness, etc.)

Crowell, K.L., and Rothstein, S.I.
1981 Clutch sizes and breeding strategies among Bermuda and North American passerines. *Ibis* 123:42–50.

Darley, J.A.
1978 Pairing in captive Brown-headed Cowbirds *(Molothrus ater). Canadian Jour. Zool.* 56:2249–2252.

Davis, D.E.
1942 Number of eggs laid by Herring Gulls. *Auk* 59:549–554. (Attempts to restrain gulls from laying the full normal clutch and to induce them to lay more.)
1955 Determinate laying in Barn Swallows and Black-billed Magpies. *Condor* 57:81–87.

Davis, J.
1960 Nesting behavior of the Rufous-sided Towhee in coastal California. *Condor* 62:434–456. (Important information on egg-laying, incubation, and related activities.)

Davis, J.; Fisler, G.F.; and Davis, B.S.
1963 The breeding biology of the Western Flycatcher. *Condor* 65:337–382. (Extensive data on nesting, including incubation periods, role of sexes, attentiveness, etc.)

Demong, N.J., and Emlen, S.T.
1975 An optical scope for examining nest contents of tunnel-nesting birds. *Wilson Bull.* 87:550–551.

De Weese, L.R.; Pillmore, R.E.; and Richmond, M.L.
1975 A device for inspecting nest cavities. *Bird-Banding* 46:162–165.

Dixon, C.L.
1978 Breeding biology of the Savannah Sparrow on Kent Island. *Auk* 95:235–246. (Low nesting success; reasons given.)

Dufty, A.M., Jr.
1982 Movements and activities of radio-tracked Brown-headed Cowbirds. *Auk* 99:316–327.

Elliott, P.F.
1980 Evolution of promiscuity in the Brown-headed Cowbird. *Condor* 82:138–141.

Emlen, J.T., Jr.
1941 An experimental analysis of the breeding cycle of the Tricolored Red-wing. *Condor* 43:209–219. (Experiments in reducing and extending the incubation cycle.)

Erickson, M.M.
1938 Territory, annual cycle, and numbers in a population of Wren-tits *(Chamaea fasciata). Univ. Calif. Publ. in Zool.* 42:247–334.

Erskine, A.J.
1959 A method for opening nesting holes. *Bird-Banding* 30:181–182. (Drilling a hole from back or side of tree. Details given.)

Finney, G., and Cooke, F.
1978 Reproductive habits in the Snow Goose: The influence of female age. *Condor* 80:147–158.

Fraga, R.
1979 Differences between nestlings and fledglings of Screaming and Bay-winged Cowbirds. *Wilson Bull.* 91:151–154.

Friedmann, H.
1929 *The cowbirds: A study in the biology of social parasitism*. Springfield, Illinois: Charles C. Thomas.
1948 *The parasitic cuckoos of Africa*. Washington, D.C.: Washington Academy of Science.
1955 *The honey-guides*. U.S. Natl. Mus. Bull. 208.
1956 Further data on African parasitic cuckoos. *Proc. U.S. Natl. Mus.* 106:377–408.
1960 The parasitic weaverbirds. *U.S. Natl. Mus. Bull.* 223.
1963 Host relations of the parasitic cowbirds. *U.S. Natl. Mus. Bull.* 233.
1964 Evolutionary trends in the avian genus *Clamator. Smithsonian Misc. Coll.* 146(4):1–127.
1968a Additional data on brood parasitism in the honey-guides. *Proc. U.S. Natl. Mus.* 124 (3648):1–8.
1968b The evolutionary history of the avian genus *Chrysococcyx. U.S. Natl. Mus. Bull.* 265.

Friedmann, H.; Kiff, L.F.; and Rothstein, S.I.
1977 A further contribution to the knowledge of the host relations of the parasitic cowbirds. *Smithsonian Contr. Zool.* No. 235.

Frith, H.J.
1962 *The Mallee-fowl: The bird that builds an incubator.* Sydney: Angus & Robertson.

Grant, G.S.
1981 Belly-soaking by incubating Common, Sandwich, and Royal Terns. *Jour. Field Ornith.* 52:244.
1982 Avian incubation: Egg temperature, nest humidity, and behaviorial thermoregulation in a hot environment. *Amer. Ornith. Union, Ornith. Monogr.* No. 30. (Several charadriiform species and Lesser Nighthawks nesting in the area of the Salton Sea, southeastern California.)

Graul, W.D.
1973 Adaptive aspects of the Mountain Plover social system. *Living Bird* 12:69–94. (Three eggs in set; female lays first set which male incubates; then she lays second set which she incubates.)

Gross, A.O.
1940 Eastern Nighthawk. In *Life histories of North American cuckoos, goatsuckers, hummingbirds and their allies*. By A.C. Bent. *U.S. Natl. Mus. Bull.* 176:206–234.

Hamilton, W.J., III, and Orians, G.H.
1965 Evolution of brood parasitism in altricial birds. *Condor* 67:361–382.

Hann, H.W.
1937 Life History of the Oven-bird in southern Michigan. *Wilson Bull.* 49:145–237. (Egg-laying; incubation; parasitism by cowbird; etc.)
1941 The cowbird at the nest. *Wilson Bull.* 53:211–221. (Fifteen observations of the female cowbird at the Ovenbird's nest.)

Harrison, C.
1978 *A field guide to the nests, eggs and nestlings of North American birds*. London and Toronto: Collins.

Harrison, H.H.
1975 *A field guide to birds' nests of 285 species found breeding in the United States east of the Mississippi River.* Boston: Houghton Mifflin.
1978 *A field guide to western birds' nests of 520 species found breeding in the United States west of the Mississippi River.* Boston: Houghton Mifflin.

Hartshorne, J.M.
1962 Behavior of the Eastern Bluebird at the nest. *Living Bird* 1:131–149.

Hayes, H., and Le Croy, M.
1971 Field criteria for determining incubation stage in eggs of the Common Tern. *Wilson Bull.* 83:425–429.

Hochbaum, H.A.
1981 *The Canvasback on a prairie marsh*. 3rd ed. Lincoln: Univ. of Nebraska Press. (Information on egg-laying; cases of different species of ducks laying in each other's nests.)

Holcomb, L.C.
1974a Incubating constancy in Red-winged Blackbird. *Wilson Bull.* 86:450–460.
1974b The influence of nest building and egg laying behavior on clutch size in renests of the Willow Flycatcher. *Bird-Banding* 45:320–325.

Howard, D.V.
1967 Variation in the breeding season and clutch-size of the Robin in the northeastern United States and the Maritime Provinces of Canada. *Wilson Bull.* 79:432–440.

Howe, M.A.
1975 Behavioral aspects of the pair bond in Wilson's Phalarope. *Wilson Bull.* 87:248–270. (Incubation, solely by male, begins after last egg is laid and pair-bond dissolves.)

Jackson, J.A.
1976 How to determine the status of a woodpecker nest. *Living Bird* 15:205–221.

Kendeigh, S.C.
1940 Factors affecting length of incubation. *Auk* 57:499–513.
1952 Parental care and its evolution in birds. 2nd corrected printing, 1955. *Illinois Biol. Monogr.* 22:i–x; 1–356. (Contains a wealth of information on incubation activities in many species.)
1963 New ways of measuring the incubation period of birds. *Auk* 80:453–461.

Kessler, F.
1962 Measurement of nest attentiveness in the Ring-necked Pheasant. *Auk* 79:702–705.

Koford, C.B.
1953 The California Condor. *Natl. Audubon Soc. Res. Rept.* No. 4, New York.

Labisky, R.F., and Jackson, G.L.
1966 Characteristics of egg-laying and eggs of yearling pheasants. *Wilson Bull.* 78:379–399. (An analysis of eggs laid by nine yearling Ring-necked Pheasants in captivity, showing variation in rate of laying, shell color, size, weight, shape, and seasonal pattern.)

Lack, D.
1947–1948 The significance of clutch-size. *Ibis,* 89:302–352; 90:25–45.
1954 *The natural regulation of animal numbers*. London: Oxford Univ. Press.
1963 Cuckoo hosts in England. *Bercly Study* 10:185–202.
1966 *Population studies of birds*. London: Oxford Univ. Press.
1968 *Ecological adaptations for breeding birds*. London: Methuen.

Lawrence, L. de K.
1953 Nesting life and behaviour of the Red-eyed Vireo. *Canadian Field-Nat.* 67:47–77. (Egg-laying, incubation, etc.)

Maclean, G.L.
1972 Clutch size and evolution in the Charadrii. *Auk* 89:299–324.

Mayfield, H.
1960 The Kirtland's Warbler. *Cranbrook Inst. Sci. Bull.* No. 40. Bloomfield Hills, Michigan.
1961 Vestiges of a proprietary interest in nests by the Brown-headed Cowbird parasitizing the Kirtland's Warbler. *Auk* 78:162–166.

McGeen, D.S., and McGeen, J.J.
1968 The cowbirds of Otter Lake. *Wilson Bull.* 80:84–93.

Murray G.A.
1976 Geographic variation in the clutch sizes of seven owl species. *Auk* 93:602–613.

Nelson, J.B.
1964 Factors influencing clutch-size and chick growth in the North Atlantic Gannet *Sula bassana*. *Ibis* 106:63–77. (Experiments showing that this species is capable of hatching two eggs and rearing two young even though it normally lays one egg.)

Nice, M.M.
1932 Observations on the nesting of the Blue-gray Gnatcatcher. *Condor* 34:18–22. (Suggestions for study of incubation rhythm.)
1937 Studies in the life history of the Song Sparrow, I. *Trans. Linnean Soc. New York* 4:i–vi; 1–247. (Chapters 11–13.)
1943 Studies in the life history of the Song Sparrow, II. *Trans. Linnaean Soc. New York* 6:i–viii; 1–328. (Chapter 18.)
1954 Problems of incubation periods in North American birds. *Condor* 56:173–197.

Nolan, V., Jr.
1979 The ecology and behavior of the Prairie Warbler *Dendroica discolor. Amer. Ornith, Union Ornith. Monogr.* No. 26. (Chapters 18–23 of particular importance.)

Norman, R.F., and Robertson, R.J.
1975 Nest-searching behavior in the Brown-headed Cowbird. *Auk* 92:610–611.

Norris, R.T.
1947 The cowbirds of Preston Frith. *Wilson Bull.* 59:83–103.

Odum, E.P.
1941 Annual cycle of the Black-capped Chickadee-2 *Auk* 58:518–535. (Egg-laying; incubation.)

Parmelee, D.F., Stephens, H.A., and Schmidt R.H.
1967 The birds of southeastern Victoria Island and adjacent small islands. *Natl. Mus. Canada Bull.* 222. (Data on nesting, incubation periods, and role of sexes in species hitherto studied very little.)

Parsons, J.
1976 Factors determining the number and size of eggs laid by the Herring Gull. *Condor* 78:481–492. (Species can compensate for early loss of its first egg by laying an additional one.)

Payne, R.B.
1973 Behavior, mimetic songs and song dialects, and relationships of the parasidic indigobirds *(Vidua)* of Africa. *Amer. Ornith. Union Ornith. Monogr.* No. 11.
1976 The clutch size and numbers of eggs of Brown-headed Cowbirds: Effects of latitude and breeding season. *Condor* 78:337–342.
1977 The ecology of brood parasitism in birds. *Ann. Rev. Ecol. Syst.* 8:1–28.

Paynter, R.A., Jr.
1949 Clutch-size and the egg and chick mortality of Kent Island Herring Gulls. *Ecology,* 30:146–166.
1951 Clutch-size and egg mortality of Kent Island eiders. *Ecology* 32:497–507.
1954 Interrelation between clutch-size, brood-size, prefledging survival, and weight in Kent Island Tree Swallows. *Bird-Banding* 25:35–58; 102–110; 136–148.

Pettingill, O.S., Jr.
1938 Intelligent behavior in the Clapper Rail *Auk* 55:411–415.
1960 The effects of climate and weather on the birds of the Falkland Islands. *Proc. XIIth Internatl. Ornith Congr.* pp. 604–614.
1963 All-day observations at a Robin's nest. *Living Bird* 2:47–55.

Phillips, C.L.
1887 Egg-laying extraordinary in *Colaptes auratus. Auk* 4:346. (Flicker induced to lay 71 eggs in 73 days. After the bird had laid two eggs, one egg was removed, the other was left as a nest-egg. This process was continued until the remarkable number was obtained.)

Picman, J., and Picman, A.K.
1980 Destruction of nests by the Short-billed Marsh Wren. *Condor* 82:176–179. (Shows "significant impact" on nesting success of other small sympatric nesting birds by destroying their eggs.)

Pitelka, F.A.
1940 Breeding behavior of the Black-throated Green Warbler. *Wilson Bull.* 52:3–18.
1941 Presentation of nesting data. *Auk* 58:608–612.

Prescott, K.W.
1964 Constancy of incubation for the Scarlet Tanager. *Wilson Bull.* 76:37–42.

Putnam, L.S.
1949 The life history of the Cedar Waxwing. *Wilson Bull.* 61:141–182. (Contains considerable information on various aspects of incubation.)

Rahn, A., and Ar, A.
1974 The avian egg: Incubation time and water loss. *Condor* 76:147–152.

Reed, C.A.
1904 *North American birds eggs.* New York: Doubleday, Page. (A "revised and unabridged republication" of this work, called "the Dover

edition," was issued in 1965 by Dover Publications, New York.)

Reilly, E.M., Jr.
1968 *The Audubon illustrated handbook of American birds*. New York: McGraw-Hill.

Robertson, R.J., and Norman, R.F.
1977 The function of evolution of aggressive host behavior towards the Brown-headed Cowbird. *Canadian Jour. Zool*. 55:508–518.

Romanoff, A.L., and Romanoff, A.J.
1949 *The avian egg*. New York: John Wiley & sons.

Rothstein, S.I.
1971 Observation and experiment in the analysis of interactions between brood parasites and their hosts. *Amer. Nat*. 105:71–74.
1975a An experimental and telenomic investigation of avian brood parasitism. *Condor* 77:250–271.
1975b Mechanisms of avian recognition: Do birds know their own eggs? *Animal Behavior* 23:268–278.
1976a Experiments on defenses Cedar Waxwings use against cowbird parasitism. *Auk* 93:675–691.
1975b Cowbird parasitism of the Cedar Waxwing and its evolutionary implications. *Auk* 93:498–509.
1978 Mechanisms of avian recognition: Additional evidence for learned components. *Animal Behavior* 26:671–677.

Ryder, J.P.
1975 Egg-laying, egg size, and success in relation to immature-mature plumage of Ring-billed Gulls. *Wilson Bull*. 87:534–542.

Scott, D.M., and Ankney, C.D.
1979 Evaluation of a method for estimating the laying rate of Brown-headed Cowbirds. *Auk* 96:483–488.
1980 Fecundity of the Brown-headed Cowbird in southern Ontario. *Auk* 97:677–683.

Schardien, B.J., and Jackson, J.A.
1979 Belly-soaking as a thermoregulatory mechanism in nesting Killdeers. *Auk* 96:604–606.

Sealy, S.G.
1978 Possible influence of food on egg-laying and clutch size in the Black-billed Cuckoo. *Condor* 80:103–104. (Large clutches in response to available forest tent caterpillars.)

Skutch, A.F.
1933a A male kingfisher incubating at night. *Auk* 50:437.
1933b Male woodpeckers incubating at night. *Auk* 50:437.
1945 Incubation and nestling periods of Central American birds. *Auk* 62:8–37.
1952 On the hour of laying and hatching of birds' eggs. *Ibis* 94:49–61.
1957 The incubation patterns of birds. *Ibis* 99:69–93.
1962 The constancy of incubation. *Wilson Bull*. 74:115–152.

Slack, R.D.
1976 Nest guarding behavior by male Gray Catbirds. *Auk* 93:292–300. (Largely during incubation. Though male does not incubate, he "spends a considerable amount of time near the nest.")

Smith, J.M.
1981 Cowbird parasitism, host fitness, and age of the host female in an island Song Sparrow population. *Condor* 83:152–161.

Smith, N.G.
1968 The advantage of being parasitized. *Nature* 219:690–694.

Spencer, O.R.
1943 Nesting habits of the Black-billed Cuckoo. *Wilson Bull*. 55:11–22. (Egg-laying and incubation.)

Stonehouse, B.
1960 The King Penguin, *Aptenodytes patagonica*, of South Georgia. 1. Breeding behaviour and development. *Falkland Islands Dependencies Surv. Sci. Repts*. No. 23.

Sutton, G.M., and Spencer, H.H.
1949 Observations at a nighthawk's nest. *Bird-Banding* 20:141–149.

Tate, J., Jr.
1967 Cowbird removes warbler nestling from nest. *Auk* 84:422.

Terres, J.K.
1980 *The Audubon encyclopedia of North American birds*. New York: Alfred A. Knopf.

Tickell, W.L.N.
1968 The biology of the Great Albatrosses, *Diomedea exulans* and *Diomedea epomophora*. *American Geophysical Union, Antarctic Res. Ser*. 12:1–55.

Tinbergen, N.
1953 *The Herring Gull's world: A study of the social behaviour of birds*. London: Collins.

Truslow, F.K.
1967 Egg-carrying by the Pileated Woodpecker. *Living Bird* 6:227–236.

Verner, J.
1965 Breeding biology of the Long-billed Marsh Wren. *Condor* 67:6–30.

von Haartman, L.
1967 Clutch-size in the Pied Flycatcher. *Proc. XIVth Internatl. Ornith. Congr.* pp. 155–164.

Walkinshaw, L.H.
1949 Twenty-five eggs apparently laid by a cowbird. *Wilson Bull.* 61:82–85.
1965 Attentiveness of cranes at their nests. *Auk* 82:465–476.

Weeden, J.S.
1966 Diurnal rhythm of attentiveness of incubating female Tree Sparrows *(Spizella arborea)* at a northern latitude. *Auk* 83:368–388.

Weller, M.W.
1956 A simple field candler for waterfowl eggs. *Jour. Wildlife Mgmt.* 20:111–113. (Use a mailing tube. Hold the eggs at one end against the sun and look through from the other. Apply the criteria in the paper for determining the age of the embryo. If the shell of the egg is too dense, the size of the air cell can be used as a criterion.)
1958 Observations on the incubation behavior of a Common Nighthawk. *Auk* 75:48–59.
1959 Parasitic egg laying in the Redhead *(Aythya americana)* and other North American Anatidae. *Ecol. Monogr.* 29:333–365.
1968 The breeding biology of the parasitic Black-headed Duck. *Living Bird* 7:169–207.

West, M.J., King, A.P., and Eastzer, D.H.
1981 Validating the female bioassay of cowbird song: Relating differences in song potency to mating success. *Animal Behavior* 29:490–501.

Weston, H.G., Jr.
1947 Breeding behavior of the Black-headed Grosbeak. *Condor* 49:54–73.

White, S.C., and Woolfenden, G.E.
1973 Breeding of the Eastern Bluebird in central Florida. *Bird-Banding* 44:110–123. (Number of broods, clutch size, nest success.)

Woolfenden, G.E.
1973 Nesting and survival in a population of Florida Scrub Jays. *Living Bird* 12:25–49. (Only females incubate; males may crouch over nests to defend contents.)

Young and Their Development

The development of young birds involves the acquisition of their physical and behavioral attributes from hatching to maturity.

Hatching

The shell of the egg gradually weakens during the course of incubation and thus becomes susceptible to the increasing pressure brought to bear by the maturing embryo. The first external evidence of hatching is a star-shaped crack that appears on the side of the egg, toward the larger end. The crack is caused by the embryo's **egg-tooth**—a calcareous, scale-like structure on the tip of the upper mandible. Muscular action, particularly by a special "hatching muscle" (Fisher, 1966), scrapes and presses the egg-tooth against the inside of the already weakened shell until a crack results. An egg is then said to be **pipped.**

An egg may remain pipped for a variable length of time, ranging from a few hours to 48 hours or more. Generally the pipped condition persists longest among species of lower orders.

Final hatching takes place in a comparatively short time. From the star-shaped crack, a fissure develops, usually around the larger end. Muscular action of the embryo, chiefly in the legs and neck, forces the shell apart at the circular fissure. The embryo on emerging becomes a **young bird.** Its egg-tooth is sloughed during the first week after hatching in some species, after the first week in others (Clark, 1961).

Eggs in a clutch do not usually hatch simultaneously, especially if incubation began before the clutch was complete. Nests may, therefore, contain young of different ages. Hatching may occur at any time during the day or night.

Condition of Development at Hatching

Young birds of different species show different degrees of development at hatching. Some are poorly developed and helpless while others are advanced in development and somewhat self-reliant. These contrasting conditions are expressed in three sets of terms:

Psilopaedic and Ptilopaedic Psilopaedic birds are either naked at hatching or have very sparse down (neossoptiles) on the dorsal region. Ptilopaedic birds at hatching are covered with down (usually dense) dorsally and ventrally.

Nidicolous and Nidifugous Nidicolous birds stay in the nest for an extended period after hatching whereas nidifugous birds leave the nest soon after hatching. Nidicolous birds may be psilopaedic (naked) or ptilopaedic (downy), but nidifugous birds are always ptilopaedic.

Altricial and Precocial Altricial birds, called **nestlings,** are fed by their parents; they are psilopaedic and nidicolous. Precocial birds, called **chicks,** feed themselves; they are nidifugous because they leave the nest to find food for themselves and ptilopaedic because they need a protective, downy covering. As a rule at hatching, nestlings have their eyes closed, chicks their eyes open.

Some young birds are neither typical nestlings nor typical chicks. For conditions of this sort, Nice (1962) proposed two terms: **semi-altricial** for young with down and staying in the nest; **semi-precocial** for young with down but soon moving from the nest to varying short distances while still fed by their parents. Essentially, Nice's "semi-altricial" means a bird that is ptilopaedic but nidicolous—and altricial in the strict sense of being fed by parents; "semi-precocial" means a bird that is ptilopaedic and nidifugous, but still altricial as it is fed by parents.

Table 7 shows the condition of development of young among the North American orders. Orders marked with one asterisk (*) are semi-altricial; those marked with two asterisks (**) are semi-precocial.

Table 7 **Condition of Development at Hatching**

	PSILOPAEDIC	PTILOPAEDIC	NIDICOLOUS	NIDIFUGOUS	ALTRICIAL	PRECOCIAL	EYES CLOSED	EYES OPEN
Gaviiformes		X		X		X		X
Podicipediformes		X		X		X		X
*Procellariiformes		X	X		X			X[1]
Pelecaniformes	X[2]		X		X		X	
*Ciconiiformes		X	X		X			X
Anseriformes		X		X		X		X
*Falconiformes		X	X		X			X
Galliformes		X		X		X		X
Gruiformes		X		X		X		X
Charadriiformes: Charadrii		X		X		X		X
**Charadriiformes: Lari		X	X			X		X
**Charadriiformes: Alcae		X	X			X		X
Columbiformes	X		X		X		X	
Psittaciformes	X		X		X		X	
Cuculiformes	X		X		X		X	
*Strigiformes		X	X		X		X	
**Caprimulgiformes		X	X			X		X
Apodiformes	X		X		X		X	
Trogoniformes	X		X		X		X	
Coraciiformes	X		X		X		X	
Piciformes	X		X		X		X	
Passeriformes	X		X		X		X	

1 Eyes closed in most petrels
2 Tropicbirds are ptilopaedic

Several other terms, commonly used to designate youth in birds, require explanation. **Fledgling** is applied broadly to an altricial bird from the time it leaves the nest until it becomes independent of its parents. A young bird actually **fledges** (or has **fledged**) when it takes flight for the first time. A **juvenile bird** (or simply **juvenile**) refers to any young bird with teleoptiles that has not attained sexual maturity and is therefore incapable of breeding. The term **juvenal** refers only to a particular plumage (see p. 162) and should not be used, as is sometimes the case, to designate youthfulness. An **immature bird** (or simply **immature**) indicates any young bird with teleoptiles that has not acquired its fully adult plumage (definitive feathering) but may nevertheless be capable of breeding.

Development of Altricial Young

The Newly Hatched Young

The altricial nestling of a passerine species immediately after hatching is ungainly and usually has an enormous head and abdomen, which are in marked contrast to very short and undeveloped limbs. The mouth is large and is conspicuous because of the **rictal flanges** that protrude laterally as if the rictal region were swollen.

The nestling is usually naked (psilopaedic), but in certain species, long, sparse down feathers (neossoptiles) are present on the dorsal and capital pterylae. The down may be of different colors in different species and may increase slightly in length after hatching. There are no macroscopic indications of developing teleoptiles, although their papillae may in some species be sufficiently large to indicate faintly the outlines of all pterylae.

The coloration of the skin is generally pinkish, darker above than below. The viscera show through the skin of the abdomen. The **soft parts**—i.e., parts of the bill and feet—are variously colored, sometimes intensively. The lining of the mouth is often strikingly colored (see Ficken, 1965). The tip of the bill, egg-tooth, and rictal flanges, however, are predominantly white, or yellowish white, in all species.

The eyes are closed, but the eyeballs show through the lids. The nestling rests on its abdomen with head slightly under the breast and the legs kept forward. Activities consist chiefly of simple grasping with the feet, for feeding, for defecating, and for keeping the body upright. The bird is responsive mainly to vibrations of low intensity.

The nestling may or may not be vocally silent. In some species it may give faint sounds even before it emerges from the egg.

The nestling is "cold-blooded" (poikilothermic), which means that its body temperature corresponds to the surrounding air temperature. It shows no ability to regulate its body temperature. The brooding parent must therefore supply the heat. Should the parent bird leave the nest for an extended period after brooding, the nestling's body temperature will drop soon to the temperature level of the surrounding air. Unless the air temperature is excessively low, the effect on the nestling will not be lethal as it can endure considerable cooling.

Length of Time in Development

The development of an altricial bird is regarded as divisible into two phases: development in the nest and development after leaving the nest. Both phases vary greatly in length among different species.

The length of time in development is commonly measured *in days* from the day of hatching (called 0 day) to the day when the young bird is independent of parental care and may range from approximately 25 days to several months. In passerine species Nice (1943) found that the majority attain flight proficiency at about 17 days and become independent of parental care at about 28 days. The end of the phase spent in the nest may not coincide with the attainment of flight, some species leaving before they are able to fly.

In most altricial species there is some correlation between the length of the incubation period and the length of the period spent in the nest. If the incubation period is a long one, so is the period of nest life. There are also two factors with which time of leaving the nest seems to be correlated: the safety of the nest and the size of the bird. Thus a bird nesting in a cavity undergoes a longer period of nest life than one nesting on the ground; a large bird takes a longer time to develop than a smaller one. The young Eastern Bluebird *(Sialia sialis)* fledges from its cavity nest in about 15 days, whereas the young Hermit Thrush *(Catharus guttatus)* leaves its ground nest in 9–10 days though it may not take its first flight until two or three days later. The young Bald Eagle *(Haliaeetus leucocephalus)* fledges in 10–12 weeks while the young of the California Condor *(Gymnogyps californianus)*, one of the largest flying land birds, may take as long as five months before first venturing on the wing (Koford, 1953). Perhaps the longest period of nest life of any bird is that of the Wandering Albatross *(Diomedea exulans)*, one of the largest flying sea birds, whose young takes nine months (278 days) to fledge (Tickell, 1968). For the ages at which many North American species take first flight, see Reilly (1968) and Terres (1980).

Weight

The weight of the nestling of most passerine birds at hatching averages approximately two-thirds of the weight of the fresh egg (Heinroth, *in* Nice, 1943) and, except in the Corvidae, is about 6–8 percent of the weight of the adult female (Nice, 1943). The weight of the nestling rapidly increases during nest life until, at the time of nest-leaving, its weight is from 20–30 percent less than that of the adult female, or sometimes as much as 50 percent less. At nest-leaving, a slight drop in weight occasionally occurs; thereafter the weight increases, though more slowly than before nest-leaving, until by the time the young bird has attained independence, its weight nearly equals that of the adult female.

In some birds, particularly nonpasserine species with a slow rate of growth, the nestling stores a vast amount of fat, thereby attaining a weight after the halfway mark in its nestling life that may greatly exceed the weight of the adult. Sea birds such as procellariiform species are notable in this respect. The young Leach's Storm-Petrel *(Oceanodroma leucorhoa)* may weigh half again as much as the adult at 40 days of age and 23–30 days before fledging (estimated from W.A.O. Gross, 1935). The young Wandering Albatross may reach maximum weight at least one-third in excess of an adult at 221–230 days of age and 50–60 days before departure (estimated from Tickell, 1968). Following this peak in weight the young of both species are fed less frequently, finally not at all, since the parents begin voluntarily to desert them. Consequently, their weight steadily declines as they use their fat stores, until at fledging their weight approximates that of the adults. Possibly fat storage by these young may be an adaptation for survival during long, stormy periods when the parents fail to obtain food from the tempestuous sea. Roberts (1940) found that young of the Wilson's Storm-Petrel *(Oceanites oceanicus)* in the Antarctic may weigh twice as much as an adult early in their long period of confinement to burrows and that their weight fluctuates markedly thereafter due, in this case, to heavy snowfalls successively covering their burrows and preventing the parents from entering to feed them. One young bird was able to survive as long as 20 days.

Plumage

Teleoptiles of the juvenal plumage first show evidence of development when the feather papillae darken and enlarge. Soon sheaths ("pinfeathers") emerge from the papillae, first on the alar, scapulohumeral, and dorsal pterylae, then on the capital pteryla, and finally on the remaining pterylae below. Later, and in this same order, the sheaths open, allowing the teleoptiles to continue growth. By the time of nest-leaving, the teleoptiles are sufficiently lengthened, unfolded, and expanded to cover all the apteria. Down feathers, when present, grow from the same papillae as the teleoptiles and are pushed out by them. They remain for a time attached to the tips of the teleoptiles. Frequently, young birds with well-developed juvenal plumage show the down feathers still adhering. Eventually the down feathers drop off.

For an excellent day-by-day account and description of the development of plumage in a passerine bird, see the paper by Banks (1959).

Body-temperature Control

In passerine species the nestling develops "warm-bloodedness" (homeothermy) early, while undergoing its most rapid growth, and attains temperature control soon after the midpoint in the period of nest life. Young House Wrens *(Troglodytes aedon)* develop temperature control by nine days of age (Baldwin and Kendeigh, 1932). Three emberizines, the Chipping, Field, and Vesper Sparrows

(*Spizella passerina*, *S. pusilla*, and *Pooecetes gramineus*), which leave the nest in 10 days or fewer, show effective temperature control by six to seven days after hatching (Dawson and Evans, 1957, 1960). The acquisition of a feather covering is not necessary for the establishment of temperature control as the nestling already has some control at moderate air temperature when the teleoptiles start emerging from their sheaths. The insulating plumage that the nestling later acquires serves mainly to prevent excessive loss of body heat when the air temperature is much lower than the body temperature. Excessively warm air tends to kill nestlings more quickly than cold.

Coloration of Soft Parts

Colors of the soft parts change gradually during early development. The whitish parts of the bill and the brilliantly colored mouth lining tend to darken as nest life advances. The iris and the feet slowly assume a coloration resembling that of the adult.

Behavior

As the young altricial bird develops in the nest, its many and complex behavioral traits of later life begin to emerge. The following descriptions apply primarily to the passerine bird.

Within minutes after hatching the nestling spontaneously lifts its head straight up and **gapes**—opens its mouth wide. This is the primordial "begging" action for food. It may be repeated spontaneously many times and will reoccur on receiving any vibratory or tactile stimulus from its parent or from a like stimulus simulated by a person at the nest. If not on the first day, then two or more days later, depending on the species, faint vocal sounds accompany the action. As the nestling develops and its eyes begin to open, it responds to visual stimuli and directs its gaping toward them. Meanwhile, increasingly sensitive to sounds of varying intensity, it begs readily on hearing its parent's "food calls" at the nest. The begging action itself gradually modifies: the nestling stands in the nest, vocalizes strongly, fans its wings, and stretches accurately toward the food source. If fed, the nestling, during its first few days, simply raises its posterior abdominal region slightly and voids the fecal matter; during its later life in the nest, it raises its abdominal region higher and more conspicuously, sometimes with a wavering action as if to draw attention to the process, and may turn so as to void the fecal material toward the edge of the nest.

When a young bird no older than two days is removed from its nest and placed on its back on a flat surface, it reacts by clutching at the air with a grasping action of its toes and attempts, albeit futilely, to turn itself over. At this early stage the nestling cannot grasp the bottom of the nest effectively and is just barely able to keep itself upright; if it rolls over on its back inadvertently, it can right itself only by persistently struggling until it manages to push against the wall of the nest or one of its nest-mates. Two or three days later the same nestling, when put on a flat surface, can quickly right itself. In the nest it retains such a firm grasp of the bottom that one can disengage its feet to lift it out only with difficulty. For exact ages at which the grasping appears in several species, see Holcomb (1966a, 1966b).

As it approaches the midpoint in its nest life, the young bird begins several maintenance activities, particularly preening and head-scratching; at the same time it displays such comfort movements as stretching, yawning, and shaking. It spends a considerable amount of time sleeping, the head forward on the nest itself or on its nest-mates. Toward the end of nest life, it stays awake for long periods during which it may indulge in numerous sessions of wing-fanning. If it happens to be larger than its nest-mates, it may be quite aggressive, pecking at them, crowding them, and taking more than its share of food from the parents. In eagles and some of the other large predators, which have only two young unequal in age and size, the larger one may early in nest life so severely harass the smaller one as to kill it.

At the start of nest life the young bird responds indiscriminately to stimuli, whether from the parent or any other source, and begs. Later, as it acquires vision, the nestling becomes increasingly more selective in its response to stimuli, finally responding only to stimuli coming from the parents. It may gape at the human intruder, but not stretch in the intruder's direction, vibrate its wings, or give vocal sounds. Eventually, near the time of nest-leaving, it will not gape at all; instead it cowers, sinking lower in the nest, sleeking the plumage, and sometimes even closing the eyes. If removed from the nest, it ordinarily "comes alive," giving "fear calls" and attempting to escape. The commotion may induce its nest-mates to come alive, burst from the nest, and disperse. Thus alarmed, neither the nestling, taken from the nest, nor its fellows which departed of their own accord, can be put back in the nest with the expectation of their staying. Therefore caution should be taken not to handle nestlings that show cowering behavior lest this cause them to leave the nest prematurely—sometimes before they are capable of faring for themselves.

The young bird leaves the nest of its own accord on the maturing of an inherent impulse or instinct to do so. The action may be triggered by any one of a number of factors, such as discomfort from high air temperature, hunger, exuberance during a session of wing-fanning, or a period of restlessness among the nest-mates. Its departure may in turn trigger the departure of its nest-mates, if they are of the same age. Parents may sometimes withhold food in view from the nest with the result that the offspring are prompted to "come and get it." What may *seem* to be a conscious intent of the parents to induce their offspring to fledge is either their hesitation, occasioned by some undetermined case in bringing the food, or their lowered feeding rate (see Lawrence, 1967). Any such break in routine increases the young birds' hunger and their consequent aggressiveness. In most species once the young bird fledges of its own accord, it rarely returns to the nest. Young eagles, however, and a few other large predators return to the nest repeatedly, using it as a "home base" for several days. Young White Storks *(Ciconia ciconia)* return to the nest to be fed and continue to use the nest as a sleeping place (Haverschmidt, 1949) until they migrate from the area.

Course of Development

For purposes of comparison with other species, it is convenient to consider the course of development of a species as divisible into **stages.** In the Song Sparrow *(Melospiza melodia),* a typical altricial species, Nice (1943) described five stages that may well serve as a guide in investigating development in other species. Her summary of these stages, in which she pointed out their duration and characteristics and indicated the time that certain activities were first manifest, follows verbatim.

> The **first** stage embraces the first 4 days of nest life; it is characterized by rapid growth and the start of feather development; the chief motor coordinations are the food response and defecation; the first food call has been heard at 2 days.
>
> The **second** stage—5 and 6 days—is one of rapid growth in weight and feathering and the establishment of temperature control; the eyes open, and the beginnings of several motor coordinations are seen—standing, stretching up of the legs, and the first intimations of preening.
>
> The **third** stage—7, 8 and 9 days—shows rapid development of motor coordinations; cowering typically appears; the birds are well covered with feathers and are able to leave the nest at any time. They stretch their wings up and sidewise, they scratch their heads and shake themselves, they fan their wings and flutter them when begging. New notes appear—the scream, the location call, and several new feeding notes. In the 9 to 10 days of its nest life the Song Sparrow has changed from a nearly naked, blind, and practically cold-blooded creature of about 1.5 grams, absolutely dependent on parents and nest for food, warmth and shelter, to a warm-blooded, well-feathered individual of 15 to 18 grams, independent of nest and nest-mates, able to care for its feathers, to move about, to escape enemies to some extent, to inform its parents of its whereabouts so they can feed it, and to respond to its parents' notes of alarm and fear.
>
> The **fourth** stage starts at the age of 9 to 10 days when the fledglings leave the nest. It is spent in retirement by the young bird away from its nest-mates, is characterized by silence (except when calling for food) and by general immobility. The chief advance is the acquisition of flight. A new method of begging for food appears: the birds no longer make themselves conspicuous, but call their parents and beg from a horizontal rather than vertical position. One brood was "psychologically" still in the nest for a day after leaving it. Sleeping in the adult position typically appears at 10 days. A beginning is made in independent feeding activities: wiping the bill, pecking at objects, picking up food, catching insects, working at grass heads, and scratching the ground. Bathing reactions appeared in several of the hand-raised birds. Singing appeared with 3 birds at 13, 14 and 15 days.

The **fifth** stage, starting with the attainment of flight and ending with independence, lasts about 10 days. At 17 days Song Sparrows in Ohio come out from retirement; they come into contact again with nest-mates and soon pursue their parents for food. They may beg from their parents until they are 5 weeks old, but the latest that I saw one fed was at 30 days. The Massachusetts Song Sparrows attempted to shell seeds at 17 days, but were not able to do so until the age of 25 days. The complete bathing technique, consisting of 3 chief motions, is attained during this stage. Sunning was first recorded at 29 days. Frolicking [play-fleeing] was first seen at 17 and 18 days. The fear note appears at 19 and 21 days. This fifth period may be the chief time when the young are conditioned by parental behavior with regard to what to fear and what not to fear. A note of antagonism was first noted at 17 days; pecking others at 18 days; threatening postures and fighting at 19 days. *Tsip*, denoting a social bond, first appeared at 19–20 days. The characteristic adult note *tchunk* was heard at 28 and 29 days.

Effects of Brood Parasitism

Before reading the following discussion, a review of "Brood Parasitism," pages 302–305, is advised.

With very few exceptions, brood parasitism lowers the output of hosts. In some cases, however, nestlings of the Giant Cowbird *(Scaphidura oryzivora)* may actually benefit their hosts by removing harmful botflies (N.G. Smith, 1968), and hosts of the Black-headed Duck *(Heteronetta atricapilla)* suffer little, if at all, because the parasite uses them solely for incubation. More typically, parasites harm their hosts in two ways: removal of one or more host eggs by the adult female and reduced survival of the remaining host progeny due to the young parasite. This second cost of parasitism may in some cases be slight if the host and parasite are similar in size and if nests have the usual single parasite—e.g., as with the Viduines and their hosts (Payne, 1973). In one study of the Song Sparrow (J.N.M. Smith, 1981) the effect of cowbird parasitism was slight, but in her study of the same species, Nice (1973) found parasitism deleterious. Parasitic young may cause the loss of host young because they have special adaptations. Nestlings of honey-guides (*Indicator* spp.) and of the Striped Cuckoo *(Tapera naevia)* in South America have sharp hooks on the tips of their mandibles, which they use to stab and kill the host young (Friedmann, 1955; Morton and Farabough, 1979). The nestling Common Cuckoo *(Cuculus canorus)* is endowed with special reflexes that enable it to move under, push up on its back, and roll or thrust over the nest's edge, any egg or other nestling. In these cases the parasites gain the full attention of their foster parents.

The Brown-headed Cowbird *(Molothrus ater)* has no such specializations. However, it does have an incubation period of 11–12 days, which is often shorter than its host's and thus gives the nestling a head start in development. Furthermore, when the cowbird parasitizes warblers and other species smaller then itself, the nestling at the outset has a distinct advantage in size with the result that, more often than not, it "swamps out" the young of its host, causing their death within two or three days after hatching by constantly trampling them and outreaching them for food brought by their parents. This is the usual fate of most nestling Kirtland's Warblers *(Dendroica kirtlandii)*, which normally hatch two days after the cowbird (Mayfield, 1960). Small flycatchers such as *Empidonax* species (Walkinshaw, 1961) and the Eastern Phoebe *(Sayornis phoebe;* Klaas, 1975) suffer even greater losses because their eggs usually do not hatch. The cowbird egg has such a shorter incubation period that these flycatchers do not normally continue to incubate their own eggs for the needed period. Instead, they prematurely switch to feeding the cowbird. Larger host species fare much better and some are probably better hosts because they provide more food for the young parasites. Nestlings of the Scarlet Tanager *(Piranga olivacea)* and the Brewer's Blackbird *(Euphagus cyanocephalus)*—species nearly as large or slightly larger than the cowbird—compete successfully with young cowbirds (Prescott, 1965; Furrer *in* Friedmann et al., 1977). Nestlings of the Red-eyed Vireo *(Vireo olivaceous)*—a species intermediate in size between the Kirtland's Warbler and Scarlet Tanager—compete poorly to only moderately well with the rapidly growing parasites. Ordinarily one nestling vireo, never more than two, survives to fledging (Southern, 1958).

The common tendency in observing brood parasitism is to abhor the parasite and pity the host as an unfortunate victim of a degenerate habit. The truth is that brood parasitism is a highly developed habit and a wholly successful mode of survival at no appreciable cost to most host species since the reproductive potential of the species as a whole is not affected. However, generalist parasites such as the Brown-headed Cowbird have the potential to depress or even to bring about the extinction of some host species. Specialized parasites limited to one or a few species are likely to become rare if they depress the population size of their hosts, thereby allowing the host species to recover. But a generalized parasite, such as the cowbird, can continue to depress the population of at least rare hosts with little effect on its own population as most of the cowbird's reproductive output may be derived from common hosts (Mayfield, 1977). Indeed, there is strong, though still inconclusive, evidence that some species with limited distributions are threatened with local extirpation or even extinction by cowbird parasitism. The best example is the rare Kirtland's Warbler, which is parasitized and suffers a roughly 40 percent reduction in reproductive output throughout its limited breeding range in Michigan. The cowbird seems to pose such a great threat of extinction to the warbler that a massive control program was initiated in the early 1970's, killing thousands of cowbirds annually (Mayfield, 1978). So far the control program may have reversed the warbler's decline, but, unfortunately, there has been no increase in the warbler.

The Brown-headed Cowbird is a remarkably successful parasite as evidenced by the long list of its host species, vast population, and expanding geographical range. Formerly an inhabitant of the grassy plains in mid-America, it has in the last 200 years penetrated the eastern forested regions and found new hosts that may lack defenses against brood parasitism (Mayfield, 1965). The situation is similar westward in North America to the Pacific Coast. The cowbird was absent or extremely rare in virtually all of California as recently as 1900, but it is now common throughout the state (Rothstein et al., 1980). This recent upsurge of cowbird parasitism in California may have caused at least nine host species, most notably the Bell's Vireo *(Vireo bellii)*, to decline or even approach extirpation from much of the state (Gaines, 1974). But all these hosts, like the Kirtland's Warbler, had small patchy populations to begin with and later habitat destruction by man may have contributed to their decline.

Development of Precocial Young

The development of the typically precocial chick of wild birds has been given much less attention than the development of the typically altricial nestling. One obvious reason for this is the great difficulty in making extended observations on a creature that, in a few hours after hatching, becomes extremely ambulatory and elusive. A résumé of the development of the typically precocial chick of North American orders follows, the emphasis being placed on characteristics that are in contrast to the development of altricial young.

The Newly Hatched Young

The newly hatched chick is more advanced in development than the altricial nestling. Its proportions more closely match those of its parents, except for the wings, which are relatively small and undeveloped. The chick's more advanced development, compared to the nestling's, is correlated with the larger supply of yolk that it had for nourishment as an embryo.

The chick is thickly covered with neossoptiles on all pterylae (ptilopaedic), the down feathers being extremely long on the upper pterylae in land chicks (i.e., chicks of Galliformes and Charadrii), and very thick on the lower pterylae of aquatic chicks (i.e., chicks of Gaviiformes, Podicipediformes, Anseriformes, and Gruiformes). The down becomes dry and fluffy within two or three hours after hatching.

The whitish egg-tooth and white tip of the lower mandible are conspicuous. The rictal region is without flanges. The lining of the mouth is no more brilliant in color than the adult's and is often less brilliant.

The eyes are open at hatching. Both vision and hearing are already developed, and the chick shows a quick response to visual and auditory stimuli. The chick rests on its tarsi; the neck is too weak to hold up the head, and consequently the head rests forward on the floor of the nest. Within minutes after hatching, the chick of a shorebird will "teeter" or "bob" if the trait is characteristic of the species.

At hatching the chick gives brood calls—i.e.,

"peeping" sounds—that can be heard earlier before it emerges from the egg.

Although the chick has body-temperature control partly established, it is essentially cold-blooded. When left unbrooded in cool weather, its body temperature soon begins to drop to the level of the surrounding air temperature with the result that it becomes increasingly immobile.

Length of Time in Development

The development of the precocial bird is divisible into two phases: the development between hatching and the attainment of flight and the development after the attainment of flight. The young bird becomes partially independent of parental care (i.e., brooding, feeding, etc.) during the first phase, but it may remain dependent socially on the family group during the second phase. Thus the two phases in the development of the precocial bird differ from the two phases in the development of the altricial bird.

The length of time before the chick attains flight ranges widely: 7–27 days in gallinaceous birds; 14–34 days in shorebirds; 4–12 weeks in most aquatic birds. See Reilly (1968) and Terres (1980) for the length of time in many species.

Weight

The weight of the precocial chick at hatching has the same proportion to the weight of the fresh egg as the weight of the altricial nestling. In other respects the weight of the precocial chick differs from the weight of the altricial nestling: (1) The weight of the precocial chick is about one to six percent the weight of the adult female, that is, the chick is somewhat smaller than altricial young in proportion to the size of the adult female. (2) The weight of the precocial chick decreases between hatching and the first feeding; thereafter it increases *steadily* throughout development. There is no retarding of weight increase during and after the attainment of flight. (3) At the time flight is attained, the chick of land birds weighs from 60–70 percent less than the adult female. In other words, the chick is much smaller than altricial young in proportion to the size of the adult female at the time of flight attainment.

Plumage

In land chicks the teleoptiles are first evident on the alar, scapulohumeral, and caudal pterylae. They appear within the first five or six days after hatching and continue well ahead of the other teleoptiles in growth. In certain species of Galliformes the teleoptiles of the alar pterylae (particularly the remiges and their respective coverts) are evident at hatching and may even be opened out of their sheaths. In aquatic chicks the teleoptiles are first evident on the ventral pteryla, then on the scapulohumeral and the caudal pterylae. They appear between the first and third weeks and continue well ahead of the other teleoptiles in growth. In general, the teleoptiles are first evident much later in Anseriformes than in other aquatic birds. In land chicks the teleoptiles last to appear are those of the capital pteryla, cervical region, and pelvic region. In aquatic chicks the teleoptiles last to appear are those of the capital pteryla, interscapular region (in Anseriformes), pelvic region, and alar pteryla. The remiges and their coverts are farthest behind in point of development.

Body-temperature Control

The newly hatched chick already has body-temperature control partly established. From hatching on, temperature regulation develops slowly, much more so than in the altricial bird. In most species the chick does not attain full temperature control until about four weeks of age.

Behavior

The activities directly adjusted to prolonged nest life do not occur in precocial young.

Most precocial young birds remain on the nest from 3–24 hours or more. During this period the down dries and the birds become increasingly active, rapidly acquiring perfection of muscular coordinations associated with standing, walking, and running or swimming. While the parent continues to brood, they peck at objects, especially those that are brightly colored or shining, fan their wings, stretch, yawn, and preen. When resting, they customarily sit on their tarsi with heels touching the ground. They give **brood calls** frequently and are soon responsive to the **food call** or **assembly call** of the parent.

After the initial period in the nest, the chick will show one of the following four patterns of behavior that have been described by Nice (1962) in the precocial species studied by her.

1. Stay in or near the nest, even though able to walk, and be fed by its parents. Characteristic of Lari (skuas, jaegers, gulls, terns, and skimmers) and Alcae (auks, murres, and puffins). For the first two or three days the chick remains in the nest where the parents bring it food. Thereafter, unless the nesting site or burrow prohibits departure, it may wander as far as the edge of the nesting territory, sometimes seeking shade under plants or other objects; but it returns to the nest for brooding and food.

2. Follow the parents from the nest and receive food from the parents. Characteristic of Gaviiformes (loons), Podicipediformes (grebes), and Gruiformes (cranes, rails, gallinules, moorhens, and coots). The chick, if a loon, grebe, coot, gallinule or moorhen, leaves the nest by swimming; if a rail or crane, by walking. Initially, the chick receives food passed to it by the parent; in a short time it is able to obtain its own. The young grebe or loon, for example, waits on the water's surface for one of the parents, searching for food below, to come up with food in the bill and deliver it; within a few days, the same young bird dives for its own food.

3. Follow the parent and be shown food. Characteristic of probably most Galliformes (grouse, pheasants, quail, turkeys). The chick leaves the nest, usually with the hen only. She directs the chick's attention to the food by picking it up and dropping it repeatedly, meanwhile giving a food call. The chick responds by taking the food dropped and swallowing it. Within a short time the chick learns to find its own food, often by scratching in the ground litter with its feet.

4. Follow the parents and find its own food. Characteristic of Anseriformes (swans, geese, and ducks) and Charadrii (shorebirds). The cygnet and gosling leave the nest with both parents, the duckling usually with the hen only; the shorebird chick leaves with one or both parents, depending on the species. At the outset the chick starts feeding without parental inducement or assistance by first nibbling indiscriminately at objects and soon learning by trial-and-error what is edible. This independence, however, is not the case among all shorebirds. At the outset the chick of certain species depends on particular foods that only its parents can obtain with their specialized bills. Two examples: the chick of the oystercatcher (*Haematopus* sp.) needs the parent to open up bivalve mollusks or to dislodge limpets for the fleshy parts and to probe in beaches for sandworms; the chick of the American Woodcock (*Scolopax minor*) needs the parent to pull up earthworms deep in soil. While oystercatcher and woodcock chicks await development of their own bills to match their parents', they may pick up some soft foods—crustaceans, insects, spiders, and plant matter—but not in sufficient quantity to sustain their rapid growth.

At sudden noises or abrupt, unfamiliar actions, most chicks scatter, hide under or beside large objects when available, and "freeze"—i.e., crouch flat and remain motionless with eyes closed. Usually, but not always, they remain frozen until the parent utters the assembly call. Aquatic chicks frequently give evidence of an ability to dive under water in the face of danger even though normally they do not dive at all or not until later in life.

A definite social bond is evident in a brood. When one chick is "lost"—i.e., out of contact with the parent and the brood—it utters the brood call loudly and shows evident distress until contact is reestablished. The family bond holds strongly until the chicks are well able to fly and require no further brooding. It is not unusual for the bond to persist into the fall and winter. Chasing, play-fighting, and pecking sometimes occur between members of a brood.

Calls are various in number among different precocial chicks. Nearly all species give the brood call just mentioned. It is uttered continually by each chick during the day while the brood is active. Presumably it functions as a means of keeping the family together. Nearly all chicks except those that stay in or near the nest give **alarm** or **warning calls,** or both, when danger approaches or when something is disturbing in appearance. Such calls serve to protect the entire brood by warning all members at once. All precocial young give **distress calls** when being harmed or handled. Some precocial young give **food calls** if they are accustomed to being fed. Some precocial young also give **contentment calls** or **conversational calls** when being brooded. These are faintly uttered and audible to the observer only when close at hand. Probably all chicks give **threat calls** when challenging a fellow member to a playful duel or when assuming dominance over another member. Very frequently, all sounds given by young, except the brood call, resemble sounds of similar purpose given by the parent. The chief difference is in the tone.

Course of Development

The course of development of the precocial young bird differs from that of the altricial in that the first three stages of development are passed in the egg; the precocial bird hatches at the beginning of the fourth stage. At this stage, both precocial and altricial young are ready to leave the nest. Both have well-developed muscular coordinations, vision, and hearing. Both are fully feathered, the precocial with neossoptiles, the altricial with teleoptiles. Both are still dependent on parental care, the precocial chiefly for brooding, the altricial for both brooding and feeding. The main difference between the two groups at this stage is that the precocial chick is still incapable of flight whereas the altricial nestling is just beginning to fly.

Development of Megapodes

Always of special interest in the study of development of precocial birds are the megapodes, gallinaceous birds whose eggs, after an exceptionally long period of incubation in mound nests (see pages 284 and 300), produce chicks no less exceptional in their precocity. Each chick, its egg-tooth already having disappeared, digs out of the mound on its own and fares alone, receiving no parental attention whatsoever and existing without the companionship of brood-mates. It is mute since brood calls can have no function. The chick of one species, the Mallee Fowl *(Leipoa ocellata)*, is able to run swiftly within two hours after emerging from the mound and can fly strongly in 24 hours (Frith, 1962). These highly precocial features are reptilelike and suggest primitiveness, yet there is no convincing evidence that this is the case. More likely, megapodes evolved from conventional galliform stock through varying selective pressures on their reproductive physiology and behavior (Clark, 1964).

Age of Maturity for Breeding

The period between the bird's attaining its independence from its parents and reaching maturity for breeding varies widely among different species. With very few exceptions, no wild North American bird breeds until about a year (9–13 months) of age. As for the exact ages at which birds of different species breed, reliable records are so few that one can only make generalizations.

The great majority of passerine species and other small land birds normally breed for the first time when just under a year old, and this is also the general rule for ducks, many gallinaceous birds, pigeons, some of the smaller owls, and presumably for grebes and the smaller herons and shorebirds. The larger herons and shorebirds probably breed when they are two years old. Birds that are almost certainly two years old, if not older, include the loons, pelecaniform birds, and geese. Some gulls, terns, and alcids breed on reaching the three-year mark, as do some swans, although most not until the four- or even the five-year mark. Puffins do not breed until reaching the five-to-six year mark. Nearly all the diurnal birds of prey breed beginning at two years of age except eagles, which breed at three or four years of age. California Condors (*Gymnogyps californianus*) first nest when six to eight years old. The age for breeding in procellariiform birds is unknown for most species—it is probably three years of age for the smaller species. For the two largest albatrosses, the Royal and Wandering (*Diomedea epomophora* and *D. exulans*), breeding is known to start at eight or nine years (see Lack, 1968, for source data), making the span of time between hatching and sexual maturity the longest among all birds.

While species do not breed generally until they have acquired the fully adult plumage (definitive feathering), this is not always the rule. For instance, the American Redstart (*Setophaga ruticilla*) breeds the first year even though its first nuptial (first alternate) plumage lacks the contrasting black and orange coloration of the successive nuptial (alternate) plumage. The Wandering Albatross may take 20–30 years or more before acquiring its definitive feathering (Tickel, 1968).

Selected Studies

Select a nest containing eggs under incubation and be prepared to observe in detail the hatching, the appearance and behavior of the young at hatching, and the physical and behavioral development of the young. When removing the young from the nest for weighing and measuring, do so as quickly as possible and return them unharmed. To hold the young in captivity is unlawful without the necessary permits. See Appendix A, pages 372–375, for procedures in obtaining permits.

Procedure Take full notes on the entire process of hatching from the time the first egg is pipped. Determine how long it takes for the pipped egg to hatch. Note time of hatching of each egg.

As soon as the young are dry, mark them so that they may be recognized as individuals. Since newly hatched young, notably those of altricial species, have small tarsi that will not hold bands, each should be temporarily marked with red nail polish. Mark one bird on the right tarsus, another on the left tarsus, another on the right outer toe, another on the left outer toe, etc. When, a few days later, the tarsi are large enough to hold bands, substitute colored bands (see Appendix A, p. 370).

If hatching was not simultaneous, observe the effect of age differences on growth and development.

Describe in full the newly hatched young bird: plumage; coloration (use color charts; see Appendix A); condition of eyes; responses; body attitudes and behavior. Weigh each bird as soon after hatching as possible.

Determine the length of time (*in days*) involved in the two phases of development. Consider the day of hatching as day 0.

Weigh each bird each day, preferably early in the morning before its first feeding. Always weigh at the same hour each day in order to get an accurate spacing of 24 hours. Follow the methods for weighing described in Appendix A. (When altricial young are weighed during the third stage of development, the resulting disturbance to the birds often causes them to leave the nest after being put back, thus preventing the determination of the total time involved in normal nestling life.) Strive to obtain weights of young after nest-leaving. Such weights of both altricial and precocial young are rare and most desirable.

Studies of the increase in length of wings, tail, tarsi, bill, body, and extent of wings are sometimes worthwhile in comparative investigations of species or in investigations of relative growth of parts of the body. Measurements should be taken according to the methods described in Appendix A.

Observe the development of plumage, noting the changes occurring in each pteryla. Sometimes it is desirable to measure the day-to-day growth of typical feathers in each pteryla. When the juvenal plumage is finally complete, describe it in full.

Note the day-to-day changes in the coloration of the fleshy parts and the time when the egg-tooth is lost. Either compare the colors directly with a color chart or record them in water-color and compare them with the chart later in the laboratory.

Make detailed observations, using a blind if necessary, of the development in activities and behavior. Extensive note-taking is necessary each day. In tracing the course of development determine the stages as described by Nice (1943) and compare the two sets of results as to time involved and characteristics.

If the nest is parasitized by cowbirds, follow the development of the young cowbirds along with the development of the hosts' young. Attempt to determine whether or not the hosts' young suffer from the presence of the young cowbirds.

The periods of development during and after nest-leaving warrant particular attention because of the dearth of existing information. Therefore, attempt to determine what triggers the young to leave; where and how far they go during the days following nest-leaving; whether they ever return to the nest or its immediate vicinity; where and how they roost; whether they remain on their parents' territory; and how long before they scatter.

If possible, take a series of photographs of young

on successive days that shows their growth and development. These will be particularly valuable to illustrate the description of plumage and the general course of development. Use the same bird for successive photographs, placing it on a dark, plain background (preferably black cloth) and always in the same position and at the same distance from the camera. Also, take photographs showing various activities, and use them as illustrations when presenting the results.

Presentation of Results Draw up a report on the results obtained from a study of young birds and their development. Prepare the manuscript as directed in Appendix B. An outline of suggested topics is given below.

Hatching Number of eggs; date and time of day when each is pipped; date and time of day when each is hatched; process of hatching; comments on time involved.

Newly Hatched Young Describe: Plumage; coloration; condition of eyes; responses; body attitudes; behavior. Give the weight. Comment on the condition of development (whether precocial or altricial, etc.); on the significance of coloration (whether cryptic or not, etc.); on the weight as compared with the weight of adult female.

Development of Young Present a day-by-day account of the development of the young birds. Include in each daily account: Weight; changes in plumage and coloration; description of activities and behavior.

Discussion After having given the above chronological account, discuss in summary form the following: *Weight:* Changes in weight during the first two phases of development. Represent the weights of each bird in a graph. See Fautin (1941) for a method used. Compare the weight at the end of each phase with the weight of the adult female. *Plumage:* General course of development of the different pterylae. Describe the complete juvenal plumage. Relation of plumage growth to activities—i.e., preening, bathing or dusting, swimming, etc.

Coloration: General changes in color and their relation to activities. *Activities and behavior:* Classify and discuss briefly the various activities observed, stating the time when each was first observed and its relation to phases of development. Note indications of learned behavior. Describe in detail the activities and behavior related to nest-leaving and the subsequent period of development. *Length of time:* Give the total length of time in the last two phases of development and compare with that given in the literature on the same species and closely related species. *Course of development:* Trace briefly the course of development and compare with Nice (1943, 1962) and other available sources.

Development of Young Cowbirds Present an account of the development of young cowbirds, stressing any effects that their presence may have had on the development of the hosts' young.

References

Austin, G.T., and Ricklefs, R.E.
1977 Growth and development of the Rufous-winged Sparrow *(Aimophila carpalis). Condor* 79: 37–50.

Baker, M.F.
1948 Notes on care and development of young Chimney Swifts. *Wilson Bull.* 60:241–242.

Baldwin, S.P., and Kendeigh, S.C.
1932 Physiology of the temperature of birds. *Sci. Publ. Cleveland Mus. Nat. Hist.* 3:i–x; 1–196. (Body temperature of nestling birds and mention of body temperature in precocial young.)

Banks, R.C.
1959 Development of nestling White-crowned Sparrows in central coastal California. *Condor* 61:96–109.

Beason, R.C., and Franks, E.C.
1973 Development of young Horned Larks. *Auk* 90:359–363.

Best, L.B.
1977 Nestling biology of the Field Sparrow. *Auk* 94:308–319.

Brackbill, H.
1947 Period of dependency in the American Robin. *Wilson Bull.* 59:114–115.

Brown, C.R.
1978a Double-broodedness in Purple Martins in Texas. *Wilson Bull.* 90:239–247.
1978b Post-fledging behavior of Purple Martins. *Wilson Bull.* 90:376–385.

Carney, W.P.
1974 Notes on raising passerine birds in the laboratory. *Bird-Banding* 45:29–32.

Clark, G.A., Jr.
1961 Occurrence and timing of egg teeth in birds. *Wilson Bull.* 73:268–278.
1964 Life histories and the evolution of Megapodes. *Living Bird* 3:149–167.
1969 Oral flanges in juvenile birds. *Wilson Bull.* 81:270–279. (Oral flanges [= rictal flanges] tend to be larger in hole-nesting passerines.)

Davis, E.
1943 A study of wild and hand reared Killdeers. *Wilson Bull.* 55:223–233. (Important observations on the development and behavior of precocial young.)

Dawson, W.R.; Bennett, A.F.; and Hudson, J.W.
1976 Metabolism and thermoregulation in hatchling Ring-billed Gulls. *Condor* 78:49–60.

Dawson, W.R., and Evans, F.C.
1957 Relation of growth and development to temperature regulation in nestling Field and Chipping Sparrows. *Physiol. Zool.* 30:315–327.
1960 Relation of growth and development to temperature regulation in nestling Vesper Sparrows. *Condor* 62:329–340.

Dawson, W.R.; Hudson, J.W.; and Hill, R.W.
1972 Temperature regulation in newly hatched Laughing Gulls *(Larus atricilla). Condor* 74:177–184.

Dunn, E.H.
1975 The timing of endothermy in the development of altricial birds. *Condor* 77:288–293.
1976a Development of endothermy and existence energy expenditure of nestling Double-crested Cormorants. *Condor* 78:350–356.
1976b The development of endothermy and existence energy expenditure in Herring Gull chicks. *Condor* 78:493–498.
1979 Age of effective homeothermy in nestling Tree Swallows according to brood size. *Wilson Bull.* 91:455–457.

Fautin, R.W.
1941 Development of nestling Yellow-headed Blackbirds. *Auk* 58:215–232.

Ficken, M.S.
1965 Mouth color of nestling passerines and its use in taxonomy. *Wilson Bull.* 77:71–75. (With few exceptions, it is distinctive of each passerine family.)

Fisher, H.I.
1966 Hatching and the hatching muscle in some North American ducks. *Trans. Illinois State Acad. Sci.* 59:305–325. (The muscle originates in the back and sides of the upper neck and inserts on the back of the skull. Reaches maximum development at hatching and serves to raise the egg-tooth against the shell. Earlier papers by the author describe the muscle in other groups of birds.)
1971 The Laysan Albatross: Its incubation, hatching, and associated behaviors. *Living Bird* 10:19–78. (Detailed, well-illustrated; based on a long-range investigation.)

Friedmann, H.
1955 The honey-guides. *U.S. Natl. Mus. Bull.* 208.

Friedmann, H.; Kiff, L.F.; and Rothstein, S.I.
1977 A further contribution to the knowledge of host relations of the parasitic cowbirds. *Smithsonian Contr. Zool.* No. 235.

Frith, H.J.
1962 *The Mallee-fowl: The bird that builds an incubator.* Sydney: Angus & Robertson.

Gaines, D.
1974 A new look at the nesting riparian avifauna of the Sacramento Valley, California. *Western Birds* 5:61–80.

Gotie, R.F., and Kroll, J.C.
1973 Growth rate and ontogeny of thermoregulation in nestling Great-tailed Grackles, *Cassidix mexicanus prosopidicola* (Icteridae). *Condor* 75:190–199.

Gross, A.O.
1938 Eider Ducks of Kent's Island. *Auk* 55:387–400. (Information, including weights, on newly hatched precocial young.)

Gross, W.A.O.
1935 The life history cycle of Leach's Petrel *(Oceanodroma leucorhoa leucorhoa)* on the Outer Sea Islands of the Bay of Fundy. *Auk* 52:382–399.

Hann, H.W.
1937 Life history of the Oven-bird in southern Michigan. *Wilson Bull.* 49:145–237. (Hatching and development of altricial young; method of leaving nest.)

Haverschmidt, F.
1949 *The life of the White Stork.* Leiden: E.J. Brill.

Hochbaum, H.A.
1981 *The Canvasback on a prairie marsh.* 3rd ed. Lincoln: Univ. of Nebraska Press. (Development of precocial young, particularly plumage.)

Hoffmeister, D.F., and Setzer, H.W.
1947 The postnatal development of two broods of Great Horned Owls *(Bubo virginianus). Univ. Kansas Publ. Mus. Nat. Hist.* 1:157–173.

Holcomb, L.C.
1966a The development of grasping and balancing coordination in nestlings of seven species of altricial birds. *Wilson Bull.* 78:57–63.

1966b Red-winged Blackbird nestling development. *Wilson Bull.* 78:283–288. (Concerned mainly with development of ability to grasp and balance.)

Howell, T.R.

1964 Notes on incubation and nestling temperatures and behavior of captive owls. *Wilson Bull.* 76:28–36.

Kendeigh, S.C.

1939 The relation of metabolism to the development of temperature regulation in birds. *Jour. Exp. Zool.* 82:419–438.

Klaas, E.E.

1975 Cowbird parasitism and nesting success in the Eastern Phoebe. *Univ. Kansas Mus. Nat. Hist., Occas. Papers* No. 41.

Koford, C.B.

1953 The California Condor. *Natl. Audubon Soc. Res. Rept.* No. 4. New York.

Lack, D.

1968 *Ecological adaptations for breeding birds.* London: Methuen.

Lack, D., and Lack, E.

1951 The breeding biology of the Swift *Apus apus. Ibis* 93:501–546. (Important correlations between weather and nestling survival.)

Lanyon, W.E., and Lanyon, V.H.

1969 A technique for rearing passerine birds from the egg. *Living Bird* 8:81–93.

Lawrence, L. de K.

1953 Nesting life and behaviour of the Red-eyed Vireo. *Canadian Field-Nat.* 67:47–77.

1967 A comparative life-history study of four species of woodpeckers. *Amer. Ornith. Union, Ornith. Monogr.* No. 5.

Mayfield, H.

1960 The Kirtland's Warbler. *Cranbrook Inst. Sci. Bull.* No. 40. Bloomfield Hills, Michigan.

1965 The Brown-headed Cowbird, with old and new hosts. *Living Bird* 4:13–28.

1977 Brown-headed Cowbird: Agent of extermination? *Amer. Birds* 31:107–113.

1978 Brood parasitism: Reducing interactions between Kirtland's Warblers and Brown-headed Cowbirds. In *Endangered birds: Management techniques for preserving threatened species,* ed. S.A. Temple. Madison: Univ. of Wisconsin Press.

McVaugh, W., Jr.

1972 The development of four North American herons. *Living Bird* 11:155–173. (Tricolored Heron, Black-crowned Night-Heron, Great Egret, Little Blue Heron.)

1975 The development of four North American herons. II. *Living Bird* 14:163–83. (Green-backed Heron, Snowy Egret, Yellow-crowned Night-Heron, Least Bittern. This and the 1972 article, both well-illustrated by the author, describe day-to-day plumage development during the first four to five weeks from hatching.)

Morton, E.S., and Farabaugh, S.M.

1979 Infanticide and other adaptations of the nestling Striped Cuckoo *(Tapera naevia). Ibis* 121:212–213.

Mosby, H.S., and Handley, C.O.

1943 *The Wild Turkey in Virginia: Its status, life history and management.* Richmond Virginia: Commission of Game and Inland Fisheries. (Information on weights, development, and behavior of precocial young.)

Nice, M.M.

1937 Studies in the life history of the Song Sparrow, I. *Trans. Linnaean Soc. New York* 4:i–vi; 1–247.

1943 Studies in the life history of the Song Sparrow, II. *Trans. Linnaean Soc. New York* 6:v–iii; 1–328. (Chapters 2–5.)

1962 Development of behavior in precocial birds. *Trans. Linnaean Soc. New York* 8:i–xii; 1–211.

Nolan, V., Jr.

1978 The ecology and behavior of the Prairie Warbler *Dendroica discolor. Amer. Ornith. Union, Ornith. Monogr.* No. 26. (Chapters 24–34.)

O'Connor, R.J.

1977 Growth strategies of nestling passerines. *Living Bird* 16:209–238. (Three representative species studied: Blue Tit for clutch size adjustment; House Sparrow for brood reduction; House Martin for resource—fat—storage.)

Odum, E.P.

1941 Annual cycle of the Black-capped Chickadee-2. *Auk* 58:518–535. (Development of altricial young, and dispersal.)

Payne, R.B.

1973 Behavior, mimetic songs and song dialects, and relationships of the parasitic indigobirds *(Vidua)* of Africa. *Amer. Ornith. Union, Ornith. Monogr.* No. 11.

Pettingill, O.S., Jr.

1936 The American Woodcock *Philohela minor* (Gmelin). *Mem. Boston Soc. Nat. Hist.* 9:167–391. (Notes on development and behavior of precocial young.)

1939 History of one hundred nests of Arctic Tern. *Auk* 56:420–428. (An account of destructive forces operating in a breeding colony.)

Pinkowski, B.C.
1975 Growth and development of Eastern Bluebirds. *Bird-Banding* 46:273–289.

Prescott, K.W.
1965 The Scarlet Tanager. *New Jersey St. Mus., Investigations* No. 2.

Rand, A.L.
1937 Notes on the development of two young Blue Jays *(Cyanocitta cristata)*. *Proc. Linnaean Soc. New York* 48:27–58.
1941 Development and enemy recognition of the Curve-billed Thrasher *Toxostoma curvirostre*. *Bull. Amer. Mus. Nat. Hist.* 78:213–242.

Reilly, E.M., Jr.
1968 *The Audubon illustrated handbook of American birds*. New York: McGraw-Hill.

Ricklefs, R.E.
1975 Patterns of growth in birds. III. Growth and development of the Cactus Wren. *Condor* 77:34–45. (One of the author's several papers on growth and development in birds.)

Ricklefs, R.E.; White, S.; and Cullens, J.
1980 Postnatal development of Leach's Storm-Petrel. *Auk* 97:768–781. (Aspects of development "resemble semi-precocial more closely than semi-altricial.")

Roberts, B.
1940 The life cycle of Wilson's Petrel *Oceanites oceanicus* (Kuhl). *Brit. Graham Land 1934–37 Sci. Repts.* 1 (2):141–194.

Rothstein, S.I.; Verner, J.; and Stevens, E.
1980 Range expansion and diurnal changes in dispersion of the Brown-headed Cowbird in the Sierra Nevada. *Auk* 97:253–267.

Schreiber, R.W.
1976 Growth and development of nestling Brown Pelicans. *Bird-Banding* 47:19–39.

Skutch, A.F.
1945 Incubation and nestling periods of Central American birds. *Auk* 62:8–37.

Smith, J.N.M.
1981 Cowbird parasitism, host fitness, and age of the host female in an island Song Sparrow population. *Condor* 83:152–161.

Smith, N.G.
1968 The advantage of being parasitized. *Nature* 219:690–694.

Southern, W.E.
1958 Nesting of the Red-eyed Vireo in the Douglas Lake Region, Michigan. *Jack-Pine Warbler* 36:105–130; 185–207.

Stoddard, H.L.
1931 *The Bobwhite Quail: Its habits, preservation and increase*. New York: Charles Scribner's Sons. (Important notes on development and behavior of precocial young in Chapters 2 and 4. Chapter 5 is devoted to the Bobwhite's vocabulary, including that of the young.)

Stonehouse, B.
1960 The King Penguin, *Aptenodytes patagonica*, of South Georgia. 1. Breeding behaviour and development. *Falkland Islands Dependencies Surv. Sci. Repts.* No. 23.

Stoner, D.
1942 Behavior of young Bank Swallows after first leaving the nest. *Bird-Banding* 13:107–110.

Terres, J.K.
1980 *The Audubon Society encyclopedia of North American birds*. New York: Alfred A. Knopf.

Tickell, W.L.N.
1968 The biology of the Great Albatrosses, *Diomedea exulans* and *Diomedea epomophora*. *American Geophysical Union, Antarctic Res. Ser.* 12:1–55.

Tinbergen, N.
1939 The behavior of the Snow Bunting in spring. *Trans. Linnaean Soc. New York* 5:1–94. (Important data on young leaving the nest.)
1953 *The Herring Gull's world: A study of the social behaviour of birds*. London: Collins. (Chapters 18, 21, and 22 deal with hatching and behavior of precocial young. Highly recommended reading.)

Walkinshaw, L.H.
1961 The effect of parasitism by the Brown-headed Cowbird on *Empidonax* flycatchers in Michigan. *Auk* 78:266–268.

Webster, J.D.
1941 The breeding of the Black Oyster-catcher. *Wilson Bull.* 53:141–156.
1942 Notes on the growth and plumages of the Black Oyster-catcher. *Condor* 44:205–211. (An excellent account of growth and development of plumages in precocial young.)

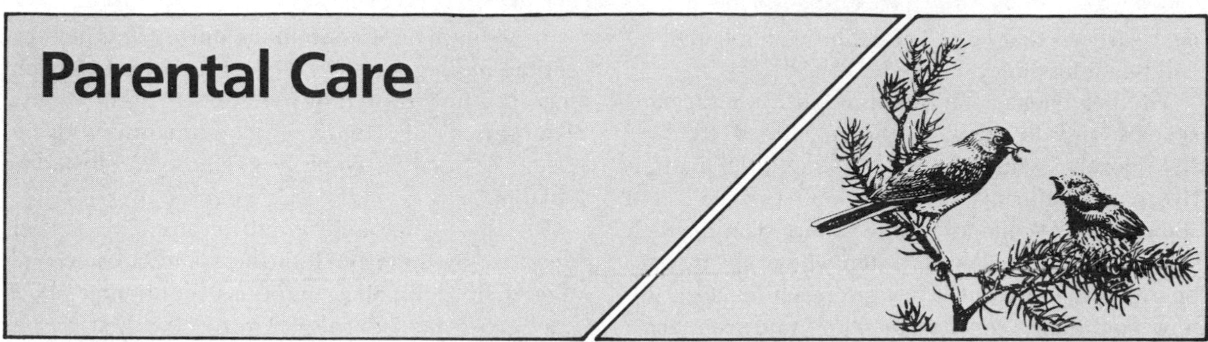

Parental Care

Numerous important problems relative to the care of young by parent birds are reviewed below. Unless otherwise indicated, statements concern the parents of both altricial and precocial young.

Removal of Eggshells

Parent birds may or may not remove the eggshells from the nest after the young hatch. Generally, in nidicolous species, the parents either eat the shells or carry them away and drop them. Some individuals may eat the shells of one egg and carry away the shells of another. Species that normally carry away shells may remove strange objects that inadvertently get into their nests. A Common Tern *(Sterna hirundo)* carried off a film carton that had blown into its nest (Pettingill, 1931). In nidifugous species the birds depart, leaving the shells in the nest.

Brooding

Brooding is fundamentally a continuation of incubation behavior through the early stages in the development of the young. Although the term, **brooding,** is sometimes applied to the process of incubating the eggs, as well as to the process of covering the young, it actually means the covering of the young only.

Start and Duration

The brooding period begins with the hatching of the first egg and continues until the parents no longer cover the young at any time, day or night.

Participation and Behavior of the Sexes

Generally the participation of the sexes in brooding is the same as in incubation. Thus, if the female alone incubates, she alone broods; if the sexes share in incubation, they also share in brooding.

The behavior of the male in passerine species often changes after the young hatch. He sings less and less each day, finally ceasing altogether; and sometimes he shows a gradual waning of aggressiveness in defense of territory.

Even though the male takes no part in incubation, he is quite likely to assist in feeding the young. In many passerine species he appears at the nest, long before the eggs hatch, carrying food with the "anticipation" of feeding the young—an act called **anticipatory food-bringing.** Nolan (1958) found that the male Prairie Warbler *(Dendroica discolor)* carried food from one to three times a day throughout the incubation period, taking it to the nest whether or not the female was present. In her absence he preoffered the food to the eggs, then either swallowed it or took it away with him. Krause (1965) observed similar behavior in the male Canada Warbler *(Wilsonia canadensis).* While this activity serves no on-the-scene purpose, Skutch (1953) suggested that, when only the female incubates, it may be useful in bringing the male's attention to the nest early so that later, when the young hatch, he will begin feeding them right away.

The behavior of the brooding bird in nidicolous species tends to follow a common pattern. The bird sits "closely," with relatively few and brief inattentive periods during the first one or two days, and shows evident concern for the young, standing high in the nest every now and then when they wriggle below. If the rays of the hot sun reach the nest, the bird commonly remains standing and frequently spreads the wings, thus exposing more of its body to heat loss and at the same time providing additional shade and ventilation for the young. For the first few days the passerine bird is secretive in its going and coming and departs and returns as deliberately as during incubation, but, as the young develop, it may fly off the nest and return openly. For this reason nests with well-developed young are easier to find; the observer has merely to watch the bold activities of the bird. Activities of the brooding bird during inattentive periods are devoted more extensively to gathering food than during incubation because the bird must find food for the young, as well as for itself. Copulation stops altogether.

In nidicolous species during the first few days the brooding bird reacts to intrusions in much the same manner as during the last part of incubation. Thereafter, it becomes more vociferous and aggressive. It flushes from the nest and remains to undertake defensive action (discussed later).

Length of Time

The number of days involved in brooding depends on several factors:

1. *The rapidity with which young develop body-temperature control and protective plumage*. The more rapidly they develop, the shorter the time they need brooding.

2. *Climatic conditions*. Excessively cool weather, heavy rains, or exposure of the nest to intense sunlight may prolong, or cause a return to, brooding.

3. *Protection of the nest*. Birds nesting in cavities, for example, where air temperatures are more uniform, require less brooding.

Brooding is not continuous during any day (except in nidifugous birds during the period of hatching). It is interrupted by periods of inattentiveness. The periods of attentiveness alternating with periods of inattentiveness compose a **brooding rhythm.**

The frequency and length of attentive periods decrease as the growth of the young advances although the climatic conditions mentioned above may cause irregularities. During the first day or two, the brooding rhythm may correspond to the incubation rhythm and one or both sexes may participate, as in incubation. In nidicolous species brooding may cease altogether before the young leave the nest; in nidifugous species brooding may cease before the young fly. Probably birds continue night brooding longer than day brooding, but actual evidence in the majority of species is lacking.

Feeding

Feeding of the young is one of the commonly studied phases in the breeding cycle of birds because it is easily observed; at the same time, it is extremely challenging because of the numerous complexities.

Start and Duration

Feeding starts after a variable lapse of time following hatching. In typically altricial species it usually begins within two or three hours after hatching, but in other species that require parental feeding the interval between hatching and the first feeding is much longer. Feeding continues until the young reach independence, long after the departure from the nest. The cessation of feeding indicates the attainment of independence.

Participation of the Sexes

In the majority of nidicolous species both sexes share in the feeding of the young. When the female does all of the brooding, the male takes a greater part in feeding, making many trips to the nest with food. If the female is brooding when he arrives, she may either fly from the nest for an inattentive period or remain on the nest, standing aside to permit the male to feed the young. In many passerine species the female may take the food from the male when he arrives at the nest and either consume it or pass it to the young. The part taken by the sexes may change as the young develop. One sex, taking an active part in feeding immediately after the hatching, may gradually or suddenly become less active, while the other sex takes over the main responsibility. Individuality of the birds may determine the role played by the sexes in feeding, especially after the brooding period has ceased.

Methods of Feeding

There are several methods by which adults feed their young. A classification of these methods follows:

I. *Indirect Feeding.* Feeding that consists merely of *attracting young to the food*, sometimes by calling, sometimes by picking up the food and holding it, and sometimes by repeatedly picking up the food, manipulating it, and dropping it, while giving food calls. Characteristic only of those nidifugous species that feed their young.

II. *Direct Feeding.* Feeding that consists of *carrying the food to the young*. Characteristic of all nidicolous species.

- A. Food is carried in the bill and placed in the opened mouths of the young. Characteristic of a wide variety of species, mostly passerine.
- B. Food is swallowed and later made available to the young in the following ways:
  1. By regurgitating (i.e., disgorging) the food:
     - a. Into the opened mouths of the young. Examples: finches and waxwings.
     - b. Into the throats of the young. Example: hummingbirds.
     - c. Into the opened mouths of the young after their mandibles have straddled the parent's bill. Example: herons.
     - d. Onto the floor of the nest or a neighboring surface, where it is picked up by the young, or held in the mouth and taken by the young. Example: gulls.
  2. By opening the mouth and allowing the young to penetrate the throat region, where they pick up the food. Examples: pelicans and cormorants.
- C. Food consists of "milk" from degenerating epithelial cells sloughed from the crop lining and regurgitated into the partially opened mouths of the young after the parent's mandibles have straddled the bills of the young. Example: pigeons.
- D. Food is carried in the talons to the nest and there shredded with the bill, then either passed to the young or taken by the young. Example: birds of prey.

Many nidicolous young receive food at first by regurgitation and later receive it directly.

Feeding trips to the nest are quick and direct. By habit the bird usually comes to the nest by the same route each time, stays at the nest edge long enough to complete feeding and, in some species, to pick up fecal material, and then either leaves hastily or perhaps remains to brood. The feeding trips become less and less secretive as the growth of the young advances and feeding becomes more frequent.

In typical passerine species the young become aware of the parent's approach to the nest by vibrations when it alights on or near the nest, by its call notes, and by its visible actions. The young respond immediately with begging behavior (described in the preceding chapter). Their brightly colored mouth linings, seen as they gape for food, provide **food targets** which "direct" the parent to place the food properly.

When the parent arrives at the nest, most often the same side each time, *all* the nestlings normally respond, although with varying degrees of intensity, depending on the state of their hunger. The parent makes no deliberate attempt to distribute the food equally, as it does not recognize the individuality of the nestlings and hence does not know which one it fed last. The nestling with the most effective response—probably the hungriest—receives the food. Then the nestling with the next most effective response—now probably the hungriest—receives the food. In this manner food is distributed by a process called **automatic apportionment.** Reed (1981) found that in a nest of Song Sparrows (*Melospiza melodia*) with five nestlings of equal size, the nestlings shifted after each feeding, competing for the position on the side of the nest where the parent usually arrived. By these regular shifts, each of the nestlings automatically received a portion during successive parental visits. Orderly as this process may be in some nests, it is not the case in others. A parent arriving at a nest in which all nestlings respond may "**trial feed**" by placing the food in one mouth. If the nestling fails to swallow the food, probably because its swallowing reflex is inhibited by an already full esophagus, the parent removes the food and puts it in another mouth. This behavior is another mode of automatic apportionment.

Rate of Feeding

In passerine species the rate of feeding increases daily because of the increasing needs of fast-growing young. In species that stay in the nest until attaining flight (as contrasted with those species that leave before attaining flight) there is an increase for the first week or ten days, after which there is no increase (Nice, 1943). The rate of feeding is increased by either making more trips per day or by supplying greater amounts of food per trip. Generally those species that customarily carry only one item of food per trip increase the rate of feeding *by making more trips*; whereas, those species that customarily either carry several items of food or feed by regurgitation increase the rate of feeding *by supplying greater amounts of food per trip*.

The feeding rate in nidicolous species, other than passerine, presumably increases as growth advances, but the increase is less evident because growth is comparatively slow.

In passerine species the number of young fed per trip often depends on the amount of food supplied and the number of young in the nest. When a bird brings one item, usually it feeds only one young. When a bird brings several items and there are three or more young in the nest, it does not usually feed them all. When the bird supplies food by regurgitation, it usually feeds all young, regardless of number, each trip. No matter how food is supplied, the individuals in a nest generally receive an equal amount of food during the course of the day, due to automatic apportionment. Exceptions to the rule of automatic apportionment occur when young are not of equal size or vigor, and the larger, or more vigorous, young obtain more food at the expense of the smaller.

Nest Sanitation

Defecation in nidicolous young occurs often immediately after feeding. Young of certain species either void fecal material over the edge of the nest (e.g., hawks and hummingbirds) or into the nest and over the immediate vicinity (e.g., herons). Supposedly all passerine species void fecal material in the form of **fecal sacs**—masses of whitish semisolid urine and darkish intestinal excreta enveloped in thick mucous. In some species parents regularly dispose of the sacs throughout nest life; in other species through approximately the first three-fourths of nest life, after which the sacs accumulate on the nest. The adults usually collect the sacs following a feeding and either eat them or carry them away and drop them. Birds seem to vary greatly in respect to the methods they use in disposing of the sacs. Certain individuals may consistently eat the sacs, others may consistently carry them away, and still others may use both methods. Although the birds usually collect fecal sacs following a feeding, they do not necessarily collect them after every feeding. In 35 studies of 28 species Nice (1943) calculated that the percentage of times the excreta were removed (in terms of number of feeding trips) ranged from 9–67 percent, the median being 25.

Nest Helpers

In nesting activities there may be **nest helpers**, birds that assist the nesting pair of their own species in one way or another. Some nest helpers may be birds-of-the-year aiding a late-nesting pair in building a nest or feeding the young; a mated adult, whose young have been recently lost, feeding the young of another pair of the same or of a different species; or an unmated adult assisting in building the nest or feeding the young of a pair. Refer to Skutch (1961) for a review and discussion of the subject.

When studying three Ohio nesting colonies of Barn Swallows (*Hirundo rustica*), all color-banded, Myers and Waller (1977) noted helpers providing a "substantial portion" of total feeding visits. Immature offspring of the first clutch fed newly hatched siblings and, later in the breeding season, a flock of "unidentified individuals" joined the parents and immatures in feeding.

Nest-helping is practically a way of life in the complex group-breeding of certain jays—e.g., the Florida Scrub Jay (*Aphelocoma c. coerulescens*) and the Brown Jay (*Cyanocorax morio*). In his study of a population of Florida Scrub Jays, all marked, Woolfenden (1975) found half the breeders nesting as "simple pairs"; the others had helpers consisting virtually of all the yearlings and about half of the two-year-olds. Rarely did any of the helpers assist other than their own parents and stepparents. Woolfenden believed that the helpers might well be the major beneficiaries of the system. In Costa Rica Lawton and Guindon (1981) found that Brown Jays generally live in flocks of six to ten. Each flock builds one nest attended by both brooding and nonbreeding members comprised of three age classes: one to two years; three years; and four or more years. The authors showed that the one- to two-year-old birds improved in nestling care, thus leading support to the suggestion by Lack (1968) that young helpers serve an apprenticeship, learning to breed successfully on their own.

Parental Carrying of Young

A few birds carry their young. Aquatic birds such as loons, grebes, and coots let their chicks ride on their backs while they are swimming, as do some waterfowl—swans, sheldgeese, and several ducks (Johnsgard and Kear, 1968). But species that intentionally pick up their young and bear them for any distance are exceptional. Rails, particularly long-billed species, may do so. Pettingill (1938) reported an instance of a parent Clapper Rail *(Rallus longirostris)* removing one chick from its brood held captive in an open-top box. On hearing the brood calling inside, the parent bird found its way to the top edge, jumped in, grasped one chick in its bill, then jumped out with it and put it on the ground. Pettingill later tested (and filmed) the same ability in a parent Virginia Rail *(Rallus limicola)* by putting its newly hatched brood in a 12-inch high (30.48 cm), open-top carton about eight feet (2.44 m) from the nest. The parent removed the chicks in the same manner, one after the other, and carried them back to the nest. The African Jacana *(Actophilornis africanus)* and another jacana, the Lotus-bird *(Irediparra gallinacea)* of Australia, carry their chicks under their wings. Hopcraft (1968) describes this procedure in the African Jacana. The parent bird first crouches low and makes a churring noise, whereupon the chicks run under its wings as if to be brooded. The parent then presses its wings firmly against the sides, stands up with the feet of the chicks dangling below the wings, and walks away with them over the broad-leaved water plants. Most extraordinary is young-carrying by the male American Finfoot *(Heliornis fulica)*, a neotropical sun-grebe, which has special "pockets" for the purpose (Alvarez del Torro, 1971). The finfoot builds its nest on a platform of sticks in a tangle of vegetation over water. There both sexes incubate the normal clutch of two eggs. After the eggs hatch the male tucks the chicks, naked and helpless, beneath the wing on either side, into a pocket consisting of a pleat of skin into which the chicks fit—one in each pocket. Side feathers overlap the pockets, further enhancing the chicks' security. The male then carries his progeny to the water and keeps them and feeds them in the pockets until the chicks gain sufficient independence to emerge on their own.

Ornithological literature contains numerous reports of various other species intentionally carrying young in one way or another, but, as in the case of the American Woodcock *(Scolopax minor)* that is said to secure a chick between its legs and carry it in flight, they are entirely without adequate proof.

Defense

Defense of young against enemies begins to appear late in incubation and reaches peak of vigor when the young leave the nest. In nidifugous species defense is most active soon after hatching; in nidicolous species, many days later, following the prolonged period of nest life.

As a rule, when both sexes participate in brooding or feeding, both sexes also participate in defense, but when one sex alone broods and feeds, the same sex may or may not defend the young. In many species the male shows a tendency to defend the young, even though he takes little or no part in feeding them. However, studies of pairs of birds of the same species show much individual variation in intensity of defensive action. In one pair, both sexes may defend the young vigorously; in another pair, only the male; in still another pair, only the female.

A classification of some of the more common methods by which parents defend their young follows. For a discussion of defense methods in birds, see the chapter "Behavior," page 218.

Direct Defense Defense by dealing directly with the enemy.

Threatening The bird gives loud sometimes piercing calls and displays the plumage, remaining defiantly in the presence of the enemy and sometimes feigning bodily attack.

Attacking The bird rushes at the intruding enemy with intent to strike with bill, or feet (sometimes with talons or tarsal spurs), or wings, or combinations of these. Frequently threat displays precede or follow attack.

Mobbing Occasionally the parents and several birds of the same or closely allied species attack an enemy together in a concerted effort to distract its attention.

Indirect Defense Defense by dealing indirectly with the enemy.

Communicating Danger The bird gives warning calls that silence and immobilize the young (sometimes also the sex-partner) and, in some species, make them hide and "freeze."

Giving Distraction Display (also called *Injury-feigning* or *Rodent-running*) The bird behaves in such a way as to detract the attention of the enemy from the nest or young.

Apparently birds show different methods of defense before different enemies. They may, for example, give threatening displays before human beings and distraction displays before small mammals. The subject of defense methods by one species against different enemies needs investigation.

Young–Parent Recognition

The young of a precocial species, once imprinted on its own parent, presumably "knows" the parent by appearance and repertoire of vocalizations—**contact calls,** such as the clucking sounds in gallinaceous birds; **rallying calls** for food; **alarm calls;** and others. The parent, in turn, learns to know its young by similar means. A strong family bond is thus established and the family integrity is maintained for the duration of the young bird's dependency, or longer. Exceptions are broods of ducks, such as mergansers and eiders, which often convene, forming mixed families under the surveillance of one or more hens.

The best example of mutual recognition can be observed in colonies of precocial aquatic birds where many families of one species exist in close association. When visiting a colony of Herring Gulls *(Larus argentatus)* with chicks, notice that the adults attack any strange chick that comes near their own, even though it is of the same age, and that the chicks respond to their parents arriving with food yet ignore other adults that come near them. Tinbergen (1953) believed that Herring Gull parents know their own young beginning about five days after hatching.

To the human observer Herring Gull chicks and adults look so much alike as to be indistinguishable. To a Herring Gull, however, each bird undoubtedly has nuances in appearance—and probably voice, too—that give it individuality. Just what the nuances are in Herring Gulls remains to be determined, but studies made in colonies of Rockhopper and Adélie Penguins *(Eudyptes crestatus* and *Pygoscelis adeliae)* revealed that subtilties in voice almost certainly play a role in mutual recognition (Pettingill, 1960; Penney, 1968).

The chicks of both species of penguins, when left alone in their colonies after two or three weeks of age, customarily gather from their nests to huddle together in **crèches.** Their parents, on returning from the sea to feed them, proceed directly to their respective nest sites and call. Immediately, their own chicks leave the crèches, come to their parents, give begging calls, and are soon fed. If other chicks respond, as is sometimes the case, they are not only refused food, but are sometimes severely pecked and driven away.

To test the ability of the chick to recognize its parent by voice, Penney (1968) tape-recorded the calls given by several parent Adélie Penguins, then played them back to their chicks huddled with others in crèches. In nine out of the ten playbacks the chick, or the chicks, of the parent whose voice was played "stood up, looked around the colony, and then moved to the edge of the crèche in an alert posture. None of the other crèche chicks changed position, although several raised their heads and looked briefly about. In seven of the nine cases the chicks kept moving and returned to their natal territory within one minute after the playback had started. These chicks remained on the territory for three to nine minutes after the sound stopped. On the territory they assumed an appearance best described as alert and anticipatory." Penney later analyzed by sound spectrograph the calls of the different parents and found that each had distinctive and consistent characters of voice sufficient to give that bird a recognizable individuality.

In altricial birds mutual recognition develops as soon as the young fledge. Lawrence (1967) found that woodpeckers recognize their own fledged young and refuse to feed any other fledged young of their own species. But woodpeckers will, she observed, sometimes react positively to begging fledglings of other woodpecker species.

Duration of Family Bond

The bond of parents to young and young to parents persists for a varying length of time, depending on the species. In many altricial birds, the bond weakens and soon disappears when the young have obtained proficiency in flight and can forage for themselves. Among a few altricial birds, such as permanent-resident titmice, families stay together through the fall and winter months. In group-nesting jays families may stay together for two or more years, the younger birds becoming nest helpers (see previous discussion). The family bond in precocial and semi-precocial birds continues for a longer time, owing to the young bird's need of brooding during its slow development of body-temperature control as well as its need of protection during its prolonged flightless period. Young gulls, fully grown, may be seen long after leaving their natal colonies, receiving food from adults, presumably their parents. Families of swans, geese, and cranes maintain their integrity for many months, migrating south in family units to their wintering grounds and even returning north before the yearling birds disperse.

Selected Studies

Select a nest containing eggs under incubation, and be prepared, as hatching approaches, to make detailed studies of the methods used by the birds in caring for their young. The study may be combined with the preceding study of young birds and their development. Before undertaking the study, consult works by Kendeigh (1952), Nice (1943, 1962), and others for important information already available.

Procedure Make careful observations during the period of hatching from the pipping of the first egg. Watch the adults *continuously* for as long as possible. Note significant activities both at the nest and in the vicinity, any changes in behavior after incubation, the part taken by the sexes in nesting affairs, and the way they remove the eggshells.

Study the activity of the sexes during brooding, ignoring no detail of behavior either at the nest or in the vicinity. Determine the length of time *in days* involved in brooding. Consider the day of hatching as 0 day. Record specific data on the brooding rhythm at different times of the day, at different stages of brooding, and under different weather conditions. Use the tabular form suggested by Pitelka (1941) for incubation. (See the directions for the study of incubation, page 305.) Attempt to determine when night brooding ceases.

In addition to studying activities connected with brooding and the brooding rhythm, study the activities associated with feeding. Determine the time of the first feeding after hatching and (if at all possible) the last feedings long after the young have left the nest. Observe carefully the role of the sexes in feeding; the methods of feeding used at the beginning and later, when the food is gathered; the route followed when adults return to the nest; the responses of the young; and the adult's reactions to the young. Gather accurate data on the rate of feeding at different ages of the young. To do this, make at least one observation each day, not less than three hours long. Record the data in tabular form suggested in part by Pitelka (1941):

Date
Age of young
Time of day; total hours and minutes of observation
Air temperature at beginning and end of observation
Wind velocity
Weather conditions

1. Total feeding visits
   Number per hour
   Average
   Extremes
   Intervals
   Number per hour
   Average length
   Maximum and minimum length
   Number of visits in which only one young was fed
   Number of visits in which only two were fed
   (Repeat as above until all young are included)
   Total number of visits in which young were fed
   Average number of young fed per visit
   Extremes
2. Total feeding visits of male
   (Repeat as in 1)
3. Total feeding visits of female
   (Repeat as in 1)

If there are size differences in the young or if there is anything peculiar about the number of young, location of nests, etc., take it into account when recording the data.

Observe the disposal of fecal sacs, recording the number removed or eaten per feeding visit, and the role of each sex in disposing of them.

Devote considerable attention to the methods of defense. Note how the adults react to humans at different times and under different circumstances. When possible, observe methods of defense in the presence of various birds and other animals. If feasible, test defense responses by deliberately placing animals (under control) near the birds. Use discretion so as not to interfere with the normal course of nesting events.

Make every effort to follow the family after nest-leaving, and strive to learn how long before the family breaks up. Even meager information is valuable.

Photographs that show brooding attitudes and methods of feeding, nest sanitation, and defense will be useful in illustrating information obtained.

Presentation of Results Prepare a report on the results obtained from a study of parental care. Prepare the manuscript as directed in Appendix B. Some suggested topics follow.

Hatching Number of eggs; dates of hatching; duration of hatching. Role and behavior of the sexes. Evident changes in behavior since incubation. Methods by which eggshells are removed (if removed). Number of unhatched eggs.

Brooding *Start and duration/Participation and behavior of the sexes:* Activities of the male and his relation to the nest; activities and behavior of the brooding bird on the nest and off; manner of approaching and of leaving the nest; relations of the sexes. *The brooding rhythm:* General description; frequency and length of attentive periods; relation of periods to climatic conditions, age of young, time of day and night.

Feeding *Start and duration/Participation of the sexes:* Methods by which the sexes take part, and variations throughout dependency of young. *Methods of feeding:* Detailed account of the way food is delivered and received throughout brooding. *Time involved in feeding:* Number of days; time of first and last feedings in relation to age of young; rate of feeding at different ages of the young, using tables suggested and discussing the results in relation to age, time of day, and climatic conditions. Comment on automatic apportionment and any data on amount of food (e.g., items of food) brought per visit. *Type of Food:* Identify food when possible.

Nest Sanitation Method of defecation; number of defecations in relation to feeding visits; part played by sexes in removal of fecal material; frequency of removal.

Nest Helpers and Parental Carrying Give whatever information is obtained.

Defense Different kinds of defense witnessed and their relation to nest and age of young; kinds of enemies stimulating defense; participation of sexes.

Young–Parent Recognition Describe any instances and give supporting evidence. Determine, if at all possible the duration of the family bond.

References

Alvarez del Toro, M.
1971 On the biology of the American Finfoot in southern Mexico. *Living Bird* 10:79–88.

Amadon, D.
1964 The evolution of low reproductive rates in birds. *Evolution* 18:105–110. (The evolutionary trend among birds has been toward producing relatively few young but bestowing on them protracted parental care. One evolutionary reason is that the complex physical skills of flight and food-getting require a certain time for maturation and perfection.)

Blair, R.H., and Tucker, B.W.
1941 Nest-sanitation. *Brit. Birds* 34:206–215; 226–235; 250–255.

Brackbill, H.
1944 Juvenile Cardinal helping at a nest. *Wilson Bull.* 56:50.

Cogswell, H.L.
1949 Alternate care of two nests in the Black-chinned Hummingbird. *Condor* 51:176–178.

Crockett, A.B., and Hansley, P.L.
1977 Coition, nesting, and postfledging behavior of Williamson's Sapsucker in California. *Living Bird* 16:7–19. (Family units "broke down quickly after fledging.")

Dexter, R.W.
1952 Extra-parental cooperation in the nesting of Chimney Swifts. *Wilson Bull.* 64:133–139.

Hann, H.W.
1937 Life history of the Oven-bird in southern Michigan. *Wilson Bull.* 49:145–237. (Extensive account of parental care.)

Hopcraft, J.B.D.
1968 Some notes on the chick-carrying behavior in the African Jacana. *Living Bird* 7:85–88.

Howell, T.R., and Dawson, W.R.
1954 Nest temperatures and attentiveness in the Anna Hummingbird. *Condor* 56:93–97.

Jenkins, M.A.
1978 Gyrfalcon nesting behavior from hatching to fledging. *Auk* 95:122–127.

Johnsgard, P.A., and Kear, J.
1968 A review of parental carrying of young by waterfowl. *Living Bird* 7:89–102.

Kendeigh, S.C.
1952 Parental care and its evolution in birds. 2nd corrected printing, 1955. *Illinois Biol. Monogr.* 22:i–x; 1–356.

Krause, H.
1965 Nesting of a pair of Canada Warblers. *Living Bird* 4:5–11.

Krekorian, C.O.
1978 Alloparental care in the Purple Gallinule. *Condor* 80:382–390. (Chicks fed and protected by juvenile helpers.)

Lack, D.
1968 *Ecological adaptations for breeding birds.* London: Methuen.

Lawrence, L. de K.
1967 A comparative life-history study of four species of woodpeckers. *Amer. Ornith. Union, Ornith. Monogr.* No. 5.

Lawton, M.F., and Guindon, C.F.
1981 Flock composition, breeding success, and learning in the Brown Jay. *Condor* 83:27–33.

Lenington, S.
1980 Bi-parental care in Killdeer: An adaptive hypothesis. *Wilson Bull.* 92:8–20. (Two parents rather than one give chicks greater protection and thus provide an adaptive advantage.)

Munro, J., and Bédard, J.
1977 Crèche formation in the Common Eider. *Auk* 94:759–771.

Myers, G.R., and Waller, D.W.
1977 Helpers at the nest in Barn Swallows. *Auk* 94:596.

Nethersole-Thompson, C., and Nethersole-Thompson, D.
1942 Egg-shell disposal by birds. *Brit. Birds* 35:162–169; 190–200; 214–223. (An important discussion of the eggshell disposal of some 186 British birds.)

Nice, M.M.
1943 Studies in the life history of the Song Sparrow, II. *Trans. Linnaean Soc. New York* 6:i–viii; 1–328. (Chapters 19–21.)
1962 Development of behavior in precocial birds. *Trans. Linnaean Soc. New York* 8:i–xii; 1–211.

Nisbet, I.C.T.; Wilson, K.J.; and Broad, W.A.
1976 Common Terns raise young after death of their mates. *Condor* 80:106–109.

Nolan, V.,Jr.
1958 Anticipatory food-bringing in the Prairie Warbler. *Auk* 75:263–278.
1962 A Catbird helper at a House Wren nest. *Wilson Bull.* 74:183–184.
1974 Notes on parental behavior and development of the young in the Wood Thrush. *Wilson Bull.* 86:144–155. (Observations of a pair in a single season.)
1978 The ecology and behavior of the Prairie Warbler

Dendroica discolor. Amer. Ornith. Union, Ecol. Monogr. No. 26. (Chapter 28 in particular.)

Odum, E.P.
1941 Annual cycle of the Black-capped Chickadee–2. *Auk* 58:518–535. (Care of the young.)

Penney, R.L.
1968 Territorial and social behavior in the Adélie Penguin. *American Geophysical Union, Antarctic Res. Ser.* 12:83–131.

Pettingill, O.S., Jr.
1931 An analysis of a series of photographs of the Common Tern. *Wilson Bull.* 43:165–172.
1936 The American Woodcock *Philohela minor* (Gmelin). *Mem. Boston Soc. Nat. Hist.* 9:167–391. (Notes on parental care and methods of defense in a precocial species.)
1937 Behavior of Black Skimmers at Cardwell Island, Virginia. *Auk* 54:237–244. (Feeding and defense of young.)
1938 Intelligent behavior in the Clapper Rail. *Auk* 55:411–415. (Instances of adults carrying young.)
1960 Crèche behavior and individual recognition in a colony of Rockhopper Penguins. *Wilson Bull.* 72:213–221.

Pinkowski, B.C.
1978 Feeding of nestling and fledgling bluebirds. *Wilson Bull.* 99:84–98.

Pitelka, F.A.
1940 Breeding behavior of the Black-throated Green Warbler. *Wilson Bull.* 52:3–18. (Detailed analysis of parental care, particularly feeding.)
1941 Presentation of nesting data. *Auk* 58:608–612.

Purdue, J.R.
1976 Thermal environment of the nest and related parental behavior in Snowy Plovers, *Charadrius alexandrinus. Condor* 78:180–185.

Ramsey, A.O.
1951 Familial recognition in domestic birds. *Auk* 68:1–16.

Reed, J.R.
1981 Song Sparrow "rules" for feeding nestlings. *Auk* 98:828–831.

Schantz, W.E.
1939 A detailed study of a family of Robins. *Wilson Bull.* 51:157–169. (Parental care, departure from the nest, etc.)

Silver, R., ed.
1977 *Parental behavior in birds. Benchmark papers in animal behavior.* Vol. 2. Stroudsburg, Pennsylvania: Dowden, Hutchinson & Ross, (Collection of 32 articles from start of breeding to posthatching behavior.)

Skutch, A.F.
1935 Helpers at the nest. *Auk* 52:257–273.
1953 How the male bird discovers the nestlings. *Ibis* 95:1–37; 505–542.
1954–1955 The parental stratagems of birds. *Ibis* 96:544–564; 97:118–142. (Various maneuvers, including distraction displays, employed by birds in protecting their nests.)
1961 Helpers among birds. *Condor* 63:198–226.
1976 *Parent birds and their young.* Austin: Univ. of Texas Press. (An indispensable reference.)

Smith, S.
1941 The instinctive nature of nest sanitation. *Brit. Birds* 35:120–124.
1943 The instinctive nature of nest sanitation. *Brit. Birds* 36:186–188.

Spencer, O.R.
1943 Nesting habits of the Black-billed Cuckoo. *Wilson Bull.* 55:11–22. (Parental care.)

Stoddard, H.L.
1931 *The Bobwhite Quail: Its habits, preservation and increase.* New York: Charles Scribner's Sons. (Parental care in Chapter 2.)

Tinbergen, N.
1953 *The Herring Gull's world: A study of the social behaviour of birds.* London: Collins.

Walkinshaw, L.H.
1937 The Virginia Rail in Michigan. *Auk* 54:464–475. (Includes description of parental carrying of young.)

Wittenberger, J.F.
1980 Feeding of secondary nestlings by polygynous male Bobolinks in Oregon. *Wilson Bull.* 92:330–340. (Feeding nestlings of second nests after primary nestlings attained specific age.)

Woolfenden, G.E.
1975 Florida Scrub Jay helpers at the nest. *Auk* 92:1–15.

Longevity, Numbers, and Populations

This book does not attempt a thorough presentation of the problems pertaining to the longevity, numbers, and populations of birds. The purpose here is merely to introduce the subjects and to suggest possible studies of populations.

Longevity

The length of life or **longevity** of birds depends on many factors, including their size. As a rule, the larger the bird, the longer it is likely to live. But this is a very broad generalization.

Some estimates of the average ages of wild birds, roughly calculated and computed from the time they are *independent* of their parents, give between one and two years for passerine birds and pigeons; and two and three years and possibly longer for some herons, some shorebirds, and gulls (data from Lack, 1954). Some of the larger birds, especially those with a long prebreeding period such as albatrosses, have a much longer life expectancy. The two largest albatrosses, the Royal and Wandering *(Diomedea epomophora* and *D. exulans)*, which require eight or nine years to reach breeding maturity, may have a lifespan averaging between 30 and 40 years (from Lack, 1954), and an occasional bird may live theoretically as long as 80 years (Westerskov, 1963; Tickell, 1968).

Do not confuse the average lifespan of a bird with its potential lifespan, which is very much longer. The length of life of a wild bird that *far* exceeds the average lifespan is rarely attained as records of banded birds show. Of the 15,054 recoveries from Ring-billed Gulls *(Larus delawarensis)* banded in the Great Lakes Region, only 10 were older than 20 years and the oldest was 31 years and 9 months (Southern, 1975). During a period of 52 years in southeastern Pennsylvania, Middleton (1974) trapped and banded 2,672 Blue Jays *(Cyanocitta cristata)* in his yard. One individual when last trapped was at least 14 years and 6 months old, two individuals were 10 years of age, and only 16 others lived to be from 9 to 7 years of age.

Among the longevity records for North American birds listed by Kennard (1975) were five species with individual records exceeding 30 years: Brown Pelican *(Pelecanus occidentalis)*, 31 years, 5 months; Osprey *(Pandion haliaetus)*, 32 years; Herring Gull *(Larus argentatus)*, 31 years, 2 months; Arctic Tern *(Sterna paradisaea)*, 34 years; and Sooty Tern *(Sterna fuscata)*, 32 years, 1 month. Among passerine species Kennard listed only three individuals, besides the Blue Jay (previously mentioned), exceeding 14 years: American Crow *(Corvus brachyrhynchos)*, 14 years, 1 month; Clark's Nutcracker *(Nucifraga columbiana)*, 17 years, 5 months; and Red-winged Blackbird *Agelaius phoeniceus)*, 14 years, 5 months. Possibly the longevity record for all wild birds in the world, excepting albatrosses, is still held by a European Oystercatcher *(Haematopus ostralegus)*. Banded as a chick, it was last trapped on its nest when 36 years old (see Nice, 1966; Terres, 1968). No doubt as time passes the returns coming in from the many more thousands of birds banded yearly will show ages surpassing existing longevity records (e.g., see Clapp et al., 1982).

Captive birds, sheltered as they are from predation and other unfavorable aspects of the natural environment, have exceeded the normal life expectancy manifold. Indeed, some captive birds have lived to astonishing ages—an Eagle Owl *(Bubo bubo)*, 68 years; Andean Condor *(Vultur gryphus)*, 65 years; one Siberian White Crane *(Grus leucogeranus)*, 62 years, another 59 years; Canada Goose *(Branta canadensis)*, 33 years; Common Pigeon *(Columba livia)*, 30 years; and House Sparrow *(Passer domesticus)*, 23 years. For further listings, see J. Fisher and Peterson (1964) and Flower (1938).

Numbers

How many birds are there? Nobody knows with any certainty. J. Fisher (1941) placed the number at 100 billion, but it could be higher. Peterson (1941, 1964) guess-estimated the number of breeding birds in conterminous United States to be "not less than" 6 billion and "probably closer" to 6 billion at the start of the breeding season and as many as 20 billion at the end. Nobody has ever disagreed in print with his figures.

Only two countries have attempted anything like an accurate estimation of their total numbers of birds. In Great Britain (88,619 square miles—229,523.21 sq. km), from bird censuses on selected habitats—cultivated and wooded lands, scrublands, swamps, gardens, etc.—the number of breeding land birds totaled 120 million (J. Fisher, *op. cit.*). In Finland (130,165 square miles—337,127.35 sq. km), from strip censuses in the breeding season through representative habitats in all parts of the country from north to south, the total number of birds was found to be about 64 million (Merikallio, 1958).

Two statewide censuses have been undertaken in the United States. In Illinois (55,400 square miles—143,486 sq. km) Graber and Graber (1963) estimated by the strip census method a summer population of 61 million birds and 54 million during the preceding winter. In North Dakota (70,665 square miles—183,022.35 sq. km) R. E. Stewart and Kantrud (1972) estimated the bird population by dividing the state into eight major strata representing the basic biotic regions and then censusing sample units allocated to each stratum, the number of sample units being in proportion to the extent of the stratum. The total breeding bird population projected for the entire state was close to 52 million birds. Passerine birds alone accounted for 79.6 percent of the total breeding bird population of all species combined.

Probably the Domestic Fowl *(Gallus gallus)* is the most numerous bird in the world, and it has been suggested that the House Sparrow and European Starling, introduced from Europe to many parts of the world, may be almost as numerous. But excluding these birds, whose numbers have been aided and abetted by man, the most abundant birds are no doubt certain marine species inhabiting vast areas of sea where the supply of food is seemingly limitless.

The guano-producing cormorant, or Guanay *(Phalacrocorax bougainvillii)*, breeding on desert islands off the Peruvian coast, may be one of the most abundant. One island alone was once estimated conservatively to have more than 5,600,000 adults and young, all of which required not less than a thousand tons (907 t) of fish per day (Murphy, 1925). The several species of alcids of the North Atlantic and North Pacific must comprise many millions of individuals. Lockley (1953) put the breeding population of just one North Atlantic species, the Atlantic Puffin *(Fratercula arctica)*, at 15 million. J. Fisher (1954) believed that the Wilson's Storm-Petrel *(Oceanites oceanicus)*, which breeds in the Antarctic area and wanders over all the oceans, is probably the most abundant species, marine or land, of any wild bird in the world.

Ornithologists are by no means certain as to the most abundant land bird in North America, but the Red-winged Blackbird may lead all other native land birds. It breeds from the northern fringes of the Arctic tundra to Central America. Although primarily a marsh bird, it nests in various open, upland situations across the continent. Wherever traveling by car in the summer, one sees it on utility wires, seldom far from its nesting site in a grassy ditch, low shrub, or field. Possibly the Red-eyed Vireo *(Vireo olivaceus)* is the most common bird in the woodlands of eastern North America and the Horned Lark *(Eremophila alpestris)* the most common in the prairie country.

Accurate information on the absolute number of any one species is sparse. Generally, the species whose numbers have been determined accurately are those whose breeding areas are sharply restricted, completely known, and accessible for direct counting. Colonies of marine birds lend themselves particularly well to direct counting because they are confined to islands or headlands where their nests can be found and the number of breeding pairs computed therefrom. J. Fisher (1954), working with a group of collaborators in 1949, censused all the colonies of the Northern Gannet *(Sula bassanus)* on the east side of the North Atlantic and came up with a total count of 82,394 nests or 164,788 birds. Westerskov (1963) and Tickell (1968) provided figures for the world's populations, respectively, of the Royal and Wandering Albatrosses, both of which breed on islands exclusively in the far southern oceans. By collecting data on the number of breeding pairs in any one year on every island where the birds are known to nest, and by making calculations based on long-term knowledge of the species' breeding cycle and mortality rate, Westerskov determined the Royal Albatross population to consist of 18,960 birds, and Tickell, the Wandering Albatross population to consist of 58,760.

The most accurate figures available on absolute numbers are those species with small populations. The Kirtland's Warbler *(Dendroica kirtlandii)*, which breeds only in a small section of the jack pine barrens in northern Lower Michigan, was systematically censused in 1951 and found to have 432 singing males or an adult population—assuming each singing male had a mate—of fewer than 1,000 (Mayfield, 1953). The population has since declined, with a total in 1981 of 232 singing males or 464 individuals (Ryel, 1981). The remaining wild and selfsustaining population of the Whooping Crane *(Grus americana)* was 78 birds in 1980–81, a number readily obtained on the Arkansas National Wildlife Refuge, coastal Texas, where it winters almost exclusively. Forty years before, in 1940–41, there were only 15.

For statewide estimates in Illinois and North Dakota of numbers of individuals comprising some of the more common species, see Graber and Graber (1963) and R.E. Stewart and Kantrud (1972), respectively. For a wide variety of methods in estimating numbers of birds, see Ralph and Scott (1981).

Populations

A **population** of birds is the total number of individuals in an area. Although the term is usually applied to single species, it is commonly used to cover all the species in the area, and since interactions between species sharing a tract of habitat are close and complex, studies often focus on the entire bird community as a unit. Bird communities vary in both composition and size, and population studies cover both the diversity and abundance aspects of their structure.

The information obtained from such studies is of practical importance as well as of general interest for the following reasons:

1. Conservationists need to know the status of rare or endangered species and what measures might be effective in protecting and encouraging existing populations where they persist.

2. The real estate business is discovering that songbirds are an important aspect of the attractiveness and sale value of home sites and developments.

3. Environmental impact statements, now required by law for approval of most land development projects, often call for evaluations of wildlife assets, including birds.

4. Sportsmen and game departments need to know how many game birds are present in a hunting preserve at the beginning of the hunting season and how many can be safely harvested without endangering the "capital stock."

5. The control of pest species such as House Sparrows, European Starlings, and occasionally others requires information on densities, flocking behavior, and appropriate management procedures.

6. And finally, an understanding of the mechanisms of natural regulation of densities within and between species populations can provide the only reliable basis for confronting a wide range of problems as they arise in this complex and often unpredictable man-dominated world.

Variations and Fluctuations in Populations

The size and structure of a bird community depends ultimately on its locaiton with respect to sources of colonization and the local hazards to survival. More directly, however, they reflect the environmental conditions at the site, the climate, and particularly the nature and richness of local food and habitat resources. These factors change from place to place and, in temperate latitudes, from season to season. Walk 100 yards (91.40 m) from a forest into an open field and the representation of birds shifts drastically; or visit a city park in February and again in May and find a different world, contrasting sharply in both the composition and abundance of its bird populations.

Such regular and predictable changes should be documented as accurately as possible, using the best censusing procedures available. Furthermore, they should be documented each year, for while essentially regular, climatic conditions do change subtly from year to year, population–environment correlations provide the only way reliable answers can be obtained for many fundamental questions of population and community behavior.

The vegetation and habitat structure of a community may also change slowly and progressively over long periods of time (see "Seral Communities," page 187) or abruptly, as in man-induced landscape developments. Such changes reveal the nature of bird responses and provide a basis for predicting the changes that will follow artificial land developments or disturbances.

Population Regulation

Factors Although sensitive to fluctuation in weather and other irregular and erratic variables, bird populations in stable environments rarely deviate appreciably from the local "norm" and tend to return quickly to that norm after a disturbance. Careful analyses of environments and the populations characteristic of them can thus provide information on the factors that control and regulate bird populations in nature. Such factors can be roughly categorized into two types: **density-related factors,** which provide a natural feedback regulation system by intensifying their effects when densities are high and relaxing when densities are low, and **density-independent factors,** which fluctuate independently of population levels, exerting their enhancing or suppressing influences without feedback effects:

Density-related Factors

- Food availability as affected by habitat productivity and depletion by competitors
- Nest site availability as affected by habitat structure and competition
- Territory availability—becoming limited when and where population densities reach a social capacity threshold
- Interference in gaining access to important resource substrates
- Predation
- Diseases and parasites

Density-independent Factors

- Weather
- Natural catastrophes and habitat alterations
- Man-induced disturbances

Mechanisms The population level attained and maintained by a species or a group of species is generally thought to reflect the resource level of the local environment—i.e., its carrying capacity for that particular species or species group. To meet the requirements of this system—i.e., to fully replace themselves in the next generation—breeding adults must lay more eggs than can be expected to survive as breeding adults. The margin of safety between number of eggs and number of breeding-age adults is generally large, especially in species where egg, juvenile, and immature mortality rates are high.

According to Lack (1954), a female will lay as many eggs as she and her mate can raise to independence under the constraints by the specialized and genetically determined life history pattern of the species. This reproductive potential is in rough balance with the mortality hazards of the species' environment. Circumstances that the young birds encounter before reaching reproductive age will determine whether total reproduction will exceed or fall short of total mortality for the year and whether the local population will emerge with a net gain or net loss in numbers. Favorable reproduction–mortality ratios over a series of years will mean progressive growth in the local population, while a series of years with unfavorable ratios will mean a progressive decline and the threat of local extinction.

In a study of 100 nests (144 eggs) of the Arctic Tern only 15.9 percent of the chicks fledged (Pettingill, 1939). In another study of 428 nests (820 eggs) of the Least Tern (*Sterna antillarum*), only 9 percent fledged (Austin, 1938). The failure of so many young birds to survive, bringing the mortality rate to about 85 percent and 90 percent for the two species of terns, is not unusual for nidicolous, precocial birds nesting close together in colonies. Anyone who visits a colony of gulls or terns, as fledging times approaches, is invariably appalled by the large number of unhatched or broken eggs and the carcasses of young at different stages of development. The normally aggressive action of adults toward one another, their nests and young, is one of the chief causes of mortality. For a description of this action as well as others in a colony, read the paper by Pettingill (1939), which traces the fate of 100 nests of the Arctic Tern.

The mortality rate of young among nidifugous species is nearly as high—about 75 percent (derived from Lack's summary, 1954)—and is not surprising when one considers the hazards to which chicks are subjected in their long out-of-the-nest period between hatching and fledging.

The mortality rate of young among altricial species is altogether much lower—roughly 45 percent—because their period between hatching and fledging is spent as nestlings with greater protection. Nice (1957) brought together data on nestling success (= young fledged) from 68 studies of altricial species, 35 of which build open nests and 33 of which nest in cavities. Her instructive summaries showed considerable variation in nesting success in both groups and, at the same time revealed a greater success among cavity-nesters (66.8 percent) than open-nesters (46.9 percent). Her study emphasized the point that the fewer the hazards to which young birds are subjected in their period of dependency, the lower their mortality rate.

Once young birds have attained independence and flight proficiency, their chances of survival rapidly improve, with the mortality rate decreasing sharply until it levels off at a low figure and remains so for a period dependent on the species' normal life expectancy.

For information on the many factors affecting populations, as well as on related problems, refer to the work by Lack (1954) and its sequel (1966), also to the excellent review by Gibb (1961). For views contrary to Lack's—that the reproductive rate is inherent and thus adapted to correspond with the largest number of young that the parents can supply with food—read Wynne-Edwards (1955) in the case of sea birds and Skutch (1949) and Wagner (1957) in the case of tropical birds.

Comparison of Populations

A comparison of populations may have at least three objectives: to show differences in their average densities, to determine differences in their average composition, to find differences in their seasonal fluctuations.

Populations with Like Habitat Frequently, a comparison of populations in different areas having similar plant associations and physiographic features can meet all three of the objectives indicated. One must, however, measure the populations by the same method, at the same time of day and season, and in areas where the vegetation complex and physiographic features are as closely alike as the minimum factors of natural variation will allow.

Populations of Dissimilar Habitat Sometimes it is worthwhile to compare populations with dissimilar habitat (e.g., beech-hemlock forest and oak-maple forest). The results serve mainly to show the differences in population density and composition. The populations must be measured by the same method at the same time of day and season.

Successional Populations Biotic communities succeed one another in a given area until the dominant or climax community is established. (For a discussion of this phenomenon, see "Distribution," page 181.)

A study of changes in bird populations, associated with community succession in one area, is largely impracticable because of the great number of years involved. However, one may analyze bird populations in a region where different stages in a succession can be found in adjoining areas at one time. Such a study is valuable principally in demonstrating the changes in species composition. Comparison of the population densities and composition of different communities is generally difficult because of the merging of communities, but it can be done if one is careful to select typical stages of succession.

Seasonal Populations in the Same Habitat A very common type of comparison is that of populations in the same habitat in different seasons of the year. If an area of known size is selected and its population studied by the same method in different seasons, the changes in population density and composition can then be determined.

Yearly Populations in the Same Habitat The analysis of populations in the same habitat in the same season, or seasons, of successive years is useful for showing any fluctuations in density or changes in composition that may have occurred. Obviously, one must employ the same methods of measurement from year to year.

Units and Methods of Measuring Populations

The ideal unit for measuring population size is **density**—the number of birds present in a given area. But birds can rarely be counted directly as humans are counted by a census-taker making rounds. Birds are aggravatingly elusive for the bird counter; in one situation they may remain silent and undetectable as the census-taker passes while in another they come forward to scold, and thus thoroughly confuse the censusing operation.

Abundance Levels The traditional solution to these problems is to settle for a system of relative levels of abundance. In early studies levels of abundance were based simply on subjective appraisals, which recognized from 5–10 grades and used descriptive terms such as abundant, common, moderately common, uncommon, and rare. But as the value of objective procedures became appreciated, indices of abundance-related counts of units of time, distance, or observer-effort became popular: birds detected per trip afield, per hour of observation, per party hour or party mile (as in the National Audubon Society's organized Christmas Bird Counts with many observers involved in a cooperative effort), or per observation period of fixed duration (as in the U.S. Fish and Wildlife Service's cooperative Breeding Bird Surveys). Such systematic, effort-related counts can provide a numerical basis for the familiar relative abundance categories: Abundant (A) = detected in 90–100 percent of the specified units; Common (C) = 65–89 percent; Moderately Common (MC) = 31–64 percent; Uncommon (UC) = 10–30 percent; Rare (R) = 1–9 percent.

Effort-related abundance indices are most valuable when used directly in comparing populations in different habitat types, different seasons, or different years. A large majority of practical problems requiring bird population data are of this type and can be satisfactorily solved by using indices of relative abundance based on units of observer-effort if care is taken to use the same procedures in each set of operations.

Numerical Levels Although not necessary for most studies comparing the numerical status of species in different sites, or years, there is considerable intellectual satisfaction in recording the number of birds per unit of area. More importantly, units of effort have different significance for species that differ in conspicuousness or detectability. Counts made on a mapped survey area or along a trail must be reduced to a common denominator before they can be summed to provide abundance values for communities. Units of area probably provide the best common denominator for studies of this type.

And finally, birds can be counted directly in a few favorable situations by using maps and plotting the location of individuals that are temporarily confined to restricted sites such as nests in a breeding colony or nesting territories on a circumscribed study area. Colonial nesters can be counted quite accurately if the nests are closely grouped on a flat terrain and precautions are taken to neither omit nor double-count any nests. Territorial nesters can provide a reasonably accurate count if all the nests can be found—a most difficult task—or if all parts of the selected survey area are covered with sufficient thoroughness to assure detection of all resident species.

Selected Studies

The following directions describe several procedures commonly used in estimating relative abundance along transects, in estimating densities along transects, and in direct counting in nesting colonies and on mapped plots.

1. Estimating Relative Abundance

While making trips, either for the purpose of census-taking or for other field work, it is worthwhile to record and compute the relative abundance of birds in various regions traversed. This information, obtained over a prolonged period of time, will yield a picture of bird distribution and fluctuations in numbers. Considering the value of the information, the amount of time and effort expended in obtaining it is relatively slight.

Relative abundance may be determined in numerous ways and under different conditions. The following two procedures are suggested, since they suit an average class trip, but they may be modified to meet particular circumstances.

Procedure (a) While walking through an area, mark on a daily field checklist the species observed. Use a new list for each walk taken. Note the date, the length of time involved on the trip, and the biotic communities passed through. Devise a method for indicating on the check-list the communities in which the birds were observed.

Tabulation of Results. At the conclusion of the season compute the results as follows:

1. To determine the relative abundance of all species observed, divide the number of trips (or days) a species is observed by the total number of trips (or days). Then list the species in the order of percentage figures from the highest to the lowest, thus giving a composite view of relative abundance.

2. To determine the relative abundance of species in similar biotic communities, divide the number of times a species is observed in a community by the total number of times the community was visited. Then list the species for each community in the order of percentage figures on another chart of like design.

If it is desirable to use relative abundance categories such as abundant, common, etc., see the previous section "Units and Methods of Measuring Populations."

Procedure (b) When automobile transportation is necessary to the point of starting field work, one can frequently record birds observed along the highway. Use a checklist for marking off the species observed. Indicate the date and the particular region traversed. Note the mileage on the speedometer at the start and conclusion of the trip and compute the total number of miles covered.

Tabulation of Results At the conclusion of the trip divide the number of times a species was observed by the number of miles covered. The results will have obvious limitations; nevertheless they may be of value in giving some conception of the abundance of the more conspicuous species. If several trips are made through the same region during the course of the season, divide the total number of times a species is observed by the total number of miles covered. This will give a somewhat more accurate indication of the abundance of conspicuous species.

2. Estimating Relative Densities

Make a census of an area representative of a major local habitat type in the area. Make it early in the morning when birds are generally active. Repeat the census on as many mornings as possible over a period of 10–12 days.

Procedure Lay out a route a mile (1.61 km) or more long and measure its total length. Slowly walk the length of the route, recording all birds detected by sight or sound within 100 feet (30.48 m) of each side of the route. For details of the procedure, see Emlen (1971).

Tabulation of Results Total the number of individuals of each species detected out to the 100-foot (30.48 m) boundary line and divide these totals by the area of the strip (length × 200-foot, 60.96 m width) multiplied by the number of traverses for a density value in terms of detection per unit of area for each species and for the total habitat. To translate detections per unit of area to birds per unit of area requires information on the proportion of birds that were making detectable sound cues while the census-taker was within detection range. A thorough familiarity with songs and call notes is obviously needed for the undertaking.

3. Direct Counting

Visit a large colony of nesting birds, preferably ground-nesting species such as pelicans, cormorants, gulls, terns, or skimmers. Select the time of year when the nests are likely to contain either complete clutches of eggs, or newly hatched young.

Procedure Members of the group line up abreast, several feet apart. (The exact spacing depends on the particular circumstances of terrain and visibility of nests. Ordinarily 10 feet (3.05 m) apart is a workable distance.) Then move back and forth through the colony until the entire nesting area has been covered. The person on the right of the line is considered the leader upon whom rests the responsibility of making certain that no area is passed over twice and that no area is slighted. As the lines move—always straight abreast—each person notes the contents and counts the number of nests that occur between the observer and the person to the left. Empty nests, destroyed eggs, and dead young should be separately recorded. If more than one species nests in the colony, keep separate records for each of them.

Tabulation of Results At the conclusion of the census the figures obtained by the group are assembled by the leader. Later they may be tabulated on the suggested chart, "Census of a Nesting Colony," page 347. If previous censuses have been conducted in the same colony, it will be worthwhile to compare the results. Undoubtedly certain fluctuations will be noted. On the basis of the figures, observations, and other information, attempt to determine the causes of the fluctuations.

4. Plot Census of Breeding Birds

Select a square plot of woods (preferably with uniform plant association) of approximately 50 acres (20.24 ha). Make at least five different counts at intervals during the breeding season between the time of territory establishment and the time when the young leave the nest. Conduct the counts early in the day when all males present are likely to be singing.

Procedure Lay out a plot in grid form by first measuring off (either by pacing or by using a measured rope) a straight line 1,456 feet (443,79 m) long. Use a compass to make certain that the line is straight. Then measure off another line of the same length, parallel to the first, 208 feet (63.40 m) away. Continue measuring off additional 1,456-foot (443.79 m) parallel lines until altogether eight such lines, 208 feet (64.40 m) apart, have been marked. Make certain that the lines are permanently indicated by conspicuous markers (see Appendix A, p. 367) conveniently spaced. Next lay out eight 1,456-foot (443.79 m) lines 208 feet (63.40 m) apart at right angles to the first eight lines, thus subdividing the plot into 49 units, each 208 feet square (19.34 sq. m). Where the lines intersect, use numbered markers, conspicuously positioned, preferably on posts.

After the plot has been laid out, make a grid map. Number intersecting lines as on the plot. Indicate topographical features, peculiarities of vegetation, and various landmarks. Show the position of the plot in relation to the points of the compass. After the map is complete, make copies available to each observer.

When ready to make the first count, walk across the plot following different parallel lines. If there are enough observers for all parallel lines in one direction, only one crossing a day is necessary. However, they must move at the same time and in the same direction. Each observer is equipped with a copy of the map. Indicate on the map the places where species are found and their breeding status or activity. Use abbreviations and letters. Abbreviate bird names thus: Black-and-white Warbler = B-w War; White-throated Sparrow = W-t Sp; Red-eyed Vireo = R-e V. Indicate observations regarding status or activity by letters in parentheses after the abbreviations of names, thus: Singing male, pair, or bird not singing but obviously paired = (P); transient individual moving over or through plot but not residing = (T); occupied nest = (N); young bird = (Y).

When the second and succeeding counts of the plot are made, use a new map and follow the same procedure.

CENSUS OF A NESTING COLONY

Species in colony:						
Location of colony:						
Estimated size (in acres) of area containing nests:	Date of census					
Length of time involved in census:	Number of persons making census:					
	Species	Species	Species	Species	Species	
Total number of nests:						
Total number with eggs only:						
with one egg:						
with two eggs						
with three eggs:						
with four eggs						
with five eggs:						
Total number with one egg and one young:						
with three combination:						
with four combination:						
with five combination:						
Total number with young only:						
with one young:						
with two young:						
with three young:						
with four young:						
with five young						
Total number of empty nests:						
Total number of eggs and young:						
Average number of eggs and young per nest (empty nests not included):						
Estimated total of breeding adults:						
Estimated number of nests per acre:						
Approximate ratio of adults to eggs and young:						
Total number of eggs found destroyed:						
Total number of young found dead:						
Comments or notes:						

PLOT CENSUS OF BREEDING BIRD POPULATION

General location of plot:					
Size of plot:					
Classification of habitat:					
General topography:					
	First Trip	Second Trip	Third Trip	Fourth Trip	Fifth Trip
Date of census trip:					
Time of day:					
Temperature:					
Precipitation:					
Wind:					
Weather rating:					
Species	First Trip	Second Trip	Third Trip	Fourth Trip	Fifth Trip
Total number of pairs:					
Number of pairs per 100 acres:					
Estimated total number of pairs based on above censuses:					
Final total of pairs per 100 acres:					

Tabulation of Results When the five counts along the parallel lines have been completed, assemble the various maps and record the results for each count with crayons of different colors on a composite map. Tabulate the results using the chart suggested on p. 347, "Plot Census of Breeding Bird Population." On this chart, list the species in the sequence of *The A.O.U. Check-list*. Record the number of pairs, transients, nests, and young in the adjoining columns. The following hypothetical examples demonstrate the method: One Great Blue Heron seen flying over the plot = 1 (T); three Red-eyed Vireos heard singing and one nest found = 3 (P) 1 (N); one adult White-throated Sparrow and one young found = 1 (P) 1 (Y).

In estimating totals, include only pairs. Count singing males as one pair, a nest as another pair, etc. However, avoid counting a singing male and a nest as two pairs when there is actual evidence that they represent only one pair of breeding adults. The matter of knowing when to count, for example, a singing male and a nest as one pair or two is left to the judgment of the census-taker.

References

Each year *American Birds* (a bimonthly journal) publishes a "Breeding Bird Census" and the "Audubon Christmas Bird Count" from reports of many hundreds of amateur and professional ornithologists in the United States and Canada. Both are rich sources of information. The journal also publishes "The Changing Seasons" from hundreds of reports on numbers of birds in spring and fall migrations and on winter bird populations.

The Colonial Bird Register, headquartered at the Cornell Laboratory of Ornithology, provides a central data base on colonial nesting birds throughout North and Central America and the Caribbean. Working with the National Audubon Society and the U.S. Fish and Wildlife Service, the Register tracks population trends among colonial nesting species by collecting data on colony locations, species composition, habitat characteristics, and numbers of breeding birds. For information on how to use the Colonial Bird Register, write to the Laboratory (Sapsucker Woods, Ithaca, New York 14850).

Students desiring procedures and methods in measuring populations more suited to their purposes than those described in this book should consult Ralph and Scott (1981).

Following are some of the more important publications dealing with longevity, numbers, and census-taking.

References

Amman, G.D., and Baldwin, P.H.
1960 A comparison of methods for censusing woodpeckers in spruce-fir forests of Colorado. *Ecology* 41:699–706.

Austin, O.L.
1938 Some results from adult tern trapping in the Cape Cord colonies. *Bird-Banding* 9:12–25.

Best, L.B.
1975 Interpretational errors in the "mapping" method as a census technique. *Auk* 92:452–460.

Breckenridge, W.J.
1935 A bird census method. *Wilson Bull.* 47:195–197. (A census of a square mile, 2.59 sq. km, area. The author worked alone, traversing the area along compass lines.)

Brewer, R.
1978 A comparison of three methods of estimating winter bird populations. *Bird-Banding* 49:252–261.

Clapp, R.B.
1976 [A review of the longevity records of world birds listed by the Polish ornithologist, W. Rydzewski, in the journal, *Ring*.] *Bird-Banding* 47:279–281.

Clapp, R.B.; Klimkiewicz, M.K.; and Kenard, J.H.
1982 Longevity records of North American birds: Gaviidae through Alcidae. *Jour. Field Ornithology* 53:81–124. (First paper in a four-part series.)

Emlen, J.T.
1971 Population densities of birds derived from transect counts. *Auk* 88:323–342.
1977 Estimated breeding season bird densities from transect counts. *Auk* 94:455–468.

Farner, D.S.
1955 Birdbanding in the study of population dynamics. In *Recent studies in avian biology*, ed. A. Wolfson. Urbana: Univ. of Illinois Press.

Fisher, H.I.
1975 Longevity of the Laysan Albatross *Diomedea immutabilis*. *Bird-Banding* 46:1–6. (Observed longevity ranged from 20 years to 30 or more.)

Fisher, J.

1941 *Watching birds*. Hammondsworth, Middlesex, England: Penguin Books.

1954 *A history of birds*. Boston: Houghton Mifflin. (Chapters 8, "Absolute Numbers of Birds"; 9, "Comparative Numbers of Birds"; 10, "Changing Bird Populations"; 11, "Bird Numbers and Man.")

Fisher, J., and Peterson, R.T.

[1964] *The world of birds*. Garden City, New York: Doubleday.

Flower, S.S.

1938 Further notes on the duration of life in animals. *Proc. Zool. Soc. London* 108 (Ser. A):195–235.

Franzreb, K.E.

1976 Comparison of variable strip transect and spot-map methods for censusing avian populations in a mixed-coniferous forest. *Condor* 78:260–262.

Gibb, J.A.

1961 Bird populations. In *Biology and comparative physiology of birds*, vol. 2, ed. A.J. Marshall. New York: Academic Press.

Graber, R.R., and Graber, J.W.

1963 A comparative study of bird populations in Illinois, 1906–1909 and 1956–1958. *Illinois Nat. Hist. Surv. Bull*. 28:377–528. (A classic example of how populations in the same areas may be compared after a lapse of many years—in this case 50.)

Hensley, M.M., and Cope, J.B.

1951 Further data on removal and repopulation of the breeding birds in a spruce-fir forest community. *Auk* 68:483–493.

Hickey, J.J.

1955 Some American population research on gallinaceous birds. In *Recent studies in avian biology*, ed. A. Wolfson. Urbana: Univ. of Illinois Press.

Howell, J.C.

1951 The roadside census as a method of measuring bird populations. *Auk* 68:334–357.

Kadlec, J.A., and Drury, W.H.

1968a Aerial estimation of the size of gull breeding colonies. *Jour. Wildlife Mgmt*. 32:287–293. (By sight and photography, but neither method "will reliably detect annual changes of less than about 25 percent.")

1968b Structure of the New England Herring Gull population. *Ecology* 49:644–676. (An analysis of age structure, population increase, and mortality rate.)

Kendeigh, S.C.

1944 Measurement of bird populations. *Ecol. Monogr*. 14:67–106.

1947 Bird Population studies in the coniferous forest biome during a spruce budworm outbreak. *Dept. Lands and Forests, Ontario, Canada. Div. Res. Biol. Bull*. No. 1.

1948 Bird populations and biotic communities in northern Lower Michigan. *Ecology* 29:101–114.

Kennard, J.H.

1975 Longevity records of North American birds. *Bird-Banding* 46:55–73.

Lack, D.

1954 *The natural regulation of animal numbers*. London: Oxford Univ. Press.

1966 *Population studies of birds*. London: Oxford Univ. Press.

Linsdale, J.M.

1936 Frequency of occurrence of summer birds in northern Michigan. *Wilson Bull*. 48:158–163.

Lockley, R.M.

1953 *Puffins*. New York: Devin-Adair.

Mayfield, H.

1953 A census of the Kirtland's Warbler. *Auk* 70:17–20.

Merikallio, E.

1958 Finnish birds: Their distribution and numbers. *Fuana Fennica*, 5.

Middleton, R.J.

1974 Fifty-two years of banding Blue Jays at Norristown, Pennsylvania. *Bird-Banding* 45:206–209.

Murphy, R.C.

1925 *Bird islands of Peru: The record of a sojourn on the west coast*. New York: G.P. Putnam's Sons.

Nice, M.M.

1927 Seasonal fluctuations in bird life in central Oklahoma. *Condor* 29:144–149.

1957 Nesting success in altricial birds. *Auk* 74:305–321.

1966 Territory defense in the Oystercatcher. *Bird-Banding* 37:132. (Review of an article by H.R. Rittinghaus, 1964.)

Odum, E.P.

1971 *Fundamentals of ecology*. 3rd ed. Philadelphia: W.B. Saunders. (Chapters 5, "Introduction to Population and Community Ecology"; 6, "Principles and Concepts Pertaining to Organization at the Species Population Level.")

Peterson, R.T.

1941 How many birds are there? *Audubon Mag*., 43:179–187.

1964 *Birds over America*. New and revised ed. New York: Dodd, Mead.

Pettingill, O.S., Jr.
1939 History of one hundred nests of Arctic Tern. *Auk* 56:420–428.

Ralph, C.J., and Scott, J.M., eds.
1981 *Estimating numbers of terrestrial birds.* Proceedings of an International Symposium held at Asilomar, California, October 26–31, 1980. Cooper Ornithological Society, Studies in Avian Biology, No. 6.

Reynolds, R.T.; Scott, J.M.; and Nussbaum, R.A.
1980 A variable circular-plot method for estimating bird numbers. *Condor* 82:309–313.

Ryel, L.A.
1981 The fourth decennial census of Kirtland's Warbler. *Jack-Pine Warbler* 59:93–95.

Skutch, A.F.
1949 Do tropical birds rear as many young as they can nourish? *Ibis* 91:430–455.

Southern, W.E.
1975 Longevity records for Ring-billed Gulls. *Auk* 92:369.

Stamm, D.D.; Davis, D.E.; and Robbins, C.S.
1960 A method of studying wild bird populations by mist-netting and bird-banding. *Bird-Banding* 31:115–130.

Stewart, P.A.
1973 Estimating numbers in a roosting congregation of blackbirds and starlings. *Auk* 90:353–358.

Stewart, R.E., and Aldrich, J.W.
1949 Breeding bird populations in the spruce region of the central Appalachians. *Ecology* 30:75–82.
1951 Removal and repopulation of breeding birds in a spruce-fir forest community. *Auk* 68: 471–482.

Stewart, R.E., and Kantrud, H.A.
1972 Population estimates of breeding birds in North Dakota. *Auk* 89:766–788.

Terres, J.K.
1968 *Flashing wings: The drama of bird flight.* Garden City, New York: Doubleday.

Tickell, W.L.N.
1968 The biology of the Great Albatrosses, *Diomedea exulans* and *Diomedea epomophora. American Geophysical Union, Antarctic Res. Ser.* 12:1–55.

Twomey, A.C.
1945 The bird population of an elm-maple forest with special reference to aspection, territorialism, and coactions. *Ecol. Monogr.* 15:173–205.

Van Ryzin, M.T., and Fisher, H.I.
1976 The age of Laysan Albatrosses, *Diomedea immutabilis,* at first breeding. *Condor* 78:1–9. (Mean age between eight and nine years.)

Wagner, H.O.
1957 Variation in clutch size at different latitudes. *Auk* 74:243–250.

Westerskov, K.
1963 Ecological factors affecting distribution of a nesting Royal Albatross population. *Proc. XIIIth Internatl. Ornith. Congr.* pp. 795–811.

White, K.A.
1942 Frequency of occurrence of summer birds at the University of Michigan Biological Station. *Wilson Bull.* 54:204–210.

Wynne-Edwards, V.C.
1955 Low reproductive rates in birds, especially seabirds. *Proc. XIth Internatl. Ornith. Congr.* pp. 540–547.

Ancestry, Evolution, and Decrease of Birds

Despite the astronomical numbers of birds that have lived and died in the passing of millions of years, the fossil record is meager when compared with other vertebrate groups. Bird bones are usually small and fragile and are therefore less likely to be preserved than the massive bones of mammals and some reptiles. Most bird bones that have been preserved as fossils are parts of legs, wings, or vertebrae, and occasionally a skull or sternum is recovered. Whole skeletons are unusual. However, in the past decade many new bird fossils, both single elements and whole skeletons, have come to light, increasing considerably the understanding of avian evolution.

Ancestry

Undeniably, birds arose from reptilian stock. In the first place, modern birds and reptiles show numerous physical features in common. For example, their skulls bear one rather than two occipital condyles for articulation with the vertebral column; their lower mandibles articulate with movable quadrate bones as in lizards and snakes instead of being hinged directly on the skull; and their ears have a single bone, the columella, for conducting vibrations from the tympanum across the middle ear to the internal ear.

In the second place, there is the highly important evidence of reptilian ancestry from *Archaeopteryx lithographica*, an ancient winged creature whose fossil remains were found at six different times in lithographic limestone near Solnhofen, Bavaria. The first discovery, in 1860, comprised the imprint of an undoubted feather. The second, the next year, consisted of an incomplete skeleton (now in the British Museum) showing particularly forearm and leg bones; a long tail with many vertebrae and the impressions of feathers, some attached to the forearms; and a pair each to most of the tail vertebrae. The third discovery, in 1877, was a skeleton (now in a Berlin museum), virtually complete except for the lower mandible, the right foot, and some of the cervical vertebrae. The fourth discovery, in 1958, was another skeleton (now at the University of Erlangen), much less complete than the two found 80 years or so earlier. The other two specimens were discovered in 1970 and 1973 in museums; both were recovered earlier, but they were misidentified as a pterosaur and a small dinosaur, *Compsognathus*.

A little larger than the Common Pigeon, *Archaeopteryx* lived in the Late Jurassic Period (see Table 8), some 140 million years ago, at a time when reptiles were dominant vertebrate animals. Without the tell-tale feathers in the fossil remains, the discoverers might have identified it as just another reptile from that remote age. But now that numerous authorities (e.g., Heilmann, 1927; de Beer, 1954) have scrutinized and evaluated the 1861 and 1877 specimens and studied the 1861 specimen by direct and indirect lighting and by ultra-violet and X-rays, they consider *Archaeopteryx* more bird than reptile with nonetheless remarkable and—from an evolutionary viewpoint—significant features intermediate between the two.

The skull has such reptilian features as an overall heavier structure, large fossae in the facial region,

Table 8 **Geologic Time Scale**

Era	Period	Epoch	Million years before present
CENOZOIC (AGE OF BIRDS AND MAMMALS)	QUATERNARY	Recent	0.01
		Pleistocene	1.5–3.5
	TERTIARY	Pliocene	7
		Miocene	26
		Oligocene	37–38
		Eocene	53–54
		Paleocene	65
MESOZOIC (AGE OF REPTILES)	CRETACEOUS	Late	100
		Early	135
	JURASSIC	Late	155
		Middle	170
		Early	180–190
	TRIASSIC		230

From Feduccia, 1980, in *The Age of Birds,* Harvard University Press, Cambridge, Massachusetts.

no bill, and pointed teeth in sockets. Also, the brain-case shows that the cerebral hemispheres were elongated, the optic lobes dorsal in position, and the cerebellum small and not extended forward to overlap the posterior parts of the cerebral hemispheres. The orbit, however, is relatively large and contains a sclerotic ring.

The vertebral column, comprised of 10 cervical, 12 thoracic and lumbar, 6 sacral, and 20 caudal vertebrae, is distinctly reptilian, with little if any fusions of vertebrae. The vertebral centra are biconcave or amphicoelous—simpler than the saddle-shaped or heterocoelous type in modern birds. The caudal vertebrae are elongated and free, with no evidence of their fusing terminally to form a pygostyle. All the ribs are unjointed, lack uncinate processes, and do not reach the sternum. The sternum itself is short, broad, and unkeeled, while posterior and lateral to it are about 12 pairs of dermal ribs or gastralia which are present in the ventral abdominal wall of some reptiles.

The pectoral girdle shows, besides a scapula and coracoid, a clavicle fusing with its fellow of the opposite side to form a distinctly avian feature, the furcula or wishbone.

The forelimb is as long as the hindlimb. Its humerus is longer than the ulna, and the ulna is a little longer and stouter than the radius, but none of the three bones is very strongly developed. In the wrist are five carpals, two (the radiale and ulnare) proximal and three distal. The hand has three metacarpals. Whereas modern birds have only the radiale and ulnare separate and the other carpals and the metacarpals fused as one bone, the carpometacarpus, *Archaeopteryx* shows all the carpals and metacarpals separate with the possible exception of the outermost distal carpal being fused with the outermost metacarpal. Articulating with the metacarpals are three separate digits or fingers (the middle one longest) terminating in conspicuously long, curved, sharply pointed claws. Each digit is comprised of more phalanges than its homologue in modern birds.

The pelvic girdle is connected to, rather than fused with, the sacral vertebrae, and its three bones—the ilium, ischium, and pubis—are separate and individually identifiable instead of being fused into one bone. The pubis is distinctly avian by being long, slender, and directed posteriorly. At the distal end it fuses with its fellow of the opposite side to form a symphysis.

In the hindlimb the fibula is a separate bone alongside the tibia and extends all the way from the femur to the heel. The fibula and tibia are longer than the femur. The second, third, and fourth metatarsals remain separate and parallel for most of their length distally; only at their proximal ends are they fused with one another and certain tarsals to form the tarsometatarsus. As in the Common Pigeon and most other modern birds capable of perching, there are four digits or toes, the first (hallux) projecting backward. The fourth or outermost digit has four phalanges instead of five.

However strikingly birdlike some of the skeletal features of *Archaeopteryx* may be, none is uniquely avian as are the feathers. Careful studies of their impressions in lithographic limestone show them to be in every way typical of modern flying birds. The forearm appears to have borne true primaries (as many as eight, according to Savile, 1957) and ten secondaries, each with overlapping coverts. Although the tail is more reptilian than avian in length and number of vertebrae, it supports rectrices—one pair with coverts attached to each 15 caudal vertebrae beginning with the sixth. Besides the feathers of the forearms and tail are the impressions of numerous other feathers which, from their positions and size, clearly suggest that the body, neck, and legs had a feather covering.

Nobody knows precisely what *Archaeopteryx* looked like in life, how it was shaped, and whether or not it was brightly or somberly colored. Numerous artists, however, have reconstructed it partly from anatomical evidence and the rest from supposition. See, for example, some very meticulous attempts in the paintings by Heilmann in the frontispiece of his book (1927), by R.T. Peterson (*in* Fisher and Peterson, 1964), and by Rudolf Freund (see Fig. 56).

While *Archaeopteryx* shows beyond a doubt that birds arose from reptilian stock (Thulborn and Hamley, 1982), it is only to a limited extent helpful in providing clues to the ancestry of birds.

Existing contemporaneously with *Archaeopteryx* were two groups of reptiles—the pterosaurs (winged reptiles) and smaller bipedal (two-footed) dinosaurs—which shared many of the skeletal features with birds. The pterosaurs, while equipped with pneumatic bones as are modern birds, depended on wings with membranes and with bones of different number and proportions. The bipedal dinosaurs had long hindlimbs on which they could run while keeping their forelimbs off the ground. Their pectoral girdles, however, lacked certain features, including a furcula, that modern birds possess. Obviously, these two groups of reptiles were no more ancestral to modern birds than *Archaeopteryx*.

To find the ancestor common to birds (including *Archaeopteryx*), pterosaurs, and dinosaurs, one must search farther back in time for a more generalized form of reptile. Speculation, based on current researches, points to the Pseudosuchia, a small group among the extinct thecodont (socket-toothed) reptiles, which existed in the Old World during the early to middle Triassic Period some 200 million years ago.

The pseudosuchians were bipedal reptiles with hindlimbs that were longer than the forelimbs and used for walking, running, and jumping. Their skulls had many of the basic features found in birds. In one form, *Euparkeria capensis* from Africa, the skull had all the bony elements present in *Archaeopteryx*, although it was more heavily constructed with a higher profile and stronger jaws. The orbits and brain-case were smaller. In another form, *Ornithosuchus woodwardi* from Scotland, the skull was more lightly constructed and thus more

Figure 56 **A Reconstruction of *Archaeopteryx***
Painting by Rudolf Freund, courtesy of Carnegie Museum.

closely resembled that of *Archaeopteryx*. The feet showed a tendency toward reduction in the fourth and fifth toes. As in other pseudosuchians, *Ornithosuchus* supported a paired row of scale-like epidermal plates on the dorsal surface of the back and tail.

Besides *Euparkeria* and *Ornithosuchus* there were many other similarly generalized forms of pseudosuchians existing at the time. It could have been from any one of these forms—or even from an older and still more generalized form which lived in the preceding Palaeozoic Era—that birds branched off, but more fossil evidence must come to light before there can be any certainty. The stock from which modern birds evolved has yet to be discovered.

Evolution

Early Evolution

If the numbers of early bipedal pseudosuchians were as vast as commonly believed, one can assume that they were subject to considerable competition for food as well as living space. Being carnivores by virtue of their socketed teeth, they were forced to seek small animals in every possible situation. At least one form of pseudosuchian and possibly more used their sharp claws to advantage in climbing trees where, through the continuing process of adaptive radiation, they developed physical attributes and techniques that enabled them to forage successfully. From an arboreal habit thus acquired, they proceeded (as suggested by Heilmann, 1927) to attain ever greater proficiency by jumping from branch to branch. This required the simultaneous development of more muscular power in their hindlimbs for leaping and of opposable first toes (halluces) for grasping.

The next probable step in development was the extension of jumping movements to gliding from higher to more distant lower branches and eventually from high in one tree down to the lower part of another. The evolutionary consequences of these feats were the gradual acquisition and elongation of scales from the trailing edge of the forelimbs and (for balance as well as sailing surface) the expansion and lateral projection of the scales along the tail. In due course the scales became lengthened and elaborated to form contour feathers.

Archaeopteryx well represents that stage in the evolution toward flight where the subject became adapted to an arboreal life and capable of extensive gliding. Its feet were suited to grasping and perching; its forelimbs, while retaining some grasping ability as evidenced by the three-clawed fingers of the hands, were elongated and more suited to supporting the body in the air. The great extent to which *Archaeopteryx* depended on gliding is reflected by a number of anatomical features—for example, the lightening of the skull; the increase in size of the orbits to accommodate larger eyes for greater distance perception and visual acuity; and the larger brain-case to accommodate the enlargement of those brain centers that control and coordinate the more complicated neuromuscular mechanisms for gliding.

With true contour feathers already formed in *Archaeopteryx,* and with no evidence in the few fossil remains of early reptiles that even remotely suggest how they formed, the origin of feathers is a fruitful field for speculation (see a brilliant paper on this subject by Parkes, 1966). The general consensus is that feathers developed first on the forelimbs and tail of an early arboreal reptile by the lengthening, broadening, and overlapping of those scales that would increase sailing surface and at the same time be lighter without sacrificing strength and durability. Since the trend toward lightness is of selective advantage to any airborne animal, it was quite natural for the modifying process to involve the remaining scales on the forelimbs and tail and eventually all the scaly covering of the body, neck, and head. A secondary advantage to feathers was no doubt to streamline the body by smoothing over angular surfaces and folds from head to tail. All of the feathers were contour feathers and developed initially as an adaptation to gliding. All other kinds of feathers found in modern birds—down feathers (both teleoptilian and neossoptilian) and filoplumes—presumably evolved from the contour feathers of the type in *Archaeopteryx*.

After *Archaeopteryx* comes a wide gap in the fossil record until the Cretaceous Period some 20 million years later. In this long interval the evolution of birds must have accelerated, for by the Cretaceous, if one may judge by the several species discovered, birds were not only well established and globally distributed but had taken divergent paths of development, many becoming aquatic and many others remaining terrestrial. Among the species were the first known representatives of Charadriiformes and representatives of other orders, now completely extinct, that are known by fossil material from sedimentary rocks laid down by the great inland sea that occupied the west-central portion of North America for some 40 million years. Nearly complete skeletons have been found for some of these birds, notably *Ichthyornis victor* and *Hesperornis regalis*.

Ichthyornis, a superficially gull-like bird about the size of a Common Pigeon, was obviously capable of sustained flight, for it had well-developed wings with a carpometacarpus fully formed, strong pectoral girdles amply supported by a furcula, and a keeled sternum for the attachment of large muscles to control the wing strokes. *Hesperornis,* a huge bird five to six feet long (1.52–1.83 m) was highly specialized for diving—perhaps the equal of loons in diving ability—and totally flightless. Far back on its elongated body the hindlimbs, with their paddlelike feet, projected at right angles to the body axis. Probably it could maneuver on land only to the extent of reaching its nest. The wings were vestigial, with no bones distal to a slender humerus; the pectoral girdles were not only weak but the clavicles failed to meet at the midline; and the sternum was flat and unkeeled. Whatever powers of flight the bird once had were long since lost.

The skeletal structure of *Ichthyornis* and *Hesperornis,* though widely divergent for purposes correlated with locomotion, shared certain similarities such as a typical avian tarsometatarsus, a small skull with bones well fused, and the reduction in the number, and the fusing of, the terminal caudal vertebrae to form a pygostyle. In *Hesperornis* and *Ichthyornis* both mandibles retained reptilian teeth, fewer in the upper mandible than the lower. *Hesperornis* had heterocoelous vertebrae characteristic of modern birds, but *Ichthyornis* had amphicoelous (bi-concave) vertebrae that were peculiarly enjoined as they are in fish. This accounts for the name *Ichthyornis*—fish bird.

Ichthyornis was undoubtedly able to fly in the manner of modern birds and, according to the fossil record of other Cretaceous birds, was not alone in having achieved this ability. Indeed, except for the skull, the skeleton of *Ichthyornis* was that of a modern bird.

Presumably, during the merging of the Late Jurassic and Early Cretaceous Periods, the feat of gliding, as exemplified by *Archaeopteryx*, was rather quickly extended for increasingly longer distances by flapping the forelimbs, thus sustaining the weight of the body against gravity. This gave the performer such selective advantages as capturing insects on the wing and more readily escaping from enemies and unfavorable aspects of the environment. In the development and achievement of flight the long tail that was advantageous only in sailing was foreshortened and its feathers directed posteriorly and spread fanlike to provide a mechanism for steering and braking; the eyes were enlarged further to give still greater visual acuity; and the teeth played a steadily decreasing role and ultimately disappeared, allowing the mandibles to be lighter in weight.

The attainment of flight was unquestionably accompanied by a heightening of body metabolism through modifications in the organ systems. Cretaceous birds such as *Ichthyornis* and *Hesperornis* must have been homeothermous (warm-blooded) but how early in their development birds acquired a thermoregulatory mechanism is problematical. Some authorities contend that *Archaeopteryx* was at least somewhat homeothermous, as were its pseudosuchian ancestors and the pterosaurs, and that the feather covering of *Archaeopteryx* developed to hold its body heat as it took to an arboreal life in which there was more exposure to lower air temperature. The pterosaurs were able to fly successfully without any such special covering. It is more likely that the feather covering developed to facilitate gliding and later flight and that, as warmbloodedness increased with the attainment of flight, the feather covering eventually assumed an insulating function.

Later Evolution

Athough the fossil record of the Cretaceous is poor, advances in the avifauna must have been great, for with the coming of the Tertiary Period toothed birds had disappeared and a host of new forms began to emerge. Among them are the earliest known representatives of a great number of modern families or orders that include the penguins (Spheniscidae), rheas (Rheidae), loons (Gaviidae), tropicbirds (Phaethontidae), anhingas (Anhingidae), cormorants (Phalacrocoracidae), herons (Ardeidae), ducks (Anatidae), vultures (Cathartidae), hawks (Accipitridae), grouse (Tetraonidae), cranes (Gruidae), rails (Rallidae), sandpipers (Scolopacidae), auks (Alcidae), cuckoos (Cuculidae), owls (Strigidae), swifts (Apodidae), trogons (Trogonidae), and the first few forms of Passeriformes.

By early Tertiary time the proliferation of birds was occurring at a very rapid rate. Gone were the dinosaurs, pterosaurs, and other early reptiles that had dominated the Cretaceous, leaving habitats more than ever available to birds. Besides the proliferation of many families of modern birds and many forms obviously ancestral to modern species, there also appeared several gigantic flightless birds that seem to have replaced the giant reptiles of the Cretaceous. They thrived until the emergence of the giant mammals, and then died out, leaving no known descendants. A notable example was *Diatryma* from the Paleocene and Eocene of North America and Europe. Specialized for terrestrial life, with massive, powerful legs, a huge head with an enormous bill, and small degenerate wings, it stood nearly seven feet tall (2.13 m).

The fossil deposits of the Oligocene in the mid-Tertiary yield, among other forms, the first known grebes (Podicipediformes), albatrosses (Diomedeidae), shearwaters (Procellariidae), storks (Ciconiidae), turkeys (Meleagridinae), limpkins (Aramidae), plovers (Charadriidae), stilts (Recurvirostridae), gulls (Laridae), pigeons (Columbidae), parrots (Psittacidae), and kingfishers (Alcedinidae), together with a few more passerine birds.

In the Miocene and Pliocene deposits, toward the close of the Tertiary, the fossil remains of birds are much more numerous and varied. While the majority represent forms already known, others are the first recorded evidences of ostriches (Struthionidae), tinamous (Tinamidae), falcons (Falconidae), oystercatchers (Haematopodidae), goatsuckers (Caprimulgidae), and several families of passerine birds.

From the Pleistocene Epoch or Ice Age come the first records of the cassowaries (Casuariidae), emus (Dromiceidae), elephant birds (Aepyornithidae), moas (Dinornithidae), kiwis (Apterygidae), ospreys (Pandioninae), jacanas (Jacanidae), phalaropes (Phalaropodinae), skuas (Stercorariidae), barn-owls (Tytonidae), hummingbirds (Trochilidae), motmots (Momotidae), and many passerine species. Undoubtedly, these birds evolved much earlier, in the Tertiary, since most all species living today and those recently extinct are believed to have been in existence at the beginning of the Pleistocene and well established in the Pleistocene. This may not have been the case with some of the "higher" passerines, which are thought to have acquired their species distinctions as a result of the Pleistocene. For example, see Mengel (1964) or a partial summation of his paper in this book, page 115.

In its abundance and variety of forms the fossil record of the Pleistocene is the richest of all the epochs. Besides 732 living species, represented by fossils, there are 270 extinct species. One, a condorlike bird, *Teratornis incredibilis*, with an estimated wingspread of 16–17 feet (4.88–5.18 m), may have been the largest bird ever to fly. Several Pleistocene deposits have yielded many hundreds of bird bones, but the most productive of any in the world are at Rancho La Brea near the center of Los Angeles, California, where during the Ice Age countless numbers of birds became entrapped in asphalt and their bones preserved in beds of tar.

Loss of Flight

In the long history of birds various forms have been characteristically flightless—i.e., unable to fly because either they lacked wings altogether or they did not have wings large enough and powerful enough for sustained locomotion in the air. Most students of paleontology agree that this condition developed secondarily in all instances from stock at one time capable of flight. All birds, whether they can fly or not, are recognizable structurally as such because they evolved essentially as flying creatures.

Hesperornis, Diatryma, and a dozen or more other genera of birds, long since extinct, were flightless. Among modern birds and some only recently extinct are a considerable number of flightless birds, two well-known groups being the ratites and the penguins.

The ratites, so called from their unkeeled and consequently raftlike sternum, comprise the ostriches (Africa), rheas (South America), emus (Australia), cassowaries (New Guinea and Australia), kiwis (New Zealand), and the recently extinct moas (New Zealand) and elephant birds (Madagascar). All are terrestrial, with strong legs and feet, and all, except the kiwis, are large and heavy bodied. Many ornithologists feel confident that the birds arose independently in their respectively isolated areas and became similar through convergence; but a few (eg., Bock, 1963) feel differently, believing that the birds arose together and then dispersed to remote areas, perhaps before losing their ability to fly. (See Feduccia, 1980, for a complete discussion of the evolution of the flightless state in birds.) Regardless of how they arose, one may postulate that their ancestors took to living in treeless country where they became grazers. The cursorial habit of walking about for food and running to escape enemies became increasingly adequate and flight decreasingly essential; hence, their legs and feet grew stronger while their wings and keeled sternum degenerated. In some instances birds, isolated on islands and free from predation by carnivores, grew to gigantic size. In New Zealand one of the 12 species of moas, *Dinornis maximus*, stood 12 feet tall (3.66 m) and may have weighed about 500 pounds (226.80 kg). Although the chicken-sized kiwis were similarly isolated in New Zealand, they never tended toward giantism; instead they reverted to a forest habitat where great size would have been only a handicap.

Unlike the ratites, penguins retained the keeled sternum and flight muscles because, in a sense, they simply readapted their locomotion from flying in the air to flying under water. Presumably, penguins evolved from flying aquatic birds—possibly stock ancestral also to the Procellariiformes (Simpson, 1946)—that reached the cool water ringing the Southern Hemisphere. Besides short, bladelike wings, or "flippers," suitable for propulsion, penguins acquired correlated structural features such as a streamlined or torpedoshaped body with legs so far back that they must stand upright when walking; feathering almost scalelike in aspect, without apteria; and a thick skin over a heavy layer of fat that provides insulation against the cool environment. A few species of penguins eventually adjusted to the more frigid water adjacent to Antarctica and soon nested exclusively on its periphery and outlying islands.

The auks, which tend to fill the niche in the Northern Hemisphere occupied by penguins in the Southern, struck a compromise in their adaptation (Storer, 1960a). From presumed gull-like ancestors, all species, with one exception, derived wings small enough for underwater swimming, yet large enough for aerial flight. The exception was the Great Auk *(Pinguinus impennis)*; like the penguins, it forsook the air altogether for submarine flight.

The Decrease of Birds

Over the ages since life began, species of animals and plants have arisen, flourished, and died out. This applies as much to birds as to other forms. From the fossil evidence and theoretical knowledge of environmental conditions that existed during the long history of birds, one can easily surmise that there were far greater numbers of species and individuals in past ages than at the present time; but, in any attempt to determine numbers one is frustrated by the meagerness of the record. Speculation is the only recourse.

The rapid multiplication of bird species in the Eocene was favored not only by the decline of reptiles and the consequent availability of their habitats but also by the persisting warm climate and the increasing development of seed plants (angiosperms) that formed immense forests, creating many new niches for occupancy. By the time of the Oligocene, mountains had begun to rise, and as the land dried in their lee, forests soon decreased and grasslands formed. Here again were new habitats—in the mountains and on the plains—for avian radiation. Through the Miocene and into the Pliocene there was a slight cooling of the climate although it stayed warm or temperate much farther north than at the present time and there were no seasonal changes. But in the Pleistocene and its succession of four glacial stages, with the long and warm interglacial periods, the relative uniformity of the climate ended and its effects on bird life were catastrophic. With each invading ice sheet, the prevailing temperature lowered. Birds were forced to shift and sometimes compress their ranges southward. Failing to cope with such radical changes, some species died out.

How many species of birds have been identified from fossil remains? Roughly 1,700, according to Austin (1961). Of this number, about 800 are still in existence and 900 are extinct. Adding the 900 extinct species to the 9,000 or so species known to be living in the world today, the total is about 9,900.

Taking a backward look at the long history of birds, one can readily see an overall increase in the number of species from the Late Jurassic to the Late Tertiary. In this great period of time the rate of increase exceeded the rate of extinction. If there was an Age of Birds, it must have been from the Miocene through the Pleistocene, a period of some 20,000,000 years. Since the Pleistocene, the rate of extinction is presumed to have exceeded the rate of increase.

The extent to which prehistoric man contributed to the decrease was probably negligible. As Greenway (1967) fancifully stated, "man and birds arranged a means of living together to the ends that no birds were extirpated." But there is no denying that in historic times man has played an awesome role in the extinction of birds.

The first species actually known to be eliminated by man is the Dodo *(Raphus cucullatus)* on the island of Mauritius in the Indian Ocean. In the 174 years following the discovery of the island by the Portuguese in 1507, men from Europeon ships, and the cats, pigs, monkeys, and rats that they brought with them, succeeded in destroying the entire population. It was nothing less than miraculous that this flightless, ground-nesting species survived as long as it did.

The Great Auk was the first species on the coast of North America that man annihilated. Breeding, probably in colonies, on rocky, coastal islands in the North Atlantic, this flightless bird was readily accessible to roving sailors and fishermen. They took its eggs for food and slaughtered it for meat, feathers, oil, and codfish bait. The last two specimens of the Great Auk were taken on Eldey, a volcanic rock off the southwest coast of Iceland, on June 3, 1884.

At about this time on mainland North America two species, the Carolina Parakeet *(Conuropsis carolinensis)* and the Passenger Pigeon *(Ectopistes migratorius)*, were on their way to extinction through excessive killing, but nobody yet realized it. Both species lived east of the Great Plains; both were gregarious, commonly existing in large flocks. The Carolina Parakeet, though perhaps never very abundant, was especially fond of fruit and consequently much despised by farmers who could kill large numbers easily. When a flock raided an apple tree, it was possible to shoot every bird because those individuals escaping the first blast from the gun hovered over those killed until they too were shot. The last specimen was killed in the wild on April 18, 1901. The Passenger Pigeon (Fig. 57), at the time of the white man's arrival in North America, may have numbered 3,000,000,000 individuals—a popultion never attained by any other bird species known—and constituted between 25 and 40 percent of the bird population in the United States (Schorger, 1955). Despite this apparent security in numbers, the species was wiped out in the course of a century, the last wild specimen being recorded with certainty between September 9 and 15, 1899. By a remarkable coincidence, both the last captive Carolina Parakeet and the last captive Passenger Pigeon died in the same place in the same month and year—in the Cincinnati Zoological Garden in September, 1914 (Greenway, 1967.)

The dramatic decline of the Carolina Parakeet and Passenger Pigeon in the last century eclipses the demise of another North American species, the Labrador Duck *(Camptorhynchus labradorius)*, the last recorded specimen of which was taken in the fall of 1875. It was never an abundant bird and it was never hunted extensively. Just what caused its extinction will never be known with any certainty.

In this century the one remaining Heath Hen *(Tympanuchus cupido cupido)*, the eastern subspecies of the Greater Prairie-Chicken, was last seen on March 11, 1932. Once prevalent along the Atlantic seaboard from New Hampshire south to Virginia, the Heath Hen became confined after 1869 to Martha's Vineyard, an island off the coast of Massachusetts. Here it survived in varying numbers, reaching a population close to 2,000 by 1916. But in the spring of that year a severe fire swept its breeding grounds, no doubt destroying many nests and nesting sites (Gross, 1928). In any case its final decline began soon thereafter and was accelerated

Figure 57 **Passenger Pigeon**

in the few remaining years by predation from cats and rats, diseases acquired from poultry, and toward the end, an excessive ratio of males to females.

All told since 1681, the last year when the Dodo was alive, no fewer than 78 species and 49 well-marked subspecies have become extinct over the world. Fisher (1964), who compiled these figures, found "fairly strong direct proof" that man destroyed nearly half of the 78 species. Although Fisher makes no statement as to man's role in the destruction of the 49 subspecies, it was undoubtedly as great, if not even greater.

Fisher attributes the causes of extinction by man to be primarily direct killing, destruction of habitat, and predation by cats, rats, and other human symbionts. Another cause, but difficult to prove, is man's introduction of competing bird species.

Worthy of note from Fisher's figures is that only nine of the 78 extinct species and two of the 49 extinct subspecies were continental; all the others lived on islands. This clearly reflects the fact that insular birds, normally with small populations, are particularly sensitive to changes. Sometimes flightless, often quite tame, and usually with quite specialized feeding habits, they tend to lack the versatility to escape from, compete with, or adjust to man-imposed modifications in their natural environment.

Besides those species and subspecies known to be extinct are a far greater number verging on extinction and, in some cases, may already be extinct. The following four North American species and two subspecies have wild populations under 100:

Species

California Condor, *Gymnogyps californianus*
Whooping Crane, *Grus americana*
Eskimo Curlew, *Numenius borealis*
Bachman's Warbler, *Vermivora bachmanii*

Subspecies

Florida Snail Kite, *Rostrhamus sociabilis plumbeus*
American Ivory-billed Woodpecker, *Campephilus p. principalis* (probably extinct)

For other bird species and subspecies in North America and elsewhere that may be nearing extinction, consult *Endangered Birds of the World: The ICBP Bird Red Data Book* (1979; reprinted in handbook form, 1981), compiled by W.B. King on behalf of the International Council for Bird Preservation and the Species Survival Commission of the International Union for Conservation of Nature and Natural Resources.

Now and then a bird, believed to be extinct, has been found still existing in some unexplored area or has simply been overlooked. For example, the Takahe, *Notornis mantelli*, a heavy, flightless gallinule that once lived over much of New Zealand's South Island was not reported after 1898 until rediscovered exactly fifty years later in a remote mountain valley (Williams, 1960). The rare Puerto Rican Whip-poor-will or Nightjar, *Caprimulgus noctitherus*, described from bones found in prehistoric cave deposits and from a single specimen collected in 1888, and seen alive only once—in 1911—thereafter went unreported by ornithologists until fifty years later, when it was rediscovered by a tape-recording of its call and from a collected specimen (Reynard, 1962). The giant race of the Canada Goose, *Branta canadensis maxima*, which breeds in the northern Great Plains, was rediscovered in 1962 after being considered extinct for three decades (Hanson, 1965).

While it is not impossible that even a few other supposedly extinct birds will some day show up, it is only wishful thinking to expect that many are still extant. The truth must be faced: most of the species declared extinct are in fact gone forever.

Will other birds become extinct? Undoubtedly. Species have apparently been decreasing since the Pleistocene, and no species can live on indefinitely. Indeed, every species has a normal life expectancy—perhaps as short as 16,000 years (calculation by Fisher, 1964; based on Brodkorb, 1960). The disturbing problem confronting ornithologists and other people interested in birds is how to blunt the human threat that may shorten life expectancy and hasten extinction.

The surging human population, now estimated at 3 billion and expected to be doubled by the year 2000, is not in itself a threat to the longevity of bird species. The threat comes in man's thoughtless abuse of the earthly environment. Promiscuous dissemination of poisons, pollution of air and water, total destruction rather than the selective use of habitats, indiscriminate creation of hazards to bird migration—all such actions, if they increase as the human population expands, can in a short time eliminate bird species by the score.

The main hope for birds lies in more aggressive conservation of natural environments. Everyone interested in ornithology should become *militant* conservationists and apply their newly acquired knowledge toward countering the threat to the early extinction of birds.

References

Austin, O.L., Jr.
1961 *Birds of the world*. New York: Golden Press.

Bock, W.J.
1963 The cranial evidence of ratite affinities. *Proc. XIIIth Internatl. Ornith. Congr.* pp. 39–54.

Brodkorb, P.
1960 How many species of birds have existed? *Bull. Florida State Mus., Biol. Sci.* 5:41–53.

de Beer, G.
1954 *Archaeopteryx lithographica: A study based upon the British Museum specimen*. London: British Museum (Natural History).
1964 Archaeopteryx. In *A new dictionary of birds*, ed. A.L. Thomson, New York: McGraw-Hill.

Feduccia, A.
1980 *The age of birds*. Cambridge: Harvard Univ. Press.

Fisher, J.
1964 Extinct birds. In *A new dictionary of birds,* ed. A.L. Thomson. New York: McGraw-Hill.

Fisher, J., and Peterson, R.T.,
[1964] *The world of birds*. Garden City, New York: Doubleday.

Greenway, J.C., Jr.
1967 *Extinct and vanishing birds of the world*. 2nd ed. New York: Dover.

Gross, A.O.
1928 The Heath Hen. *Mem. Boston Soc. Nat. Hist.* 6:487–588.

Hanson, H.C.
1965 *The Giant Canada Goose*. Carbondale: Southern Illinois Univ. Press.

Heilmann, G.
1927 *The origin of birds*. New York: D. Appleton.

Mengel, R.M.
1964 The probable history of species formation in some Northern Wood Warblers (Parulidae). *Living Bird* 3–43.

Millener, P.R.
1982 And then there were twelve: The taxonomic status of *Anomalopteryx eweni* (Aves: Dinornithidae). *Notornis* 29:165–170.

Parkes, K.C.
1966 Speculations on the origin of feathers. *Living Bird* 5:77–86.

Reynard, G.B.
1962 The rediscovery of the Puerto Rican Whip-poor-will. *Living Bird* 1:51–60.

Savile, D.B.O.
1957 The primaries of *Archaeopteryx*. *Auk* 74: 99–101.

Schorger, A.W.
1955 *The Passenger Pigeon: Its natural history and extinction*. Madison: Univ. of Wisconsin Press.

Simpson, G.G.
1946 Fossil penguins. *Bull. Amer. Mus. Nat. Hist.* 87:1–99.

Storer, R.W.
1960a Evolution in the diving birds. *Proc. XIIth Internatl. Ornith. Congr.* pp. 694–707.
1960b Adaptive radiation in birds. In *Biology and comparative physiology of birds,* vol. 1, ed. A.J. Marshall. New York: Academic Press.

Swinton, W.E.
1960 The origin of birds. In *Biology and comparative physiology of birds,* vol. 1, ed. A.J. Marshall. New York: Academic Press.
1964 Fossil birds. In *A new dictionary of birds,* ed. A.L. Thomson. New York: McGraw-Hill.

Thulborn, R.A., and Hamley, T.L.
1982 The reptilian relationship of *Archaeopteryx*. *Australian Jour. Zool*. 30, 611–634.
1959 Birds of the Pleistocene in North America. *Smithsonian Misc. Coll*. 138:1–24.

Wetmore, A.
1955 Paleontology. In *Recent studies in avian biology,* ed. A. Wolfson. Urbana: Univ. of Illinois Press.
1956 A Check-list of the fossil and prehistoric birds of North America and the West Indies. *Smithsonian Misc. Coll*. 131 (5):1–105.

Williams, G.R.
1960 The Takahe *(Notornis mantelli,* Owen, 1848): A general survey. *Trans. Royal Soc. New Zealand* 88:235–258.

Ornithological Methods

APPENDIX A

Among the numerous methods for facilitating the study of birds in the field, some of the more important are presented in the following pages.

Observing and Photographing Birds

Blinds

Blinds, or **hides,** are frequently necessary for close-up observation and photography. Although they must sometimes be specially constructed to suit particular places, the following specifications can be adapted to a wide variety of situations.

The blind should be self-supporting, portable, and easy to set up, move, or take down. For the single observer or photographer the recommended size is five feet (1.52 mm) in height and four feet (.37 sq. m) square—large enough to allow room for sitting on a campstool, standing up, and equipment.

Framework The material must be strong and sturdy, preferably metal. Ideal are ¾-inch aluminum conduits, four cut in 5-foot (1.52 m) lengths for the upright corner supports and four cut in 4-foot (1.22 m) lengths for the sides of the square roof. These conduits are then assembled to form the frame by slipping each one into a triple-opening elbow, one for each corner, and securing with the aid of small screws.

Covering The covering must be a tough fabric, preferably waterproof, nontranslucent, and dark green or brown. After the fabric has been cut into five pieces of proper size, the four sides should be sewed to the square top so that the covering can be laid over the frame, with the sides hanging down, and tied tightly together at the corners. At each of the four roof corners there should be a loop, or "eye," for the attachment of a guy to some distant object. If the blind is on an elevated platform, the guys may be tied down to the floor. The front covering—i.e., the covering facing the direction for observation or photography—should have three vertical slits, zippered, the middle one longer. It is also desirable to have at least one zippered vertical slit on each of the other three sides to be opened for occasional observation if not for ventilation.

Placing and Using Blinds

When placing blinds in the vicinity of nests, certain precautions must be taken. Always put the framework together far from the nests in order to prevent disturbance. For *small birds 12 inches (30.48 cm) in length or less,* quickly place the blind framework about 10 feet (3.05 cm) from the nest, put on the covering and tie well so as to prevent excessive flapping in the wind, attach guys, then leave the vicinity until the next day. If the blind appears to be "accepted" after this lapse of time, move the blind closer to the nest (seldom nearer than five feet–1.52 m), and again leave the vicinity. It is accepted on the following day, it may be used. It is sometimes necessary to take a longer time to condition birds to elevated blinds since the structures are apt to be conspicuous and, therefore, frightening. In this case place the framework in position 10 feet (3.05 m) from the nest without covering for a day. Put on the covering the next day and move the blind closer the third day. For *large birds over 12 inches (30.48 cm)* place the completely erected and covered blinds at greater distances from the nests and take a longer time before moving the blinds closer. A preliminary knowledge of the habits and reactions of large birds is often necessary before one can place blinds in their presence with successful results.

When entering blinds, either try to choose a time when the nesting birds are not present, or have a second person walk away as a deflector of the birds' attention. Even though birds may condition themselves to the blind itself, they may not behave normally if they are aware of its being occupied.

Cameras and Equipment

Cameras ordinarily used for photographs to illustrate birds or to supplement bird observtions are the single lens reflex, 35 mm models that have, or can accommodate, the following equipment:

Reflex finder A mirror behind the lens that reflects through the viewfinder the exact field and focus and frequently activates an exposure meter.

"Fast" lenses Lenses with large diameters in relation to focal length and designated by small "*f*" numbers. Because birds are quick moving and often in areas with poor lighting, such lenses are highly desirable. The standard 50 mm lens is recommended for photographing birds when close and should be *f*1.8, *f*.9, or *f*2.8. The shutter speed used will depend on the amount of light (determined by a light meter) and the particular emulsion speed of the film in the camera.

Telephoto Lenses Lenses designed to give large images of distant subjects. In all cases the smaller the *f* number, the better. Even though blinds will allow the photographer to get closer to birds, frequently they are not close enough with the 50 mm lens for large images. Telephoto lenses should have a maximum aperture of *f*5.6 and permit satisfactory exposure at 1/250 second in good light. Any telephoto lens with an aperature less than *f*8 is of little use in bird photography. Zoom lenses of good quality—capable of producing *sharp* images—with focal lengths from 70–210, 90–230, or 100–300 are advantageous for a quick change of magnification without loss of time in changing lenses.

Extension Tubes, or Extenders Increase the focal length by mounting lenses farther out from the film to obtain larger and more detailed images. Since their use requires wider apertures for more light, they have the disadvantages of requiring slower exposure and reducing depth of focus.

Flash Equipment Used for close-up pictures of birds in poor light or for high-speed action shots. Flash units are of several varieties, all synchronized with the camera shutter. They must have a duration of no more than 1/1000 second to stop most birds in motion on the ground or in slow flight. Flash units are rated by their brightness numbers: the larger their guide numbers, the brighter their flashes. Units with a guide number approximating 65, mounted within three feet (0.91 m) of the subject, will usually produce satisfactory pictures. A flash unit can be connected to the camera by a long cable, making it possible to photograph a nesting bird from a blind with the unit outside and the cable running to the camera in the blind. Before the bird returns, the unit is mounted at the proper distance from the nest. Once the bird is back on the nest and the camera in focus, the photographer is ready to press the shutter release for the picture.

Tripod Supports the camera when photographing from a blind, when using a telephoto lens of 300 mm or more, and when taking pictures of nearby objects at slow speed.

Gunstock Mount Holding the camera when obtaining high-speed shots of birds in action, especially in flight. This type of mount allows the photographer to follow bird movements more readily. Under most circumstances, a camera on a tripod cannot be moved rapidly enough to keep up with the action.

Film Recommended for the best results is positive slide film, which has sharper definition and more accurate color value than negative film and is thus much preferred for reproductions.

This is essentially an outline of basic equipment and operational procedure. For more detailed information on the construction and use of blinds, photographic equipment, and techniques in photography, consult *The Audubon Society Handbook for Birders* by S.W. Kress (Charles Scribner's Sons, New York, 1981). Also see *A Portable Observation Tower-Blind,* with design features, construction, transport, and assembly, by N.L. Rodenhouse and L.B. Best (Wildlife Society Bull. 11:293–297; 1983).

Recording Bird Vocalizations

Recording bird songs and other vocalizations suitable for spectrographic analysis requires careful choice of equipment together with a few simple techniques in its application.

The equipment should have a higher quality and fidelity than one can expect of a home recorder for music and human speech, and it should be designed to respond *uniformly* to a wider range of sound frequencies (sound waves) in Hertz or cycles per second. Furthermore, it must not only be portable and rugged for use in the field but also include special accessories for picking up sounds efficiently at a distance with a minimum of distortion and environmental noises.

Microphones and Accessories

The microphone is responsible for converting the acoustical energy into electrical energy—i.e., sound waves into electrical impulses. It should withstand rough use, be uniformly responsive to sound waves ranging from 30 to at least 10,000 Hertz (i.e., have a "flat" response), and have a high electrical output and usually low impedance to match the tape recorder being used. Some years ago the "dynamic" microphone would have been preferred, but "condenser" microphones have improved substantially and should be considered unless the recording environment is very humid.

The ultradirectional "shotgun" microphones, while often rather expensive, are very convenient to use and therefore quite popular among professional recordists. The microphone is encased within an elongated "interference tube," looking much like a shotgun barrel, so that sounds arriving off axis are canceled; thus, sounds are recorded by merely aiming the microphone at the subject. Advantages in using a shotgun microphone include convenience and a response to the sound energy, which is "flat" throughout the frequency range of most bird songs. The microphone, however, can respond only to the energy that actually strikes its small diaphragm, so the environment must be relatively noise free or the bird must be quite close to the microphone.

The advantages and disadvantages of a shotgun microphone are exactly reversed with the parabola. The **parabola,** a dishlike device fitted with an omnidirectional microphone, can collect many times more sound energy than the microphone alone. Sound waves that are parallel to the axis of the parabola are reflected to its focal point. A microphone placed precisely at this focal point serves as a "parabolic microphone," which is extremely directional. If the microphone is placed on the axis, though slightly removed from the focal point of the parabola, some gain may be sacrificed, but the ultradirectionality of the system is reduced, making it easier to obtain good recordings even if the parabola is not aimed exactly at the subject; in addition, at the high frequencies the beam of sound energy is widened and therefore involves more of the microphone's diaphragm in producing the electrical signal.

The effectiveness of a parabola depends largely upon its size. Obviously, the larger the parabola, the greater its effective gain—and unwieldiness. Parabolas that are 40, 24, and 13 inches (101.60, 60.96, and 33.02 cm) in diameter will collect, respectively, about 1,600, 600, and 170 times more sound energy from the point at which they are directed than would the microphone alone at the same distance. Unfortunately, the smaller diameter reflectors are less effective in reflecting and recording the lower frequencies. Small, clear plastic parabolas (13–18 inches, 33.02–45.72 cm, in diameter) are available commercially and are relatively inexpensive; they are very light and easy to aim at the bird. Larger parabolas are usually aluminum, thus more durable, and because of their increased size, they are more effective in recording lower frequencies since frequency response is flatter. The smaller models are certainly very adequate for amateurs, but professionals who want tape recordings to match as closely as possible the energy spectrum in bird sounds might prefer a parabola as large as 36 inches (91.44 cm) in diameter.

An additional feature that must be considered with the parabola is the position of the focal point; the curvature of the parabola will determine whether the focal point is within or outside the actual cavity of the reflector. If it is within the dish, the microphone will be shielded to some extent from the wind, but the resonance within the cavity of the parabolic dish will increase. Resonance is not a problem with the microphone outside the dish, but wind velocity increases outside, of course, and at increasing distances from the dish the system becomes more unwieldy.

To help counteract the effects of wind, a **windscreen** is a useful accessory that may be fitted like a cap over the receiving end of the microphone. Properly designed, windscreens can prevent low frequency fluttering in the recording that results from wind striking the microphone directly. However, during a moderate to strong wind, the windscreen is of no value in keeping the wind from causing serious interference.

A parabola of aluminum is noisy and will amplify vibrations from the slightest touch. The handle should be wrapped with rubber or soft cloth, and the back should be coated with a "sound-deadener" used on car bodies. To reduce the shininess of the front, lest it startle the birds whose voices are being recorded, a light coat of dull-colored, nonglossy paint may be applied. While a parabola is usually hand held, it should nonetheless bear a bracket so that it can be set on a camera tripod for occasional uses in a stationary position.

Recorders

The ideal machine is a portable magnetic reel-to-reel tape recorder, powered by dry-cell ("flashlight") batteries and weighing between 8 and 10 pounds (3.63 kg–4.54 kg). Unfortunately, the "ideal" equipment is usually very expensive. Marked improvements have been made in less expensive cassette tape recorders. Even though their slower tape speeds (1⅞ inches, 47.63 mm, per second) and narrower tape width (0.15 inch, 3.81 mm, versus 0.25 inch, 6.35 mm) noticeably reduce the quality of the recording, their availability, convenience, and price are making them increasingly popular, even among some professionals. The recorder should have the following features and accessories:

1. Besides the recording head and amplifier, a set of headphones for testing the sounds before recording them and for monitoring the sounds when they are being recorded.

2. A playback head and amplifier for determining whether or not the recordings are meeting the desired standards and an erase head for "wiping" the recordings if they are unsatisfactory.

3. On reel-to-reel machines, recording speeds of at least 7.5 (19.05 cm) and perhaps even 15 inches (38.10 cm) per second. The slower speed is satisfactory for sounds of low to intermediate frequency, but the higher speed is particularly better for the sounds of high frequency.

4. A cable to the microphone of the best quality. It should contain one or two wires, depending on the type of microphone (balanced or unbalanced), in a woven-wire shield that is grounded to the recorder and covered with rubber or plastic.

5. A durable carrying case with an over-the-shoulder strap.

When recording, the operator ordinarily carries the recorder slung from the shoulder, wears headphones plugged into the recorder, manages the recorder's controls with one hand, and holds the parabola or shotgun microphone with the other. Through the headphones the operator can determine when the microphone is aimed at the vocalizing bird and whether or not to correct the quality of the sound being received by manipulating the controls on the recorder.

One of the serious problems in recording is to avoid extraneous sounds such as wind and all the noises emanating from human activities. Birds are very active during early morning when there is apt to be less wind and less human activity. This is the best time for recording bird songs with a minimum of interference. If conditions are very quiet, it is possible with larger parabolas to record songs at distances of 300 feet (91.44 m) or more.

This information comes primarily from the following publications. For further information on tape recorder metering systems, choices of magnetic tape, and other topics related to bird-sound recording, consult them.

Gulledge, J.L.
1977 Recording bird sounds. *Living Bird* 15: 183–204.
Kellogg, P.P.
1960 Considerations and techniques in recording sound for bio-acoustics studies. In *Animal sounds and communication*, eds. W.E. Lanyon and W.N. Tavolga. Washington, D.C.: American Institute of Biological Sciences.
1962 Bird-sound studies at Cornell. *Living Bird* 1:37–48.
Wickstrom, D.C.
1982 Factors to consider in recording avian sounds. In *Acoustic communication in birds*, eds. D.E. Kroodsma and E.H. Miller. New York: Academic Press.

Recording Colors of Birds and Bird Eggs

In describing birds and bird eggs, or in recording the colors of particular birds, alive or preserved, it is highly desirable to have some color standard to follow so that the names of different hues will have reference value. The best color standard available is F.B. Smithe's *Naturalist's Color Guide: Part I, the Color Guide; Part II, the Color Guide Supplement* (American Museum of Natural History, New York, 1975). Part I is a looseleaf pocket notebook with marginal color chip pages and a pocket containing a neutral gray card with cutouts for isolating colors. Part II, paperbound, has descriptive data for comparative use.

Another color standard is *Atlas de los Colores* by C. Villalobos-Dominguez and J. Villalobos; English text by A.M. Homer (El Ateneo, Florida 340, Buenos Aires, Argentina, 1947). This work contains a table converting the symbols, which the authors use, to the nomenclature in R. Ridgway's *Color Standard and Nomenclature* (1912), once the accepted standard for ornithologists but long out of print and difficult to obtain.

Lacking a copy of Villalobos, follow the more simplified version of the Villalobos standard, with a double-page color chart prepared under the supervision of J. Villalobos, in the introduction to the *Handbook of North American Birds*, Volume 1, edited by R.S. Palmer (Yale University Press, New Haven, 1962). The choice of color terms available is sufficiently wide for most descriptions.

Measuring Elevation of Nests and Flight Paths

When nests are in trees or other situations too high for accurate determination with measuring stick or tape, a hand-held and hand-operated finder may be useful. Variously called rangefinder, clinometer, altimeter, or height meter, depending on the model, it can be obtained from any one of several companies specializing in equipment for work in forestry.

This same instrument, usually available at modest cost, can also be applied to determining within 100–150 feet (30.48–45.72 m) the elevation of flight paths taken by birds above land or water in their local maneuvers. It can also be applied to measuring straight line distances such as the extent of a bird's territory, the width of a river, and others.

Marking Nest Sites, Boundaries of Territories, and Census Lines

Useful for marking nest sites, territorial boundaries, and lines to be followed in census work is a vinyl plastic ribbon called flagging, available in an assortment of bright colors. This can be obtained in rolls from companies that supply equipment for foresters. It is easy to tie and easy to see; it lasts indefinitely and its color is permanent. Because it lasts so long, remove all flagging when through with the study lest it become litter.

Capturing Wild Birds

Frequently, it is necessary to capture wild adult birds for banding, marking, weighing, or examination. There are several devices and techniques commonly used. See page 372 for permits required.

Traps for Nesting Birds Birds nesting on the ground may be captured in a **pull-string** trap, which consists of a wooden or metal frame covered with hardware cloth on the sides and top. It is placed over the nest and tilted upward at a 45-degree angle by a trip-stick to which is attached a string leading to the trapper hidden at a distance. When the bird enters the nest, the trip-stick is quickly pulled from under the trap, which falls over the nest and imprisons the bird. The bird is removed by reaching either under the trap or through a door constructed for the purpose.

Birds nesting in trees may also be captured by a trap specially constructed to suit the particular nesting site. It is made entirely of hardware cloth, cut to fit around and below the nest and supporting branch, and held in place either by the branch or a prop coming from the ground below. The top, which is several inches above the nest rim, is hinged, weighted, and held partially open by a trip-stick to which a string is attached. When the bird settles on the nest, after entering through the top (the only opening), the trip-stick is pulled out and the weighted top falls to hold the bird a prisoner. A more simple method of capturing tree-nesting adults is a net made of several hair nets sewed together and attached to a wire hoop. This is placed over the nest at a 45-degree angle and held in place by strings. After settling on the nest, the bird is abruptly flushed and it flies into the net where it becomes entangled.

Birds nesting in cavities may be captured in two ways: by a mist net (description following) placed in such a manner in front of the entrance to the cavity that the bird flies into it when leaving the nest or returning; or by a net on a long-handled pole applied to the opening when the bird is inside, as at night. Either way, the nets should not be left unattended.

Mist Nets Made of fine nylon or silk and dyed to make them practically invisible against almost any dark background, mist nets are excellent for capturing flying birds in quantity. The nets, about 30–38 feet (9.14–11.58 m) long and 3–7 feet (.91–2.13 m) high, are available in different mesh sizes for birds ranging from kinglets and warblers to grouse, large shorebirds, and medium-sized ducks and hawks. Each net consists of five horizontal lines or "trammels," which are stretched taut between two vertical poles, and the netting itself, which hangs loose from the trammels to form bags. The nets are positioned strategically across flight paths commonly taken by birds. Birds fail to see the netting, strike it, and fall into the bags where they become entangled and held. Special pulleys and guys can be used to raise the nets up to 40 feet (12.19 m) or more for capturing birds in the forest canopy. The person operating the nets should be in constant attendance and not let the birds be captive longer than the time necessary to remove them. If the birds become badly entangled, as is sometimes the case, great patience and skill are required to free them.

Bait Traps Birds that normally feed on the ground may be captured in traps suitably baited. The variety of traps is infinite, ranging from single-cell types, which automatically capture one bird at a time, to large "decoy" enclosures, which can imprison hundreds at one setting. The simple trap previously described for capturing ground-nesting birds can be easily adapted by setting it up over bait instead of a nest. Another simple trap is a wire-cloth, bottomless box placed flat on the ground over the bait. This box has one or more entrances, each consisting of a funnel that has its big opening flush with the side of the box and its small opening far inside. Birds, entering through a funnel, fail to exit since they overlook the small inner opening of the funnel. The birds are removed through a covered opening in the top of the trap. Raptors can be readily taken in a "bal-chatri" trap consisting of a chicken wire cage holding a bird or rodent as bait. To the top of the trap are attached many monofilament nooses that entangle the talons of the raptors attacking the lure.

Cannon Nets Gregarious species such as blackbirds, doves, and gulls; waterfowl that are easily attracted to bait; and shorebirds feeding together on a beach may be captured effectively by large cannon or rocket nets. Each net is attached to three or more heavy projectiles that are inserted in "cannons," or mortars, loaded with a propellant. When simultaneously detonated electrically by an operator hidden some distance away, the projectiles instantly carry the net over the feeding birds. As there is an element of danger involved in setting up and firing cannon nets, obtain prior instruction from an experienced operator before using them.

Corral Nets and Drift Traps Waterfowl, while flightless during the simultaneous molt of their remiges, and young of a few colony-nesting species may be driven into enclosures or corrals, constructed either of hardware cloth or netting. Long fences flare from the entrance to form a funnel into which the birds are first herded and ultimately guided into the trap. "Leads" of sea drift placed across feeding areas of rails and shorebirds may passively direct the birds into funnel traps.

Hand Nets Used with Spotlights Shorebirds, such as woodcock, and many aquatic birds may be handnetted at night after first being "blinded" by a powerful spotlight. The operation usually requires at least two persons side by side, one to direct the spotlight and the other to wield a large circular net on a long pole. Dark, overcast nights are necessary. If the bird is on the ground, the night-lighters walk quietly toward it, all the while holding the beam on the bird until it can be reached with the net. Should the bird flush, it can sometimes be brought down by "dazzling" it with the spotlight. If the bird is on the water, the night-lighers can follow the same procedures from a boat by quietly maneuvering their craft toward the bird until they can take it in the net.

Chemicals Birds may be captured by bait treated with a tranquilizing or stupefying agent. Since most such chemicals are invariably lethal in heavy doses, do not apply any agent of this sort without first being thoroughly familiar with its potentialities, knowing precisely how to administer it, and understanding how to handle doped birds and effect their recovery.

The following references will provide much detailed information on methods of capturing wild birds. For agencies supplying traps and nets, peruse the "Banders' Marketplace" in *North American Bird Bander,* quarterly publication of the Eastern, Inland, and Western Bird Banding Associations (Eldon Publishing Company, P.O. Box 446, Cave Creek, Arizona 85331).

Berger, D.D., and Mueller, H.C.
1959 The Bal-chatri: A trap for birds of prey. *Bird-Banding* 30:18–26. (Bal-chatri comes from the East Indian name for "boy's umbrella.")

Day, G.I.; Schemnitz, S.D.; and Taber, R.D.
1980 Capturing and marking wild animals. In *Wildlife management techniques manual*. 4th ed., ed. S.D. Schemnitz. Washington, D.C.: The Wildlife Society.

Edgar, R.L.
1968 Catching colonial seabirds for banding. *Bird-Banding* 39:41–43. (Using an adjustable noose on a pole.)

Greenlaw, J.S., and Swinebroad, J.
1967 A method for constructing and erecting aerial-nets in a forest. *Bird-Banding* 38:114–119.

Heimerdinger, M.A., and Leberman, R.C.
1966 The comparative efficiency of 30 and 36 mm mesh in mist nets. *Bird-Banding* 37:280-285.

Humphrey, P.S.; Bridge, D.; and Lovejoy, T.E.
1968 A technique for mist-netting in the forest canopy. *Bird-Banding* 39:43–50.

Johns, J.E.
1963 A new method of capture utilizing the mist net. *Bird-Banding* 34:209–213. (Net stretched horizontally just above shallow water and shore, then dropped by release string, captures shorebirds successfully).

Johnston, D.W.
1965 An effective method for trapping territorial male Indigo Buntings. *Bird-Banding* 36:80–83. (Equipment: a mist net, a stuffed male in breeding plumage, and a recording of species' song. Set-up must be *in* an occupied territory or between contiguous territories.)

Hussell, D.J.T., and Woodford, J.
1961 The use of a Heligoland trap and mist-nets at Long Point, Ontario. *Bird-Banding* 32:115–125.

Lacher, J.R., and Lacher, D.D.
1964 A mobile cannon net trap. *Jour. Wildlife Mgmt*. 28:595–597. (Cannons mounted on the front of a jeep.)

Licinsky, S.A., and Bailey, W.J., Jr.
1955 A modified shorebird trap for capturing woodcock and grouse. *Jour. Wildlife Mgmt*. 19:405–408. (Using funnel nets strategically placed in feeding and resting areas.)

Loftin, H.
1960 Use of decoys in netting shorebirds. *Bird-Banding* 31:89–90. ("Decoys used are simply profiles cut from plywood or even pasteboard, with a heavy wire rod attached for sticking into the sand.")

Low, S.H.
1957 Banding with mist nets. *Bird-Banding* 28:115–128. (The classic paper on the subject.)
Martin, S.G.
1969 A technique for capturing nesting grassland birds with mist nets. *Bird-Banding* 40:233–237.
McCamey, F.
1961 The chickadee trap. *Bird-Banding* 32:51–55.
Nolan, V., Jr.
1961 A method of netting birds at open nests in trees. *Auk* 78:643–645.
Reeves, H.M.; Geis, A.D.; and Kniffen, F.C.
1968 *Mourning Dove capture and banding*. U.S. Fish and Wildlife Service, Special Sci. Rept., Wildlife No. 117.
Schemnitz, S.D., ed.
1980 *Wildlife management techniques manual*. 4th ed. Washington, D.C.: The Wildlife Society. (Comprehensive coverage; indispensable.)
Sheldon, W.G.
1960 A method of mist netting woodcocks in summer. *Bird-Banding* 31:130–135. (Nets placed on spring singing fields and operated for a few minutes after sunset.)
Thompson, M.C., and DeLong, R.L.
1967 The use of cannon and rocket-projected nets for trapping shorebirds. *Bird-Banding* 38:214–218.
Vandenbergh, J.G.
1960 A bird holding cage. *Bird-Banding* 31:221–222.
Woodford, J., and Hussell, D.J.T.
1961 Construction and use of Heligoland traps. *Bird-Banding* 32:125–141. (Contains a wealth of detailed information and very useful suggestions.)

Banding Wild Birds

There are many schemes for banding (called ringing in Europe) wild birds throughout the world. All involve the use of metal leg bands, systematically numbered and carrying an address to which the recoverer of the banded bird may report.

In North America the United States and Canadian Governments—i.e., the U.S. Fish and Wildlife Service and the Canadian Wildlife Service—cooperate in a single, continent-wide program. Each band bears a number and directions "Advise Bird Band, Washington, D.C." After the band is attached to the bird's leg, a record is sent to the Bird Banding Laboratory at the Office of Migratory Bird Management, Laurel, Maryland, where it is placed on file. If at some future date a person kills the bird, finds it dead, or captures it, that person reports the number to the address on the band, thus revealing its whereabouts. In North America well over 30,000,000 birds have been banded and approximately 1,000,000 more are banded each year. Over 2,000,000 have been "recovered" and reported to the Bird Banding Laboratory.

Any person wishing to band birds must first obtain the necessary permits (see p. 372). The Bird Banding Laboratory then provides, free of charge, all bands, forms for reporting, postage-free envelopes, and a *Bird Banding Manual*. The bander is expected to provide traps, nets, and other equipment.

The data from recoveries sent to the Bird Banding Laboratory are computerized, and then made available to students and other persons qualified to undertake research. In order to protect the interests of banders, many of whom are engaged in long-term research projects and who have invested significant amounts of money in the accumulation of data, the Bird Banding Laboratory places certain limitations on the publication of information from its files. Requests for data from the files should be forwarded to the Office of Migratory Bird Management, Laurel, Maryland 20708.

Marking Wild Birds

It is sometimes desirable to mark birds for individual recognition. See page 372 for necessary permits.

Colored Plastic Leg Bands Various combinations of colored plastic bands enable the identification of individual birds visually without having to recapture them. By devising a coding system with strong, easy-to-see colors (e.g., green band on the *left* leg of one bird, green band on the *right* leg of another, green and yellow bands on the *left* leg of still another, etc.) it is possible to recognize individuals in a great number of birds. Caution: While it is all right to place two or more plastic bands on the same leg, never use two or more metal bands on the same leg as their constant tapping together produces a sharp "flange" that can seriously injure or even sever the leg.

Neckbands and Tags A wide variety of markers have been devised for large birds to enable their recognition from a distance. Most such markers consist of plastic material in different colors attached to the bird by a preformed band or collar with overlapping ends or by noncorrosive clips. Two laminated plastic strips of contrasting colors can be formed into neckbands for large birds, with the outer color separated to expose the inner color in the form of letters or numbers. The use of wing (patagial) tags of varying colors has proven successful. Caution: Any marker can be a handicap to a bird, or even a cause of death, unless properly designed and applied to suit the bird's habits and habitat. No marker should impair flight, be so loose as to become entangled in vegetation, or be heavy enough to erode the feathers or to chafe the skin.

Bill Markers Several types of disks may be used on the bills of larger birds, primarily waterfowl. Made of plastic in different colors, the markers are attached on either side of the bill or saddled over the bill by a metal or nylon pin through the nostrils. Their chief detriment is their tendency to become entangled in vegetation or debris while the birds are feeding. Under certain weather conditions ice will form on the plastic marker causing birds to become incapacitated. This occurs occasionally to waterfowl wintering in northern states and provinces. See Greenwood and Blair (1974).

Feather Marking and Coloring Birds may be **feather-marked** by cementing white feathers that have been dyed in different colors to their upper tail coverts. Birds may also have their flight feathers or other parts of their plumage colored by applying either an aniline dye or a quick-drying lacquer (e.g., the so-called airplane dope used for model airplanes). Considerable care must be taken, however, to apply the substance lightly and to make sure that it is dry when the birds are released.

When it is desirable to mark birds that cannot for one reason or another be captured, it is sometimes possible to spray them with a quick-drying dye or lacquer shot from a water pistol or to place fresh coloring such as paint or powdered dye on objects—e.g., the edge of a nest-opening—with which the birds will come in contact. Either method will give the birds random markings sufficient for individual identification.

The Bird Banding Laboratory maintains a central file containing records of all color-marking schemes throughout North America. This prevents two or more research workers from duplicating each other's schemes and enables the Laboratory to respond to reports of sightings of color-marked birds and to notify the research workers of the birds' whereabouts.

Use of Radiotelemetry When a bird cannot be followed visually, it may be equipped with a miniaturized radio transmitter, held in place by a harness slipped over the bird's neck or wings or stuck with an adhesive to the skin of the back between the wings. The transmitter is then monitored by a receiver and the movements of the bird charted.

The following references contain further information with regard to marking wild birds.

Amlaner, C.J., and Macdonald, D.W., eds.
1979 *A handbook of radiotelemetry and radio tracking*. New York: Oxford Univ. Press.

Baumgartner, A.M.
1938 Experiments in feather marking Eastern Tree Sparrows for territory studies. *Bird Banding* 9:124–135. (Consult for methods of marking by dyeing and attaching feathers.)

Cochran, W.W.
1980 Wildlife telemetry. In *Wildlife management techniques manual*. 4th ed., ed. S.D. Schemnitz. Washington, D.C.: The Wildlife Society.

Craven, S.R.
1979 Some problems with Canada Goose neckbands. *Wildlife Soc. Bull*. 7:268–273.

Day, I.D.; Schemnitz, S.D.; and Taber, R.D.
1980 Capturing and marking wild animals. In *Wildlife management techniques manual*. 4th ed., ed. S.D. Schemnitz. Washington, D.C.: The Wildlife Society.

Emlen, J.T., Jr.
1954 Territory, nest building, and pair formation in the Cliff Swallow. *Auk* 71:16–35. (Consult for methods of marking birds that cannot be captured.)

Fankhauser, D.
1964 Plastic adhesive tape for color-marking birds. *Jour. Wildlife Mgmt*. 28:594.

Gullion, G.W.; Eng, R.L.; and Kupa, J.J.
1962 Three methods for individually marking Ruffed Grouse. *Jour. Wildlife Mgmt*. 26:404–407. (Color banding, dyeing, and back-tagging.

Greenwood, R.J., and Blair, W.C.
1974 Ice on waterfowl markers. *Wildlife Soc. Bull.* 2:130–134.

Havlin, J.
1968 Wing-tagging ducklings in pipped eggs. *Jour. Wildlife Mgmt.* 32:172–174.

Hester, A.E.
1963 A plastic wing tag for individual identification of passerine birds. *Bird-Banding* 34:213–217.

Hewitt, O.H., and Austin-Smith, P.J.
1966 A simple wing tag for field-marking birds. *Jour. Wildlife Mgmt.* 30:625–627. (For use with small birds.)

Marion, W.R., and Shamis, J.D.
1977 An annotated bibliography of bird marking techniques. *Bird-Banding* 48:42–61.

Moseley, L.J., and Mueller, H.C.
1975 A device for color-marking nesting birds. *Bird-Banding* 46:341–342. (For Least Terns and other ground-nesting birds.)

Sherwood, G.A.
1966 Flexible plastic collars compared to nasal discs for marking geese. *Jour. Wildlife Mgmt.* 30:853–855. (Collars superior in visibility, retention, and ease of placement.)

Southern, W.E.
1965 Biotelemetry: A new technique for wildlife research. *Living Bird* 4:45–58.

Stiles, F.G., and Wolf, L.L
1973 Techniques for color-marking hummingbirds. *Condor* 75:244–245.

Sugen, L.G., and Poston, H.J.
1968 A nasal marker for ducks. *Jour. Wildlife Mgmt.* 32:984–986.

Swank, W.G.
1952 Trapping and marking of adult nesting doves. *Jour. Wildlife Mgmt.* 16:87–90. (Large feathers on wings and tail the best surfaces for marking.)

Wadkins, L.A.
1948 Dyeing birds for identification. *Jour. Wildlife Mgmt.* 12:388–391.

Wallace, M.P.; Parker, P.G.; and Temple, S.A.
1980 An evaluation of patagial markers for Cathartid Vultures. *Jour. Field Ornith.* 51:309–427.

Collecting Birds

If specimens cannot be captured by a net or trap, they may be collected with a gun. The weapon commonly used is either a 12-gauge, or 16-gauge, double-barreled shot-gun equipped with an auxiliary .410 barrel that may be inserted in one of the regular barrels. The regular barrel is used for large birds, the auxiliary barrel for small birds. The shells and desired loads may be secured through commercial channels. Usually the .410 shell is loaded with No. 12 ("dust") shot for birds under 10 inches (25.4 cm). The 12, or 16, shells are loaded with No. 10 shot for birds of medium size and with Nos. 7½ and 6 shot for large birds.

Permits are required for collecting (see the following section). Everyone holding them should regard themselves as specially privileged and personally responsible to the granting agencies and to the institution they represent. Callous or indiscriminate killing of birds is not only unwarranted but can create bad public relations. To avoid misunderstandings and objections, heed these admonitions.

1. Advise local conservation officers and constabulary about collecting activities contemplated.

2. Obtain the permission of land owners before collecting on their property.

3. *Never* carry firearms or shoot birds in the presence of persons uninformed about the purpose or value of the procedure.

Permits for Capturing, Banding, Marking, and Collecting Birds

With few exceptions, all species of wild North American birds are fully protected by federal regulations under the provisos of treaties between the governments of the United States, Canada, and other countries. In most states and Canadian provinces additional species in their avifauna are fully protected by regulations not afforded by the federal governments.

In accordance with federal regulations it is therefore illegal to capture, hold, transport, or kill any protected species and to take or otherwise destroy its nest or eggs. This means that the following acts involving *a protected species* are illegal.

1. To be in the possession of any live specimen, even one that is sick or injured.

2. To pick up or otherwise be in the possession of a dead specimen or the parts (including the feathers) of a specimen.

3. To be in the possession of the eggs or nests, including *old* nests long since abandoned by the builders.

Federal regulations in the United States and Canada provide for the issuance of various permits that authorize qualified students and bona fide research workers to capture, mark, and hold live specimens; to "salvage" dead specimens; and to collect specimens, their nests and eggs, of protected species. Special restrictions, however, are placed on any study activities involving eagles or other species designated "endangered," or involving the use of mist nets or chemicals.

Students or research workers who wish to capture wild migratory birds for the purpose of banding or marking must first obtain a federal scientific marking permit. Each applicant must be at least 18 years of age, show evidence of a sound ornithological background, and provide the names of at least three licensed banders or recognized ornithologists who will vouch for the applicant's knowledge and ability.

Applications for United States banding permits should be forwarded to Chief, Bird Banding Laboratory, Office of Migratory Bird Management, Laurel, Maryland 20708; for Canadian permits to Director, Canadian Wildlife Service, Environmental Management Service, Department of the Environment, Ottawa, Ontario K1A OE7, Canada.

The federal permit is invalid unless the permittee also possesses any required state or provincial permits or licenses. Therefore, *after* obtaining the federal permit, the permittee must apply for a state or provincial license. Applications for state permits should be forwarded to the appropriate agencies listed at the end of this section. Applicants for provincial licenses, if needed, may obtain the necessary advice and addresses from the Chief, Canadian Wildlife Service.

Students or research workers who wish to collect (live or dead), propagate, hold in captivity for "special use," or take the nests and eggs of one or more specimens of wild migratory birds, may obtain the necessary permits. The collecting of protected birds, their nests, and their eggs is warranted only when particularly significant problems are being studied. Permits to collect a variety of species are not generally granted.

Applications for permits to collect, propagate, or hold wild migratory birds in the United States must first be directed to the U.S. Fish and Wildlife Service through the appropriate regional office in the following listings. Each applicant must be at least 18 years of age, give a careful description of the nature and ultimate purpose of the study to be undertaken, and later give a report stating the exact number of specimens taken and the manner of their disposal.

1. **Pacific Region:** California, Hawaii, Idaho, Nevada, Oregon, Washington
   Lloyd 500 Building
   Suite 1692
   500 Northeast Multnomah Street
   Portland, Oregon 97232
2. **Southwest Region:** Arizona, New Mexico, Oklahoma, Texas
   P.O. Box 1306
   Albuquerque, New Mexico 87103
3. **North-Central Region:** Illinois, Indiana, Iowa, Michigan, Minnesota, Missouri, Ohio, Wisconsin
   Bishop Henry Whipple Federal Building, Fort Snelling
   Twin Cities, Minnesota 55111
4. **Southeast Region:** Alabama, Arkansas, Florida, Georgia, Kentucky, Louisiana, Mississippi, North Carolina, South Carolina, Tennessee
   Richard B. Russell Federal Building, Room 1200
   75 Spring Street, Southwest
   Atlanta, Georgia 30303
5. **Northeast Region:** Connecticut, Delaware, Maine, Maryland, Massachusetts, New Hampshire, New Jersey, New York, Pennsylvania, Rhode Island, Vermont, Virginia, West Virginia
   One Gateway Center
   Suite 700
   Newton Corner, Massachusetts 02158
6. **South-Central Region:** Colorado, Kansas, Montana, Nebraska, North Dakota, South Dakota, Utah, Wyoming
   P.O. Box 25486
   Denver Federal Center
   Denver, Colorado 90225
7. **Alaska Region:** Alaska
   1011 East Tudor Road
   Anchorage, Alaska 99501

Anyone who picks up a sick or injured wild migratory bird and wants to nurse it back to health may obtain temporary legal possession by writing the appropriate regional office and requesting a permit to retain it. The permittee will be required to release the bird when it has recovered.

As in the case of federal permits to capture and mark wild migratory birds, no permit to collect, propagate, or hold (except sick or injured birds) is valid until the permittee obtains another permit or license from the state agency or provincial office (through the Canadian Wildlife Service) that has jurisdiction over the area where the project is undertaken.

Applications for the state permits should be addressed to the officer of the one of the following appropriate agencies.

Alabama: Director, Department of Conservation and Natural Resources, 64 North Union Street, Montgomery 36130

Alaska: Commissioner, Department of Fish and Game, Support Building, Juneau 99801

Arizona: Director, Game and Fish Department, 2222 West Greenway Road, Phoenix 85023

Arkansas: Director, Game and Fish Commission, No. 2 Natural Resources Drive, Little Rock 72205

California: Director, Department of Fish and Game, 1416 Ninth Street, Sacramento 95814

Colorado: Director, Division of Wildlife, 6060 Broadway, Denver 80216

Connecticut: Commissioner, Department of Environmental Protection, State Office Building, 165 Capital Avenue, Hartford 06115

Delaware: Secretary, Department of Natural Resources and Environmental Control, P.O. Box 1401, Dover 19901

Florida: Director, Game and Fresh Water Fish Commission, 620 South Meridian, Tallahassee 32301

Georgia: Commissioner, Department of Natural Resources, 270 Washington Street SW, Atlanta 30334

Hawaii: Director, Division of Forestry and Wildlife, Department of Land and Natural Resources, 1151 Punchbowl Street, Honolulu 96813

Idaho: Director, Fish and Game Department, 600 South Walnut Street, P.O. Box 25, Boise 83707

Illinois: Director, Department of Conservation, Lincoln Tower Plaza, 524 South Second Street, Springfield 62706

Indiana: Director, Department of Natural Resources, 608 State Office Building, Indianapolis, 46204

Iowa: Director, State Conservation Commission, Wallace State Office Building, Des Moines 50319

Kansas: Director, Fish and Game Commission, Box 54A, RR 2, Pratt 67124

Kentucky: Commissioner, Department of Fish and Wildlife Resources, No. 1 Game Farm Road, Frankfort 40601

Louisiana: Director, Department of Wildlife and Fisheries, 400 Royal Street, New Orleans 70130

Maine: Commissioner, Department of Inland Fisheries and Wildlife, State Office Building, 284 State Street Station No. 41, Augusta 04333

Maryland: Secretary, Department of Natural Resources, Rome Boulevard and Taylor Avenue, Annapolis 21401

Massachusetts: Director, Division of Fisheries and Wildlife, 100 Cambridge Street, Boston 02202

Michigan: Director, Department of Natural Resources, Box 30028, Lansing 48926

Minnesota: Commissioner, Department of Natural Resources, 300 Centennial Office Building, 658 Cedar Street, St. Paul 55155

Mississippi: Director of Conservation, Department of Wildlife Conservation, Southport Mall, P.O. Box 451, Jackson 39205

Missouri: Director, Department of Conservation, P.O. Box 180, Jefferson City 65102

Montana: Director, Department of Fish, Wildlife and Parks, 1420 East Sixth, Helena 59601

Nebraska: Director, Game and Parks Commission, 2200 North 33rd Street, P.O. Box 30370, Lincoln 68503

Nevada: Director, Department of Wildlife, Box 10678, Reno 89520

New Hampshire: Director, Fish and Game Department, 34 Bridge Street, Concord 03301

New Jersey: Director, Division of Fish, Game, and Wildlife, Department of Environmental Protection, P.O. Box 1809, Trenton 08625

New Mexico: Director, Game and Fish Department, Villagra Building, Santa Fe 87503
New York: Director, Division of Fish and Wildlife, Department of Environmental Conservation, 50 Wolf Road, Albany 12233
North Carolina: Executive Director, Wildlife Resources Commission, Archdale Building, 512 North Salisbury Street, Raleigh 27611
North Dakota: Commissioner, State Game and Fish Department, 2121 Lovett Avenue, Bismarck 58505
Ohio: Director, Department of Natural Resources, Fountain Square, Columbus, 43224
Oklahoma: Director, Department of Wildlife Conservation, P.O. Box 53465, Oklahoma City 73152
Oregon: Director, Department of Fish and Wildlife, P.O. Box 3503, Portland 97208
Pennsylvania: Executive Director, Game Commission, P.O. Box 1567, Harrisburg 17120
Rhode Island: Director, Department of Environmental Management, 83 Park Street, Providence 02903
South Carolina: Director, Wildlife and Marine Resources Department, Rembert C. Dennis Building, P.O. Box 167, Columbia 29202
South Dakota: Director, Game, Fish and Parks Department, Sigurd Anderson Building, Pierre 57501
Tennessee: Executive Director, Wildlife Resources Agency, P.O. Box 40747, Ellington Building, Agricultural Center, Nashville 37204
Texas: Executive Director, Parks and Wildlife Department, 4200 Smith School Road, Austin 78744
Utah: Director, Division of Wildlife Resources, 1596 West North Temple, Salt Lake City 84116
Vermont: Commissioner, Fish and Game Department, Montpelier 05602
Virginia: Executive Director, Commission of Game and Inland Fisheries, 4010 West Broad Street, P.O. Box 11104, Richmond 23230
Washington: Director, Department of Game, 600 North Capitol Way, Olympia 98504
West Virginia: Director, Department of Natural Resources, 1800 Washington Street East, Charleston 25305
Wisconsin: Secretary, Department of Natural Resources, Box 7921, Madison 53707
Wyoming: Director, Game and Fish Department, Box 1589, Cheyenne 82002

Preparing Specimens

Skins

Study skins of birds have great value in research on birds in the field as well as in taxonomic investigations. Read *The Contribution of Museum Collections to Knowledge of Living Birds* by K.C. Parkes (Living Bird, 2: 121–130, 1963) for highly instructional information along this line.

The preparation of birds as study skins requires experience that can be gained only through practice. For a description of procedures in bird-skinning, cataloguing, and labeling, consult the following: *Perserving Birds for Study* by E.R. Blake (Fieldiana: Technique No. 7, Chicago Nat. Hist. Mus., 1949); *The Preparation of Birds for Study* by J.P. Chapin (Science Guide No. 58, Amer. Mus. Nat. Hist., 1946); "Making Birdskins" in *Handbook of Birds of Eastern North America* by F.M. Chapman (D. Appleton and Company, New York, 1932; Dover reprint available); *Collecting and Preparing Study Specimens of Vertebrates* by E.R. Hall (Misc. Publ. No. 30, Univ. Kansas Mus. Nat. Hist., 1962). See also *A Bird Skin Drying Form for Field Use* by G.E. Watson (Bird-Banding, 33: 95–96, 1962).

In preparing study skins bear in mind that it is important not only to make skins that neatly and properly preserve each bird's general form and feather arrangement but also to record all necessary data pertaining to each specimen. This means placing on each bird skin label full information as to locality and date of collection; name of collector; sex, age and weight of specimen; amount of fat; condition of gonads; presence or absence of incubation patch; and the perishable colors of the soft parts. A skin, no matter how good it is, has little value for study purposes without data of this sort. For details as to just what information should be placed on labels and a discussion of the importance of full data, read *Principles and Practices in Collecting and Taxonomic Work* by J. Van Tyne (Auk,

69:27–33, 1952). Also instructive is *An Aspect of Collectors' Technique* by T.T. McCabe (Auk, 60:550–558, 1943) in which suggestions are given for recording cyclic conditions of gonads, fat, and incubation patch.

For comprehensive information on the preservation and storage of bird skins and birds in the flesh (including embryos), as well as illustrated descriptions of the preparation of bird skins and bird eggs, refer to *The Preservation of Natural History Specimens,* Vol 2, by R. Wagstaffe and J.H. Fidler (Philosophical Library, New York, 1968).

In preserving small birds in quantity—e.g., large kills from television towers—it may not be feasible to make regular skins of them; yet the specimens may be necessary for later study and analysis. See *A New Method for Preserving Bird Specimens* by R.A. Norris (Auk, 78:436–440, 1961) wherein skins may be mounted flat on large cards with ample room for data. The procedure is time-saving and the space required for storage is economical.

Skeletal Material

The preparation of skeletal material, as well as specimens to be kept in the flesh, requires special care. For helpful information, read *Suggestions Regarding Alcoholic Specimens and Skeletons of Birds* by A.J. Berger (Auk, 72:300–303, 1955) and *Anatomical Prepartions* by M.Hildebrand (Univ. of California Press, Berkeley, 1968), a standard manual for all vertebrate anatomical preparation techniques, with bibliographic references to pertinent papers.

Skeletons may be cleaned by putting them in a colony of dermestid beetles maintained for this purpose. For details on the method, read *Dermestid Beetles for Cleaning Skulls and Skeletons in Small Quantities* by R. Hardy (Turtox News, 23:69–70, 1945) and *Defleshing of Skulls by Beetles* by E.J. Coleman and J.R. Zbijewska (Turtox News, 46: 204–205, 1968). Or skeletons may be cleaned by macerating them in a solution of ammonium hydroxide. For details read *Method for the Preparation and Preservtion of Articulated Skeletons* by C.P. Egerton (Turtox News, 46:156–157, 1968).

Determining the Sex and Age of Live Birds

When coloration or other features of a captured bird's exterior do not reveal its sex or age, its cloaca may be examined for the presence, absence, or state of development of the following structures, which will provide the information desired. (Before examining a cloaca, first understand its general anatomy by reading the directions for the dissection of the pigeon's cloaca, page 82.)

Penis

The ventral wall of the male proctodeum may be modified to form a penis. In waterfowl (swans, geese, and ducks) the penis is a thickening which, when erected and protruded through the vent, acts as an intromittent organ during copulation. A groove on its dorsal surface conveys the spermatozoa into the cloaca of the female. In adult ducks—and presumably all waterfowl—the penis, except when protruded, is covered by a sheath that appears as a grayish white fold on the left wall of the proctodeum. But in immature ducks—i.e., ducks that are no older than six months—the penis is merely a short prominence just inside the lip of the vent and has no sheath.

By examining the cloaca of a duck, it is possible to determine the sex and, in case the individual is a male, the age. The technique used in making the examination briefly follows: have the bird held with the posterior abdomen up and the tail pointing away from the examiner; force the tail back with the forefingers; with the thumbs, part the feathers over the vent and press down and away from the two sides of the vent simultaneously, spreading the opening of the vent and exposing the cloaca within. If the duck is a male, a penis will protrude; if the male is an adult, the penis will be sheathed, but if it is an immature bird, the penis will be small and unsheathed.

A penis like the waterfowl's occurs in ostriches, rheas, cassowaries, emus, kiwis, tinamous, and guans. A rudimentary form of the same structure also persists in herons and flamingos. Males of a few other kinds of birds often show a slight modification of the cloacal wall for copulatory purposes.

For example, the Domestic Fowl *(Gallus),* Turkey *(Meleagris),* and Ring-necked Pheasant *(Phasianus colchicus)* have paired thickenings with erectile tissue on the ventral wall. Males of most kinds of birds, however, lack a penis.

Cloacal Protuberance

The posterior wall of the male proctodeum may swell and protrude posteriorly, bearing the encircled vent with it. Called the cloacal protuberance, it contains chiefly two bodies of compactly coiled tubules (vasa deferentia), which are believed to act as storage reservoirs for spermatozoa. So far as known, the protuberance occurs in certain passerine species only.

The protuberance increases in size as the breeding season advances, reaching its maximum size at the height of the season. After the breeding season it is very small, if not entirely absent. The presence of the protuberance is an unquestionable indicator of the male sex, and its size shows the breeding condition. A protuberance is easily looked for by holding the bird with posterior abdomen up and parting the feathers over the vent. If there is any pronounced swelling around the vent, the bird is a male; if the area around the vent is greatly distended, the male is very near or in breeding condition, but if the swelling is slight, the male has not neared or has definitely passed breeding condition.

Bursa of Fabricius

This is an outpocketing of the dorsal wall of the proctodeum and is connected to the proctodeum by an orifice. It is present in all young birds, but as birds approach maturity it disappears by involution and its orifice, at the same time, becomes smaller and finally closes. An orifice in the dorsal cloacal wall means that there is a bursa behind it and is thus positive proof of the individual's immaturity. With larger species of birds a cloacal examination is practicable, for the orifice is big enough to be readily seen, but with smaller birds an examination requires too great precision to be recommended. The examination procedure briefly follows: under strong light, have the bird held with posterior abdomen up and the tail pointing toward the examiner; with the outer fingers of one hand force back the tail; with the thumb and index finger of the same hand part the feathers over the vent, grasp the dorsal lip of the vent and pull it toward the tail; with the thumb and index finger of the other hand grasp the ventral lip of the vent and pull it away from the tail. The cloaca will now be sufficiently open to view. Look for the orifice in the dorsal wall of the cloaca. Sometimes the orifice, even though present, may be covered by a membranous fold of the adjacent cloacal wall. Therefore, probe the area gently with a blunt instrument in order to uncover the orifice, should it be there.

Certain investigators, by measuring the depth of bursae with scaled instruments, have been able to determine the ages of young birds by the stages of bursal involution.

For additional means of determining the age and sex of live birds, refer to the *North American Bird Banding Manual* (Volume 1, 1976; Volume 2, 1980), published by the U.S. Fish and Wildlife Service and the Canadian Wildlife Service, and the following references.

Davis, D.E.
1947 Size of bursa of Fabricius compared with ossification of skull and maturity of gonads. *Jour. Wildlife Mgmt.* 11:244–251. (Includes information on many species of tropical birds.)

Horwich, R.H.
1966 Feather development as a means of aging young mockingbirds *(Mimus polyglottos). Bird-Banding* 37:257–267.

Kessel, B.
1951 Criteria for sexing and aging European Starlings *(Sturnus vulgaris). Bird-Banding* 22:16–23.

Kirkpatrick, C.M.
1944 The bursa of Fabricius in Ring-necked Pheasants. *Jour. Wildlife Mgmt.* 8:118–129. (Includes a description of the technique for determining ages by the depth of bursae.)

Larson, J.S., and Taber, R.D.
1980 Criteria of sex and age. In *Wildlife management techniques manual*, 4th ed., ed. S.D. Schmenitz. Washington, D.C.: The Wildlife Society.

Martin, F.W.
1964 Woodcock age and sex determination from wings. *Jour. Wildlife Mgmt.* 28:287–293.

Miller, A.H.
1946 A method of determining the age of live pas-

serine birds. *Bird-Banding* 17:33–35. (In dead birds it is possible to tell age by the condition of the skull. Thus if the specimen is immature, its skull is uniformly pinkish; if it is an adult, its skull is whitish and finely speckled. The author suggests a simple operation on a live bird that will reveal the condition of the skull.)

Miller, W.J. and Wagner, F.H.
1955 Sexing mature Columbiformes by cloacal characters. *Auk* 72:279–285.

Norris, R.A.
1961 A modification of the Miller method of aging live passerine birds. *Bird-Banding* 32:55–57.

Palmer, W.L.
1959 Sexing live-trapped juvenile Ruffed Grouse. *Jour. Wildlife Mgmt.* 23:111–112.

Parks, G.H.
1962 A convenient method of sexing and aging the Starling. *Bird-Banding* 33:148–151. (Use of the Kessel, 1951, method.)

Petrides, G.A.
1950 Notes on determination of sex and age in the Woodcock and Mourning Dove. *Auk* 67:357–360. (The bursa is useful for age determination in the American Woodcock only by dissection, and in the live Mourning Dove during the breeding season.)

Salt, W.R.
1954 The structure of the cloacal protuberance of the Vesper Sparrow *(Pooecetes gramineus)* and certain other passerine birds. *Auk* 71:64–73.

Scott, D.M.
1967 Postjuvenal molt and determination of age of the Cardinal. *Bird-Banding* 38:37–51.

Swank, W.G.
1955 Feather molt as an ageing technique for Mourning Doves. *Jour. Wildlife Mgmt.* 19:412–414.

Wolfson, A.
1952 The cloacal protuberance—A means for determining breeding condition in live male passerines. *Bird-Banding* 23:159–165.
1954 Notes on the cloacal protuberance, seminal vesicles, and a possible copulatory organ in male passerine birds. *Bull. Chicago Acad. Sci.* 10(1):1–23.

Wood, M.
1969 *A bird-bander's guide to determination of age and sex of selected species*. University Park: College of Agriculture, Pennsylvania State Univ. (Spiral binding. Includes Passeriformes and "a few other species commonly handled by bird-banders" in northeastern United States.)

Weighing Birds

Weights of birds have three main uses: to aid systematic and physiological studies within species; to ascertain differences or similarities between sexes; to record the growth of young.

The study of bird weights reveals a wide variety of significant problems such as seasonal fluctuations in weight within a species, or in individuals; weight in relation to migration; weight in relation to time of day or night; weight in relation to age; weight in relation to the type and amount of food consumed. See *The Biological Significance of Bird Weights* by M.M. Nice (Bird-Banding, 9:1–11, 1938) for a discussion of some of these problems. See also "Weights of 151 Species of Pennsylvania Birds Analyzed by Month, Age, and Sex," by M.H. Clench and R.C. Leberman (*Bull. Carnegie Mus. Nat. Hist.*, No. 5, 1978) and "Body Weights of Birds: A Review" by G.A. Clark, Jr. (*Condor*, 81:193–202, 1979).

Weights are taken in grams on scales sensitive to one-tenth of a gram. Scales recommended for field and laboratory use should have the following specifications: for **small birds,** triple-beam, agate-bearing scales with weighing capacity of 0.01 to 111 grams without separate, auxiliary weights and up to 201 grams with auxiliary weights; for **large birds,** triple-beam, agate-bearing trip scales with weighing capacity of 610 grams without separate, auxiliary weights and up to 2,610 grams with auxiliary weights. When the scales are used in the field, they should be housed in specially constructed carrying cases with handles. The cases should be so designed that they may be entirely opened on one side by hinged doors. Birds may thus be weighed within the case where, if the opened side faces leeward, there is a minimum of interference from wind.

When birds are too active to stay on the scales while being weighed, place them in a cloth sack or paper bag. The weight of the sack or bag should be determined separately and deducted from the total. If the sack or bag is to be used over a period of days, the weight should be determined each day, since it is liable to fluctuate under different atmospheric conditions.

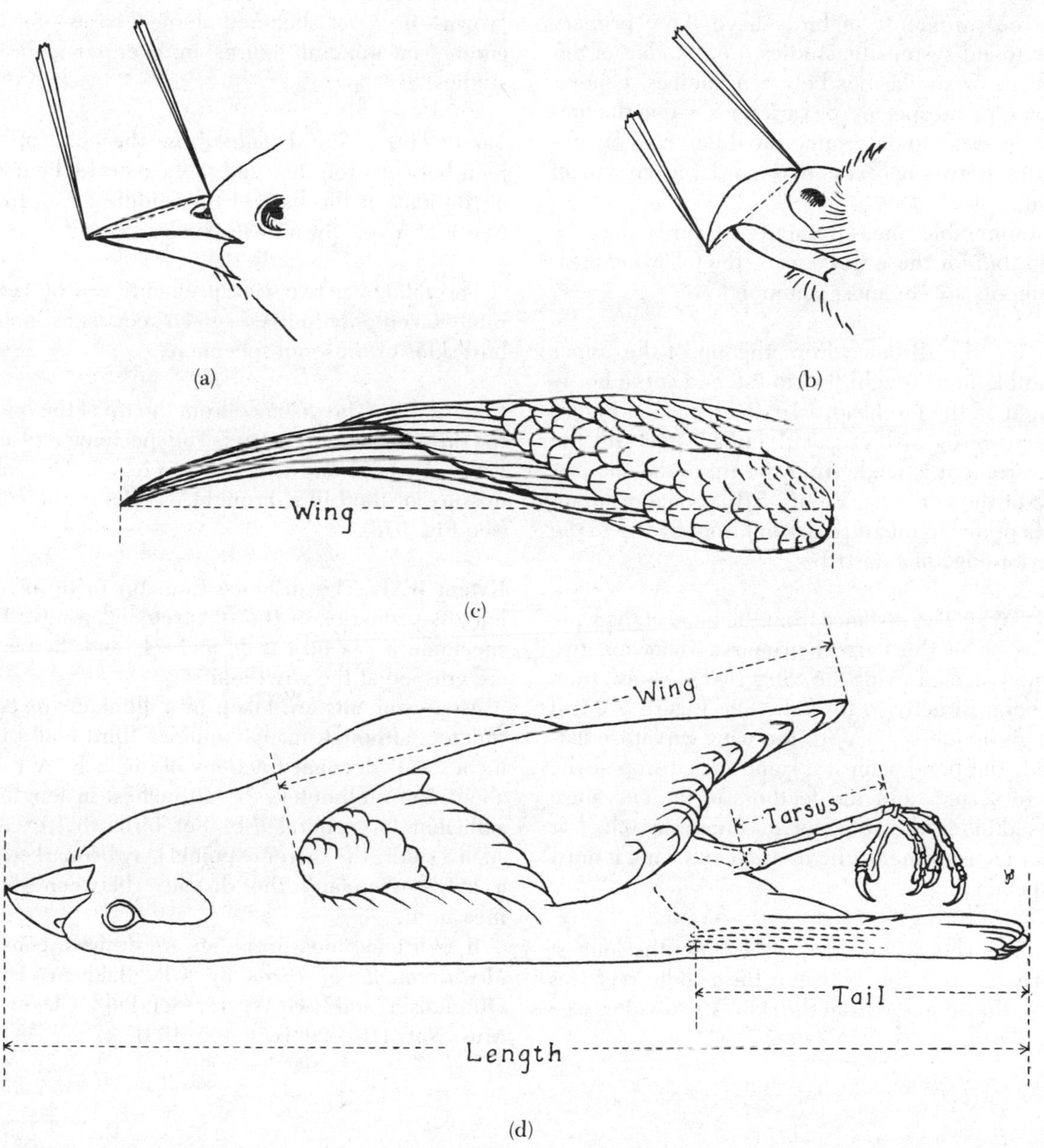

Figure 58 **Measurements of Birds**

Measuring Birds

The measurements of birds have three primary uses: to aid systematic studies (i.e., studies of differences or similarities between families, genera, species, or subspecies, or variations within the limits of species); to determine the differences or similarities between sexes; to record the growth of young.

Innumerable measurements of birds may be taken, but for these three uses, the following measurements are the most common.

Bill (B) the distance from the tip of the upper mandible in a straight line to the end of the horny culmen on the forehead. Use dividers. (See Figure 57a). In birds with cere, or similar structure, the measurement is made from the tip to the anterior edge of the cere (see Figure 57b). Some ornithologists prefer to measure the bill from the tip to the anterior edge of a nostril.

Wing (W) The distance from the bend of the wing to the tip of the longest primary. There are two methods of measuring the wing: (1) The chord, from the bend directly to the tip. (See Figure 57c and d). Use dividers. (2) With the wing curvature flattened, the bend against a right-angled stop at the end of a ruler and the feathers of the curvature pressed along the ruler for maximum length. For either measurement, the method for taking it must be specified.

Tail (T) The distance from the tip of the longest rectrix to the point between the middle rectrices where they emerge from the skin. Use dividers (see Fig. 57d).

When measuring the wing and tail be sure that the feathers included in the measurements are fully grown—i.e., not sheathed at their bases—for accurate, meaningful figures in later comparative studies.

Tarsus (Tar) The distance from the point of the joint between the tibia and metatarsus to the point of the joint at the base of the middle toe in front (see Fig. 57d). Always use dividers.

The following two measurements can be accurately taken only in live or fresh specimens, not in bird skins or museum specimens.

Length (L) The distance from the tip of the bill to the tip of the longest rectrix. The specimen is placed flat on its back and is gently stretched. The commissure of the bill is brought parallel to the ruler (see Fig. 57d).

Extent (Ex) The distance from tip to tip of the longest primaries of the outstretched wings. The specimen is placed flat on its back, and the wings are grasped at the wrist joints.

Measurements are taken in millimeters or centimeters although many "popular" bird books use inches and decimal fractions of an inch. A ruler about 250 millimeters (= 10 inches) in length is sufficient for general use. For birds that exceed such a ruler, the extreme points may be marked on a flat surface and the distance between them measured.

If additional measurements are desired, consult *Measurements of Birds* by S.P. Baldwin, H.C. Oberholser, and L.G. Worley (Sci. Publ. Cleveland Mus. Nat. Hist. 2:i–ix; 1–165, 1931).

Preparation of a Paper

APPENDIX B

The following instructions may aid in the preparation of a paper for a class requirement or for publication.

Preparation of the Manuscript

The manuscript must be typewritten on paper of good quality, preferably bond (size 8½ x 11 inches), with lines double-spaced. On each page allow a left-hand margin 1 to 1½ (2.54 cm–3.81 cm) inches wide. Do not use erasable, light-weight, or mimeo bond. Never fold the manuscript. Keep at least one copy.

Number the pages consecutively, beginning with the first page *after* the table of contents.

Underscore all words to be printed in italics.

Do not abbreviate geographical names or names of days and months except in lists and tables. Do not abbreviate "Figure," "Plate," etc., when referring to illustrations in the paper. Thus: "Figure 2," not "Fig. 2."

Abbreviate designations of measurements after the first time given. Thus: "... millimeters" the first time, "... mm" the second time, without a period after the abbreviation.

Write out ordinal numbers except in tables: "First, second," etc., not "1st, 2nd."

Write out all numbers below 10 except for pages, tables, plates, etc., or when giving numbers in a series.

Fasten the pages together with clips—do not staple them. Place tables, figures, plates, and maps on separate sheets of paper and number them, but do not include them in the pagination.

Use capitals for designating various items. Thus: "Bird A"; "Nest 3"; "Plate 6."

Capitalize all English names of birds when designating a species, but not otherwise. Thus: "Song Sparrow," but "sparrow"; "Eastern Meadowlark," but "meadowlark." Follow *The A.O.U. Check-list* (latest edition) for the correct scientific and English names of North American birds.

Do not capitalize English names of species other than of birds.

Underscore all technical names of genera, species, and subspecies, but not technical names of other taxonomic categories.

Underscore or enclose in quotation marks the phonetic representations of bird sounds.

Give the technical name of a species the first time mentioned in the paper, but not again unless in the summary.

Use parentheses for technical names after their English names and for citations to literature. Otherwise keep parenthetical interruptions to a minimum.

Follow the latest edition of one standard dictionary for spelling, use of hyphens, etc. For abbreviations of statistical terms and mensural units, conform to the *Council of Biology Editors Style Manual*, Fourth edition, 1978. Published by the C.B.E. Available from the American Institute of Biological Sciences, 1401 Wilson Boulevard, Arlington, Virginia 22209.

If the paper is for publication, consult the editorial directions inside the journal to which the paper will be submitted.

Organization

Title Page

The title page includes the title of the study in as few words as possible, the name and address of the author, and the date when the paper is submitted. If the paper is for publication, omit the title page and incorporate the information on the first and last pages of the paper in accordance with the custom of the journal to receive the paper for consideration.

Table of Contents

A table of contents helps to guide the instructor as well as an editor who may review the paper for publication. The table follows the title page and should look somewhat like the following:

TABLE OF CONTENTS

Text

Introduction Set forth the scope and purposes of the paper, including the English and scientific names of the species studied, dates of the study, the precise political and geographical location of the study, the number of birds studied, the number of days involved in field and laboratory work, the number of hours involved in observations. Include under a subheading the methods and techniques used.

Observations or Results Before preparing this part of the paper, review any information that may have been published on the subject in ornithological journals (see Appendix C) and elsewhere. Cite the information if it corroborates, differs from, fills gaps in, or serves to explain any findings. A great aid in seaching the ornithological literature is *Aves* from *The Zoological Record* compiled annually by the staff of the Zoological Society of London. Each issue has an author index, subject index, geographical index, and systematic index. Normally the issue covering a particular year does not reach publication until about five years thereafter. Issues of *Aves* are available in the United States from BioSciences Information Service, 2100 Arch Street, Philadelphia, Pennsylvania 19103.

This is the body of the paper. Here correlate all the essential information and present it in a logical sequence. Sift out data irrelevant to the main theme. Always keep the reader oriented as the theme unfolds and posted on dates and places. If this part of the paper is long, use subheadings so that the reader will not lose track of the theme. Make sure that these subheadings are given in the Table of Contents. Avoid long lists of numbers. If many must be included, incorporate them in tables or graphs.

Discussion Here discuss certain facts and suppositions, particularly when they are at variance with known information, or need explanation, interpretation, and analysis. While it may be appropriate to discuss some of the results in those sections of the paper where the facts were given, it is usually better to reserve the principal discussion for this part of the paper. The discussion is second in importance only to the summary and is often the part of the paper of greatest interest to the reader.

Summary Review very briefly the essential results of the study. Do not repeat methods, techniques, and descriptions. Since the summary is often the first part of the paper read by the instructor or editor, it is frequently the basis for judging the value of the paper. Therefore use just one or more short pargraphs concisely worded.

Acknowledgments Give full credit to the persons and organizations contributing directly or indirectly to the study. When in doubt about acknowledging someone's assistance, give it rather than risk the loss of good will.

References If 10 or more publications are mentioned in the text, they should be listed by authors alphabetically under the caption "Literature Cited." All titles must be quoted verbatim and the words spelled exactly as in the original. Only the proper names and the English names of bird species are capitalized unless the journal to receive the paper for consideration requires another style of capitalization. If necessary to quote a citation without having seen the original, state "Original publication not examined" or "Title quoted from (name or source)." Do not include under "Literature Cited" titles, which are not referred to in the text. Several types of publications to show style follow.

Feduccia, A.
1980 The age of birds. Harvard University Press, Cambridge, Massachusetts. *(Note how the author's name is reversed and placed by itself on one line. Only the author's last name is spelled out. There is no period after the date of publication.)*

Bailey, A.M., and R.J. Niedrach
1965 Birds of Colorado. Volume 2. Denver Museum of Natural History, Denver, Colorado. *(When any book or article is authored by more than one person, only the name of the first author is reversed in most American journals. Note the Arabic numeral for volume, followed by a period. Do not use Roman numerals in citing volumes, numbers, etc.)*

Rockwell, E.D.
1982 Intraspecific food robbing in Glaucous-winged Gulls. Wilson Bull. 94:282–288. *(Note the punctuation, also the abbreviation, of Bulletin. For regularly used abbreviations of journals, peruse citation lists in current ornithological journals.)*

Holcomb, L.C.
1966a The development of grasping and balancing coordination in nestlings of seven species of altricial birds. Wilson Bull., 78:57–73.
1966b Red-winged Blackbird nestling development. Wilson Bull. 78:283–288. *(When two or more articles are published by the same author in the same year, the dates of publication are distinguished by small letters in series. Page numbers include the entire article, not the part cited.)*

All citations in the text refer to the publications under "Literature Cited." Note the following variety of citations:

"Feduccia (1980:206) stated that...."
"In Colorado, Bailey and Niedrach (1965:585) reported...."
"One authority (Rockwell, 1982:288) has this to say...."

The numbers after the colon refer to the exact page of the quotation from the publications. If the publications are mentioned in a general way, give only their dates. Example: "How Holcomb (1966b) studied nestling development deserves attention...."

Some journals require the following style of listing citations:

Feduccia, A. 1980. The age of birds. Harvard University Press, Cambridge, Massachusetts.
Rockwell, E.D. 1982. Intraspecific food robbing in Glaucous-winged Gulls. Wilson Bull. 94:282–288.

The style has the advantage of saving space and the great disadvantage of obscuring the dates of publication for quick reference. It is not recommended for student papers.

If there are fewer than 10 publications, they may be cited parenthetically in the body of the paper. For example: (A. Feduccia, *The Age of Birds,* Harvard University Press, Cambridge, Massachusetts, 1980); E.D. Rockwell (Wilson Bull. 94:282–288, 1982). Note that the authors' names are not reversed and the title of the book is capitalized and underscored. Also note the different punctuation and arrangement of dates.

Tables, Figures, Plates, and Maps Tables are generally lists of data or related items arranged for ease of reference or comparison, often in parallel columns. Figures are generally graphs, diagrams, and drawings; plates are generally photographs and paintings.

All tables, figures, plates, and maps must be separately titled, numbered consecutively in separate series (e.g., Table 1, Table 2, etc.; Figure 1, Figure 2, etc.) with their captions on attached separate sheets of paper. All photographs not taken by the author must be credited, either by a credit line or in "Acknowledgments."

Tables, figures, plates, and maps should *supplement* the text, not take the place of it. Refer to each in the text by number; if necessary, indicate its significance.

Style of Writing

Use a simple style. Avoid wordiness. Make precise statements. Whenever possible, *use the first person* in giving personal observations and opinions—the most effective means of indicating the source. Never use contractions—e.g., write "the bird was not," instead of "the bird wasn't."

After carefully organizing all data in a logical sequence and writing the paper, review it for style, keeping in mind some of the following common pitfalls that all too frequently show up in scientific reports.

The deadly passive voice: Example: "The worms *were eaten* by the robin." Make the verb active and the paper live. Revise: "The robin ate the worms."

The ambiguous observer: Example: "Fifteen Evening Grosbeaks were seen on January 1." Always let the reader know who made the observation. "[ ] saw 15 Evening Grosbeaks on January 1."

Meaningless adjectives: Examples: "An *interesting* behavior, a *beautiful* forest, an *exciting* experience." The adjectives are lazy words and tell nothing. Describe the behavior, the forest, and the experience and let the reader judge the quality of the behavior, the forest, and the experience.

Unnecessary use of "would": Examples: "When the rain came, the robin *would* cover her eggs." "The hawk *would* take off at the slightest disturbance." Write instead: "When the rain came, the robin covered her eggs." "The hawk took off at the slightest disturbance." Avoid the past perfect tense unless an imprecise statement is intended.

The superfluous "located": Example: "The nest *located* at the edge of the road." Strike out "located" and the statement conveys the same meaning.

Too many adjectival nouns: Examples: "Bird population decline, pesticide identification techniques, forest habitat preference." The use of words in this manner, applied in the name of economy of space or to suggest the scientific approach, impairs rather than facilitates readability. "The decline in the population of birds, the techniques for identifying pesticides, the preference for habitat in a forest" takes little more space and detracts nothing whatsoever from the stature of the writer.

The unidentified "it": Question every "it" in the paper. Is "it" necessary? If so, make sure that "it" has a definite antecedent. Many times one can avoid "it." Examples: "It is possible that...." "It is the opinion of the writer that...." Write instead: "The possibility that...." "My opinion is that...." Or "I believe that...."

"Utilization" and other pompous words: Example: "Utilization of habitat." This says nothing more than "the uses of habitat," but it has a high sounding ring, chosen to impress the reader. Utilization is synonymous with use. When given a choice, always use (not utilize!) the simpler word.

A final word of advice: Before attempting to write the paper, or later when polishing the text before typing the final copy, take time to read *The Elements of Style* by W. Strunk, Jr. *With Revisions, an Introduction, and a Chapter on Writing* by E.B. White. 3rd ed. in paperback. Macmillan Publishing Company, 1979. No little book "in shorter space, with fewer words, will help any writer more."

If the paper is to be revised later for publication, first thoroughly read *How to Write and Publish a Scientific Paper,* 2nd ed. by R.A. Day, ISI Press, Philadelphia, 1983. Available in hardcover or paperback from ISI Press, 3501 Market Street, University City Science Center, Philadelphia, Pennsylvania 19104. There is no better book for practical instructions on all aspects of scientific writing and publishing.

Current Ornithological Journals

APPENDIX C

Following are the titles and sponsoring organizations of current journals that either are entirely ornithological in content or quite frequently publish articles dealing with birds.

United States and Canada

The principal journals published in the United States and Canada are given in the following two lists. The first contains the five leading journals—publications essential for any student of birds.

The Auk (quarterly).
American Ornithologists' Union.
American Birds (bimonthly).
National Audubon Society.
The Condor (quarterly).
Cooper Ornithological Society.
Journal of Field Ornithology (formerly **Bird-Banding**; quarterly).
Northeastern Bird-Banding Association.
The Wilson Bulletin (quarterly).
Wilson Ornithological Society.

Alabama Birdlife (quarterly).
Alabama Ornithological Society.
Animal Behaviour (quarterly).
Association for the Study of Animal Behaviour and Animal Behavior Society (co-sponsor).
Animal Kingdom (bimonthly).
New York Zoological Society.
Audubon (bimonthly).
National Audubon Society.
Birding (bimonthly).
American Birding Association.
The Bluebird (quarterly).
The Audubon Society of Missouri.
The Blue Jay (quarterly).
Saskatchewan Natural History Society.
Bulletin of the Oklahoma Ornithological Society (quarterly).
Bulletin of the Texas Ornithological Society (irregular).
The Canadian Field-Naturalist (quarterly).
Ottawa Field-Naturalists' Club.
Canadian Journal of Zoology (monthly).
National Research Council, Ottawa.
Cassinia (biennial).
Delaware Valley Ornithological Club.
The Chat (quarterly).
Carolina Bird Club.
The Colorado Field Ornithologists' Quarterly.
Colorado Field Ornithologists.
Connecticut Audubon (bimonthly).
Connecticut Audubon Society.
Delmarva Ornithologist (irregular).
Delmarva Ornithological Society. Delaware.
Elepaio (quarterly).
Hawaii Audubon Society.
Florida Field Naturalist (quarterly).
Florida Ornithological Society.
Habitat (bimonthly).
Maine Audubon Society.
The Hermit Thrush (bimonthly).
Green Mountain Audubon Society.
Illinois Audubon Bulletin (quarterly).
Illinois Audubon Society.

The Indiana Audubon Quarterly.
Indiana Audubon Society.
International Wildlife (bimonthly).
National Wildlife Federation.
Iowa Bird Life (quarterly).
Iowa Ornithologists' Union.
The Jack-Pine Warbler (quarterly).
Michigan Audubon Society.
The Journal of Wildlife Management (quarterly).
The Wildlife Society.
Kansas Ornithological Society Bulletin (quarterly).
The Kentucky Warbler (quarterly).
Kentucky Ornithological Society.
The Kingbird (quarterly).
Federation of New York State Bird Clubs.
The Living Bird (annual).
Laboratory of Ornithology, Cornell University.
The Living Bird Quarterly
Laboratory of Ornithology, Cornell University.
The Loon (quarterly).
Minnesota Ornithologists' Union.
Maryland Birdlife (quarterly).
Maryland Ornithological Society.
The Migrant (quarterly).
Tennessee Ornithological Society.
The Mississippi Kite (irregular).
Mississippi Ornithological Society.
The Murrelet (triannually).
Pacific Northwest Bird and Mammal Society.
National Wildlife (bimonthly).
National Wildlife Federation.
Natural History (monthly).
American Museum of Natural History.
The Nebraska Bird Review (quarterly).
Nebraska Ornithologists' Union.
New Jersey Audubon (quarterly).
New Jersey Audubon Society.
New Mexico Ornithological Bulletin (quarterly).
North American Bird Bander (quarterly).
Eastern, Inland, and Western Bird-Banding Associations.
Ontario Bird Banding (quarterly).
Ontario Bird Banding Association.
Ontario Birds (biannual).
Ontario Field Ornithologists.
The Oriole (quarterly).
Georgia Ornithological Society.
The Passenger Pigeon (quarterly).
Wisconsin Society for Ornithology.
The Raven (quarterly).
Virginia Society of Ornithology.
The Redstart (quarterly).
Brooks Bird Club, West Virginia.
Sialia (quarterly).
North American Bluebird Society.
South Dakota Bird Notes (quarterly).
South Dakota Ornithologists' Union.
Vermont Natural History (annual).
Vermont Institute of Natural Science.
Western Birds (quarterly).
Western Field Ornithologists, California.
Wildlife Society Bulletin (quarterly).
The Wildlife Society.

Other Countries

Acta Ornithologica (irregular).
Polska Akademia Nauk, Instytut Zoologii. Poland.
Alauda (quarterly).
La Société d'Etudes Ornithologiques. France.
Angewandte Ornithologie (irregular).
Internationale Union für Angewandte Ornithologie. West Germany.
Aquila (annual).
Instituti Ornithologici Hungarici. Hungary.
Ardea (biannual).
Tijdschrift der Nederlandse Ornithologische Unie. The Netherlands.
Ardeola (annual).
Sociedad Española de Ornitología. Spain.
The Australian Bird Watcher (eight times yearly).
Bird Observers Club. Australia.
Australian Wildlife Research (triannual).
The Commonwealth Scientific and Industrial Research Organization, Melbourne, Australia.
Avicultural Magazine (quarterly).
The Avicultural Society. England.
Behaviour (monthly).
The Netherlands: E.J. Brill, Publisher.
Beiträge zur Vogelkunde (quarterly).
Akademische Verlagsgesellschaft. West Germany.
Bird Study (triannual).
British Trust for Ornithology. England.
Birds (quarterly).
The Royal Society for the Protection of Birds. England.

Bokmakierie (quarterly).
South African Ornithological Society. South Africa.

British Birds (monthly).
British Birds Limited. England.

Bulletin of the British Ornithologists' Club (quarterly).
England.

Corella (five times a year).
Australian Bird Study Association. Australia.

Dansk Ornithologisk Forenings Tidsskrift (quarterly).
Dansk Ornithologisk Forening. Denmark.

Dutch Birding (quarterly).
Stichting Dutch Birding Association. The Netherlands.

Egretta (biannual).
Vogelkundliche Nachrichten aus Österreich. Austria.

The Emu (quarterly).
Royal Australasian Ornithologists Union. Australia.

Vår Fågelvärld (bimonthly).
Sveriges Ornitogiska Forëning. Sweden.

Le Gerfaut (quarterly).
L'Institut Royal des Sciences Naturelles de Belgique. Belgium.

Gli Uccelli d'Italia (quarterly).
Societá Ornitologica Italiana. Italy.

Honeyguide (quarterly).
Ornithological Society of Zimbabwe.

El Hornero (irregular).
La Asociación Ornitólogica del Plata. Argentina.

The Ibis (quarterly).
British Ornithologists' Union. England.

Irish Bird Report (annual).
Irish Ornithologists' Club. Ireland.

Irish Birds (annual).
The Irish Wildbird Conservancy. Ireland.

Journal für Ornithologie (quarterly).
Deutsche Ornithologen-Gesellschaft. West Germany.

The Lammergeyer (irregular).
Journal of the Natal Parks, Game and Fish Preservation Board. South Africa.

Larus.
Annual of the Institute of Ornithology of the Yugoslav Academy of Sciences and Arts, Zagreb. Yugoslavia.

Limosa (quarterly).
Nederlandse Ornitologische Unie. The Netherlands.

Malimbus (biannual).
Journal of the West African Ornithological Society. Nigeria.

Moscow Society of Naturalists (bimonthly).
Bulletin. (English summaries.) Russia.

Nos Oiseaux (bimonthly).
La Société Romande pour l'Étude et la Protection des Oiseaux. Switzerland.

Notornis (quarterly).
Ornithological Society of New Zealand. New Zealand.

L'Oiseau et la Revue Française d'Ornithologie (quarterly).
La Société Ornithologique de France. France.

Ökologie der Vögel (biannual).
Published by Kuratorium für avifaunistische Forschung in Baden-Württemberg e.V. West Germany.

Ornis Fennica (quarterly).
Ornitologiska Foreningen i Finland. Finland.

Ornis Scandinavica (quarterly).
Scandinavian Ornithologists' Union, an association of the national societies of Denmark, Finland, Norway, Sweden.

Der Ornithologische Beobachter (bimonthly).
Schweizerische Gesellschaft für Vogelkunde und Vogelschutz. Switzerland.

Ornithologische Mitteilungen (monthly).
Monatsschrift für Vogelkunde und Vogelschutz. West Germany.

Ornitologiya (annual).
Moscow University. (English summaries.) Russia.

The Ostrich (quarterly).
South African Ornithological Society. South Africa.

Pavo (irregular).
The Indian Journal of Ornithology. India.

The Ring (quarterly).
International Ornithological Bulletin, Polish Zoological Society. Poland.

Rivista Italiana di Ornitologia (irregular).
Societá Italiana di Scienze Naturali Museo Civico di Storia Naturale. Italy.

The Sandgrouse (annual).
Ornithological Society of the Middle East (c/o

Royal Society for the Protection of Birds, England).

Scopus (quarterly).
Ornithological Sub-committee of the East African Natural History Society. Kenya.

Scottish Birds (quarterly).
Scottish Ornithologists' Club. Scotland.

The South Australian Ornithologist (biannual).
South Australian Ornithological Association. Australia.

Sterna (irregular).
Tidsskrift utgitt av Norsk Ornitogisk Forening og Stavanger Museum. Norway.

The Sunbird (quarterly).
Queensland Ornithological Society. Australia.

Tori (quarterly).
Ornithological Society of Japan. Japan.

Wildfowl.
Annual Journal of the Wildfowl Trust. England.

Verhandlungen der Ornithologischen Gesellschaft (irregular).
Transactions of the Ornithological Society in Bavaria. West Germany.

Die Vogelwarte (quarterly).
Vogelwarten Helgoland und Radolzell. West Germany.

Die Vogelwelt (bimonthly).
Zeitschrift für Vogelkunde und Vogelschutz. West Germany.

Journal of the Yamashina Institute for Ornithology (triannual).
Japan.

Zeitschrift für Tierpsychologie (monthly).
West Germany: Paul Parey Scientific Publishers.

Zoologichesky Zhurnal (monthly).
Leningrad Academy of Sciences. (Many bird articles with English summaries.) Russia.

Books for General Information

APPENDIX D

The following books provide wide sources of information on birds, bird life, and bird study from historical times to the present.

Alden, P., and Gooders, J.

1981 *Finding birds around the world*. Boston: Houghton Mifflin.

Allen, E.G.

1951 The history of American ornithology before Audubon. *Trans. Amer. Phil. Soc.* 41:387–591. (Republished by Russell and Russell, New York, 1969.)

Armstrong, E.A.

1958 *The folklore of birds: An enquiry into the origin and distribution of some magico-religious traditions*. London: Collins.

1975 *The life and lore of the bird in nature, art, myth, and literature*. New York: Crown.

Austin, O.L., Jr.

1961 *Birds of the world: A survey of the twenty-seven orders and one hundred and fifty-five families*. Reprinted in 1983. New York: Golden Press.

Bellrose, F.C.

1980 *Ducks, geese & swans of North America*. 3rd ed. Harrisburg, Pennsylvania: Stackpole.

Bent, A.C.

1919 Life histories of North American diving birds. *U.S. Natl. Mus. Bull.* 107. Reprinted 1946. New York: Dodd, Mead.

1921 Life histories of North American gulls and terns. *U.S. Natl. Mus. Bull.* 113. Reprinted 1947. New York: Dodd, Mead.

1922 Life histories of North American petrels and pelicans and their allies. *U.S. Natl. Mus. Bull.* 121.

1923 Life histories of North American wild fowl. Part 1. *U.S. Natl. Mus. Bull.* 126.

1925 Life histories of North American wild fowl. Part 2. *U.S. Natl. Mus. Bull.* 130.

1926 Life histories of North American marsh birds. *U.S. Natl. Mus. Bull.* 135.

1927 Life histories of North American shore birds. Part 1. *U.S. Natl. Mus. Bull.* 142.

1929 Life histories of North American shore birds. Part 2. *U.S. Natl. Mus. Bull.* 146.

1932 Life histories of North American gallinaceous birds. *U.S. Nat. Mus. Bull.* 162.

1937 Life histories of North American birds of prey. Part 1. *U.S. Natl. Mus. Bull.* 167.

1938 Life histories of North American birds of prey. Part 2. *U.S. Natl. Mus. Bull.* 170.

1939 Life histories of North American woodpeckers. *U.S. Natl. Mus. Bull.* 174.

1940 Life histories of North American cuckoos, goatsuckers, hummingbirds, and their allies. *U.S. Natl. Mus. Bull.* 176.

1942 Life histories of North American flycatchers, larks, swallows, and their allies. *U.S. Natl. Mus. Bull. 179.*

1946 Life histories of North American jays, crows, and titmice. *U.S. Natl. Mus. Bull.* 191.

1948 Life histories of North American nuthatches, wrens, thrashers, and their allies. *U.S. Natl. Mus. Bull.* 195

1949 Life histories of North American thrushes, kinglets, and their allies. *U.S. Natl. Mus. Bull.* 196.

1950 Life histories of North American wagtails, shrikes, vireos, and their allies. *U.S. Natl. Mus. Bull.* 197.

1953 Life histories of North American wood warblers. *U.S. Natl. Mus. Bull.* 203.

1958 Life histories of North American blackbirds, orioles, tanagers, and allies. *U.S. Natl. Mus. Bull.* 211.

1968 Life histories of North American cardinals, grosbeaks, buntings, towhees, finches, sparrows, and allies. Parts 1,2 and 3. Compiled and edited by O.L. Austin, Jr. *U.S. Natl. Mus. Bull.* 237.

(Dover reprints of the above Bent series are available.)

Berger, A.J.
1961 *Bird study.* New York: John Wiley & Sons.

Blake, E.R.
1977 *Manual of neotropical birds, Volume 1.* Spheniscidae to Laridae. Chicago: Univ. of Chicago Press.

Bramwell, M., project ed.
1974 *The Mitchell Beazley world atlas of birds.* London: Mitchell Beazley.

Brooks, P.
1980 *Speaking for nature: How literary naturalists from Henry Thoreau to Rachel Carson have shaped America.* Boston: Houghton Mifflin.

Brown, L.
1976 *Birds of prey: Their biology and ecology.* London and New York: Hamlyn.

Brown, L., and Amadon, D.
1968 *Eagles, hawks and falcons of the world.* 2 vols. New York: McGraw-Hill.

Brush, A.H., and Clark, G.A., Jr., eds.
1983 *Perspectives in ornithology.* New York: Cambridge Univ. Press. (Collection of 13 essays, presented at the Centennial of the American Ornithologists' Union, dealing with areas of contemporary research in ornithology.)

Burton, J.A., ed.
1973 *Owls of the world: Their evolution, structure, and ecology.* New York: E.P. Dutton.

Cade, T.
1982 *The falcons of the world.* Ithaca, New York: Cornell Univ. Press.

Cantwell, R.
1961 *Alexander Wilson: Naturalist and pioneer.* Philadelphia: J.B. Lippincott. (The definitive biography of "The Father of American Ornithology.")

Clark, R.J.; Smith D.G.; and Kelso, L.H.
1978 *Working bibliography of owls of the world: With summaries of current taxonomy and distributional status.* National Wildlife Federation Scientific/Technical Series No. 1.

Cramp, S., and Simmons, K.E.L., eds.
1977–1982 *The birds of the Western Palearctic.* 3 vols. Oxford: Oxford Univ. Press.

Cruickshank, H., compiler
1964 *Thoreau on birds.* New York: McGraw-Hill. (Selections from Thoreau's writings with authoritative comments by Mrs. Cruickshank.)

Cutright, P.R., and Brodhead, M.J.
1981 *Elliott Coues: Naturalist and frontier historian.* Urbana: Univ. of Illinois Press. (One of the greatest ornithologists of all time.)

Delacour, J.
1954–1964 *The waterfowl of the world.* 4 vols. London: Country Life. (Reprinted in 1974 by Arco Publishing Company, New York.)
1977 *The pheasants of the world.* 2nd ed. Hindhead, England: Spur.

Delacour, J., and Amadon, D.
1973 *Curassows and related birds.* New York: American Museum of Natural History.

Farber, P.L.
1982 *The emergence of ornithology as a scientific discipline:* 1760–1850. Dordrecht, Holland: D. Reidel. Available: Kluwer Boston, 190 Old Derby Street, Hingham, Massachusetts. (How ornithology, rooted in early literary activity, was tranformed into a specialized scientific discipline.)

Farner, D.S., ed.
1973 *Breeding biology of birds. Proceedings of a symposium on breeding behavior and reproductive physiology in birds.* Washington, D.C.: National Academy of Science.

Farner, D.S.; King, J.R.; and Parkes, K.C., eds.
1971–1983 *Avian biology.* 7 vols.New York: Academic Press.

Feduccia, A
1980 *The age of birds.* Cambridge, Massachusetts: Harvard Univ. Press.

Fisher, J., and Lockley, R.M.
1954 *Sea-birds: An introduction to the natural history of the sea-birds of the North Atlantic.* Boston: Houghton Mifflin.

Fisher, J., and Peterson, R.T.
[1964] *The world of birds.* Garden City, New York: Doubleday.

Forshaw, J.M.
1978 *Parrots of the world.* 2nd (revised) ed. Melbourne, Australia: Lansdowne Editions.

Gillard, E.T.
1958 *Living birds of the world.* Updated, 3rd printing, 1967. Garden City, New York: Doubleday.
1969 *Birds of paradise and bower birds.* London: Weidenfeld & Nicolson.

Goodwin, D.
1967 *Pigeons and doves of the world*. London: British Museum (Natural History).
1976 *Crows of the world*. London: British Museum (Natural History), and Ithaca, New York: Cornell Univ. Press.
1978 *Birds of man's world*. London: British Museum (Natural History), and Ithaca, New York: Cornell Univ. Press.
1982 *Estrildid Finches of the world*. Ithaca, New York: Cornell Univ. Press.

Grzimek, B., ed.
1972–1973 *Grzimek's animal life encyclopedia, Volumes* 7–9 (Birds I–III). New York: Van Nostrand Reinhold. (Many articles written by experts.)

Hancock, J., and Elliot, H.
1978 *The herons of the world*. New York: Harper & Row.

Harrison, C.J.O., ed.
1978 *Bird families of the world*. New York: Harry N. Abrams.

Heinroth, O., and Heinroth, K.
1958 *The birds*. Ann Arbor: Univ. of Michigan Press.

Herrick, F.H.
1938 *Audubon the naturalist: A history of his life and time*. 2nd ed. New York: D. Appleton-Century. (The definitive biography of John James Audubon.)

Hickey, J.J.
1975 *A guide to bird watching*. New York: Dover. (Reprint of the original book, published in 1943. The annotated list of books in Appendix D revised in 1974.)

Hume, E.E.
1942 *Ornithologists of the United States Army Medical Corps: Thirty-six biographies*. Baltimore: Johns Hopkins Press. (One of the most important biographic works in American ornithology.)

Johnsgard, P.A.
1973 *Grouse and quails of North America*. Lincoln: Univ. of Nebraska Press.
1978 *Ducks, geese, and swans of the world*. Lincoln: Univ. of Nebraska Press.
1981 *The plovers, sandpipers, and snipes of the world*. Lincoln: Univ. of Nebraska Press.
1983 *Hummingbirds of North America*. Washington, D.C.: Smithsonian Institution Press.
1983 *Grouse of the world*. Lincoln: Univ. of Nebraska Press.
1983 *Cranes of the world*. Bloomington: Indiana Univ. Press.

Kear, J., and Duplaix-Hall, N., eds.
1975 *Flamingos*. Berkhamsted, England: T & A.D. Poyser. (Authored by 29 persons from 15 countries.)

King, A.S., and McLelland, J., eds.
1979–1981 *Form and function in birds*. 2 vols. New York: Academic Press.

Kress, S.W.
1981 *The Audubon Society handbook for birders. A guide to locating, observing, identifying, recording, photographing and studying birds*. New York: Charles Scribner's Sons.

Lack, D.
1968 *Ecological adaptations for breeding in birds*. London: Methuen.
1971 *Ecological isolation in birds*. Oxford: Blackwell.
1976 *Island biology: Illustrated by the land birds of Jamaica*. Berkeley: Univ. of California Press.

Long, J.L.
1981 *Introduced birds of the world: The worldwide history, distribution and influence of birds introduced to new environments*. New York: Universe Books.

Marshall, A.J., ed.
1960–1961 *Biology and comparative physiology of birds*. 2 vols. New York: Academic Press.

Murton, R.K., and Westwood, N.J.
1977 *Avain breeding cycles*. Oxford: Clarendon Press.

Nelson, J.B.
1978 *The Sulidae: Gannets and boobies*. London: Oxford Univ. Press.

Newton, A.
1896 *A dictionary of birds*. London: Adam and Charles Black. (Still an indispensable reference.)

Owen, M.
1980 *Wild geese of the world*. London: B.T. Batsford.

Palmer, R.S.
1967 Species accounts. In *The shorebirds of North America*. Ed. and sponsor, G.D. Stout. New York: Viking Press.

Palmer, R.S., ed.
1962–1976 *Handbook of North American birds. Volume 1 (1962, Loons through flamingos); Volumes 2–3 (1976, Waterfowl)*. New Haven, Connecticut: Yale Univ. Press.

Pasquier, R.F.
1977 *Watching birds: An introduction to ornithology.* Boston: Houghton Mifflin.

Paynter, R.A., Jr., ed.
1974 Avian energetics. *Nuttall Ornith. Club Publ.* No. 15. Cambridge, Massachusetts.

Perrins, C.
1976 *Birds: Their life, their ways, their world*. New York: Harry N. Abrams.

Peterson, R.T., and the Editors of LIFE
1963 *The birds. Life nature library*. New York: Time Incorporated.

Pinowski, J., and Kendeigh, S.C., eds.
1977 *Granivorous birds in ecosystems*. Cambridge: Cambridge Univ. Press.

Rand, A.L.
1967 *Ornithology: An introduction*. Reprint ed. in 1969. New York: W.W. Norton.

Reilly, E.M., Jr.
1968 *The Audubon illustrated handbook of American birds*. New York: McGraw-Hill.

Ripley, S.D.
1977 *Rails of the world: A monograph of the family Rallidae*. Boston: David R. Godine.

Scott, P., and the Wildfowl Trust
1972 *The swans*. Boston: Houghton Mifflin.

Short, L.L.
1982 *Woodpeckers of the world*. Greenville: Delaware Museum of Natural History.

Simpson, G.G.
1976 *Penguins: Past and present, here and there*. New Haven, Connecticut: Yale Univ. Press.

Snow, D.W.
1976 *The web of adaptation: Bird studies in the American tropics*. New York: Quadrangle, New York Times Book Company.
1982 *The cotingas: bellbirds, umbrellabirds, and other species*. British Museum (Natural History) and Ithaca, New York: Cornell Univ. Press.

Stonehouse, B., ed.
1975 *The biology of penguins*. Baltimore: Univ. Park Press.

Stresemann, E.
1975 *Ornithology from Aristotle to the present*. Cambridge, Massachusetts: Harvard Univ. Press.

Temple, S.A.
1977 *Endangered birds: Management techniques for preserving threatened species*. Madison: Univ. of Wisconsin Press.

Terres, J.K.
1980 *The Audubon Society encyclopedia of North American birds*. New York: Alfred A. Knopf.

Thomson, A.L., ed.
1964 *A new dictionary of birds*. New York: McGraw-Hill.

Todd, F.S.
1979 *Waterfowl: Ducks, geese & swans of the world*. New York: Harcourt Brace Jovanovich.

Van Tyne, J., and Berger, A.J.
1976 *Fundamentals of ornithology*. 2nd ed. New York: John Wiley & Sons.

Walkinshaw, L.
1973 *The cranes of the world*. New York: Winchester Press.

Wallace, G.J., and Mahan, H.D.
1975 *An introduction to ornithology*. 3rd ed. New York: Macmillan.

Welker, R.H.
1955 *Birds and men: American birds in science, art, literature, and conservation, 1800–1900*. Cambridge, Massachusetts: Harvard Univ. Press.

Welty, J.C.
1982 *The life of birds*. 3rd ed. Philadelphia: Saunders.

Index

A 5
B 6
C 7
D 8
E 9
F 0
G 1